Textbook of Robotic Liver Surgery

Mathieu D'Hondt • Iswanto Sucandy
Editors

Textbook of Robotic Liver Surgery

Springer

Editors
Mathieu D'Hondt
Department of Hepatobiliary
and Pancreatic Surgery
Groeninge Hospital
Kortrijk, Belgium

Iswanto Sucandy
Hepatopancreatobiliary (HPB)
and Robotic Surgery
AdventHealth Tampa – Digestive Health
Institute
Tampa, FL, USA

ISBN 978-3-031-76535-3 ISBN 978-3-031-76536-0 (eBook)
https://doi.org/10.1007/978-3-031-76536-0

Illustrations by: Stephanie Philippaerts

This Springer imprint is published by the registered company Springer Nature Switzerland AG
The registered company address is: Gewerbestrasse 11, 6330 Cham, Switzerland

Foreword: The Rise of Robotic Liver Surgery

We have observed a marked transformation in the field of minimally invasive liver surgery. Twenty years ago, the thought of a robotic donor right hepatectomy for live donor liver transplantation or a robotic liver resection for a Klatskin tumor was unthinkable. Perhaps, it was a lack of imagination or possibly it was the stubbornness of experienced transplant and hepatobiliary surgeons who are resistant to change.

At the first international laparoscopic liver resection consensus conference, hosted in Louisville in 2008, robotic liver surgery was an afterthought. In a world review of nearly 3000 minimally invasive liver resections published in 2009, only 3 cases (0.1%) were performed with a robotic approach. These innovative robotic liver surgeries resection operations are being performed at multiple centers throughout the world, and the field is expanding at a rapid pace.

The International Laparoscopic Liver Society (ILLS) was established in 2016 and convened its first biannual World Congress in 2017. The quality, complexity, and volume of robotic liver resection cases and videos have grown substantially at each of the subsequent ILLS congresses. The past two years have witnessed a dramatic increase in academic conferences, either entirely focused on robotic liver surgery or featuring robotics as a large component of broader minimally invasive liver surgery programs. These include the ILLS fourth World Congress convened in Rome in September 2023, the ILLS and iLDLT Joint Conference on Minimally Invasive Donor Hepatectomy and Liver Transplantation in Naples in September 2023, the HPB Consensus Conference in Paris in December 2023, and the International European Guidelines Meeting on Minimally Invasive Liver Surgery (IEGUMILS) in Brescia in May 2024. Upcoming meetings include the ILLS Single Topic Conference on Robotic Liver Resection in Kyoto in October 2024, the Minimally Invasive Organ Transplant Consensus Conference in Riyadh in December 2024, and the fifth World Congress of ILLS in Seoul in June 2025.

The co-editors, Iswanto Sucandy, M.D., and Mathieu D'Hondt, M.D., Ph.D., are leading the field of robotic liver surgery in the USA and Europe, and they have developed a masterpiece: the *Textbook of Robotic Liver Surgery*. This scholarly work is timely, and the editors have assembled a renowned list of global contributors. In the 32 chapters, the authors provide a comprehensive and detailed description of anatomic robotic liver resection techniques, accompanied by illustrations, including robotic living donor hepatectomy. Specific outcomes for HCC, CRLM, and biliary tract cancers are provided. Learning curve, training programs, cost analysis, and registry outcomes are presented. The readers will be thoroughly delighted by a broad overview of the rapidly evolving field.

What will the future hold? Augmented and mixed virtual reality imaging are already being applied in some settings. As computing power, software, 3D imaging, and robotic technology continue to advance, surgeons' enthusiasm to adopt these innovative approaches in complex liver surgery will likely grow and disseminate. It is plausible to see the presence of artificial intelligence (AI) guiding robotic liver surgery and helping surgeons perform safer and more complex liver surgeries in the near future.

Division of Hepatobiliary and Pancreatic Surgery David A. Geller
UPMC Liver Cancer Center
University of Pittsburgh
Pittsburgh, PA, USA

Preface

The introduction of laparoscopy in the 1980s revolutionized many surgical domains, reducing postoperative complications and enhancing patient outcomes. Since the turn of the century, the robotic approach has become predominant in the urologic field. Over the last ten years, a significant increase in the use of robotic approaches has been noted at the expense of open and laparoscopic surgery in colorectal, pancreatic, esophageal, and hernia surgery. Although recent years have shown an increasing incidence of the robotic approach in the field of minimally invasive liver surgery, practice change remains very slow. Despite all the advances in minimally invasive liver surgery, most hepatectomies continue to be performed via open surgery. More than two years ago, I decided together with my good friend Dr. Iswanto Sucandy to compile a textbook on robotic liver surgery.

This unique textbook has been structured to provide the reader with a "start-to-finish" understanding of robotic liver surgery, enabling them to incorporate the robotic liver surgery into their own HPB practice. This book is the result of a combined effort of numerous world-renowned experts in the field of robotic liver surgery. The format of this book is such that the reader can quickly review a specific robotic liver procedure, with an emphasis on practical tips and tricks to be used while performing a specific procedure. Many key surgical steps have been illustrated with intraoperative pictures, medical illustrations, or diagrams. A part of this book is also dedicated to training in robotic liver surgery and the learning curve for the adoption of robotic liver surgery.

This book is stratified into seven main sections that cover all aspects of robotic liver surgery, including patient preparation, anesthesia, training, learning curves, and technical aspects. The final section covers the financial impact, registries, and the future directions of robotic liver surgery.

It has been a pleasure and a privilege to edit this first edition of the *Textbook of Robotic Liver Surgery* with my friend Dr. Iswanto Sucandy, and we hope to edit more editions of the *Textbook of Robotic Liver Surgery* in the future.

Kortrijk, Belgium | Mathieu D'Hondt

Contents

Part I
Introduction

Minimally Invasive Liver Surgery: Where Are We Now?

1

Mohammad Abu Hilal, Jasper P. Sijberden, and Tijs J. Hoogteijling

1.1 Introduction

Liver resection is considered one of the most complex and challenging visceral surgical procedures, requiring a strong mastery of anatomical knowledge, pathophysiological understanding, and advanced surgical skills. In general, in addition to technical skills and scientific knowledge on the topic, surgeons should have a thorough understanding of the historical evolution of their speciality to continue to be on top of new scientific discoveries and innovation and to be able to identify valid novel innovative techniques to embrace. Even though surgery is considered by many to be a modern and contemporary speciality, it traces its roots back to the ancient civilisations of Egypt, Greece, and earlier. Surgery has been practised throughout human history in various forms and has been motivated by different philosophies. Since the earliest surgical procedures, our understanding of illness and treatment has dramatically improved. This process ultimately led to the novel, advanced surgical practice of today. Even if modern surgery does not resemble what it did thousands of years ago, our historical predecessors set the groundwork for all the breakthroughs detailed in this book. This chapter provides an overview of the innovations and developments that have led to the implementation of minimally invasive liver surgery.

1.2 The Evolution of Minimally Invasive Liver Surgery

1.2.1 Historical Breakthroughs

One of the most important and successful innovations in surgery has been the rise of minimally invasive techniques. As we shall see, the global implementation of minimally invasive surgery covers only the last three centuries, but the first steps that are crucial to the introduction of any new technology were made long before that. It is beyond the scope of this chapter to give an in-depth overview of every scientific breakthrough and name every pioneer involved in the history of minimally invasive surgery. Therefore, we will solely give a brief review of some of the most important developments that have led to the wide global implementation of minimally invasive surgery. It is important to note that whilst the pioneers listed in this chapter mostly used technology that was already available to them, it is often the combination of several different technologies that leads to scientific discovery.

The evolution of minimally invasive surgery started in 1805 when the first-ever endoscopic procedure was credited to Philip Bozzini [1, 2].

M. Abu Hilal (✉) · J. P. Sijberden · T. J. Hoogteijling
Department of Surgery, Poliambulanza Foundation Hospital, Brescia, Italy

M. D'Hondt, I. Sucandy (eds.), *Textbook of Robotic Liver Surgery*,
https://doi.org/10.1007/978-3-031-76536-0_1

He combined his anatomical knowledge and the existing technology to produce a device that could be used to successfully inspect body cavities. The 'Lichtleiter', or light conductor as it was called, was built using a series of tubes, mirrors, and a candle and allowed physicians to identify and later operate on tumours in the vaginal canal and rectum [3]. This revolutionary invention could have paved the way for others to further develop endoscopic surgery, but it would take another century before the next endoscopic procedure was reported. In 1907, the German surgeon George Kelling performed a series of successful celioscopies on dogs. In his experiments, he introduced a cystoscope in the abdominal cavity after establishing a pneumoperitoneum [4]. Three years later, Swedish internist Hans Christian Jacobaeus published his series of successful endoscopic experiments and introduced the term 'laparoscopy' [5]. Although these innovations were all essential parts of the evolution of minimally invasive surgery, they were discredited in their time as being too experimental, controversial, and even 'insane'.

In the surgical community, most remained sceptical of these novel techniques. Despite this, there were several important developments in endoscopy in the years between the two world wars. Most notably, the German gastroenterologist Heinz Kalk and the American internist John Ruddock both used forward-facing endoscopes, trocars, and laparoscopic instruments [6]. Being internists rather than surgeons, they mainly used these instruments to perform solid organ biopsies. Both men served in the military medical service of their respective countries, which halted discovery and innovation during the Second World War. When the war was over, both Kalk and Ruddock published more about their initial experience with these innovative procedures, but they were still not widely adopted in general surgical practice [7]. In the subsequent time period, discoveries in the fields of fibre optics, cameras, and television paved the way for gynaecologist Camran Nezhat to develop video laparoscopy in the late 1970s [1, 8]. This pairing of a camera and an endoscope is the foundation upon which all endoscopic minimally invasive surgery is based today. Video laparoscopy started in gynaecology, mostly for tubal sterilisations, but eventually expanded to other indications and specialities. In general surgery, its implementation took more time and suffered critique by some of the most senior members of the surgical community. Despite this criticism, laparoscopic surgery started gaining momentum, and in 1987 the French surgeon Philippe Mouret successfully performed the first video-assisted laparoscopic cholecystectomy. Thereafter, the first report of a series of video-assisted laparoscopic cholecystectomies appeared in 1989 [9, 10]. After this suitably named 'second French revolution', the initially slow-paced advance of minimally invasive surgery suddenly started moving fast. Media attention led to a broad general interest in laparoscopic surgery and soon patients demanded that their operations be performed laparoscopically.

After the gallbladder, other organs and structures soon followed. Laparoscopic (partial) colectomy, splenectomy, and nephrectomy were all reported within the following few years [11]. Interestingly, the developments of minimally invasive techniques are not confined to thoracic and abdominal surgery. In cardiology and vascular surgery, the evolution of transluminal or endovascular procedures has developed parallel to that of endoscopic surgery [12]. The end of the twentieth century marked a paradigm shift in medical and surgical practice, changing the physician's approach towards the patient. The interests and medical arguments of physicians were balanced by the patient's desire to have the smallest scar possible following a surgical procedure. In line with the general trend towards 'patient-centred healthcare' and 'shared decision-making', the appeal to reduce surgical trauma and enhance postoperative recovery sparked a wave of research and innovation in minimally invasive approaches.

1.2.2 The First Laparoscopic Hepatectomies

It didn't take long after the sudden and rapid adoption of laparoscopy in general surgery before the first reports of laparoscopic hepatectomies

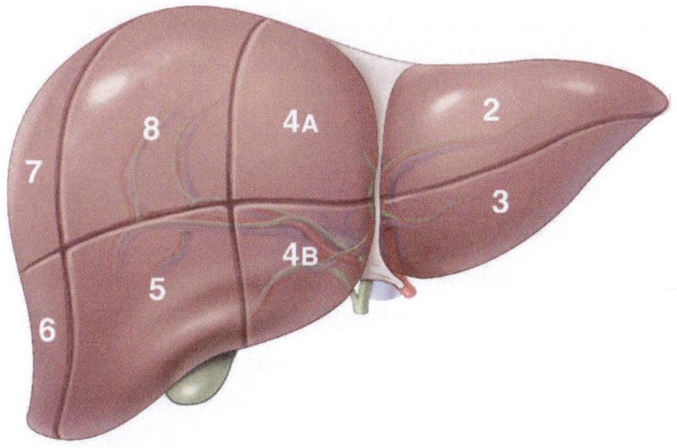

Fig. 1.1 The relatively easily accessible anterolateral part of the liver (segments II–VI) and the posterosuperior segments (segments I, IVa, VII, and VIII)

were published. The first laparoscopic hepatectomy was performed only four years after the first laparoscopic cholecystectomy by Harry Reich and published in *Obstetrics and Gynaecology* in 1991 [13]. In his article, he described the cases of two women who had superficial benign neoplasms on the liver edge, discovered during a diagnostic laparoscopy for gynaecological symptoms. In the following years, a larger series of laparoscopic liver resections performed by general surgeons were published [14–16]. Despite this initial success, international implementation was slow in the late 1990s, with just a few pioneering surgeons performing laparoscopic hepatectomies. Also, resections were initially only performed for lesions in the relatively easily accessible anterolateral part of the liver (segments II–VI) (Fig. 1.1). Nevertheless, the beginning of the twenty-first century marked a turning point as an increasing number of surgeons embraced minimally invasive techniques.

1.3 Worldwide Implementation of Minimally Invasive Liver Surgery

In the first decade of the twenty-first century, laparoscopic liver surgery was gaining more and more popularity around the world. By 2008, laparoscopic liver resections were performed in many American, European, and Asian specialist centres [17]. A growing body of evidence showed that laparoscopic minor hepatectomies were safe and effective [18–21]. A consensus meeting was deemed necessary, and therefore the first consensus meeting was held in Louisville, Kentucky, in 2008 [22]. An international panel of hepatobiliary surgery experts analysed the existing literature on minimally invasive liver resection and through a process of discussion and voting produced a consensus statement. The experts agreed that lesions smaller than 5 cm in diameter located in the anterolateral segments could be resected safely by laparoscopy. Furthermore, laparoscopic left lateral sectionectomy (resection of segments II and III) was characterised as standard practice. In addition to the consensus statement, the Louisville meeting also produced an enormous enthusiasm towards minimally invasive approaches in liver surgery. It sparked a wave of studies comparing the outcomes after laparoscopic and conventional open liver surgery [23].

For both hepatocellular carcinoma and colorectal liver metastasis, studies associating laparoscopic liver surgery with, compared to open liver surgery, improved short-term and similar oncological outcomes were being published [24, 25]. Some pioneering surgeons started performing more technically complex procedures laparoscopically, such as hemi hepatectomies or resections for lesions in the posterosuperior segments (segments I, IVa, VII, and VIII) (Fig. 1.1) [26].

The rapid increase in available evidence meant that the statements of 2008 became outdated, and the second international consensus meeting on laparoscopic liver surgery was held in 2014 in Morioka, Japan [27]. In contrast to the previous meeting, this one was divided into domains for which thorough literature reviews were conducted by expert panels. One of the conclusions was that the overall quality of the available evidence was low. Furthermore, the jury concluded that laparoscopic minor liver resections have become standard practice and major liver resections should be implemented carefully as they were still considered to be in the exploratory phase. Finally, the concept of parenchyma-sparing liver resection was recommended as a safe and feasible alternative to anatomical resections.

After Morioka, further progress was made regarding the technical aspects of laparoscopic liver surgery. One of the major advancements was the introduction of the term 'technically major' for resections of lesions in the posterosuperior segments, dividing the heterogeneous group of major liver resections into two [26, 28, 29]. After this subclassification was introduced, specific techniques to perform these technically complex resections were developed. These standardised techniques included, among others, the 'diamond technique', named after the diamond-like shape of the resection specimen, and the parenchyma-sparing anatomical resection. The latter technique uses extensive preoperative imaging-guided liver mapping to perform the smallest possible anatomical resection through Glissonean dissection and subsegmentectomy [30–32].

The third consensus meeting was held in Southampton, UK, in 2017 in the form of the first European Guidelines Meeting on Laparoscopic Liver Surgery (EGMLLS) [33]. Whilst the meetings of Louisville and Morioka were mainly focused on feasibility and non-inferiority, the aim of the EGMLLS was to develop clinical guidelines which stated in which context laparoscopic liver surgery was superior to the traditional open approach. Another unique feature was the strong synergetic combination of different methodologies to achieve a validated consensus and the strict adherence to validated methodologies. In fact, the EGMLLS adopted the SIGN, GRADE, Delphi, and AGREE methodologies to produce guidelines of the highest level of consensus [34, 35]. Interestingly, the meta-analyses performed in the preparation of EGMLLS revealed superior short-term outcomes after laparoscopic liver surgery compared to open liver surgery, whilst maintaining comparable oncological outcomes for all indications. Furthermore, in previously regarded difficult clinical indications such as cirrhosis and hepatocellular carcinoma (HCC), laparoscopy was deemed superior to open surgery when performed by expert surgeons at high-volume centres. Indications for laparoscopic hepatectomy were widened further as elderly and obese patients were no longer considered a contraindication for the use of a minimally invasive approach. Finally, the strongest domain of EGMLLS in the eyes of many was the domain of future implementation. It was considered a great achievement to reach a consensus on the need to have at least two competent laparoscopic liver surgeons in each liver unit. The quality of evidence, however, was still relatively low. Most studies included in the meta-analyses were retrospective cohort studies and randomised trials were lacking.

However, the need for higher quality evidence was met a year after the Southampton guidelines meeting. In 2018, the first randomized controlled trial (RCT) comparing laparoscopic with conventional open liver surgery was published. The OSLO-COMET trial found significantly fewer postoperative complications and similar R0 resection rates after laparoscopic liver surgery in patients with colorectal liver metastases [36]. In a recently published meta-analysis of RCTs comparing laparoscopic and open hepatectomies, 13 RCTs were included, mostly solely assessing the outcomes of minor liver resections in the anterolateral segments [37]. In this meta-analysis, laparoscopy, compared to open surgery, was associated with less blood loss, lower rates of overall and severe morbidity, and a shorter time to functional recovery and length of stay. Although these high-quality studies are definitely in favour of the laparoscopic approach in selected patients, results of randomised studies comparing the laparoscopic and open approach for resections in the posterosuperior segments or of major extent are not yet available. The ORANGE SEGMENTS and ORANGE II PLUS trials have been designed to fill in this evidence gap, and accrual for both trials was recently completed [38]. The published preliminary data of the ORANGE 2 PLUS trial, comparing laparoscopic and open left and right hemi hepatectomies, shows a benefit of laparoscopy, over open surgery, in terms of time to functional recovery and length of stay [39].

1.4 Pushing Boundaries

Whilst gaining more and more experience with minimally invasive liver surgery and supported by improving technology, pioneers and early adopters have been constantly pushing boundaries. This process has ultimately led to a wider implementation of the minimally invasive approach for challenging procedures [40]. Several patients' disease-related or anatomical factors can make a procedure 'challenging' [41–43]. An important anatomical factor is the location of the tumour. Patients with lesions situated in the posterosuperior segments of the liver were initially deemed poor candidates for a minimally invasive resection due to the difficulty in gaining adequate access to this location, limited visualisation, and difficulty in bleeding control [22]. However, the first reports demonstrating the feasibility of these procedures, in expert hands, were published in the years prior to the Louisville consensus conference [44, 45]. Since then, several standardised surgical techniques for these resections have been proposed, and in this day and age laparoscopic surgery can offer several benefits over open surgery in this setting, in terms of less blood loss, postoperative complications, analgesics requirements, a shorter hospital stay, and time to initiation of postoperative chemotherapy [32, 46–49].

As stated, liver resections, especially for colorectal liver metastases, are increasingly performed in a parenchyma-sparing manner, supported by encouraging results [50, 51]. Utilising this approach, several separate resections are often needed to clear the liver of disease. While parenchymal-sparing resections can be safely performed by a minimally invasive approach, further reducing the surgical stress, some surgeons remain reluctant to allocate patients with multiple lesions to this approach [31]. This reluctance mainly seems to be caused by the possible long operative time, need for more ports, and need to repeatedly perform ultrasonography, parenchymal transection, and possible need to change patient positioning. Despite these, a recent report has indicated that multiple separate resections can be safely performed minimally invasive [52]. Nevertheless, it has been suggested that the risk of positive surgical margins increases when multiple lesions are resected, both in laparoscopic and open surgery [53].

Aside from the number of lesions, large lesion(s) can also be challenging to resect using the minimally invasive approach [42]. When large lesions are present, laparoscopic mobilisation in the confined space of the closed abdomen is often more difficult. Additionally, large lesions can distort the anatomy of the liver, are often close to major blood vessels, and rupture more easily during surgical manoeuvres. Also, an extensive liver resection and sizable extraction site are often needed, which made surgeons question the overall benefit of a minimally invasive approach in these cases. For these reasons, large lesions were initially considered a contraindication for the use of a minimally invasive approach [22]. However, more recent studies have suggested that in expert hands and through adopting specific techniques, minimally invasive resections of large and giant lesions are feasible, safe, and oncologically efficient [25, 54–59]. Nevertheless, resection for large lesions has been associated with longer operative times, more blood loss, transfusions, morbidity, and mortality [54, 60]. Hence, a cautious surgical strategy is still warranted in these cases.

In most types of malignant liver disease, disease recurrence after a curative resection is common. Fortunately, repeat resections of the liver are now often feasible and associated with similar long-term outcomes as primary liver resections in patients without extrahepatic disease [61, 62]. Nevertheless, repeat liver resections can be challenging from a technical point of view due to adhesions, distortion of the liver anatomy, and the possible previous use of other treatment modalities (e.g. extensive chemotherapy, thermal ablation). One can imagine that these procedures can be especially difficult to perform minimally invasive, and most minimally invasive surgeons have also suggested that repeat liver resections are associated with a higher degree of technical difficulty and a greater risk of conversion [63]. This has led to the formulation of the statement that repeat liver resections should be avoided in the early phase of the learning curve in the Southampton consensus guidelines [33]. When this learning curve has been surpassed, the minimally invasive approach can still offer benefits over the open approach in this technically complex setting, and it may facilitate additional repeat resections or a salvage liver transplantation at a later stage [64–69].

It has been suggested that the latter benefits are also present in the setting of two-stage hepatectomies. Nevertheless, this procedure can be very complicated as it is per definition performed for patients with extensive disease, and in the second stage the surgeon can encounter extensive adhesions, anatomical changes, and a liver damaged by chemotherapy [70]. Additionally, it might be needed to modulate the future liver remnant, which can lead to hypertrophy of the contralateral lobe and perivascular inflammation. Therefore, these procedures are probably only performed in centres with huge experience in minimally invasive liver surgery. Nevertheless, evidence from major centres does so far suggest that this procedure can be safely performed in selected patients and is not associated with inferior oncological outcomes compared to open surgery [70–72]. Recently, several authors working at major centres have also described their initial, limited but favourable, experience with the minimally invasive approach to perform associating liver partition and portal vein ligation for staged hepatectomy (ALPPS) [73]. In the setting of these technically complex procedures, more data and larger studies are still needed before a wider implementation of the minimally invasive approach can be recommended.

Another important topic is the role of the minimally invasive approach in patients with synchronous colorectal liver metastases that are eligible for a resection of both the primary and the liver metastases. In these patients, the optimal timing strategy for both resections remains a matter of debate. Nevertheless, simultaneous resections have gained traction over the past decades. While initially thought to be possibly unsafe, recent studies have associated simultaneous resections with non-inferior survival, morbidity, and mortality rates with a shorter length of hospital stay compared to two-staged procedures [74, 75]. Simultaneous resections can obviously also provide socio-economic benefits as patients are only admitted once and undergo one surgical procedure to clear their disease. Nevertheless, patient selection seems to be of vital importance in this context [76]. These procedures can be lengthy and complex, which has also led to an initial reluctance to use a minimally invasive approach in this setting [63]. Nevertheless, due to the growing adoption and experience with the minimally invasive approach for both colorectal and liver resections, the role of minimally invasive simultaneous resections also seems to expand. One could hypothesise that the minimally invasive approach could mitigate surgical stress in the setting of these extensive procedures, thereby further reducing its potential risks. A recently published meta-analysis supports this hypothesis as the minimally invasive approach was associated with several advantages over the open approach in terms of less intraoperative blood loss, a lower postoperative morbidity rate, and a shorter length of hospital stay [77].

In conclusion, the exploratory work of pioneers and early adopters in minimally invasive liver surgery has gradually extended the indications for the minimally invasive approach, further widening the implementation and expansion of this approach.

1.5 The Future Prospects: Robotic Liver Surgery

In the present day, the minimally invasive approach has embedded itself in the surgical treatment of benign and malignant hepatic lesions. Technological advances in artificial intelligence, (contrast enhanced) imaging, and 3D modelling have enabled minimally invasive surgeons to perform the most technically challenging procedures through a minimally invasive approach [78]. Since Reich's first-ever laparoscopic hepatectomy three decades ago, minimally invasive liver surgery has evolved into the state-of-the-art practice of today. Now, surgery is on the brink of entering a new phase. Following in the footsteps of urologists and gynaecologists, general surgeons are using robotic approaches for a wide range of indications, making robotic liver resection (RLR) the next possibly game-changing evolution in surgical practice [79].

In sharp contrast to the lengthy and arduous evolution of laparoscopy, robotic surgery was quickly embraced since its introduction in the early 1990s. Although the earliest devices suffered from the low-resolution monitors and relatively crude digital technology of that time, robotic surgery profited from the wave of enthusiasm for minimally invasive techniques. Military medical branches across the world harboured a particular interest in robotic surgery platforms due to their potential for telesurgery, meaning the robot could be positioned in a field hospital while the surgeon remotely controls it from the safety of the military base [80]. In fact, at the end of the twentieth century, the US military invested greatly in the development of the early robotic platforms [81]. Even though a military application of surgical robots has not yet been realised thus far, the technique has quickly become popular with civilian doctors and patients.

The basic principle of surgery robots is that of a 'middleman' or a 'master slave' [82]. The robot functions as a large, complex, and indeed expensive piece of equipment, which is controlled through a console by an autonomous surgeon. Courtesy of its machinated arms, the robot can filter out tremors, accentuate or refine movements, and allow the surgeon control over multiple instruments at the same time [83]. The endo-articulating arms and 3D optics provide superior dexterity and vision, which are both highly desirable traits in complex surgical procedures such as a hepatectomy. These traits, together with the generally more receptive attitude of the surgical community towards new developments in these modern times, have facilitated a much faster implementation of robotic surgery compared with laparoscopic surgery. Within 30 years after the development of the first surgical robot, the DaVinci® platform found its way into the most specialised, modern operating theatres around the world [84].

Promising early results of comparative cohort studies have facilitated this rapid implementation. Since the first robotic hepatectomy was published in 1993 [85], there have been several studies that have shown non-inferior or even superior outcomes after RLR compared to laparoscopic resections [86–90]. Most notably, less intraoperative blood loss and lower conversion rates during technically challenging resections demonstrate the merits of the robot [91, 92]. On a critical note, these results are often subject to an inevitable degree of selection bias, which underlines the need for randomised controlled trials. Furthermore, the robotic approach has some important limitations in the field of liver surgery. The absence of standardised techniques, the incompatibility of specific energy devices like the Cavitron Ultrasonic Surgical Aspirator (CUSA), and the lack of textile feedback through the console might negatively affect the quality of the parenchymal transection [83] or result in an excess of tissue stress [83, 93]. One could speculate that if the proper instrumentation is introduced, more elegant and precise dissections can be performed, leading to a much quicker expansion of the robotic surgical approach that has been seen so far.

In conclusion, more than two centuries of discoveries and development of minimally invasive liver surgery have led to the cutting-edge practice of today. Highly trained surgeons are performing every type of liver resection, from relatively straightforward to extremely complex, while simultaneously trying to reduce the surgical trauma inflicted on the patient. The recent introduction of robotic liver surgery marks the beginning of the next phase in minimally invasive liver surgery and will continue to be one of the main focusses of surgical research. The following chapters will give a comprehensive overview of the history and implementation of robotic liver surgery, as well as a review of its current surgical techniques, merits, pitfalls, and financial aspects.

Box 1.1 Segmental Anatomy

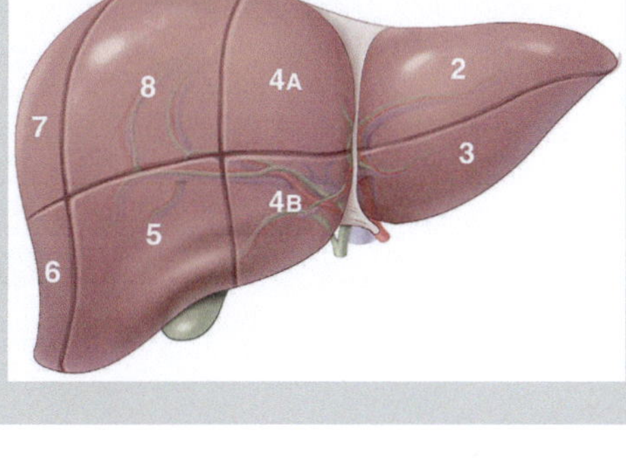

Performing surgical procedures on the liver has always been considered a hazardous undertaking, associated with a high risk of postoperative mortality, mostly due to bleeding. The liver is one of the most perfused organs in the body due to its filtering function. The liver is perfused via *arterial and portal venous branches and drains* via *the hepatic veins. Neither of these vascular structures are visible nor palpable from the liver surface. Therefore, dissection of the liver parenchyma can lead to catastrophic bleeding, leading to postoperative mortality rates of around 20% in the 1970s* [94, 95]. *A breakthrough in the anatomical understanding of the liver came in 1956 when Claude Couinaud published* 'The liver, anatomical studies and surgical studies'. [95, 96] *Based on liver casts, he identified eight segments fed by their own portal or Glissonean pedicle consisting of a hepatic artery branch, a portal branch, and a biliary duct. His anatomical findings and classification are, to this day, used in liver surgery. By identifying and selectively ligating a Glissonean pedicle, it is possible to resect an individual segment or even perform a subsegmental anatomical resection. However, due to the inferior imaging technologies of the time, segmentectomies were seldom performed. Smaller tumours confined to one segment could rarely be identified by radiography, meaning that liver resections were usually performed for larger lesions. These typically necessitate a larger liver resection such as a hemi hepatectomy. In the following decades, advancements in medical imaging led to the earlier detection of smaller liver tumours. The use of pre- and intraoperative imaging techniques as well as (3D) liver mapping and navigation has become of the utmost importance when performing complex liver surgery* [33]..

Box 1.2 The Diamond Technique

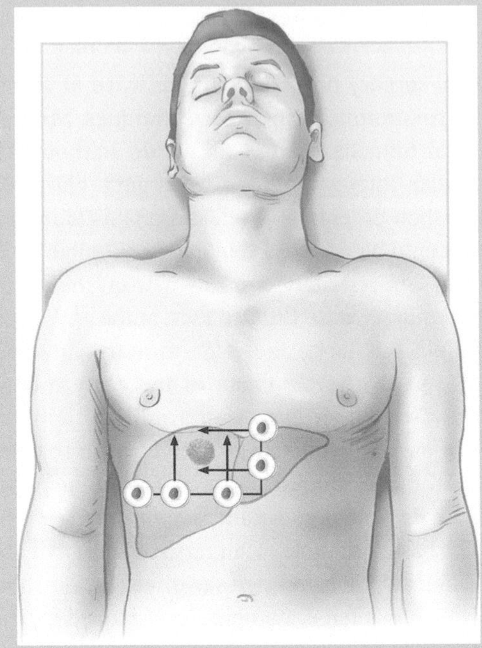

(with photo/image of the reverse L-shape port placement and 'diamond')

Laparoscopic anatomical resection of tumours located in the posterosuperior segments is considered the most technically challenging. Shielded by the diaphragm and lower ribs, and with their Glissonean pedicle buried deep within the liver parenchyma, laparoscopically resecting the posterosuperior segments is no easy task. With the emergence of parenchyma-sparing resections for metastatic (and carefully selected primary) liver tumours came the development of the 'diamond technique'. This technique is based on eight pillars which, if adhered to, ensure a parenchyma-sparing resection without sacrificing oncological safety. The eight pillars are

1. *The use of straight transection lines, minimalising the risk of an irradical resection.*
2. *Adequate port positioning in a reverse L-shape, ensuring parallel resection planes and facilitating pillar one.*
3. *Maintaining an adequate resection margin, determined and assessed by ultrasound.*
4. *Continuous use of intraoperative ultrasound assessment.*
5. *Thorough mobilisation of the liver and the use of gravity through patient positioning to ensure optimal exposure of the posterosuperior segments.*
6. *Fine and meticulous dissection using both ultrasonic and CUSA energy devices.*
7. *Gaining access to the posterior planes using cardinal stitches along the edges of the resection lines, facilitating the completion of the resection.*
8. *Employing intraoperative strategies to create a dry operative field (central venous pressure 2–5 mmHg, intermittent Pringle manoeuvre, haemostatic products) helps identify and control structures during parenchymal transection.*

A retrospective analysis of patients undergoing resection of posterosuperior segments using the diamond technique revealed non-inferior oncological and in fact superior short-term surgical outcomes when performed by highly specialised open and laparoscopic surgeons [31, 97, 98].

References

1. Nezhat C. Nezhat's history of endoscopy: a historical analysis of endoscopy's ascension since antiquity. Endo-Press; 2011.
2. Hargest R. Five thousand years of minimal access surgery: 3000 BC to 1850: early instruments for viewing body cavities. J R Soc Med. 2020;113(12):491–6.
3. Ramai D, Zakhia K, Etienne D, Reddy M. Philipp Bozzini (1773–1809): The earliest description of endoscopy. J Med Biogr. 2018;26(2):137–41.
4. Hargest R. Five thousand years of minimal access surgery: 1850 to 1990: Technological developments. J R Soc Med. 2021;114(1):19–29.
5. Litynski GS. Laparoscopy—the early attempts: spotlighting Georg Kelling and Hans Christian Jacobaeus. Jsls. 1997;1(1):83–5.
6. Morgenstern L. No surgeon he. John C. Ruddock, M.D., F.A.C.P., pioneer in laparoscopy. Surg Endosc. 1996;10(6):617–8.
7. Litynski GS. Laparoscopy between the world wars: the barriers to trans-atlantic exchange. Spotlighting Heinz Kalk and John C. Ruddock. JSLS. 1997;1(2):185–8.
8. Kelley WE Jr. The evolution of laparoscopy and the revolution in surgery in the decade of the 1990s. JSLS. 2008;12(4):351–7.
9. Dubois F, Berthelot G, Levard H. [Cholecystectomy by coelioscopy]. Presse Med. 1989; 18(19):980–2.
10. Wickham JE. The new surgery. Br Med J (Clin Res Ed). 1987;295(6613):1581–2.
11. Berci G, Forde KA. History of endoscopy: what lessons have we learned from the past? Surg Endosc. 2000;14(1):5–15.
12. Parodi JC, Palmaz JC, Barone HD. Transfemoral intraluminal graft implantation for abdominal aortic aneurysms. Ann Vasc Surg. 1991;5(6):491–9.
13. Reich H, McGlynn F, DeCaprio J, Budin R. Laparoscopic excision of benign liver lesions. Obstet Gynecol. 1991;78(5 Pt 2):956–8.
14. Gagner M. Laparoscopic partial hepatectomy for liver tumor. Surg Endosc. 1992;6:97–8.
15. Azagra JS, Goergen M, Gilbart E, Jacobs D. Laparoscopic anatomical (hepatic) left lateral segmentectomy-technical aspects. Surg Endosc. 1996;10(7):758–61.
16. Kaneko H, Takagi S, Shiba T. Laparoscopic partial hepatectomy and left lateral segmentectomy: technique and results of a clinical series. Surgery. 1996;120(3):468–75.
17. Aghayan DL, Kazaryan AM, Fretland ÅA, Røsok B, Barkhatov L, Lassen K, Edwin B. Evolution of laparoscopic liver surgery: 20-year experience of a Norwegian high-volume referral center. Surg Endosc. 2022;36(5):2818–26.
18. Koffron AJ, Auffenberg G, Kung R, Abecassis M. Evaluation of 300 minimally invasive liver resections at a single institution: less is more. Ann Surg. 2007;246(3):385–92; discussion 392–384.
19. Vibert E, Perniceni T, Levard H, Denet C, Shahri NK, Gayet B. Laparoscopic liver resection. Br J Surg. 2006;93(1):67–72.
20. Gigot JF, Glineur D, Santiago Azagra J, Goergen M, Ceuterick M, Morino M, Etienne J, Marescaux J, Mutter D, van Krunckelsven L, et al. Laparoscopic liver resection for malignant liver tumors: preliminary results of a multicenter European study. Ann Surg. 2002;236(1):90–7.
21. Cherqui D, Husson E, Hammoud R, Malassagne B, Stéphan F, Bensaid S, Rotman N, Fagniez PL. Laparoscopic liver resections: a feasibility study in 30 patients. Ann Surg. 2000;232(6):753–62.

22. Buell JF, Cherqui D, Geller DA, O'Rourke N, Iannitti D, Dagher I, Koffron AJ, Thomas M, Gayet B, Han HS, et al. The international position on laparoscopic liver surgery: The Louisville Statement, 2008. Ann Surg. 2009;250(5):825–30.
23. Abu Hilal M. Why do we need guidelines in laparoscopic liver surgery? HPB (Oxford). 2017;19(4):287–8.
24. Kazaryan AM, Marangos IP, Røsok BI, Rosseland AR, Villanger O, Fosse E, Mathisen O, Edwin B. Laparoscopic resection of colorectal liver metastases: surgical and long-term oncologic outcome. Ann Surg. 2010;252(6):1005–12.
25. Ai JH, Li JW, Chen J, Bie P, Wang SG, Zheng SG. Feasibility and safety of laparoscopic liver resection for hepatocellular carcinoma with a tumor size of 5–10 cm. PLoS One. 2013;8(8):e72328.
26. Di Fabio F, Samim M, Di Gioia P, Godeseth R, Pearce NW, Abu Hilal M. Laparoscopic major hepatectomies: clinical outcomes and classification. World J Surg. 2014;38(12):3169–74.
27. Wakabayashi G, Cherqui D, Geller DA, Buell JF, Kaneko H, Han HS, Asbun H, O'Rourke N, Tanabe M, Koffron AJ, et al. Recommendations for laparoscopic liver resection: a report from the second international consensus conference held in Morioka. Ann Surg. 2015;261(4):619–29.
28. Morise Z, Wakabayashi G. First quarter century of laparoscopic liver resection. World J Gastroenterol. 2017;23(20):3581–8.
29. Wakabayashi G. What has changed after the Morioka consensus conference 2014 on laparoscopic liver resection? Hepatobiliary Surg Nutr. 2016;5(4):281–9.
30. Berardi G, Igarashi K, Li CJ, Ozaki T, Mishima K, Nakajima K, Honda M, Wakabayashi G. Parenchymal sparing anatomical liver resections with full laparoscopic approach: description of technique and short-term results. Ann Surg. 2021;273(4):785–91.
31. Cipriani F, Shelat VG, Rawashdeh M, Francone E, Aldrighetti L, Takhar A, Armstrong T, Pearce NW, Abu Hilal M. Laparoscopic parenchymal-sparing resections for nonperipheral liver lesions, the diamond technique: technical aspects, clinical outcomes, and oncologic efficiency. J Am Coll Surg. 2015;221(2):265–72.
32. Abu Hilal M, Tschuor C, Kuemmerli C, López-Ben S, Lesurtel M, Rotellar F. Laparoscopic posterior segmental resections: how I do it: tips and pitfalls. Int J Surg. 2020;82S:178–86.
33. Abu Hilal M, Aldrighetti L, Dagher I, Edwin B, Troisi RI, Alikhanov R, Aroori S, Belli G, Besselink M, Briceno J, et al. The Southampton Consensus Guidelines for Laparoscopic Liver Surgery: From Indication to Implementation. Ann Surg. 2018;268(1):11–8.
34. Network SIG. Sign 50: a guideline developer's handbook. Healthcare Improvement Scotland; 2014.
35. Brouwers MC, Kho ME, Browman GP, Burgers JS, Cluzeau F, Feder G, Fervers B, Graham ID, Grimshaw J, Hanna SE, et al. AGREE II: advancing guideline development, reporting and evaluation in health care. CMAJ. 2010;182(18):E839–42.
36. Fretland ÅA, Dagenborg VJ, Bjørnelv GMW, Kazaryan AM, Kristiansen R, Fagerland MW, Hausken J, Tønnessen TI, Abildgaard A, Barkhatov L, et al. Laparoscopic versus open resection for colorectal liver metastases: the OSLO-COMET randomized controlled trial. Ann Surg. 2018;267(2):199–207.
37. Haney CM, Studier-Fischer A, Probst P, Fan C, Müller PC, Golriz M, Diener MK, Hackert T, Müller-Stich BP, Mehrabi A, et al. A systematic review and meta-analysis of randomized controlled trials comparing laparoscopic and open liver resection. HPB (Oxford). 2021;23(10):1467–81.
38. Kuemmerli C, Fichtinger RS, Moekotte A, Aldrighetti LA, Aroori S, Besselink MGH, D'Hondt M, Díaz-Nieto R, Edwin B, Efanov M, Ettorre GM, Menon KV, Sheen AJ, Soonawalla Z, Sutcliffe R, Troisi RI, White SA, Brandts L, van Breukelen GJP, Sijberden J, Pugh SA, Eminton Z, Primrose JN, van Dam R, Hilal MA. Laparoscopic versus open resections in the posterosuperior liver segments within an enhanced recovery programme (ORANGE Segments): study protocol for a multicentre randomised controlled trial. Trials. 2022;23:206.
39. Fichtinger RS, Aldrighetti L, Troisi R, Abu Hilal M, Sutcliffe R, Besselink M, Aroori S, Menon K, Edwin B, D'Hondt M, et al. 384O Laparoscopic versus open hemihepatectomy: the ORANGE II PLUS multicenter randomized controlled trial. Ann Oncol. 2021;32:S531.
40. Berardi G, Van Cleven S, Fretland ÅA, Barkhatov L, Halls M, Cipriani F, Aldrighetti L, Abu Hilal M, Edwin B, Troisi RI. Evolution of Laparoscopic liver surgery from innovation to implementation to mastery: perioperative and oncologic outcomes of 2,238 patients from 4 European Specialized Centers. J Am Coll Surg. 2017;225(5):639–49.
41. Ban D, Tanabe M, Ito H, Otsuka Y, Nitta H, Abe Y, Hasegawa Y, Katagiri T, Takagi C, Itano O, et al. A novel difficulty scoring system for laparoscopic liver resection. J Hepatobiliary Pancreat Sci. 2014;21(10):745–53.
42. Halls MC, Berardi G, Cipriani F, Barkhatov L, Lainas P, Harris S, D'Hondt M, Rotellar F, Dagher I, Aldrighetti L, et al. Development and validation of a difficulty score to predict intraoperative complications during laparoscopic liver resection. Br J Surg. 2018;105(9):1182–91.
43. Kawaguchi Y, Fuks D, Kokudo N, Gayet B. Difficulty of laparoscopic liver resection: proposal for a new classification. Ann Surg. 2018;267(1):13–7.
44. Yoon YS, Han HS, Choi YS, Jang JY, Suh KS, Kim SW, Lee KU, Park YH. Total laparoscopic right posterior sectionectomy for hepatocellular carcinoma. J Laparoendosc Adv Surg Tech A. 2006;16(3):274–7.
45. Cho JY, Han HS, Yoon YS, Shin SH. Feasibility of laparoscopic liver resection for tumors located in the posterosuperior segments of the liver, with a special reference to overcoming current limitations on tumor location. Surgery. 2008;144(1):32–8.

46. Scuderi V, Barkhatov L, Montalti R, Ratti F, Cipriani F, Pardo F, Tranchart H, Dagher I, Rotellar F, Abu Hilal M, et al. Outcome after laparoscopic and open resections of posterosuperior segments of the liver. Br J Surg. 2017;104(6):751–9.
47. Kazaryan AM, Røsok BI, Marangos IP, Rosseland AR, Edwin B. Comparative evaluation of laparoscopic liver resection for posterosuperior and anterolateral segments. Surg Endosc. 2011;25(12):3881–9.
48. Xiao L, Xiang LJ, Li JW, Chen J, Fan YD, Zheng SG. Laparoscopic versus open liver resection for hepatocellular carcinoma in posterosuperior segments. Surg Endosc. 2015;29(10):2994–3001.
49. D'Hondt M, Tamby E, Boscart I, Turcotte S, Parmentier I, Pottel H, Lapointe R, Ovaere S, Vansteenkiste F, Vandenbroucke-Menu F. Laparoscopic versus open parenchymal preserving liver resections in the posterosuperior segments: a case-matched study. Surg Endosc. 2018;32(3):1478–85.
50. Torzilli G, McCormack L, Pawlik T. Parenchyma-sparing liver resections. Int J Surg. 2020;82S:192–7.
51. Gold JS, Are C, Kornprat P, Jarnagin WR, Gönen M, Fong Y, DeMatteo RP, Blumgart LH, D'Angelica M. Increased use of parenchymal-sparing surgery for bilateral liver metastases from colorectal cancer is associated with improved mortality without change in oncologic outcome: trends in treatment over time in 440 patients. Ann Surg. 2008;247(1):109–17.
52. Nassar A, Tribillon E, Marchese U, Faermark N, Bonnet S, Beaussier M, Gayet B, Fuks D. Feasibility and outcomes of multiple simultaneous laparoscopic liver resections. Surg Endosc. 2022;36(4):2466–72.
53. Benedetti Cacciaguerra A, Görgec B, Cipriani F, Aghayan D, Borelli G, Aljaiuossi A, Dagher I, Gayet B, Fuks D, Rotellar F, et al. Risk Factors of Positive Resection Margin in Laparoscopic and Open Liver Surgery for Colorectal Liver Metastases: A New Perspective in the Perioperative Assessment: A European Multicenter Study. Ann Surg. 2022;275(1):e213–21.
54. Shelat VG, Cipriani F, Basseres T, Armstrong TH, Takhar AS, Pearce NW, AbuHilal M. Pure laparoscopic liver resection for large malignant tumors: does size matter? Ann Surg Oncol. 2015;22(4):1288–93.
55. Yang TH, Chen JL, Lin YJ, Chao YJ, Shan YS, Hsu HP, Su ZM, Chou CC, Yen YT. Laparoscopic surgery for large left lateral liver tumors: safety and oncologic outcomes. Surg Endosc. 2018;32(10):4314–20.
56. Kwon Y, Han HS, Yoon YS, Cho JY. Are large hepatocellular carcinomas still a contraindication for laparoscopic liver resection? J Laparoendosc Adv Surg Tech A. 2015;25(2):98–102.
57. Nomi T, Fuks D, Louvet C, Nakajima Y, Gayet B. Outcomes of laparoscopic liver resection for patients with large colorectal liver metastases: a case-matched analysis. World J Surg. 2016;40(7):1702–8.
58. Levi Sandri GB, Spoletini G, Vennarecci G, Francone E, Abu Hilal M, Ettorre GM. Laparoscopic liver resection for large HCC: short- and long-term outcomes in relation to tumor size. Surg Endosc. 2018;32(12):4772–9.
59. Eunmi Gil M, Kwon CHD, Kim JM, Gyu-Seong Choi M, Heo JS, Cho W, Gwak MS, Geum-Youn Gwak M, Joh J-W. Laparoscopic liver resection of hepatocellular carcinoma with a tumor size larger than 5 cm: review of 45 cases in a tertiary institution. J Laparoendosc Adv Surg Tech A. 2017;27(8):799–803.
60. Kabir T, Syn N, Koh YX, Teo JY, Chung AY, Chan CY, Goh BKP. Impact of tumor size on the difficulty of minimally invasive liver resection. Eur J Surg Oncol. 2022;48(1):169–76.
61. Adam R, Bismuth H, Castaing D, Waechter F, Navarro F, Abascal A, Majno P, Engerran L. Repeat hepatectomy for colorectal liver metastases. Ann Surg. 1997;225(1):51–60; discussion 60–52.
62. Sa Cunha A, Laurent C, Rault A, Couderc P, Rullier E, Saric J. A second liver resection due to recurrent colorectal liver metastases. Arch Surg. 2007;142(12):1144–9.
63. Halls MC, Cherqui D, Taylor MA, Primrose JN, Abu Hilal M. Are the current difficulty scores for laparoscopic liver surgery telling the whole story? An international survey and recommendations for the future. HPB (Oxford). 2018;20(3):231–6.
64. van der Poel MJ, Barkhatov L, Fuks D, Berardi G, Cipriani F, Aljaiuossi A, Lainas P, Dagher I, D'Hondt M, Rotellar F, et al. Multicentre propensity score-matched study of laparoscopic versus open repeat liver resection for colorectal liver metastases. Br J Surg. 2019;106(6):783–9.
65. Levi Sandri GB, Lai Q, Ravaioli M, Di Sandro S, Balzano E, Pagano D, Magistri P, Di Benedetto F, Rossi M, Gruttadauria S, et al. The role of salvage transplantation in patients initially treated with open versus minimally invasive liver surgery: an intention-to-treat analysis. Liver Transpl. 2020;26(7):878–87.
66. Morise Z, Aldrighetti L, Belli G, Ratti F, Belli A, Cherqui D, Tanabe M, Wakabayashi G. Laparoscopic repeat liver resection for hepatocellular carcinoma: a multicentre propensity score-based study. Br J Surg. 2020;107(7):889–95.
67. Levi Sandri GB, Colasanti M, Aldrighetti L, Guglielmi A, Cillo U, Mazzaferro V, Dalla Valle R, De Carlis L, Gruttadauria S, Di Benedetto F, et al. Is minimally invasive liver surgery a reasonable option in recurrent HCC? A snapshot from the I Go MILS registry. Updates Surg. 2022;74(1):87–96.
68. Shelat VG, Serin K, Samim M, Besselink MG, Al Saati H, Gioia PD, Pearce NW, Abu Hilal M. Outcomes of repeat laparoscopic liver resection compared to the primary resection. World J Surg. 2014;38(12):3175–80.
69. Chen JF, Fu XT, Gao Z, Shi YH, Tang Z, Liu WR, Zhang X, Gao Q, Ding GY, Song K, et al. Laparoscopic vs. open repeat hepatectomy for recurrent liver tumors: a propensity score-matched study and meta-analysis. Front Oncol. 2021;11:646737.
70. Görgec B, Suhool A, Al-Jarrah R, Fontana M, Tehami NA, Modi S, Abu Hilal M. Surgical technique and clinical results of one- or two-stage laparoscopic right hemihepatectomy after portal vein embolization in patients with initially unresectable colorectal liver metastases: a case series. Int J Surg. 2020;77:69–75.

71. Taillieu E, De Meyere C, D'Hondt M. The role of the laparoscopic approach in two-stage hepatectomy for colorectal liver metastases: a single-center experience. Surg Endosc. 2022;36(1):559–68.
72. Okumura S, Goumard C, Gayet B, Fuks D, Scatton O. Laparoscopic versus open two-stage hepatectomy for bilobar colorectal liver metastases: a bi-institutional, propensity score-matched study. Surgery. 2019;166(6):959–66.
73. Melandro F, Giovanardi F, Hassan R, Larghi Laureiro Z, Ferri F, Rossi M, Mennini G, Pawlik TM, Lai Q. Minimally Invasive Approach in the Setting of ALPPS Procedure: a Systematic Review of the Literature. J Gastrointest Surg. 2019;23(9):1917–24.
74. Gavriilidis P, Sutcliffe RP, Hodson J, Marudanayagam R, Isaac J, Azoulay D, Roberts KJ. Simultaneous versus delayed hepatectomy for synchronous colorectal liver metastases: a systematic review and meta-analysis. HPB (Oxford). 2018;20(1):11–9.
75. Boudjema K, Locher C, Sabbagh C, Ortega-Deballon P, Heyd B, Bachellier P, Métairie S, Paye F, Bourlier P, Adam R, et al. Simultaneous Versus Delayed Resection for Initially Resectable Synchronous Colorectal Cancer Liver Metastases: A Prospective, Open-label, Randomized, Controlled Trial. Ann Surg. 2021;273(1):49–56.
76. Ghiasloo M, Pavlenko D, Verhaeghe M, Van Langenhove Z, Uyttebroek O, Berardi G, Troisi RI, Ceelen W. Surgical treatment of stage IV colorectal cancer with synchronous liver metastases: a systematic review and network meta-analysis. Eur J Surg Oncol. 2020;46(7):1203–13.
77. Pan L, Tong C, Fu S, Fang J, Gu Q, Wang S, Jiang Z, Juengpanich S, Cai X. Laparoscopic procedure is associated with lower morbidity for simultaneous resection of colorectal cancer and liver metastases: an updated meta-analysis. World J Surg Oncol. 2020;18(1):251.
78. Kokudo N, Takemura N, Ito K, Mihara F. The history of liver surgery: achievements over the past 50 years. Ann Gastroenterol Surg. 2020;4(2):109–17.
79. Cadière MDPDG-B, Himpens MDJ, Germay O, Izizaw R, Degueldre MDM, Vandromme MDJ, Capelluto MDE, Bruyns MDJ. Feasibility of robotic laparoscopic surgery: 146 cases. World J Surg. 2001;25(11):1467–77.
80. Marino MV, Shabat G, Gulotta G, Komorowski AL. From illusion to reality: a brief history of robotic surgery. Surg Innov. 2018;25(3):291–6.
81. Tim L. A short history of robotic surgery. Ann R Coll Surg Engl. 2018;100(6_sup):5–7.
82. Shimizu A, Ito M, Lefor AK. Laparoscopic and robot-assisted hepatic surgery: an historical review. J Clin Med. 2022;11(12):3254.
83. Bozkurt E, Sijberden JP, Hilal MA. What is the current role and what are the prospects of the robotic approach in liver surgery? Cancers (Basel). 2022;14(17):4268.
84. Zwart MJW, Görgec B, Arabiyat A, Nota CLM, van der Poel MJ, Fichtinger RS, Berrevoet F, van Dam RM, Aldrighetti L, Fuks D, et al. Pan-European survey on the implementation of robotic and laparoscopic minimally invasive liver surgery. HPB (Oxford). 2022;24(3):322–31.
85. Giulianotti PC, Coratti A, Angelini M, Sbrana F, Cecconi S, Balestracci T, Caravaglios G. Robotics in general surgery: personal experience in a large community hospital. Arch Surg. 2003;138(7):777–84.
86. Kamarajah SK, Bundred J, Manas D, Jiao L, Hilal MA, White SA. Robotic versus conventional laparoscopic liver resections: a systematic review and meta-analysis. Scand J Surg. 2021;110(3):290–300.
87. Kim JK, Park JS, Han DH, Choi GH, Kim KS, Choi JS, Yoon DS. Robotic versus laparoscopic left lateral sectionectomy of liver. Surg Endosc. 2016;30(11):4756–64.
88. Fruscione M, Pickens R, Baker EH, Cochran A, Khan A, Ocuin L, Iannitti DA, Vrochides D, Martinie JB. Robotic-assisted versus laparoscopic major liver resection: analysis of outcomes from a single center. HPB (Oxford). 2019;21(7):906–11.
89. Gumbs AA, Lorenz E, Tsai TJ, Starker L, Flanagan J, Benedetti Cacciaguerra A, Yu NJ, Bajul M, Chouillard E, Croner R, et al. Study: International Multicentric Minimally Invasive Liver Resection for Colorectal Liver Metastases (SIMMILR-CRLM). Cancers (Basel). 2022;14(6):1379.
90. Gumbs AA, Croner R, Lorenz E, Cacciaguerra AB, Tsai TJ, Starker L, Flanagan J, Yu NJ, Chouillard E, Abu Hilal M. Survival Study: International Multicentric Minimally Invasive Liver Resection for Colorectal Liver Metastases (SIMMILR-2). Cancers (Basel). 2022;14(17):4190.
91. Kadam P, Sutcliffe RP, Scatton O, Sucandy I, Kingham TP, Liu R, Choi GH, Syn NL, Gastaca M, Choi SH, et al. An international multicenter propensity-score matched and coarsened-exact matched analysis comparing robotic versus laparoscopic partial liver resections of the anterolateral segments. J Hepatobiliary Pancreat Sci. 2022;29(8):843–54.
92. Crippa J, Grass F, Achilli P, Mathis KL, Kelley SR, Merchea A, Colibaseanu DT, Larson DW. Risk factors for conversion in laparoscopic and robotic rectal cancer surgery. Br J Surg. 2020;107(5):560–6.
93. Rehan M, Saleem MM, Tiwana MI, Shakoor RI, Cheung R. A soft multi-axis high force range magnetic tactile sensor for force feedback in robotic surgical systems. Sensors (Basel). 2022;22(9):3500.
94. Almersjö O, Bengmark S, Hafström L. Liver resection for cancer. Acta Chir Scand. 1976;142(2):139–44.
95. Bismuth H, Eshkenazy R, Arish A. Milestones in the evolution of hepatic surgery. Rambam Maimonides Med J. 2011;2(1):e0021.
96. Couinaud C. [Liver lobes and segments: notes on the anatomical architecture and surgery of the liver]. Presse Med. (1893) 1954;62(33):709–12.
97. Martínez-Cecilia D, Fontana M, Siddiqi NN, Halls M, Barbaro S, Abu-Hilal M. Laparoscopic parenchymal sparing resections in segment 8: techniques for a demanding and infrequent procedure. Surgical Endoscopy. 2018;32(4):2012–9.
98. Coles SR, Besselink MG, Serin KR, Alsaati H, Di Gioia P, Samim M, Pearce NW, Abu Hilal M. Total laparoscopic management of lesions involving liver segment 7. Surg Endosc. 2015;29(11):3190–5.

History and Worldwide Diffusion of Robotic Liver Surgery

2

Rubén Ciria, Rafael Calleja, and Javier Briceño

2.1 Introduction

Minimally invasive liver surgery (MILS) is one of the most important advances in liver surgery in the last decades. Traditionally, liver surgeons have been reluctant to consider minimally invasive approach as an option due to some considerations: first, liver is a hollow organ in which planes are difficult to be observed; second, bleeding has been a major concern and fear to achieve adequate and fast control during laparoscopic approach has been a limitation to its implementations; and third, concerns were raised in the beginning about oncological safety in regards to MILS due to the absence of tactile sensation, misorientation, and difficulties during intraoperative laparoscopic ultrasonography.

The first reports on laparoscopic liver resections (LLR) were dated 25 years ago in which Azagra JS et al. [1, 2] and Kaneko H et al. [3] showed pioneering left lateral sectionectomies using just energy and stapling devices. At present, even though minimally invasive liver surgery is not fully implemented with rates not higher than 40–50%, clear advances have been made. In this context, robotic liver resection (RLR) has experienced a gradual but unstoppable increase, reaching up to the field of living donation or perihilar cholangiocarcinoma [4, 5].

2.2 The First Pioneering Series on Robotic Liver Surgery

Since the first Expert Consensus Meeting held in 2009 in Louisville [6], several meetings and landmarks have happened. In this first meeting, there was no mention of robotic approach to liver surgery. It was only later during the Morioka Consensus [7] that a recommendation was made about RLR. This claimed for "both ongoing institutional ethical approval and a reporting registry of all cases in place before beginning to perform this procedure." It was also stated that a "close evaluation of results by registry and comparative studies using higher quality methodology" was needed. Also, concerns about cost/benefit were raised.

The earliest experiences on anatomical RLR were performed on left lateral sectionectomy. Vasile et al. [8] and Choi et al. [9] both reported three cases each using the Da Vinci S robotic system in 2008. In these early experiences, Vasile et al. [8] reported

R. Ciria (✉)
Unit of Hepatobiliary Surgery and Liver Transplantation, University Hospital Reina Sofia, Cordoba, Spain

R. Calleja · J. Briceño
Unit of Hepatobiliary Surgery and Liver Transplantation, University Hospital Reina Sofia, Cordoba, Spain

Maimonides Biomedical Research Institute of Cordoba (IMIBIC), University Hospital Reina Sofia, Cordoba, Spain

M. D'Hondt, I. Sucandy (eds.), *Textbook of Robotic Liver Surgery*,
https://doi.org/10.1007/978-3-031-76536-0_2

pedicles and hepatic veins sealing using Atlas robotic sealing device, contrary to Choi et al. [9], who used endoGIA. Both teams reported excellent outcomes, no need for conversion or transfusion, and, interestingly, no need for Pringle. The first comparative series was reported by Berber et al. [10] from Cleveland, in which over a 1-year period 9 patients underwent RLR of peripherally located malignant lesions measuring <5 cm and compared prospectively with 23 patients who underwent LLR of similar tumors at the same institution. The authors reported similar operative times (259 mins for the RLR versus 234 mins for the LLR) and no difference between estimated blood loss (EBL) and resection margin status. Conversion to an open operation was necessary in one patient in the robotic group. Complications were also not statistically significant. The authors also added a brief oncological follow-up of 14 months reporting equivalent disease-free survival (DFS) rates.

The first large series on RLR was reported by Giulianotti et al. [11] In 2011, where the authors reported a set of 70 cases performed from 2002 to 2009 by a single surgeon. Interestingly, and contrary to what happened with initial LLR, malignancies were the main indication for resections in up to 60% of the patients. Also, contrary to initial LLR phases, major liver resections were performed in almost 40% of the cases. The authors reported four conversions to open surgery (5,7%). The median operative time was 313 mins and 198 mins for major and minor RLR, respectively. The mortality rate was 0%, and the overall rate of complications was 21%. Major morbidity occurred in four patients in the major hepatectomies group (14.8%) and in four patients in the minor hepatectomies group (9.3%). All complications were managed conservatively and none required reoperation. With this preliminary and early experience, the authors demonstrated that robotic surgery can be used safely for liver resections, thus opening the field of a new technical option for MILS. Soon after this first large series, complex cases were reported worldwide [12]. The first series of major resections were reported with excellent results. Giulianotti et al. [13] first reported 24 totally robotic right hepatectomies with only one conversion to open (4,2%) and no perioperative deaths or reoperations. Morbidity was 25% (six patients), including transitory liver failure in two patients and pleural effusion, bile leak, fluid collection, and deep venous thrombosis in one patient each. Casciola et al. [14] first reported the specific technique and outcomes of 10 RLR cases (47.8%) in which lesions were located in the so-called 'complex segments' (posterosuperior segments). In this series, the authors reported the feasibility of this procedure in these segments with no and only one case of biliary leakage. Complexity was also reported early by authors performing hilar cholangiocarcinoma with biliary reconstruction in 2010 and 2014 [15, 16]. In the first case [16], an extended right hepatectomy with a left hepaticojejunostomy was performed. Operative time was 540 minutes, with an intraoperative blood loss of 800 mL. The postoperative course was uneventful, and the patient was discharged at postoperative day 11. The second case [15] was a complex and non-standard three-step Da Vinci approach. This case started with the drainage of left hepatic duct, followed by right-hepatic vascular control and ended up in a right hepatectomy with left cholangioanastomosis. Similarly, postoperative course was uneventful and 1-year follow up was free of disease. Another important landmark happened in 2012 when the first robot-assisted right lobe donor hepatectomy in an entirely minimally invasive fashion was done [17]. The liver graft was safely extracted through a limited lower abdominal incision. Both the donor and the recipient recovered well without postoperative complications. However, upon routine CT scan 6 months after the procedure, a short stenosis of the main trunk of the donor portal vein was noticed. The authors considered angulation of the portal vein as caused by hepatic regeneration or the formation of a band of scar tissue given the normal CT angiogram at 1 month. The stenosis was dilated through a percutaneous transhepatic approach.

2.3 Contemporary Series and Meta-Analyses

Recent series have proven that robotic approach is well-settled and may offer advantages mostly in difficult cases. Several meta-analyses have been published (Table 2.1) comparing both laparoscopic and robotic techniques. Early meta-analyses were consistent with increased operative times, estimated blood loss (EBL), and cost related to the robotic platform. On the contrary, conversion to open surgery was reported to be lower in the RLR. Interestingly, a recent meta-analysis reported reduced operative time when performed on major-only liver resections [18]. Cost-analyses were consistent in 6 out of 11 meta-analyses with up to USD 5.000 increased cost per procedure compared to laparoscopic procedures [12, 19–23]. It is interesting to state that, however, while the major meta-analyses in laparoscopic surgery included 9527 patients [24], the largest robotic cohort in a meta-analysis was 1144 patients [19] (Fig. 2.1).

Propensity score matching (PSM) series and high-quality analyses have also been reported (Table 2.2). Interestingly, most of the studies are consistent with improved short-term outcomes, mostly conversion to open surgery and EBL. These PSM studies have been reported on different etiologies and positions as observed in Table 2.2, so consistent conclusions may be difficult to obtain. The only study reporting significantly increased operative times is Rho et al. [25], in which up to a 2 hour increase was reported analyzing only right lobe living liver donation.

Similarly to what happened with the laparoscopic liver approach, the increase is gradual and the creation of dedicated societies and meetings is important. The promotion of mentoring programs, "hands-on" or workshop courses, as well as expert meetings favor the implementation of these new devices according to the most recent scientific evidence available.

Table 2.1 Meta-analyses of robotic (RLR) versus laparoscopic liver resections (LLR)

Author	Year	Journal	Type of resection	LLR	RLR	Favors robot	Against robot
Ziogas et al. [37]	2021	Surg Endosc	Major-only	300	225	–	–
Hu et al. [38]	2021	Asian J Surg	Malignancies	748	345	–	Op time Conv Right hep
Ciria et al. [12]	2020	JHBPS	All + subanalyses		2728	Major: Op time LOS	Op time EBL Costs
Zhao et al. [39]	2020	Updates Surg	All	1010	1249	–	Op time EBL
Kamarajah et al. [40]	2020	Scand J Surg	All	1680	950	EBL Readmissions	Op time
Gavriilidis et al. [19]	2020	Eur J Surg Oncol	All	4678	1144	–	Op time Pringle Costs
Zhang et al. [20]	2020	PLoS ONE	All	1312	2232	Conversion	Op time Transf Costs
Guan et al. [21]	2019	Asian J Surg	All	503	435	Conversion	Op time EBL Costs
Hu et al. [22]	2018	Asian J Surg	All	902	487	–	Op time EBL First intake Costs
Qiu et al. [23]	2016	Surg Endosc	All	522	254	–	Op time Costs
Montalti et al. [41]	2015	World J Gastroent	All	479	215	–	Op time EBL

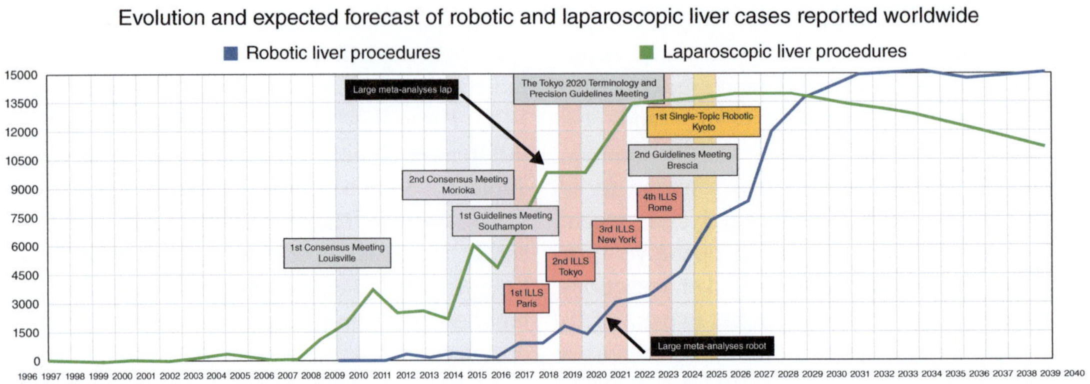

Fig. 2.1 Evolution and expected forecast of robotic and laparoscopic liver cases reported worldwide. Laparoscopic and robotic liver procedures are represented by green and blue lines, respectively. Relevant consensus and international societies meetings related to minimally invasive liver surgery are displayed throughout the frame time

Table 2.2 Propensity score matching series comparing robotic (RLR) versus laparoscopic liver resections (LLR)

Author	Year	Journal	Type of resection	LLR	RLR	Favors robot	Against robot
Kadam et al. [42]	2022	JHBPS	Anterolateral segments	296	296	Conversion Blood loss	–
Chong et al. [43]	2022	JAMA Surg	Right hepatectomy and exterior right	220	220	Conversion Hospital stay	
Rho et al. [25].	2022	Ann Surg	Right lobe LDLT	118	52	EBL	Op time (2 h)
Fagenson et al. [44]	2021	J Gastroint Surg	All	240	240	Conversion	Op time
Beard et al. [45]	2020	Wold J Surg	CRLM	115	115	–	–
Lim et al. [46]	2019	World J Surg	Malignancies	55	55	–	–
Montalti et al. [47]	2015	Surg Endosc	Posterosuperior segments	72	36	–	CCI Pringle

2.4 Concerns During the Implementation Phases of Robotic Liver Surgery

In the early studies about RLR, the following limitations were cautiously reported: (1) high cost, (2) inadequate coverage by medical insurance, (3) lack of tactile sense, (4) lack of training systems, (5) heavy robotic arms and equipment, (6) time-consuming set-up, and (7) difficulty in converting to open surgery.

Some of these limitations have been consistent during the implementation phases, but most of them are not a limitation any more. Certainly, high cost is inherent to this platform, and it has been speculated that it does not afford advantages over traditional laparoscopy and represents a nonjustifiable cost burden [26, 27]. In our opinion, this does not consider the surgeon's ability to perform complex hepatobiliary procedures in a minimally invasive fashion as it may solely take into account postoperative benefits. In fact, the trend is that as accessibility to these devices increases, the number of procedures will grow exponentially, surpassing even the laparoscopic approach (Fig. 2.1).

Training is a critical issue in regard to the development of new skills, especially in this platform. Regarding this, the lack of training systems was argued as another potential limitation [28]. This problem has certainly been overcome with simulation devices integrated within the robotic platform and an increasing number of meetings and dedicated courses over the last few years. Experts in the Pan-European survey on the implementation of robotic and laparoscopic minimally invasive liver surgery recently stated that 21–30 procedures/year is the minimum number of MILS procedures in a center to be considered suitable for training (98% respondents) [29]. Based on the CUSUM methodology, a meta-analysis of 40 studies performed by Chua D et al. [30] reported 50 cases (range 25–58) and 25 cases (16–50), respectively, as the number of cases necessary to complete the learning curve for both laparoscopic and robotic liver resections. The difference in the number of cases is remarkable, but in the authors' opinion, differences in previous open liver surgery experience or differences in MILS training may influence a shorter learning curve. In this analysis, the authors did not consider difficulty scoring systems, which may constitute a significant selection and procedure bias. Regarding this, a meta-analysis recently reported the impact of these difficulty scoring systems (DSS) in MILS, showing that there was no DSS superior to the others and only the Ban and Iwate DSS were externally validated for RLR [31]. Krenzien F et al. [32] recommended to be cautious, and, after an IWATE-DSS adjusted calculation, the learning curve has been observed to increase by up to four times. Hence, future studies should discriminate learning curves adjusted by difficulty scoring systems.

Robotic platforms may potentially overcome some of the inherent limitations of conventional laparoscopy and flatten the learning curve to enable more surgeons to confer the proven benefits of minimally invasive surgery to a broader population of patients. It has been widely proven, but it is remarkably noticeable in the setting of living liver donation. This is a difficult surgical approach with ethical considerations, and special concerns regarding donor safety have been raised. A recent meta-analysis showed lower EBS but an increase in operative time comparing robotic living donor major hepatectomy versus open approach. However, there is still insufficient evidence regarding donor outcomes in robotic versus laparoscopic approaches [18].

2.5 Innovation and Robotic Liver Surgery

Parallel to the advances in minimally invasive surgery, different innovative tools have been developed to improve the surgical preoperative planification and the surgical procedure itself. These tools may soon be easily integrated within the robotic platforms and constitute a real new environment for liver surgeons. Some of these noteworthy technologies are artificial intelligence (AI), augmented reality (AR), and image-guided surgery (IS). The popularity of AI algorithms is increasing. AI-based models with different clinical aims have been explored, with promising results in the field of liver diseases [33]. However, there are no specific publications related to RLS, although its expansion has probably contributed to improving digital surgery systems. The role of the digital era in MILS is represented mainly by AR and IS. AR allows preoperative visualization of lesions in 3D, assisting in preoperative surgical strategy to ensure correct margins and avoid injury to adjacent structures. This technology can be integrated during surgery, as proposed by Diana M et al. [34] when performing an AR-guided robotic cholecystectomy. In the field of RLS, this system has already been integrated into robotic devices. Pessaux P [35]. et al. performed three AR-guided robotic segmentectomies (two segmentectomy V and one segmentectomy VI) with assistance during the liver resection as well as for robotic ports placement. However, the need for standardized algorithms, greater accessibility to the devices, and more evidence is necessary for real applicability.

The use of intraoperative ultrasound (IUS) and/or the application of indocyanine green (ICG) are some examples of IS. At present, these tools are essential for hepatobiliary and pancreatic (HBP) surgery and their application has reached its maximum development thanks to the laparoscopic approach. The TilePro™ function of the Da Vinci Surgical System® allows for a direct visualization of IUS images and 3D reconstructions showing different video sources simultaneously on the same screen. The use of the "firefly mode" also allows the ICG view to be integrated within the 3D image. Chiow AK et al. [36] reported their experience as a safe way to perform living donor robotic resections by using these integrated tools.

2.6 Conclusions

Robotic liver surgery is the ultimate symbiosis between MILS and technological innovation. Current devices offer optimized tools and ergonomics, which may be the start of a new era in hepatobiliary surgery. These new tools are very likely to be integrated with digital navigation systems in the near future. Since its introduction, it has been demonstrated that robotic liver resections can be performed safely in the patient, allowing noninferior oncological resections even in highly complex cases. Scientific evidence is increasing exponentially. Robotic liver surgery is expected to be a technological revolution in patient care and outcomes in liver surgery. Although cost is still the main drawback of this technology, surely the future may see a drastic drop in the cost associated with these platforms and highlight a wide and extensive set of new opportunities in the robotic era of surgery.

References

1. Azagra JS, Goergen M, Brondello S, Calmes MO, Philippe P, Schmitz B. Laparoscopic liver sectionectomy 2 and 3 (LLS 2 and 3): towards the “gold standard”. J Hepato-Biliary-Pancreat Surg. 2009;16(4):422–6. https://doi.org/10.1007/s00534-009-0117-3.
2. Azagra JS, Goergen M, Gilbart E, Jacobs D. Laparoscopic anatomical (hepatic) left lateral segmentectomy-technical aspects. Surg Endosc. 1996;10(7):758–61. https://doi.org/10.1007/bf00193052.
3. Kaneko H, Otsuka Y, Takagi S, Tsuchiya M, Tamura A, Shiba T. Hepatic resection using stapling devices. Am J Surg. 2004;187(2):280–4. https://doi.org/10.1016/j.amjsurg.2003.11.005.
4. Broering D, Sturdevant ML, Zidan A. Robotic donor hepatectomy: a major breakthrough in living donor liver transplantation. Am J Transplant. 2022;22(1):14–23. https://doi.org/10.1111/ajt.16889.
5. Brolese A, Rigoni M, Pasquale A, Viel G, Brolese M, Ciarleglio FA. The role of robotic surgery for the treatment of hilar cholangiocarcinoma: a systematic review. Front Oncol. 2022;12:1001838. https://doi.org/10.3389/fonc.2022.1001838.
6. Buell JF, Cherqui D, Geller DA, O’Rourke N, Iannitti D, Dagher I, et al. The international position on laparoscopic liver surgery: the Louisville statement, 2008. Ann Surg. 2009;250(5):825–30. https://doi.org/10.1097/sla.0b013e3181b3b2d8.
7. Wakabayashi G, Cherqui D, Geller DA, Buell JF, Kaneko H, Han HS, et al. Recommendations for laparoscopic liver resection: a report from the second international consensus conference held in Morioka. Ann Surg. 2015;261(4):619–29. https://doi.org/10.1097/sla.0000000000001184.
8. Vasile S, Sgarbură O, Tomulescu V, Popescu I. The robotic-assisted left lateral hepatic segmentectomy: the next step. Chirurgia (Bucur). 2008;103(4):401–5.
9. Choi SB, Park JS, Kim JK, Hyung WJ, Kim KS, Yoon DS, et al. Early experiences of robotic-assisted laparoscopic liver resection. Yonsei Med J. 2008;49(4):632–8. https://doi.org/10.3349/ymj.2008.49.4.632.
10. Berber E, Akyildiz HY, Aucejo F, Gunasekaran G, Chalikonda S, Fung J. Robotic versus laparoscopic resection of liver tumours. HPB. 2010;12(8):583–6. https://doi.org/10.1111/j.1477-2574.2010.00234.x.
11. Giulianotti PC, Coratti A, Sbrana F, Addeo P, Bianco FM, Buchs NC, et al. Robotic liver surgery: results for 70 resections. Surgery. 2011;149(1):29–39. https://doi.org/10.1016/j.surg.2010.04.002.
12. Ciria R, Berardi G, Alconchel F, Briceño J, Choi GH, Wu YM, et al. The impact of robotics in liver surgery: a worldwide systematic review and short-term outcomes meta-analysis on 2,728 cases. J Hepatobiliary Pancreat Sci. 2022;29(2):181–97. https://doi.org/10.1002/jhbp.869.
13. Giulianotti PC, Sbrana F, Coratti A, Bianco FM, Addeo P, Buchs NC, et al. Totally robotic right hepatectomy: surgical technique and outcomes. Arch Surg. 2011;146(7):844–50. https://doi.org/10.1001/archsurg.2011.145.
14. Casciola L, Patriti A, Ceccarelli G, Bartoli A, Ceribelli C, Spaziani A. Robot-assisted parenchymal-sparing liver surgery including lesions located in the posterosuperior segments. Surg Endosc. 2011;25(12):3815–24. https://doi.org/10.1007/s00464-011-1796-9.
15. Zhu Z, Liu Q, Chen J, Duan W, Dong M, Mu P, et al. Robotic surgery twice performed in the treatment of hilar cholangiocarcinoma with deep jaundice: delayed right hemihepatectomy following the right-hepatic vascular control. Surg Laparosc Endosc Percutan Tech. 2014;24(5):e184–90. https://doi.org/10.1097/SLE.0b013e31828f708b.
16. Giulianotti PC, Sbrana F, Bianco FM, Addeo P. Robot-assisted laparoscopic extended right hepatectomy with biliary reconstruction. J Laparoendosc Adv Surg

Tech A. 2010;20(2):159–63. https://doi.org/10.1089/lap.2009.0383.
17. Giulianotti PC, Tzvetanov I, Jeon H, Bianco F, Spaggiari M, Oberholzer J, et al. Robot-assisted right lobe donor hepatectomy. Transpl Int. 2012;25(1):e5–9. https://doi.org/10.1111/j.1432-2277.2011.01373.x.
18. Ziogas IA, Kakos CD, Moris DP, Kaltenmeier C, Tsoulfas G, Montenovo MI, et al. Systematic review and meta-analysis of open versus laparoscopy-assisted versus pure laparoscopic versus robotic living donor hepatectomy. Liver Transpl. 2023;29:1063. https://doi.org/10.1097/lvt.0000000000000115.
19. Gavriilidis P, Roberts KJ, Aldrighetti L, Sutcliffe RP. A comparison between robotic, laparoscopic and open hepatectomy: a systematic review and network meta-analysis. Eur J Surg Oncol. 2020;46(7):1214–24. https://doi.org/10.1016/j.ejso.2020.03.227.
20. Zhang L, Yuan Q, Xu Y, Wang W. Comparative clinical outcomes of robot-assisted liver resection versus laparoscopic liver resection: a meta-analysis. PLoS One. 2020;15(10):e0240593. https://doi.org/10.1371/journal.pone.0240593.
21. Guan R, Chen Y, Yang K, Ma D, Gong X, Shen B, et al. Clinical efficacy of robot-assisted versus laparoscopic liver resection: a meta analysis. Asian J Surg. 2019;42(1):19–31. https://doi.org/10.1016/j.asjsur.2018.05.008.
22. Hu L, Yao L, Li X, Jin P, Yang K, Guo T. Effectiveness and safety of robotic-assisted versus laparoscopic hepatectomy for liver neoplasms: a meta-analysis of retrospective studies. Asian J Surg. 2018;41(5):401–16. https://doi.org/10.1016/j.asjsur.2017.07.001.
23. Qiu J, Chen S, Chengyou D. A systematic review of robotic-assisted liver resection and meta-analysis of robotic versus laparoscopic hepatectomy for hepatic neoplasms. Surg Endosc. 2016;30(3):862–75. https://doi.org/10.1007/s00464-015-4306-7.
24. Ciria R, Cherqui D, Geller DA, Briceno J, Wakabayashi G. Comparative short-term benefits of laparoscopic liver resection: 9000 cases and climbing. Ann Surg. 2016;263(4):761–77. https://doi.org/10.1097/sla.0000000000001413.
25. Rho SY, Lee JG, Joo DJ, Kim MS, Kim SI, Han DH, et al. Outcomes of robotic living donor right hepatectomy from 52 consecutive cases: comparison with open and laparoscopy-assisted donor hepatectomy. Ann Surg. 2022;275(2):e433–e42. https://doi.org/10.1097/sla.0000000000004067.
26. Vanounou T, Steel JL, Nguyen KT, Tsung A, Marsh JW, Geller DA, et al. Comparing the clinical and economic impact of laparoscopic versus open liver resection. Ann Surg Oncol. 2010;17(4):998–1009. https://doi.org/10.1245/s10434-009-0839-0.
27. Bhojani FD, Fox A, Pitzul K, Gallinger S, Wei A, Moulton CA, et al. Clinical and economic comparison of laparoscopic to open liver resections using a 2-to-1 matched pair analysis: an institutional experience. J Am Coll Surg. 2012;214(2):184–95. https://doi.org/10.1016/j.jamcollsurg.2011.10.020.
28. Hashizume M, Tsugawa K. Robotic surgery and cancer: the present state, problems and future vision. Jpn J Clin Oncol. 2004;34(5):227–37. https://doi.org/10.1093/jjco/hyh053.
29. Zwart MJW, Görgec B, Arabiyat A, Nota CLM, van der Poel MJ, Fichtinger RS, et al. Pan-European survey on the implementation of robotic and laparoscopic minimally invasive liver surgery. HPB (Oxford). 2022;24(3):322–31. https://doi.org/10.1016/j.hpb.2021.08.939.
30. Chua D, Syn N, Koh Y-X, Goh BKP. Learning curves in minimally invasive hepatectomy: systematic review and meta-regression analysis. Br J Surg. 2021;108(4):351–8. https://doi.org/10.1093/bjs/znaa118.
31. Linn Y-L, Wu AG, Han H-S, Liu R, Chen K-H, Fuks D, et al. Systematic review and meta-analysis of difficulty scoring systems for laparoscopic and robotic liver resections. J Hepatobiliary Pancreat Sci. 2023;30(1):36–59. https://doi.org/10.1002/jhbp.1211.
32. Krenzien F, Benzing C, Feldbrügge L, Ortiz Galindo SA, Hillebrandt K, Raschzok N, et al. Complexity-adjusted learning curves for robotic and laparoscopic liver resection: a word of caution. Annals of Surgery Open. 2022;3(1):e131. https://doi.org/10.1097/as9.0000000000000131.
33. Spann A, Yasodhara A, Kang J, Watt K, Wang B, Goldenberg A, et al. Applying machine learning in liver disease and transplantation: a comprehensive review. Hepatology. 2020;71(3):1093–105. https://doi.org/10.1002/hep.31103.
34. Diana M, Soler L, Agnus V, D'Urso A, Vix M, Dallemagne B, et al. Prospective evaluation of precision multimodal gallbladder surgery navigation: virtual reality, near-infrared fluorescence, and X-ray-based intraoperative cholangiography. Ann Surg. 2017;266(5):890–7. https://doi.org/10.1097/sla.0000000000002400.
35. Pessaux P, Diana M, Soler L, Piardi T, Mutter D, Marescaux J. Towards cybernetic surgery: robotic and augmented reality-assisted liver segmentectomy. Langenbeck's Arch Surg. 2015;400(3):381–5. https://doi.org/10.1007/s00423-014-1256-9.
36. Chiow AKH, Rho SY, Wee IJY, Lee LS, Choi GH. Robotic ICG guided anatomical liver resection in a multi-Centre cohort: an evolution from "positive staining" into "negative staining" method. HPB (Oxford). 2021;23(3):475–82. https://doi.org/10.1016/j.hpb.2020.08.005.
37. Ziogas IA, Giannis D, Esagian SM, Economopoulos KP, Tohme S, Geller DA. Laparoscopic versus robotic major hepatectomy: a systematic review and meta-analysis. Surg Endosc. 2021;35(2):524–35. https://doi.org/10.1007/s00464-020-08008-2.
38. Hu Y, Guo K, Xu J, Xia T, Wang T, Liu N, et al. Robotic versus laparoscopic hepatectomy for malignancy: a systematic review and meta-analysis. Asian J Surg. 2021;44(4):615–28. https://doi.org/10.1016/j.asjsur.2020.12.016.

39. Zhao Z, Yin Z, Li M, Jiang N, Liu R. State of the art in robotic liver surgery: a meta-analysis. Updat Surg. 2021;73(3):977–87. https://doi.org/10.1007/s13304-020-00906-3.
40. Kamarajah SK, Bundred J, Manas D, Jiao L, Hilal MA, White SA. Robotic versus conventional laparoscopic liver resections: a systematic review and meta-analysis. Scand J Surg. 2021;110(3):290–300. https://doi.org/10.1177/1457496920925637.
41. Montalti R, Berardi G, Patriti A, Vivarelli M, Troisi RI. Outcomes of robotic vs laparoscopic hepatectomy: a systematic review and meta-analysis. World J Gastroenterol. 2015;21(27):8441–51. https://doi.org/10.3748/wjg.v21.i27.8441.
42. Kadam P, Sutcliffe RP, Scatton O, Sucandy I, Kingham TP, Liu R, et al. An international multicenter propensity-score matched and coarsened-exact matched analysis comparing robotic versus laparoscopic partial liver resections of the anterolateral segments. J Hepatobiliary Pancreat Sci. 2022;29(8):843–54. https://doi.org/10.1002/jhbp.1149.
43. Chong CC, Fuks D, Lee KF, Zhao JJ, Choi GH, Sucandy I, et al. Propensity score-matched analysis comparing robotic and laparoscopic right and extended right hepatectomy. JAMA Surg. 2022;157(5):436–44. https://doi.org/10.1001/jamasurg.2022.0161.
44. Fagenson AM, Gleeson EM, Pitt HA, Lau KN. Minimally invasive hepatectomy in North America: laparoscopic versus robotic. J Gastrointest Surg. 2021;25(1):85–93. https://doi.org/10.1007/s11605-020-04703-6.
45. Beard RE, Khan S, Troisi RI, Montalti R, Vanlander A, Fong Y, et al. Long-term and oncologic outcomes of robotic versus laparoscopic liver resection for metastatic colorectal cancer: a multicenter, propensity score matching analysis. World J Surg. 2020;44(3):887–95. https://doi.org/10.1007/s00268-019-05270-x.
46. Lim C, Salloum C, Tudisco A, Ricci C, Osseis M, Napoli N, et al. Short- and long-term outcomes after robotic and laparoscopic liver resection for malignancies: a propensity score-matched study. World J Surg. 2019;43(6):1594–603. https://doi.org/10.1007/s00268-019-04927-x.
47. Montalti R, Scuderi V, Patriti A, Vivarelli M, Troisi RI. Robotic versus laparoscopic resections of posterosuperior segments of the liver: a propensity score-matched comparison. Surg Endosc. 2016;30(3):1004–13. https://doi.org/10.1007/s00464-015-4284-9.

Part II

Robotic Systems, Patient Preparation and Anesthesia in Robotic Liver Surgery

Robotic Systems, Equipment, Instrumentation, and Troubleshooting

3

Jeroen Hagendoorn

As in any surgical procedure, correct instrumentation is a key component of successful robotic liver resection. Whereas the number of robotic hepatectomies has steeply increased over the past years, equipment and instrumentation designed *specifically* for robotic liver surgery are still lacking. Thus, surgical techniques reflect the current state of exploration and adaptation of instruments designed for robotic or minimally invasive surgery *in general* to the emerging field of robotic liver surgery [22]. Therefore, in this chapter, we will summarize various alternative instruments and devices reported in the literature.

3.1 Prerequisites for a Robotic Liver Surgery Program

Several general conditions must be met prior to starting a robotic liver surgery program. *First*, a successful program is a team effort. The team consists not only of (one or several) HPB surgeons, but also incorporates anesthesiologists, OR staff, and robotic staff working in concert to support building the program. Simultaneous introduction of robotic techniques in both liver and pancreas procedures, as well as two-surgeon procedures in the initial stages of the program, can be advantageous for exposure and learning curve of the team members [6, 14]. However, preferences may differ according to local possibilities. *Second*, equipment and supportive expertise are needed. These consist of access to a robotic platform and availability of the robotic systems commensurate with target operative volume. In addition, it is helpful if there are existing robotic programs at the institution as they may support setting up a robotic liver surgery program. *Third*, training, proctoring, and volume are mandatory in the initial and later phases of the program. As clinical implementation of robotic liver surgery has progressed through its development and exploration phases, there is increasing opportunity for case observations and proctoring, which is considered essential. Training in robotic liver surgery is discussed elsewhere in this volume (Chap. 8). At the same time, safe execution of a broad spectrum of partial hepatectomies, including major and technically major procedures, necessitates adequate practice volume, although exact evidence-based cutoff volume values remain the subject of discussion [19].

3.2 Robotic Systems

3.2.1 DaVinci

The DaVinci robotic platform (Intuitive Surgical, Mountain View, CA, USA) remains the gold standard for complex robotic abdominal surgery.

J. Hagendoorn (✉)
Department of Surgical Oncology, University Medical Center Utrecht, Utrecht, The Netherlands
e-mail: j.hagendoorn-3@umcutrecht.nl

M. D'Hondt, I. Sucandy (eds.), *Textbook of Robotic Liver Surgery*,
https://doi.org/10.1007/978-3-031-76536-0_3

Currently, four models are operational worldwide. They function through the same conceptual approach that consists of the patient-side "robot" with multiple instrument arms that are "wristed" at the instrument tips and are operated by the surgeon via a console that is physically apart. These components are interconnected with a computer cart, thus allowing features such as adaptable 3D, 10x magnified view, scaling of motion, tremor filtration, and camera/instrument "clutch" that enables the surgeon to always work in a neutral position. The *DaVinci Xi* is the most commonly used, modern model. The Xi has an "overhead" architecture, where the arms are mounted on an adaptable boom. It is capable of the most advanced features such as table motion and guided targeting. In robotic liver surgery, DaVinci Xi is usually positioned on the patient's right side. The *DaVinci X* has a similar slim arm architecture and employs the same devices as the Xi, except that the arms are mounted on a static side cart. In robotic liver surgery, DaVinci X is usually positioned over the patient's head. This can be cumbersome in relation to anesthesia and also makes the system less flexible to adapt to patient size and table positioning. Nevertheless, despite the system being slightly less versatile, it is economically more accessible and robotic liver surgery is well described on both systems (e.g., a national case series of robotic hepatectomy in the Netherlands described the use of Davinci X in 22% of cases [8]). The *DaVinci Si* is currently being phased out by Intuitive Surgical, whereas it has been widely employed by centers pioneering robotic liver surgery [7]. In fact, the DaVinci Si has the same architectural design as DaVinci X, with the instrument arms mounted on a side cart. In comparison, DaVinci Si has shorter and more bulky instrument arms and less advanced features. For instance, it does not allow switching the camera and instrument arms, making the system inherently less flexible. Finally, *DaVinci SP* is Intuitive's most recent single-port system. It provides single-port access of the instrument arms that unfold and triangulate intra-abdominally. Use of the SP system has been described in minor liver resection [10].

3.2.2 Novel Systems

Several novel robotic platforms for use in abdominal surgery are now available for clinical practice. The most notable of the commercially available systems at the time of writing is Medtronic's *Hugo*™ *RAS* system (Figs. 3.1 and 3.2).

This robot consists of four independent, mobile instrument arms and an "open" surgical console. The system is modular in that the arms can be used and positioned independently. Many of Medtronic's laparoscopic devices (e.g., for coagulation, sealing, and stapling) can potentially be applied to the system, thus enhancing its versatility in combination with extensive cloud-based digital auxiliary features. Another novel system on this market is the *hinotori*™, introduced by the Japanese firm Medicaroid (a joint venture started by industrial robot producer Kawasaki Heavy Industries and Sysmex) (Fig. 3.3). This system works with a surgeon "cockpit," a compact and docking-free "operation unit," and a "vision unit." At the time of writing, several clinical case reports employing the *hinotori*™ system are available, including distal pancreatectomy [9], but liver surgery has so far not been described and the system is not yet CE-marked. Versuis®, produced by CMR Surgical Ltd., is another cart-based robotic system that has been introduced clinically, with first reports so far in the field of urology.

3.2.3 Future Considerations

Based on the literature, there is ample support for the DaVinci robotic platform in minimally invasive liver surgery. Whereas robotic liver surgery may allow a larger portion of hepatectomies being performed minimally invasively by more hospitals [8], it remains to be investigated where its added value is most pronounced for hospitals that have established large laparoscopic liver surgery programs [12]. Anyway the prospects of robotic liver surgery are very promising, with the advent of new techniques such as fluorescence guidance, augmented reality image overlay, and

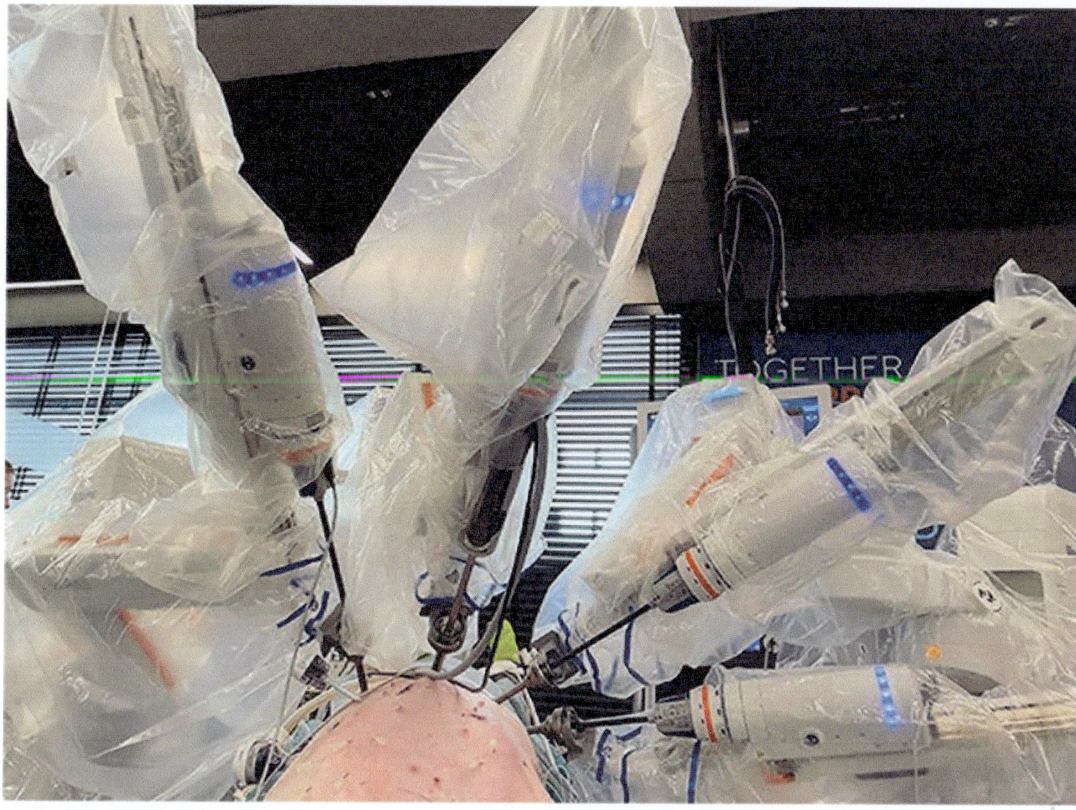

Fig. 3.1 Medtronic's *Hugo*™ *RAS* system docked for training on a porcine model

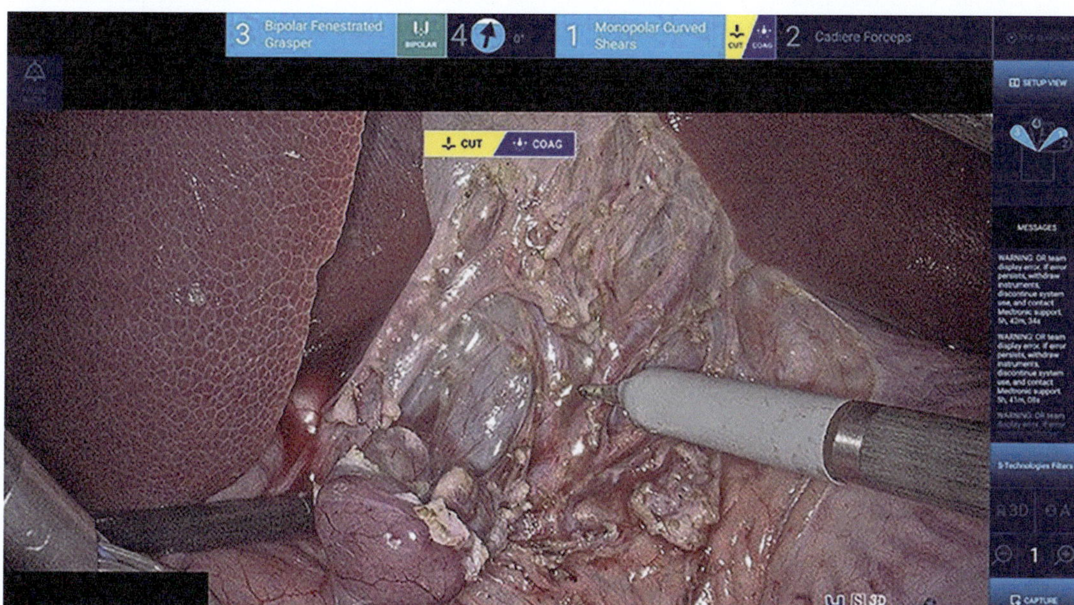

Fig. 3.2 Robotic hilar dissection in a porcine model using Medtronic's *Hugo*™ *RAS* system

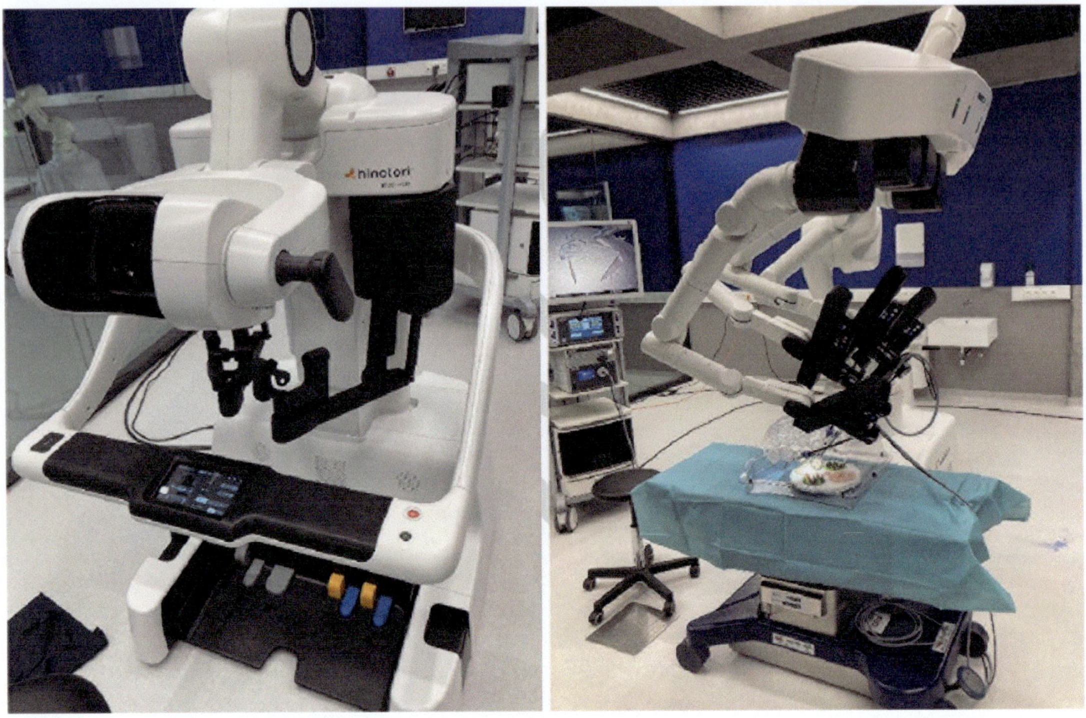

Fig. 3.3 *Hinotori*™, introduced by the Japanese firm Medicaroid

enhanced digital support features [3]. Moreover, design and eventual commercial availability of devices specifically designed for liver surgery (e.g., robotic cusa and other parenchymal transection devices) may give robotics the edge to help liver surgery definitively enter the minimally invasive era.

3.3 Instrumentation

3.3.1 Operating Room

Mandatory operating room equipment in robotic liver surgery includes the possibility of intraoperative *ultrasound*. Perioperative ultrasound is widely used to delineate oncological transection planes. In "conventional" laparoscopic liver surgery, curved-array laparoscopic transducers are employed. However, although they can be used also in robotic surgery, these have distinct disadvantages, including reduced freedom of movement in between the robotic arms, inability to reach far-away regions of the liver (such as posterior segments), and disconnection from the operating surgeon at the console. These problems are overcome with "robotic" ultrasound probes, or so-called "drop-in" probes that are attached to a cable only and can be grasped with robotic instruments such as Prograsp [17, 18]. Via Tilepro, ultrasound images are projected in the surgeon console, thus allowing full control by the operating surgeon (Fig. 3.4).

An *ultrasonic surgical aspirator* (CUSA, Integra Life Sciences; HEPACCS, Söring) is widely used for parenchymal transection in open liver surgery. While laparoscopic variants are available, unfortunately, these devices have so far remained nonrobotic. Therefore, they are applied and found of value by some institutions, especially in central resections or for dissecting closely to major hepatic veins and are in such cases operated by a tableside surgeon or assistant. Other nonrobotic devices deemed helpful in robotic liver surgery include laparoscopic *bulldog clamps* suitable for robotic use (Scanlan International). These can be used for inflow control as well as clamping in case of inadvertent

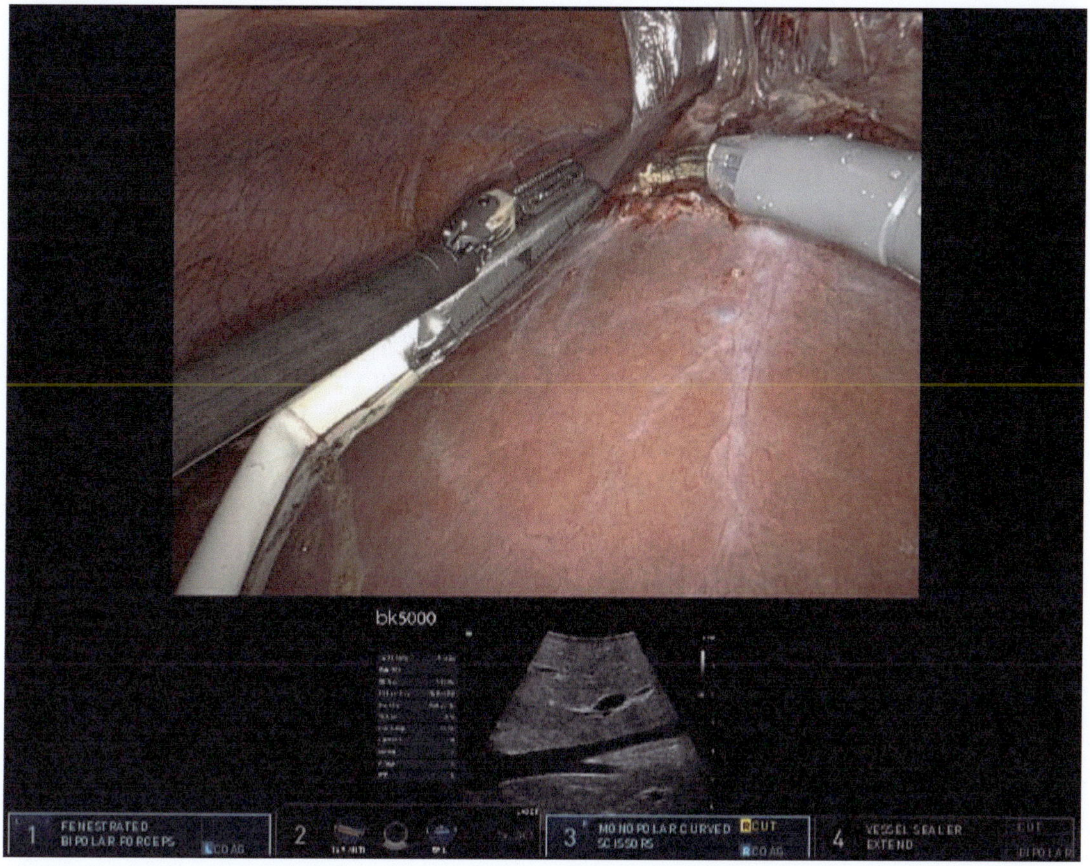

Fig. 3.4 "Robotic" ultrasound probes, or so-called "drop-in" probes that are attached to a cable only and can be grasped with robotic instruments

bleeding during the resection phase. A table-mounted, laparoscopic *liver retractor* (Mediflex Lapro-Flex or Nathanson) can be useful in case of extensive hilar dissection where the third robotic arm is needed elsewhere than elevating the liver toward the diaphragm. A (long!) laparoscopic *suction-irrigation device* is usually employed by the tableside assistant.

3.3.2 DaVinci Instrumentation

DaVinci Settings Usually, we carry out dissection with hand control motion scaling setting at "Normal" for the dissection. Consider setting it at "Fine" for reconstructive procedures, for example, hepatojejunostomy. Standard power settings and limits can be adapted for liver surgery according to the individual surgeon's preference.

Hilar Dissection Hilar dissection can be carried out with a permanent cautery hook or monopolar curved scissors in the surgeon's dominant hand and a fenestrated bipolar forceps in the other, allowing easy coagulation if small bleedings arise. A Maryland bipolar forceps can be useful for fine dissection or encircling structures such as portal vein or hepatic duct. When a laparoscopic liver retractor is not used, the liver is usually "lifted" by the third robotic arm. We always apply a rolled 10 x 10 gauze pad between the instrument (e.g., Cadiere forceps or small Graptor) and the liver surface to avoid injury to the liver.

Parenchymal Transection Various techniques for parenchymal transection have been reported; however, standardized approaches and comparative analyses are so far lacking. A number of authors of larger case series describe the use of the Vessel sealer and Vessel sealer extend (e.g., see [4]; [15]). In general, this wristed sealing-cutting device is used as a clamp-crush transection method and vessels/bile ducts should be visualized before they are sealed and cut ([11]; [16]). With the introduction of the SynchroSeal, an alternative, less bulky device has become available [4], although it cuts with an electrode instead of a knife. The field awaits a smaller version Vessel sealer specifically designed for liver surgery [22]. A second approach to parenchymal transection is the use of a wristed bipolar device such as Maryland of fenestrated bipolar combined with a robotic (but nonwristed) Harmonic ACE (e.g., see [20]). A third approach consists of conventional laparoscopic CUSA operated by a tableside surgeon or assistant, combined with robotic bipolar instruments for fine dissection and sealing (e.g., see [2]). This hybrid, two-surgeon technique in between conventional laparoscopic and robotic liver resection was proposed to enable accurate and safe hepatectomy even in very complex cases. In all these approaches, robotic as well as laparoscopic auxiliary instruments may be applied, including robotic hemolocks for taking larger vessels (main portal branches, segmental bile ducts), robotic large (suture cut) needle driver when stitches are needed (e.g., in case of bleeding or closure of bile ducts), and robotic staplers (e.g. EndoWrist curved-tip 30 with white reload for applying to pedicles or hepatic veins). When stapling is needed, this can of course also be done by a tableside surgeon or assistant. In liver transection, small diameter 8 mm staplers can be very useful.

ICG-Firefly Fluorescence imaging using indocyanine green (ICG) and Firefly mode is a helpful integrated feature of DaVinci in robotic liver surgery. The Firefly mode can be activated in the DaVinci console menu or by applying finger clutch while pressing the camera foot pedal. When administered 24 hours before surgery in patients with colorectal liver metastases, ICG accumulates in peritumor hepatic tissue creating a fluorescent around the lesion, thus allowing easy identification of (deeper) resection margins [1] (Fig. 3.5). ICG-Firefly imaging can also be employed for demarcation transection lines as laparoscopic liver surgery [13], (extra)hepatic perfusion in hepatic arterial infusion pump placement [5], and detecting liver-draining lymph nodes [21].

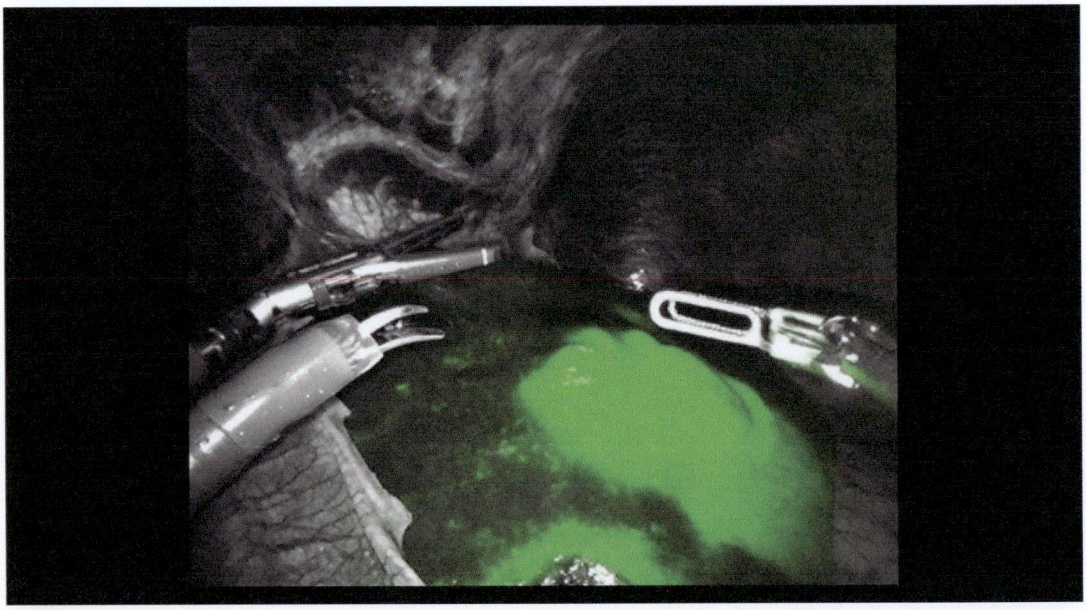

Fig. 3.5 ICG accumulates in peritumor hepatic tissue creating a fluorescent around the lesion as shown in this picture of a large colorectal liver metastasis in segment 3

VisFirefly mode can be activated in the DaVinci console menu or by applying finger clutch while pressing the camera foot pedal when administered 24 hours before surgery in patients with colorectal liver metastases.

Considerations As robotic liver surgery is an emerging field, "gold standard" approaches are still lacking and surgical technique will depend on an individual surgeon's preference, experience, and proctoring. Nevertheless, patterns of procedural steps and application of instruments are becoming apparent in the literature, as, for example, in parenchymal transection. These will need to be further evaluated scientifically in terms of safety, efficacy, versatility, and cost. Ultimately, however, the field will most benefit from robotic instrument specifically designed for liver surgery.

3.4 Troubleshooting

Collision of Instruments and Exposure Trocar positioning is a key element in multiport robot surgery. In liver surgery, differences in working depth may be substantial (e.g., from hilar dissection to mobilization of the diaphragm, etc.). Therefore, careful planning of trocar positioning relative target working area will help prevent external collision of robotic arms. We sometimes find it helpful to estimate trocar-to-target area distances on coronal or sagittal CT images. Exposure of the liver hilum and transection plane is usually performed with the third arm. A more forceful grasper (e.g., Cadiere or small Graptor) with a rolled 10 x 10 cm gauze pad can facilitate correct positioning of the liver tissues.

Bleeding The causes of bleeding obstructing a clear overview of the operative field are multifactorial. As in open liver surgery, inflow control (portal vein clamping or Pringle maneuver) is an essential tool in case of blood loss. Several minimally invasive techniques are available that are described elsewhere in this volume. At the same time, it is common clinical practice to employ low central venous pressure in major liver surgery. Fluid restriction, permissive hypotension, reduction of positive end expiratory pressure, and nitroglycerine administration are all tools that are used to try and reduce blood loss from the liver parenchyma. They apply to laparoscopic and robotic liver surgery as they do to open surgery, with the caveat of strict monitoring of potential air embolism due to pneumoperitoneum. In case of major bleeding from a major portal or hepatic vein branch, however, a technical approach is mandated by suturing of the defect. It is advisable to switch to a needle driver in the dominant hand for this purpose.

Conversion It is essential to have a conversion protocol for the operative team in place, in case of emergent reasons for conversion to open surgery such as persistently inadequate exposure or bleeding that cannot be properly controlled. The conversion protocol should include what open instrumentation is necessary, patient position/repositioning, and the procedure to get the robot "out of the way." The robotic instruments need to be released and extracted before the robotic system is undocked and retracted. Thus, in case of bleeding, compression needs to be applied by the tableside assistant, for example, using a gauze pad and a laparoscopic instrument, until laparotomy can be performed.

References

1. Achterberg FB, Sibinga Mulder BG, Meijer RPJ, Bonsing BA, Hartgrink HH, Mieog JSD, Zlitni A, Park SM, Farina Sarasqueta A, Vahrmeijer AL, Swijnenburg RJ. Real-time surgical margin assessment using ICG-fluorescence during laparoscopic and robot-assisted resections of colorectal liver metastases. Ann Transl Med. 2020;8(21):1448.
2. Aldrighetti L, Catena M, Ratti F. Maximizing Performance in Complex Minimally Invasive Surgery of the Liver: the RoboLap Approach. J Gastrointest Surg. 2022;26(8):1811–3. https://doi.org/10.1007/s11605-022-05340-x. Epub 2022 May 9. PMID: 35534750.
3. Bijlstra OD, Broersen A, Oosterveer TTM, Faber RA, Achterberg FB, Hurks R, Burgmans MC, Dijkstra J, Mieog JSD, Vahrmeijer AL, Swijnenburg RJ. Integration of three-dimensional liver models in a multimodal image-guided robotic liver sur-

gery cockpit. Life (Basel). 2022;12(5):667. https://doi.org/10.3390/life12050667. PMID: 35629335; PMCID: PMC9144252.
4. D'Hondt M, Devooght A, Willems E, Wicherts D, De Meyere C, Parmentier I, Provoost A, Pottel H, Verslype C. Transition from laparoscopic to robotic liver surgery: clinical outcomes, learning curve effect, and cost-effectiveness. J Robot Surg. 2022;17(1):79–88. https://doi.org/10.1007/s11701-022-01405-w. Epub ahead of print. PMID: 35322342.
5. van Dorst RWJJ, Ten Haaft BHEA, Franssen S, Borel Rinkes IHM, Groot Koerkamp B, Swijnenburg RJ, Hagendoorn J. Indocyanine green fluorescence perfusion testing in robot-assisted hepatic arterial infusion pump placement. Surg Endosc. 2024;38(9):5108–13. https://doi.org/10.1007/s00464-024-11010-7. Epub 2024 Jul 17. PMID: 39017958.
6. Emmen AMLH, Görgec B, Zwart MJW, Daams F, Erdmann J, Festen S, Gouma DJ, van Gulik TM, van Hilst J, Kazemier G, Lof S, Sussenbach SI, Tanis PJ, Zonderhuis BM, Busch OR, Swijnenburg RJ, Besselink MG, for HPB-Amsterdam. Impact of shifting from laparoscopic to robotic surgery during 600 minimally invasive pancreatic and liver resections. Surg Endosc. 2022;37(4):2659–72. https://doi.org/10.1007/s00464-022-09735-4. Epub ahead of print.
7. Giulianotti PC, Bianco FM, Daskalaki D, Gonzalez-Ciccarelli LF, Kim J, Benedetti E. Robotic liver surgery: technical aspects and review of the literature. Hepatobiliary Surg Nutr. 2016;5(4):311–21. https://doi.org/10.21037/hbsn.2015.10.05. PMID: 27500143; PMCID: PMC4960413.
8. Görgec B, Zwart M, Nota CL, Bijlstra OD, Bosscha K, de Boer MT, de Wilde RF, Draaisma WA, Gerhards MF, Liem MS, Lips DJ, Marsman HA, Mieog JSD, Molenaar QI, Nijkamp M, Te Riele WW, Terkivatan T, Vahrmeijer AL, Besselink MG, Swijnenburg RJ, Hagendoorn J. Implementation and outcome of robotic liver surgery in the Netherlands: a nationwide analysis. Ann Surg. 2022;277(6):e1269–77. https://doi.org/10.1097/SLA.0000000000005600. Epub ahead of print. PMID: 35848742.
9. Ide T, Ito K, Tanaka T, Noshiro H. Robotic distal pancreatectomy using a docking-free system (the hinotori™ surgical robot system). Surg Oncol. 2023;50:101974. https://doi.org/10.1016/j.suronc.2023.101974. Epub 2023 Jul 11. PMID: 37459677.
10. Kim WJ, Park PJ, Choi SB, Kim WB. Case report of pure single-port robotic left lateral sectionectomy using the da Vinci SP system. Medicine (Baltimore). 2021;100(51):e28248. https://doi.org/10.1097/MD.0000000000028248. PMID: 34941098; PMCID: PMC8701933.
11. Lafaro KJ, Stewart C, Fong A, Fong Y. Robotic liver resection. Surg Clin North Am. 2020;100(2):265–81. https://doi.org/10.1016/j.suc.2019.11.003. Epub 2020 Feb 12. PMID: 32169180.
12. Nevermann N, Feldbrügge L, Krenzien F, Ortiz Galindo S, Knitter S, Haber P, Malinka T, Lurje G, Schöning W, Schmelzle M, Pratschke J. Robotic liver surgery: technical advantages over laparoscopic technique based on parameters of surgical complexity and perioperative outcomes. J Laparoendosc Adv Surg Tech A. 2022; https://doi.org/10.1089/lap.2022.0129. Epub ahead of print. PMID: 35877812.
13. Nomi T, Hokuto D, Yoshikawa T, Matsuo Y, Sho M. A novel navigation for laparoscopic anatomic liver resection using indocyanine green fluorescence. Ann Surg Oncol. 2018;25(13):3982.
14. Nota CL, Molenaar IQ, Te Riele WW, van Santvoort HC, Hagendoorn J, Borel Rinkes IHM. Stepwise implementation of robotic surgery in a high volume HPB practice in the Netherlands. HPB (Oxford). 2020a;22(11):1596–603.
15. Nota CL, Molenaar IQ, te Riele WW, van Santvoort HC, Rinkes IHMB, Hagendoorn J. Parenchymal transection in robotic liver resection: results of 70 resections using the Vessel Sealer. Mini-invasive Surg. 2020b;4:74. https://doi.org/10.20517/2574-1225.2020.57.
16. Nota CL, Molenaar IQ, Borel Rinkes IH, Hagendoorn J. Robotic liver resection of segment 7: a step-by-step description of the technique. Surg Oncol. 2020c;34:206–7. https://doi.org/10.1016/j.suronc.2020.04.028. Epub 2020 May 4. PMID: 32891331.
17. Nota CL, Rinkes IHB, Hagendoorn J. Setting up a robotic hepatectomy program: a Western-European experience and perspective. Hepatobiliary Surg Nutr. 2017;6(4):239–45. https://doi.org/10.21037/hbsn.2016.12.05. PMID: 28848746; PMCID: PMC5554759.
18. Pesi B, Moraldi L, Guerra F, Tofani F, Nerini A, Annecchiarico M, Coratti A. Surgical and oncological outcomes after ultrasound-guided robotic liver resections for malignant tumor. Analysis of a prospective database. Int J Med Robot. 2019;15(4):e2002. https://doi.org/10.1002/rcs.2002. Epub 2019 May 27. PMID: 31022774.
19. van der Poel MJ, Fichtinger RS, Bemelmans M, Bosscha K, Braat AE, de Boer MT, CHC D, Doornebosch PG, Draaisma WA, Gerhards MF, Gobardhan PD, Gorgec B, Hagendoorn J, Kazemier G, Klaase J, WKG L, Liem MS, Lips DJ, Marsman HA, JSD M, Molenaar QI, Nieuwenhuijs VB, Nota CL, Patijn GA, Rijken AM, Slooter GD, MWJ S, Swijnenburg RJ, Tanis PJ, Te Riele WW, Terkivatan T, van den Tol PM, van den Boezem PB, van der Hoeven JA, Vermaas M, Abu Hilal M, van Dam RM, Besselink MG, Dutch Liver Collaborative Group. Implementation and outcome of minor and major minimally invasive liver surgery in the Netherlands. HPB (Oxford). 2019;21(12):1734–43.

20. Rojas AE, Paterakos P, Choi SH. Robotic central bisectionectomy for centrally located hepatic malignant tumor. Ann Surg Oncol. 2022; https://doi.org/10.1245/s10434-022-11646-6. Epub ahead of print. PMID: 35357613.
21. Zhang Y, Zhang Y, Zhu J, Tao H, Liang H, Chen Y, Zhang Z, Zhao J, Zhang W. Clinical application of indocyanine green fluorescence imaging in laparoscopic lymph node dissection for intrahepatic cholangiocarcinoma: a pilot study (with video). Surgery. 2022;171(6):1589–95.
22. Zwart MJW, Görgec B, Arabiyat A, Nota CLM, van der Poel MJ, Fichtinger RS, Berrevoet F, van Dam RM, Aldrighetti L, Fuks D, Hoti E, Edwin B, Besselink MG, Abu Hilal M, Hagendoorn J, Swijnenburg RJ. Dutch Liver Collaborative Group and E-AHPBA Innovation & Development Committee. Pan-European survey on the implementation of robotic and laparoscopic minimally invasive liver surgery. HPB (Oxford). 2022;24(3):322–31.

4 Patient Preparation

Luca Chiarella and Riccardo Memeo

4.1 Operating Room Setup

The organisation of the operating theatre represents a fundamental step in order to contain the robotic platform and all the stuff necessary to correctly perform the surgical procedure, including all the equipment and necessary devices [1–4]. The space should be organised to allow the contemporary presence of surgeons, circulating nurses, and anaesthesiologists with all their materials and instruments necessary to allow the surgical procedure, considering the necessity in some cases to divide the usual space available for anaesthesiologists with the space occupied for the docking of the robot [5] (Fig. 4.1).

All these structures require the possibility of contemporary work and maintain sterile fields and space for surgery.

Depending on the space available, some time-consuming procedures such as draping and setting up could be quicker in case of adequate space in the operating room (Fig. 4.2).

A standardisation of the procedure with a structured organisation of the operating theatre could reduce the time necessary to complete the procedures [6, 7]. The setup of the room dedicated to robotic surgery should consider the necessity to contemporarily perform the arrival of the patient in the operating room and the preparation of instruments and robotic platform, maintaining the sterile field (Fig. 4.3).

All these considerations are related to the possibility of having a workspace adapted to the operating room with enough space for all operations (Fig. 4.4).

L. Chiarella · R. Memeo (✉)
Unit of Hepato-Pancreato-Biliary Surgery, General Regional Hospital "F. Miulli",
Acquaviva delle Fonti BA, Italy

M. D'Hondt, I. Sucandy (eds.), *Textbook of Robotic Liver Surgery*,
https://doi.org/10.1007/978-3-031-76536-0_4

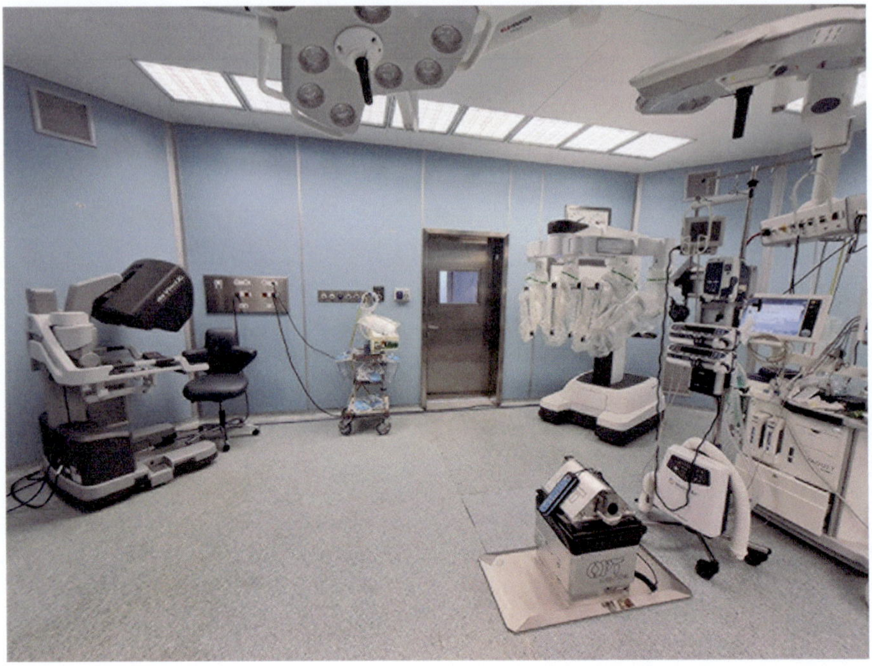

Fig. 4.1 Operating theatre setting

Fig. 4.2 Draping of the robot and the patient

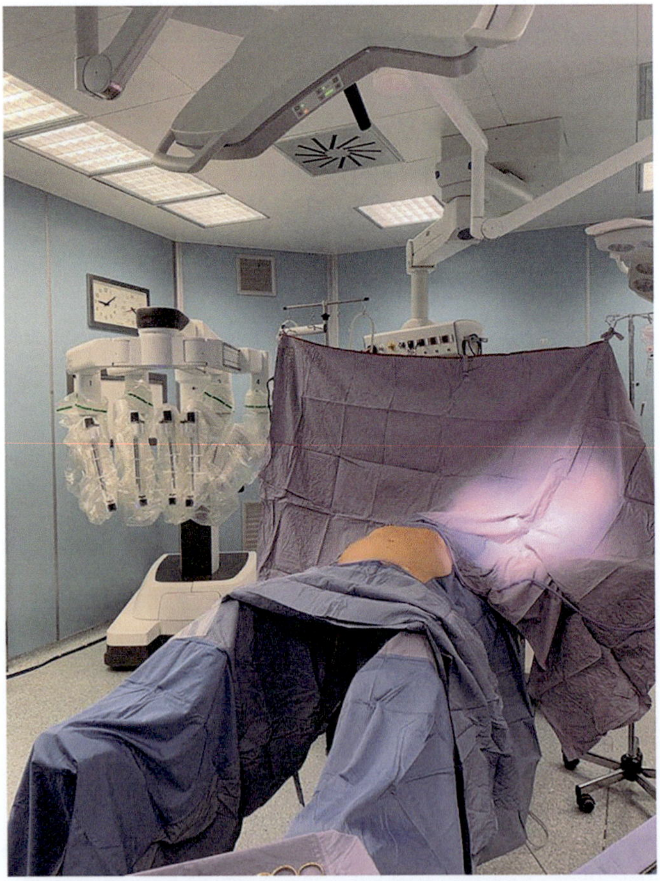

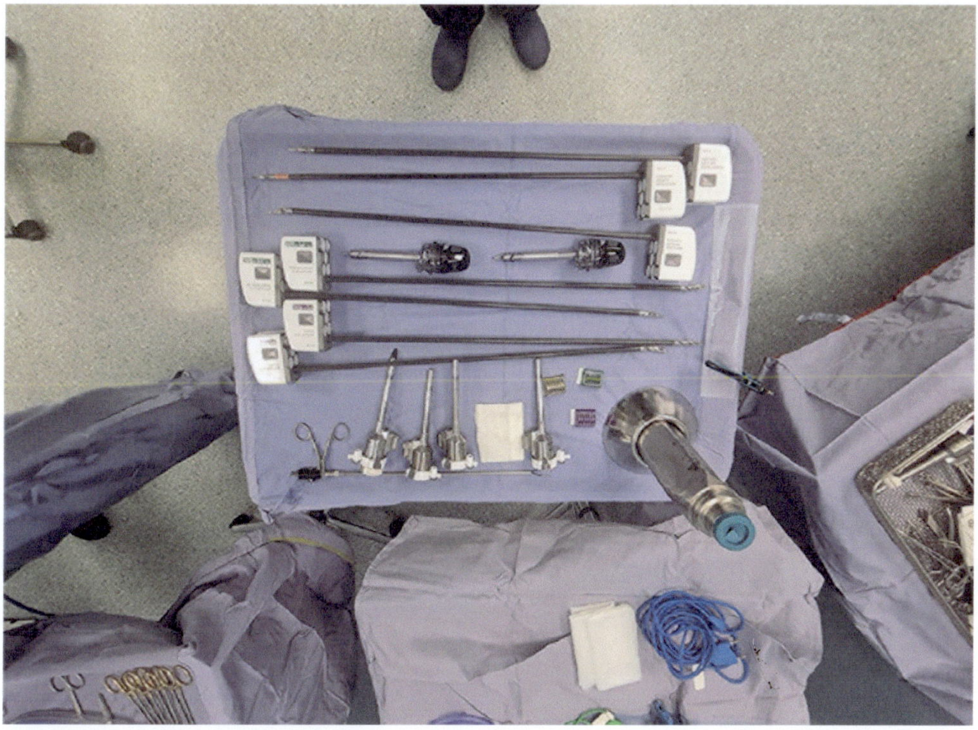

Fig. 4.3 Preparation of instruments for the robotic platform

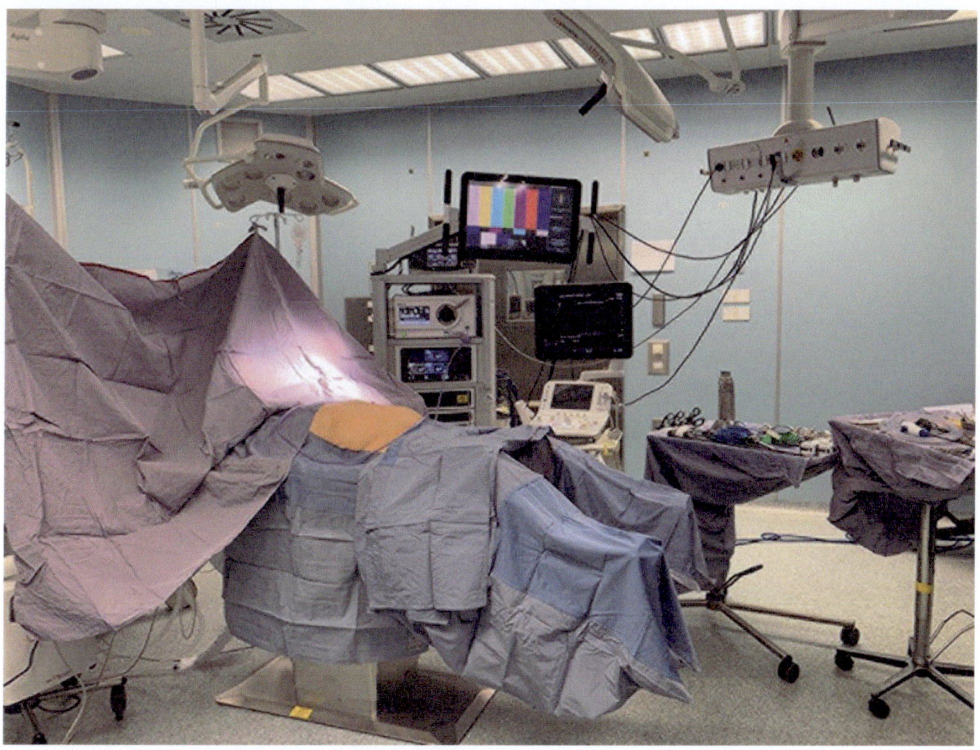

Fig. 4.4 Operating room setting before docking of the robot

4.2 Patient Installation

In most of the liver resections, the patient is placed in a supine position with parted legs in a 21-degree reverse Trendelenburg (Fig. 4.5).

The assistant is placed among the patient's leg. In case of necessity to have access to posterosuperior segments, some surgeons could prefer a left lateral decubitus, with a 30-degree reverse Trendelenburg, with the positioning of the 'bump'under the right shoulder and a pillow placed in the patient's arms. The assistant is placed on the left side of the patient. In order to avoid patient movement in Trendelenburg and reverse Trendelenburg position, a special mat could be used to reduce the risk during surgery. The port positioning varies in order to achieve the best positioning in the operating field, avoiding the collision among robot arms (Fig. 4.6).

For this reason, the body habitus of the patient should be always taken into consideration in order to avoid difficulties during the surgical procedure and facilitate the role of the assistant in surgical operation. A safe access to the head and neck of the patient must be preserved to allow the anaesthesiologist to correctly monitor the vital function of the patient.

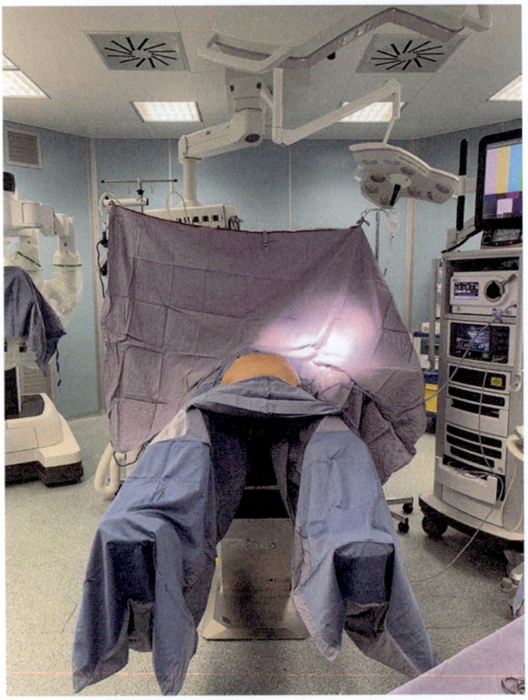

Fig. 4.5 Patient position

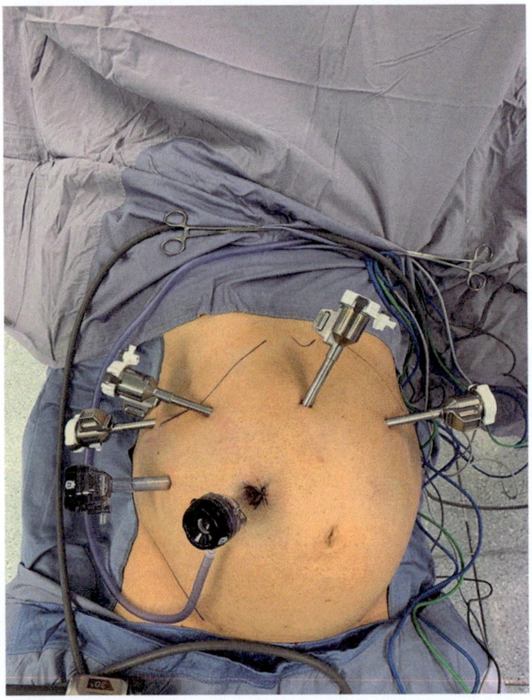

Fig. 4.6 Port positioning

4.3 Considerations for Docking

Robot docking and patient positioning are fundamental to having a simple approach to all anatomical segments of the liver. Docking should be completed in order to guarantee the correct movement of the robotic cart to correctly arrive on the patient without conflict with the anaesthesiologic field and instrumentalist carts (Fig. 4.7).

Considering the seven degrees of movement, it is extremely important to avoid conflict among robotic arms and at the same time with the patient's rib and iliac crest.

In the case of combined liver and colic resection, it could be necessary to redock the robot to perform a synchronous resection to allow a multiquadrant resection. In some cases, we can have the possibility to associate the robot with the 'table motion system' that allows integration with the operating table in synchronous with the robot, avoiding in selected cases changing decubitus in a patient who underwent multiquadrant resection.

The robot cart should be mobilised close to the patient, allowing the laser green cross pointed at the camera port. The nurse driving the robot can rotate manually the boom or the surgeon could mobilise to correctly place the robot in position and complete the connection between the patient and the robot.

Once the robot is connected to the port, the docking must be completed with the targeting of the part of the abdomen in which we should perform surgery in order to reduce the conflict during the procedure (Fig. 4.8).

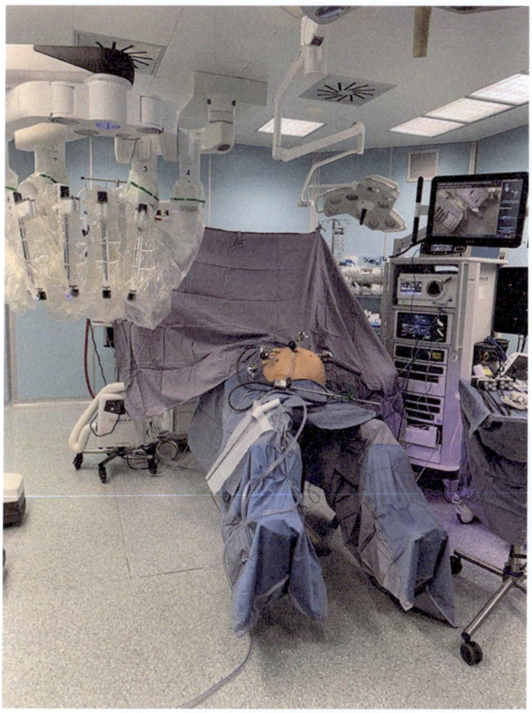

Fig. 4.7 Docking procedure

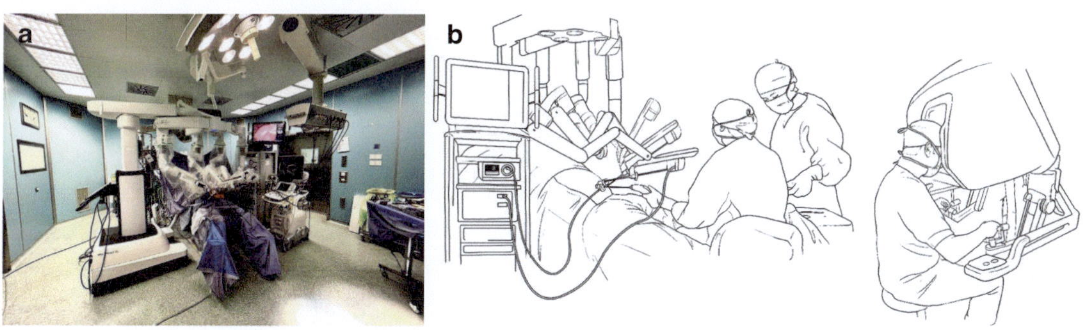

Fig. 4.8 (**a** and **b**) Final views of the operating theatre

References

1. Emmen AMLH, et al. Impact of shifting from laparoscopic to robotic surgery during 600 minimally invasive pancreatic and liver resections. Surg Endosc. 2023;37:2659–72.
2. Nota CL, et al. Developing a robotic pancreas program: the Dutch experience. J Vis Surg. 2017;3:106.
3. Farrugia A, et al. Proposed training pathway with initial experience to set up robotic hepatobiliary and pancreatic service. J Robot Surg. 2022;16:65–71.
4. Brodie A, Vasdev N. The future of robotic surgery. Ann R Coll Surg Engl. 2018;100:4–13.
5. Nota CL, Rinkes IHB, Hagendoorn J. Setting up a robotic hepatectomy program: a Western-European experience and perspective. Hepatobiliary Surg Nutr. 2017;6:23945–245.
6. Eriksen JD, et al. Implementation of a standardized surgical technique in robot-assisted restorative rectal cancer resection: a single center cohort study. BMC Surg. 2022;22:1–10.
7. Knab LM, Zureikat AH, Zeh HJ, Hogg ME. Towards standardized robotic surgery in gastrointestinal oncology. Langenbeck's Arch Surg. 2017;402:1003–14.

Anesthesia for Robotic Liver Surgery

5

Pieter Jan Steelant

While robotic liver surgery (RLS) and laparoscopic liver surgery (LLS) differ substantially from a surgical point of view, both techniques are very similar in terms of anesthesia management. In this chapter, we discuss the main points of interest for anesthesia in RLS, paying particular attention to the differences with LLS. We also address some current discussion points about anesthesia in minimally invasive liver surgery.

Minimally invasive liver surgery (MILS) includes both LLS and RLS. MILS is gaining popularity worldwide. Despite initial concerns about the safety profile, both short-term and long-term, the benefits of MILS have become clear in recent years. LLS for colorectal liver metastases has proven to be safe, both surgically and oncologically, and is associated with reduced postoperative morbidity [1, 2].

Although a robotic approach to liver surgery was first described in 2003, there is almost no published literature discussing the challenges in RLS with regard to anesthesia. The main differences between RLS and LLS are a longer operating time and more risk of gas emboli, but, on the other hand, less risk of conversion to open surgery and less blood loss [3]. As in any hepatic surgery, the outcome depends on good preparation and a shared mental model by the surgical team.

P. J. Steelant (✉)
Department of Anesthesiology, Groeninge Hospital, Kortrijk, Belgium
e-mail: pieterjan.steelant@azgroeninge.be

5.1 Preoperative Optimization

Every patient should be evaluated preoperatively for timely recognition of increased surgical risk and to make adjustments if necessary. Preoperative counseling is useful only if it makes a difference in outcome. While most patients presenting for RLS are in good general health, it is important to screen for comorbidities that might influence the anesthetic and surgical plan of action. Standard preoperative investigations are sufficient (ECG, hemoglobin, chemistry, renal function, coagulation). Extensive cardiac and pulmonary testing is only indicated if the medical history requires it.

Advanced age per se is no contraindication for MILS.

5.1.1 Cardiovascular Disease

Standard cardiological risk stratification can be used to screen patients prior to MILS.

Compared with open hepatectomy, MILS does not pose an increased perioperative risk in cardiac patients if certain precautions are taken. In patients with a higher cardiovascular risk, it is

M. D'Hondt, I. Sucandy (eds.), *Textbook of Robotic Liver Surgery*,
https://doi.org/10.1007/978-3-031-76536-0_5

advised to gradually increase intra-abdominal pressure, starting with low-flow before switching to high-flow insufflation. When cardiac stability is reached, reversed Trendelenburg can gradually be applied. This gentle approach can help minimize the impact of pneumoperitoneum and positioning on cardiovascular function [4]. Performing hypovolemic phlebotomy and vascular occlusion techniques should be done with caution in high-risk patients.

5.1.2 Pulmonary Disease

Postoperative pulmonary complications are common in liver surgery but can be reduced by preoperative optimization of patients with preexisting lung disease or smoking.

Compared with open liver surgery (OLS), MILS decreases pulmonary complications in patients undergoing major liver resections. This positive effect is due to decreased pain with faster rehabilitation, the absence of transection of abdominal muscles, and fewer pleural effusions [5].

Patients with chronic obstructive pulmonary disease who are "well," meaning they have normal oxygen saturation and no recent infection, have a similar perioperative risk as patients without lung disease. Preoperative stratification should therefore identify these patients who do not fulfill these conditions to be optimized preoperatively. Patients with obstructive sleep apnea are more prone to postoperative pulmonary complications and will need more intensive postoperative monitoring [6].

Current smokers and ex-smokers have an increased risk of postoperative pulmonary complications. Preoperative smoking cessation is beneficial. The effects of carbon monoxide and nicotine on cardiovascular function will dissipate in a few hours, but normalization of mucus production and clearance will take several weeks. A longer period of cessation will produce a more substantial reduction in postoperative complications [7]. Smoking cessation should therefore be discussed with the patient.

5.1.3 Liver Disease

Most patients who present for LLS do not have generalized or advanced liver disease. Nevertheless, some patients will have preexisting liver disease.

There is robust evidence that patients with cirrhosis who require anesthesia and surgery have higher rates of morbidity and mortality. Whether liver surgery is—more than other types of surgery—an independent risk factor predisposing cirrhotic patients to increased mortality is controversial. One retrospective review claims that cardiac surgery and intra-abdominal surgery, including liver surgery, are contraindicated in class B cirrhotic patients [8]. Another multivariate analysis demonstrated that liver surgery was no independent variable associated with increased morbidity and mortality in cirrhotic patients [9].

Several systems exist to assess preoperative liver function and postoperative mortality risk.

The Child–Turcotte–Pugh score (CTP) [10] has long been the most widely used instrument to assess operative risk in cirrhotic patients. Patients are stratified in three groups (A, B, and C) according to total bilirubin level, serum albumin level, international normalized ratio (INR), volume of ascites, and hepatic encephalopathy. Surgery is tolerated well in patients with class A cirrhosis. Those in classes B and C are at high risk. Mortality after nonhepatic abdominal surgery in CTP classes A, B, and C was found to be 10%, 17%, and 63%, respectively [4, 11].

The MELD score was developed in 2000 [12]. After a few modifications, it was used as the most important tool in assigning the need for a liver transplant. The MELD score can also predict outcomes in nontransplant surgery [13, 14]. In contrast to the CTP score, it is solely based on objective factors (serum creatinine, bilirubin, and INR).

Patients with liver disease who undergo surgery should be optimized preoperatively. Managing the altered coagulation in cirrhosis is not straightforward. Patients with cirrhosis have a decrease in both procoagulant and anticoagulant factors with a rebalanced hemostasis.

Conventional coagulation tests cannot reliably predict coagulation status in cirrhotic patients. Viscoelastic tests such as thrombo-elastography (TEG) and rotation thrombo-elastometry (ROTEM) are more useful to guide interventions in liver surgery, including cirrhotic patients [15, 16]. Ascites could be managed by a low sodium diet, diuretics, or in severe cases preoperative paracentesis with replacement by albumin. Many patients with liver disease are malnourished and should see a nutritionist preoperatively.

5.1.4 Renal Disease

Preexisting chronic renal failure is a strong predictor of postoperative acute renal failure.

Perioperative fluid restriction, used in liver surgery to lower the central venous pressure (CVP), leads to an increased risk of acute kidney injury (AKI) [17]. The incidence of AKI after liver resection is 12–15% and is associated with renal replacement therapy and mortality [18]. Slankamenac et al. used a multivariate model to develop a prediction score for postoperative acute renal failure after liver resection. Four major factors were identified that could accurately predict postoperative AKI. Preoperatively elevated alanine–transaminase, preexisting cardiovascular disease, chronic renal failure, and diabetes were the strongest predictors of AKI. Using this information could lead to preoperative identification of patients who would benefit most from protective kidney interventions perioperatively [19].

Unfortunately, few of these interventions exist. A Cochrane review looked for interventions to protect renal function in the perioperative period. Many pharmaceutical agents were investigated, such as dopamine and its analogs, diuretics, calcium channel blockers, ACE inhibitors, N-acetylcysteine, atrial natriuretic peptide, sodium bicarbonate, antioxidants, and erythropoietin or hydration fluids such as colloids and hypertonic saline [20].

None of the studied interventions during surgery prevented kidney damage. While none of the selected studies involved liver surgery, there is no reason to suspect that the results are not valid for MILS.

If no pharmacological interventions can aid us in preventing acute renal failure, we should aim for other treatable and modifiable parameters such as cardiac dysfunction, preoperative kidney dysfunction, diabetes, and preoperative cholestasis. Many intraoperative parameters influencing postoperative renal function have been identified [21–23].

A low CVP technique is nearly universally used in an effort to limit perioperative blood loss. Low intraoperative CVP and low arterial blood pressure are linked with postoperative acute renal failure. In a review article on normotensive ischemic acute renal failure, Abuelo points out the conditions that cause low-perfusion states and increased afferent arteriolar resistance. These risk factors for renal ischemia are often treatable and thus should be identified and dealt with, even perioperatively. If possible, blood pressure that is on the lower end of the normal range should be increased by correction of any hypovolemia [24]. Anesthesia for MILS is about finding the optimal balance between a low CVP strategy and adequate intraoperative renal perfusion. After all, blood loss and subsequent transfusion are major risk factors for short-term and long-term morbidity and mortality.

Renal insufficiency may be present in patients with liver disease, leading to hyponatremia. This should be slowly corrected over the course of several days to at least 130 mmol/L before surgery [11]. Excessive use of diuretics to reduce ascites or pulmonary effusions in cirrhotic patients can also worsen renal function.

5.2 Anesthetic Technique

Although an arterial catheter is not considered necessary by all experts, it seems prudent to place it in all but the most minor RLS resections. Access to the patient during RLS is restricted and the provoked relative hypovolemia quickly leads to hypotension. The arterial catheter can additionally be used to perform the phlebotomy. Moreover, there is always the possibility of a gas embolism in which case an invasive beat-to-beat measurement of blood pressure is indispensable. In contrast, a central venous catheter is certainly not always necessary. As shown further below, measurement of CVP is not useful. Administration of noradrenaline can also safely be done through a reliable peripheral infusion [25].

The choice of anesthetic agents presumably does not make much difference. Both vapor and total intravenous techniques can be used. Total intravenous anesthesia was shown to offer oncological advantages over inhalational anesthesia in vitro. A meta-analysis comparing propofol to sevoflurane could not confirm these results in real life [26]. Adequate muscle relaxation, measured using a continuous train-of-four monitor, is essential to achieve a calm surgical field and minimize insufflation pressures. Measurement of body temperature with continuous attention to prevention of hypothermia is a good standard practice.

5.2.1 Prevention of Blood Loss

Blood loss should be prevented as much as possible. It is clear that anemia, but certainly transfusion of allogeneic blood, is associated with increased morbidity and mortality, worse oncological outcome, and tumor-free survival [27–29]. Even transfusion of one unit of packed red cells is an independent predictor for worse outcome, and this effect is dose-dependent. Much of the research done on blood loss prevention in liver surgery has been done in OLS. Nevertheless, many lessons learned from OLS can be transferred to MILS. Luckily, RLS seems to be associated with less blood loss and transfusion than LLS.

The most important factors rest in the hands of the surgeon: preparation of a thoughtful surgical plan, use of new dissection devices, and topical fibrin sealants (Fig. 5.1). Discussion of hepatic inflow occlusion techniques goes beyond the scope of a chapter about anesthesia, but these techniques can be very important in the prevention of blood loss. Good communication within the team is essential as a Pringle maneuver can cause clinically relevant hypotension. In addition, hepatic ischemia is associated with lung damage that can lead to postoperative pulmonary complications [30].

Two Cochrane reviews on cardiopulmonary [31] and pharmacological interventions [32] to prevent blood loss in liver surgery concluded that only aprotinin, tranexamic acid, and hemodilution, with or without hypotension, lead to a reduced need for transfusion. The serine protease inhibitor aprotinin has an antifibrinolytic effect, but has been surrounded by controversy after

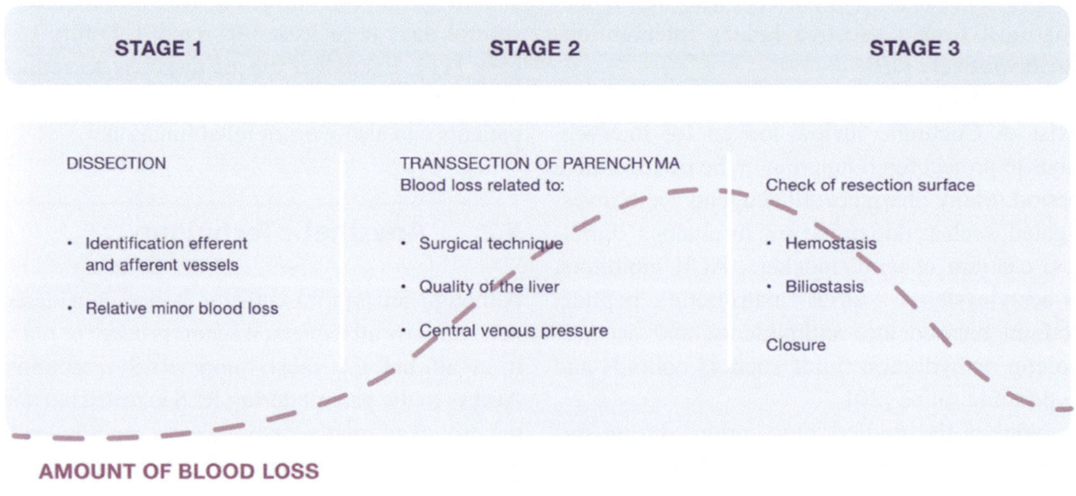

Fig. 5.1 Phases of hepatectomy and the risk of bleeding during the different stages of hepatectomy

temporary withdrawal from the market because of a poor safety profile. In contrast, the plasmin inhibitor tranexamic acid (TXA), also an antifibrinolytic, offers clearly proven added value over a wide range of clinical situations. In smaller studies, TXA reduced the risk of transfusion by up to 90% with no increased effect on thromboembolic events or mortality [33]. Since TXA is an inexpensive and readily available product, it found a place in many hepatectomy protocols in a wide range of dosing regimens [33]. The HeLiX trial [34] is a large, multicenter randomized trial that proved that TXA did not reduce blood loss or blood transfusion in oncological liver resection, but led to more perioperative complications. It seems reasonable to stop using TXA routinely in liver surgery.

Many cardiovascular strategies have been developed to reduce blood loss. Traditionally, low CVP is associated with reduced blood loss, although a Cochrane review could not confirm this. Autologous blood donation and hypoventilation have no proven benefit. The effect of PEEP has been the subject of experimental and clinical studies that show it has no effect on hepatic pressures or perioperative blood loss [35, 36]. Only hemodilution has a proven added value. First used with good results in OLS [37], it was proven to be useful in MILS. Several groups use a variation on classical normovolemic hemodilution by performing a hypovolemic phlebotomy [38, 39]. This involves drawing an amount of blood before parenchymal resection and not compensating it with intravenous fluids. Around 0.7% of body weight is drawn after which a restrictive fluid policy is applied. This blood is returned after the parenchymal transection phase or earlier if necessary. Once the parenchymal transection is over, fluid resuscitation should be done to ensure hemodynamic stability. Theoretically, this technique has the potential to cause postoperative renal insufficiency, but this is rarely seen [40].

5.2.2 Hemodynamic Management

There is no reason to guide hemodynamic management by measuring CVP. While a relative hypovolemic state prevents blood loss, an absolute measurement of CVP is not reliable and not helpful for this purpose [41, 42]. Especially in minimally invasive surgeries with pneumoperitoneum and reverse Trendelenburg, CVP has no added value. Other dynamic parameters offer a possible alternative: use of both stroke volume variation (SVV) and pulse pressure variation (PPV) demonstrated promising results in liver surgery [43].

A low-dose continuous vasopressor is often necessary to maintain adequate end-organ tissue perfusion during the hypovolemic phase when liberal fluid administration is contraindicated.

Despite all preventive measures, bleeding will occasionally still occur during MILS. At the time of bleeding, patients are typically hypovolemic due to all CVP-lowering techniques. It is important to promptly compensate all blood losses with crystalloids and/or colloids. Transfusion of homologous blood should only be considered when there is a clinical need to do so. Overtransfusion in major gastrointestinal and cancer surgery is frequent: up to half of the patients are overtransfused [44, 45]. Keeping in mind the harmful effects of even a single unit of autologous blood, this should be avoided.

Fluid resuscitation after liver resection is vital to prevent postoperative renal failure through hypoperfusion [24]. Overcorrection of the hypovolemic state is equally dangerous: perioperative fluid overload is one of the strongest risk factors for pulmonary complications [46, 47]. Dynamic fluid responsiveness monitors could be useful to find the correct balance between adequate fluid resuscitation while at the same time preventing overload and tissue edema. Several goal-directed fluid therapy protocols are proposed, mostly based on SVV and PPV [41].

5.2.3 Gas Embolism

The combination of pneumoperitoneum, hypovolemia, and reverse Trendelenburg position produces excellent working conditions for the surgeon. Even when large veins are opened, there is often minimal bleeding. While this is beneficial with regard to blood loss and prevention of transfusion, these conditions might create the potential for gas embolism. Entry of carbon dioxide through open veins leads to venous gas embolism (Fig. 5.2).

Accumulation of gas in the pulmonary arterioles leads to its pathophysiological effects as shown in Fig. 5.3. Large amounts of carbon dioxide can lead to a "gas lock" with cardiac failure and ultimately cardiovascular collapse. Porcine models of LLS demonstrate a high incidence of carbon dioxide embolism, causing respiratory and hemodynamic disturbances [48, 49]. Reported frequencies of gas embolism in human series are low (around 1%), with most episodes not leading to postoperative complications [50]. Still, clinically significant venous air embolism in laparoscopic surgery has a high mortality rate [51, 52].

The surgical team should have a high level of suspicion for carbon dioxide embolism during MILS. The most sensitive monitoring device is transesophageal echocardiography, but given its invasiveness and safety profile we cannot recom-

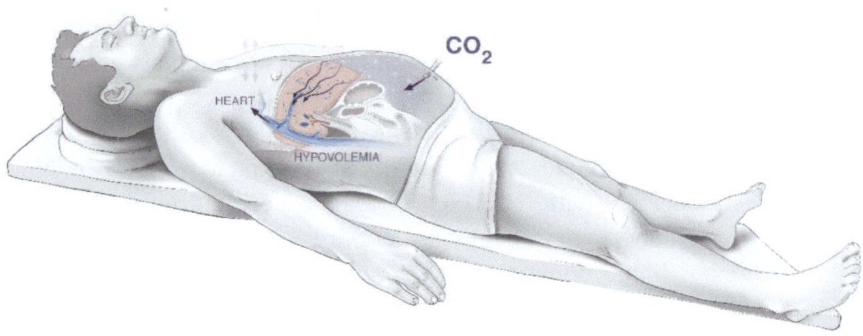

Fig. 5.2 The combination of a pneumoperitoneum, hypovolemia, and Trendelenburg position might create the potential for gas embolism

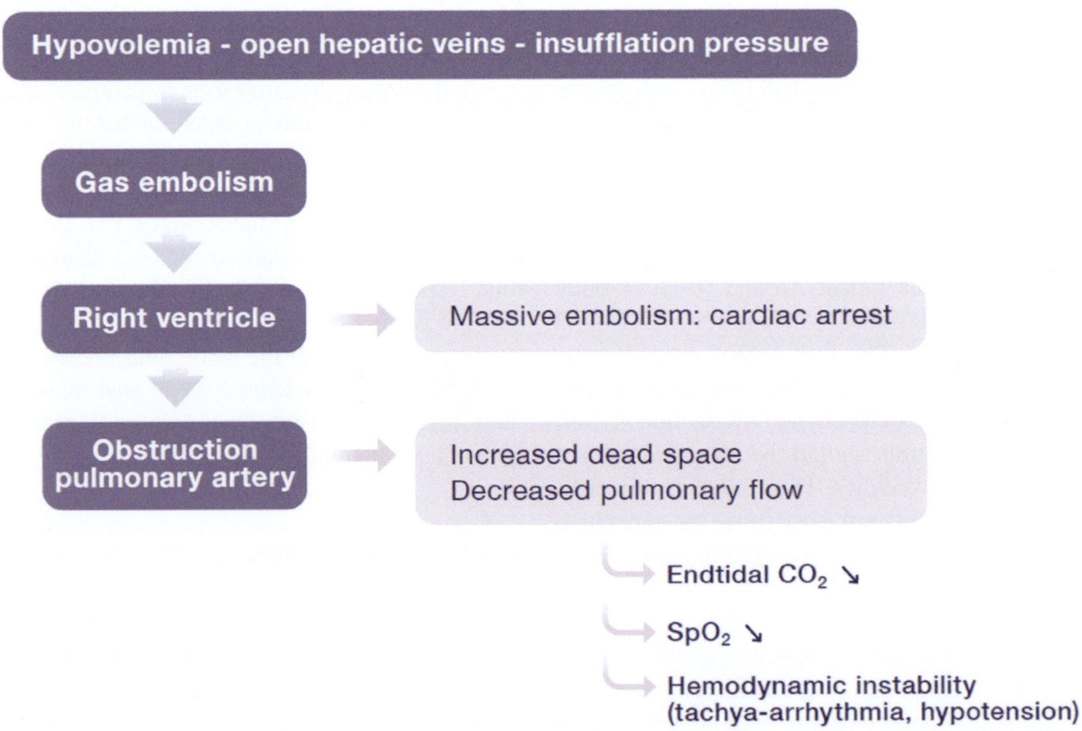

Fig. 5.3 Pathophysiology of gas embolism

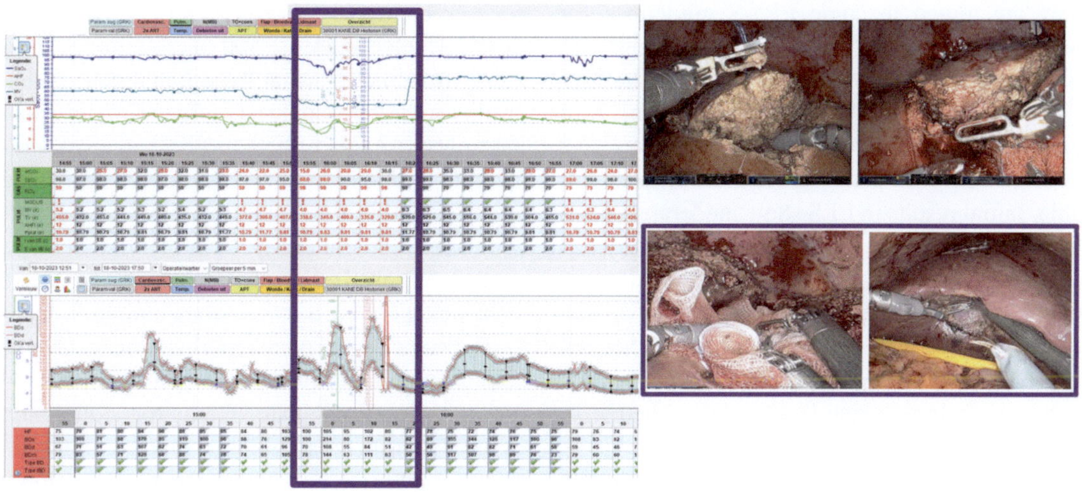

Fig. 5.4 The transection plane is covered with gauzes in a patient with signs of gas embolism

mend its use in every MILS case. Precordial Doppler ultrasound is noninvasive and very sensitive in detecting gas emboli. Standard American Society of Anesthesiologist monitoring will also detect relevant gas embolism in an early stage if paired with a vigilant anesthesiologist. Textbook management of serious gas embolism includes deflating the pneumoperitoneum, hyperventilation, application of 100% O_2, and positioning the patient in left lateral decubitus and Trendelenburg position, combined with standard resuscitative measures. Not all of these can be used in RLS: the position of the operating table is fixed during robotic surgery and deflating the abdomen will not help in solving the issue.

We suggest creating a management algorithm to deal with serious venous gas embolism during RLS, analogous to a useful algorithm developed for robotic prostatectomy [53]. In liver surgery, temporary apnea may be useful to prevent further gas embolization through open veins, until the operation field is flooded or exposed veins are covered with a moist sponge (Fig. 5.4).

5.3 Pain Management

Adequate pain management is an integral part of good anesthesia. Good pain management can also avoid chronification of postoperative pain.

In terms of pain management, results of studies in OLS with a laparotomy cannot be simply extrapolated to MILS. The PROSPECT group looked at the ideal analgesia technique in liver surgery, but excluded MILS from their analysis because at the time their guidelines were drafted, there existed only three randomized clinical trials that examined pain management in LLS [54]. Since then, more work has been done on analgetic techniques in LLS. So far, there is no published controlled study on optimal analgesia in RLS.

It seems logical that classical multimodal analgesia (paracetamol, nonsteroidal anti-inflammatory drugs, weak opiates) is the basis of pain management in any protocol. Many other pharmacological interventions can also be chosen as part of an enhanced recovery after surgery (ERAS) protocol, such as low-dose ketamine, intravenous lidocaine, dexamethasone, clonidine, and magnesium sulfate. Each of these products will act on different targets in the pain pathway to avoid postoperative pain. In addition to these, some supplementary locoregional techniques can improve postoperative pain management.

Single-shot infiltration of the trocar sites and specimen extraction wound by the surgeon is a simple, efficient technique.

Interfascial plane blocks have gained tremendous popularity over the last decade. Although they are touted in some ERAS guidelines for laparoscopic surgery, there is a lot of skepticism within the anesthesia community about the usefulness of these blocks [55]. The mechanism of action is unclear, and the efficacy on top of a good multimodal analgesia regimen is questionable. Initial results in LLS show moderately positive results in the first 24–72 hours postoperatively. Generally, single-shot techniques are used instead of continuous catheter techniques, with an opiate-sparing effect and better pain scores. Several approaches have been described in LLS: transversus abdominis plane blocks [56], whether or not combined with rectus sheath blocks [57], erector spinae plane blocks [58, 59], and transmuscular quadratus lumborum blocks [60]. The main disadvantage of all these techniques is the relatively short duration of action. Simultaneous administration of dexamethasone [61] or dexmedetomidine [62] moderately prolongs the duration of action of interfascial blocks. The future will have to show whether these techniques add a clinically important value to MILS.

Some surgical teams choose to place a thoracic epidural catheter in some LLS patients when the risk of conversion is not insignificant. Placing an epidural catheter in liver surgery used to be controversial because of the risk of postoperative coagulopathy, but this risk is so small in most hepatectomies that it is usually not an objection. On the other hand, it is unnecessary to expose patients to the risks of an epidural catheter when it offers no added value. In this regard, RLS may offer an advantage: initial meta-analyses show a reduced conversion rate for RLS compared with LLS [63], so a prophylactic epidural catheter can be avoided.

An elegant intermediate solution may be a single-shot lumbar intrathecal injection with morphine. Not only in OLS [64] but also in LLS [39, 65], spinal analgesia with intrathecal morphine produces better pain scores than intravenous opiates. Since the peak effect of intrathecal morphine occurs after about 6 hours, it should be placed preoperatively, and, to obtain a faster effect, bupivacaine may be associated. The main concern is an increased risk of respiratory depression, although in practice this does not appear to be a problem [65]. Recommended doses range from 100 micrograms to 500 micrograms of morphine. It is unclear what the optimal dose is that produces the best analgesic effect without causing unwanted side effects (pruritus, nausea, bradycardia, respiratory depression). A single intrathecal injection is not inferior to patient-controlled epidural analgesia in the first 48 hours postoperatively [64] and removes many of the disadvantages of a continuous epidural technique, such as delayed mobilization, problems with coagulopathy, dislocated catheters, and need for a specialized pain team for follow-up. Compared with interfascial techniques, intrathecal morphine has the advantage of being more reliable and having a more complete analgesia profile because both the somatic and visceral components are covered.

5.4 ERAS

Clinical pathways or fast-track pathways are multimodal ERAS protocols. These are bundles of different evidence-based interventions, each one with a small or non-measurable effect, resulting in reduced postoperative morbidity and hospital stay. Conceived in the late 1990s, they are nowadays accepted worldwide and across most surgical specialties [66–68].

Ever since the first reported fast-track program for LLS [69], several protocols have been published [70]. A meta-analysis of eight trials using ERAS in LLS concluded that the application of these programs is safe and effective [71]. We have also concluded that these programs are not only feasible, but can be financially advantageous [72]. Although a literature search did not produce a specific RLS-targeted ERAS program, there is no reason why existing MILS programs would not be applicable to RLS.

Recurring elements in ERAS protocols for MILS include reduced fasting time, preoperative clear carbohydrate drinks [73], intraoperative normothermia, removal of the nasogastric tube at the end of surgery, single-dose antimicrobial prophylaxis, antithrombotic prophylaxis, prevention of nausea and vomiting, early mobilization, and early oral intake [69, 71, 74, 75]. All indwelling catheters should be removed as soon as reasonably possible.

The success of a clinical pathway is highly dependent on a multidisciplinary approach, including not only surgeons and anesthetists, but equally preoperative intake staff, nursing staff, physical therapists, nutritionists, and the hospital's social services. It is important to inform patients of their expected preoperative and postoperative course.

5.5 Conclusion

Robotic liver surgery is an exciting and developing discipline, offering interesting challenges for the anesthetic team. Delivering optimal care to patients undergoing RLS requires close communication with the rest of the surgical team. Standardization and use of evidence-based interventions are probably best achieved by assembling a dedicated anesthesia team for liver surgery. Several groups around the world have described positive results with dedicated anesthesia teams in various subdisciplines, including liver transplant surgery [76, 77].

Equally important is a shared mental model. This concept derives from other industries and reflects the extent to which team members acquire a similar understanding of the situation leading to coordinated efficient action. A shared mental model leads to better team performance and improves making high-quality decisions [78, 79]. One example is a large variation in transfusion practices, with up to half of liver surgery patients being overtransfused in some series [44, 45] and large differences in intradepartmental transfusion practices [80], emphasizing the importance of standardized protocols with well-defined transfusion triggers.

Despite a huge increase in published literature on MILS, there is still an information gap in large studies on anesthesia techniques, hemodynamic management, and analgesia. Further research is definitely welcomed, both in basic physiology and in clinical applications.

This chapter discusses the key elements in perioperative anesthetic management for robotic liver surgery. Appropriate patient preparation, standardization of methods to avoid blood loss and transfusion, applying evidence-based interventions to prevent perioperative complications, and using the newest insights in analgesic techniques and enhanced recovery may help make robotic liver surgery a new standard.

References

1. Fretland AA, Dagenborg VJ, Bjørnelv GMW, Kazaryan AM, Kristiansen R, Fagerland MW, et al. Laparoscopic versus open resection for colorectal liver metastases. Ann Surg. Lippincott Williams and Wilkins. 2018;267:199–207.
2. Hilal MA, Aldrighetti L, Dagher I, Edwin B, Troisi RI, Alikhanov R, et al. The Southampton consensus guidelines for laparoscopic liver surgery: from indication to implementation. Ann Surg. Lippincott Williams and Wilkins. 2018;268:11–8.
3. Ayabe RI, Azimuddin A, Tran Cao HS. Robot-assisted liver resection: the real benefit so far. Langenbecks Arch Surg. Springer Science and Business Media Deutschland GmbH. 2022;407:1779–87.
4. Egger ME, Gottumukkala V, Wilks JA, Soliz J, Ilmer M, Vauthey JN, et al. Anesthetic and operative considerations for laparoscopic liver resection. Surgery (United States). Mosby Inc. 2017;161:1191–202.
5. Fuks D, Cauchy F, Fteriche S, Nomi T, Schwarz L, Dokmak S, et al. Laparoscopy decreases pulmonar y complications in patients undergoing major liver resection a propensity score analysis. Ann Surg. Lippincott Williams and Wilkins. 2016;263:353–61.
6. Lumb AB. Pre-operative respiratory optimisation: an expert review. Anaesthesia. . Blackwell Publishing Ltd. 2019;74:43–8.
7. Hawn MT, Houston TK, Campagna EJ, Graham LA, Singh J, Bishop M, et al. The attributable risk of smoking on surgical complications. Ann Surg. 2011;254:914–20.
8. Millwala F, Nguyen GC, Thuluvath PJ. Outcomes of patients with cirrhosis undergoing non-hepatic surgery: risk assessment and management. World J Gastroenterol. 2007;13:4056–63.
9. Cho HC, Jung HY, Sinn DH, Choi MS, Koh KC, Paik SW, et al. Mortality after surgery in patients with liver cirrhosis: comparison of Child-Turcotte-Pugh, MELD and MELDNa score. Eur J Gastroenterol Hepatol. 2011;23:51–9.
10. Pugh RNH, Murray-Lyon IM, Dawson JL, Pietroni MC, Williams R. Transection of the oesophagus for bleeding oesophageal varices. Br J Surg. 2005;60:646–9.

11. Leung CCH, Young KK. Clinical aspects of hepatic disease. Anaesth Intensive Care Med. 2015;16:11–3.
12. Malinchoc M, Kamath PS, Gordon FD, Peine CJ, Rank J, ter Borg PCJ. A model to predict poor survival in patients undergoing transjugular intrahepatic portosystemic shunts. Hepatology. W.B. Saunders. 2000;31:864–71.
13. Morimoto N, Okada K, Okita Y. The model for end-stage liver disease (MELD) predicts early and late outcomes of cardiovascular operations in patients with liver cirrhosis. Ann Thorac Surg. 2013;96:1672–8.
14. Befeler AS. The safety of intra-abdominal surgery in patients with cirrhosis. Arch Surg. 2005;140:650.
15. Mpaili E, Tsilimigras DI, Moris D, Sigala F, Frank SM, Hartmann J, et al. Utility of viscoelastic coagulation testing in liver surgery: a systematic review. HPB. Elsevier B.V. 2021;23:331–43.
16. Massicotte L, Hevesi Z. How to reduce bleeding and blood transfusion during liver transplantation. In: Anesthesia for hepatico-pancreatic-biliary surgery and transplantation. Cham: Springer International Publishing; 2021. p. 225–33.
17. Ham SY, Kim EJ, Kim TH, Koo BN. Comparison of perioperative renal function between epidural and intravenous patient-controlled analgesia after living-donor hepatectomy: a retrospective study. Transplant Proc. Elsevier USA. 2018;50:1365–71.
18. Tomozawa A, Ishikawa S, Shiota N, Cholvisudhi P, Makita K. Facteurs de risque périopératoires de lésion rénale aiguë après résection hépatique: une étude de cohorte historique. Can J Anesth. Springer New York LLC. 2015;62:753–61.
19. Slankamenac K, Breitenstein S, Held U, Beck-Schimmer B, Puhan MA, Clavien PA. Development and validation of a prediction score for postoperative acute renal failure following liver resection. Ann Surg. 2009;250:720–7.
20. Zacharias M, Mugawar M, Herbison GP, Walker RJ, Hovhannisyan K, Sivalingam P, et al. Interventions for protecting renal function in the perioperative period. Cochrane Database Syst Rev. 2013;(3):CD003590.
21. Schroeder RA, Collins BH, Tuttle-Newhall E, Robertson K, Plotkin J, Johnson LB, et al. Intraoperative fluid management during orthotopic liver transplantation. J Cardiothorac Vasc Anesth. 2004;18:438–41.
22. Melendez JA, Arslan V, Fischer ME, Wuest D, Jarnagin WR, Fong Y, et al. Perioperative outcomes of major hepatic resections under low central venous pressure anesthesia: blood loss, blood transfusion, and the risk of postoperative renal dysfunction. J Am Coll Surg. 1998;187:620–5.
23. Jones RM, Moulton CE, Hardy KJ. Central venous pressure and its effect on blood loss during liver resection. Br J Surg. 2003;85:1058–60.
24. Abuelo JG. Normotensive ischemic acute renal failure. N Engl J Med. 2007;357:797–805.
25. Pancaro C, Shah N, Pasma W, Saager L, Cassidy R, van Klei W, et al. Risk of major complications after perioperative norepinephrine infusion through peripheral intravenous lines in a multicenter study. Anesth Analg. Lippincott Williams and Wilkins. 2020;131:1060–5.
26. Chang CY, Wu MY, Chien YJ, Su IM, Wang SC, Kao MC. Anesthesia and long-term oncological outcomes: a systematic review and meta-analysis. Anesth Analg. Lippincott Williams and Wilkins. 2021;132:623–34.
27. Hallet J, Kulyk I, Cheng ESW, Truong J, Hanna SS, Law CHL, et al. The impact of red blood cell transfusions on perioperative outcomes in the contemporary era of liver resection. Part of this study was presented as a poster presentation at the 2014 Canadian Surgery Forum of the Canadian Association of General Surgeons, in Vancouver, BC. Surgery (United States). Mosby Inc. 2016;159:1591–9.
28. Hallet J, Tsang M, Cheng ESW, Habashi R, Kulyk I, Hanna SS, et al. The impact of perioperative red blood cell transfusions on long-term outcomes after hepatectomy for colorectal liver metastases, vol. 22. Ann Surg Oncol. Springer New York LLC; 2015. p. 4038–45.
29. Lyu X, Qiao W, Li D, Leng Y. Impact of perioperative blood transfusion on clinical outcomes in patients with colorectal liver metastasis after hepatectomy: a meta-analysis. Oncotarget. 2017;8:41740–8.
30. Lepere V, Vanier A, Loncar Y, Lemoine L, Vaillant JC, Monsel A, et al. Risk factors for pulmonary complications after hepatic resection: role of intraoperative hemodynamic instability and hepatic ischemia. BMC Anesthesiol. BioMed Central Ltd. 2017;17:84.
31. Gurusamy KS, Li J, Vaughan J, Sharma D, Davidson BR. Cardiopulmonary interventions to decrease blood loss and blood transfusion requirements for liver resection. Cochrane Database Syst Rev. 2012;2012(5):CD007338.
32. Gurusamy KS, Li J, Sharma D, Davidson BR. Pharmacological interventions to decrease blood loss and blood transfusion requirements for liver resection. Cochrane Database Syst Rev. 2009;(4):CD008085.
33. Koh A, Adiamah A, Gomez D, Sanyal S. Safety and efficacy of tranexamic acid to minimise perioperative bleeding in hepatic surgery: a systematic review and meta-analysis. World J Surg. Springer Science and Business Media Deutschland GmbH. 2022;46:441–9.
34. Karanicolas PJ, Lin Y, McCluskey S, Tarshis J, Thorpe KE, Wei A et al. Tranexamic Acid in Patients Undergoing Liver Resection. The HeLiX Randomized Clinical Trial. JAMA. 2024; 332:1080–9.
35. Neuschwander A, Futier E, Jaber S, Pereira B, Eurin M, Marret E, et al. The effects of intraoperative lung protective ventilation with positive end-expiratory pressure on blood loss during hepatic resection surgery. Eur J Anaesthesiol. Lippincott Williams and Wilkins. 2016;33:292–8.
36. Sand L, Rizell M, Houltz E, Karlsen K, Wiklund J, Odenstedt Hergès H, et al. Effect of patient position and PEEP on hepatic, portal and central venous pressures during liver resection. Acta Anaesthesiol Scand. 2011;55:1106–12.

37. Hashimoto T, Kokudo N, Orii R, Seyama Y, Sano K, Imamura H, et al. Intraoperative blood salvage during liver resection: a randomized controlled trial. Ann Surg. 2007;245:686–91.
38. Ryckx A, Christiaens C, Clarysse M, Vansteenkiste F, Steelant PJ, Sergeant G, et al. Central venous pressure drop after hypovolemic phlebotomy is a strong independent predictor of intraoperative blood loss during liver resection. Ann Surg Oncol. Springer New York LLC. 2017;24:1367–75.
39. Nguyen M, Vandenbroucke F, Roy JD, Beaulieu D, Seal RF, Lapointe R, et al. Evaluation of the addition of bupivacaine to intrathecal morphine and fentanyl for postoperative pain management in laparascopic liver resection. Reg Anesth Pain Med. 2010;35:261–6.
40. Desmet F, D'Hondt M, Pottel H, Carlier S, Hoste E, Kellum J, et al. Association between urinary TIMP-2 and IGFBP7 as early biomarkers of AKI and oliguria during liver surgery: a prospective pilot study. Crit Care. 2015;19:P291.
41. Correa-Gallego C, Tan KS, Arslan-Carlon V, Gonen M, Denis SC, Langdon-Embry L, et al. Goal-directed fluid therapy using stroke volume variation for resuscitation after low central venous pressure-assisted liver resection: a randomized clinical trial. J Am Coll Surg. Elsevier Inc. 2015;221:591–601.
42. Dunki-Jacobs EM, Philips P, Scoggins CR, McMasters KM, Martin RCG. Stroke volume variation in hepatic resection: a replacement for standard central venous pressure monitoring. Ann Surg Oncol. 2014;21:473–8.
43. Agarwal V, Divatia J, v. Enhanced recovery after surgery in liver resection: current concepts and controversies. Korean J Anesthesiol. Korean Society of Anesthesiologists. 2019;72:119–29.
44. Ejaz A, Spolverato G, Kim Y, Frank SM, Pawlik TM. Variation in triggers and use of perioperative blood transfusion in major gastrointestinal surgery. Br J Surg. John Wiley and Sons Ltd. 2014;101:1424–33.
45. Ranganathan P, Ahmed S, Kulkarni AP, Divatia J, v. Appropriateness of perioperative blood transfusion in patients undergoing cancer surgery: a prospective single-centre study. Indian J Anaesth. 2012;56:234–7.
46. Pirat A, Özgur S, Torgay A, Candan S, Zeyneloğlu P, Arslan G. Risk factors for postoperative respiratory complications in adult liver transplant recipients. Transplant Proc. Elsevier USA. 2004;36:218–20.
47. O'Brien JD, Ettinger NA. Pulmonary complications of liver transplantation. Clin Chest Med. 1996;17:99–114.
48. Fors D, Eiriksson K, Arvidsson D, Rubertsson S. Gas embolism during laparoscopic liver resection in a pig model: frequency and severity. Br J Anaesth. Oxford University Press. 2010;105:282–8.
49. Schmandra TC, Mierdl S, Bauer H, Gutt C, Hanisch E. Transoesophageal echocardiography shows high risk of gas embolism during laparoscopic hepatic resection under carbon dioxide pneumoperitoneum. Br J Surg. 2002;89:870–6.
50. Otsuka Y, Katagiri T, Ishii J, Maeda T, Kubota Y, Tamura A, et al. Gas embolism in laparoscopic hepatectomy: what is the optimal pneumoperitoneal pressure for laparoscopic major hepatectomy? J Hepatobiliary Pancreat Sci. 2013;20:137–40.
51. Park EY, Kwon JY, Kim KJ. Carbon dioxide embolism during laparoscopic surgery. Yonsei Med J. 2012;53:459–66.
52. Mirski MA, Lele AV, Fitzsimmons L, Toung TJK, Warltier DC. Diagnosis and treatment of vascular air embolism. Anesthesiology. 2007;106:164–77.
53. Vidovich C, Laserna A, Karan SB. Venous gas embolism during radical robotic prostatectomy: a case report and evidence-based management algorithm. Cureus. 2021;13(8):e17296.
54. Dieu A, Huynen P, Lavand'homme P, Beloeil H, Freys SM, Pogatzki-Zahn EM, et al. Pain management after open liver resection: Procedure-Specific Postoperative Pain Management (PROSPECT) recommendations. Reg Anesth Pain Med. BMJ Publishing Group. 2021;46:433–5.
55. Jones JH, Aldwinckle R. Interfascial plane blocks and laparoscopic abdominal surgery: A narrative review. Local Reg Anesth. Dove Medical Press Ltd. 2020;13:159–69.
56. Zhang J, Liu T, Zhou H, Fei Y, Yu X. The safety and efficacy of ultrasound-guided bilateral dual transversus abdominis plane (BD-TAP) block in ERAS program of laparoscopic hepatectomy: a prospective, randomized, controlled, blinded, clinical study. Drug Des Devel Ther. Dove Medical Press Ltd. 2020;14:2889–98.
57. Lu X, Yu P, Ou C, Wang J, Zhou Z, Lai R. The postoperative analgesic effect of ultrasound-guided bilateral transversus abdominis plane combined with rectus sheath blocks in laparoscopic hepatectomy: a randomized controlled study. Ther Clin Risk Manag. Dove Medical Press Ltd. 2020;16:881–8.
58. Huang X, Wang J, Zhang J, Kang Y, Sandeep B, Yang J. Ultrasound-guided erector spinae plane block improves analgesia after laparoscopic hepatectomy: a randomised controlled trial. Br J Anaesth. Elsevier Ltd. 2022;129:445–53.
59. Kim D, Kim JM, Choi GS, Heo G, Kim GS, Jeong JS. Ultrasound-guided erector spinae plane block for postoperative analgesia in laparoscopic liver resection: a prospective, randomised controlled, patient and observer-blinded study. Eur J Anaesthesiol. NLM (Medline). 2021;38:S106–12.
60. Pang M, Sun G, Yao W, Zhou S, Shen N, Liao H, et al. Ultrasound-guided transmuscular quadratus lumborum block reduced postoperative opioids consumptions in patients after laparoscopic hepatectomy: a three-arm randomized controlled trial. BMC Anesthesiol. 2021;21:45.
61. Chen Q, An R, Zhou J, Yang B. Clinical analgesic efficacy of dexamethasone as a local anesthetic adjuvant for transversus abdominis plane (TAP) block: a meta-analysis. PLoS One. Public Library of Science. 2018;13:e0198923.
62. Wang Q, Li H, Wei S, Zhang G, Ni C, Sun L, et al. Dexmedetomidine added to ropivacaine for ultrasoundguided erector spinae plane block prolongs

analgesia duration and reduces perioperative opioid consumption after thoracotomy: a randomized, controlled clinical study. Clin J Pain. Lippincott Williams and Wilkins. 2022;38:8–14.
63. Zhang L, Yuan Q, Xu Y, Wang W. Comparative clinical outcomes of robot-assisted liver resection versus laparoscopic liver resection: a meta-analysis. PLoS One. 2020;15:e0240593.
64. de Pietri L, Siniscalchi A, Reggiani A, Masetti M, Begliomini B, Gazzi M, et al. The use of intrathecal morphine for postoperative pain relief after liver resection: a comparison with epidural analgesia. Anesth Analg. Lippincott Williams and Wilkins. 2006;102:1157–63.
65. Ban M, Choi YS, Koo B-N. Analgesic effect of intrathecal morphine combined with low-dose bupivacaine on postoperative analgesia after liver resection: a randomized controlled study. J Pers Med. 2022;12:211.
66. Kehlet H. Multimodal approach to control postoperative pathophysiology and rehabilitation. Br J Anaesth. Oxford University Press. 1997;78:606–17.
67. Kehlet H. Enhanced Recovery After Surgery (ERAS): good for now, but what about the future? Can J Anesth. Springer Science and Business Media, LLC. 2015;62:99–104.
68. Hoffmann H, Kettelhack C. Fast-track surgery-conditions and challenges in postsurgical treatment: a review of elements of translational research in enhanced recovery after surgery. Eur Surg Res. 2012;49:24–34.
69. Stoot JH, van Dam RM, Busch OR, van Hillegersberg R, de Boer M, Olde Damink SWM, et al. The effect of a multimodal fast-track programme on outcomes in laparoscopic liver surgery: a multicentre pilot study. HPB. John Wiley and Sons Inc. 2009;11:140–4.
70. Schultz NA, Larsen PN, Klarskov B, Plum LM, Frederiksen HJ, Christensen BM, et al. Evaluation of a fast-track programme for patients undergoing liver resection. Br J Surg. 2013;100:138–43.
71. He F, Lin X, Xie F, Huang Y, Yuan R. The effect of enhanced recovery program for patients undergoing partial laparoscopic hepatectomy of liver cancer. Clin Transl Oncol. Springer-Verlag Italia Srl. 2015;17:694–701.
72. Ovaere S, Boscart I, Parmentier I, Steelant PJ, Gabriel T, Allewaert J, et al. The effectiveness of a clinical pathway in liver surgery: a case-control study. J Gastrointest Surg. Springer New York LLC. 2018;22:684–94.
73. Lobo DN, Hendry PO, Rodrigues G, Marciani L, Totman JJ, Wright JW, et al. Gastric emptying of three liquid oral preoperative metabolic preconditioning regimens measured by magnetic resonance imaging in healthy adult volunteers: a randomised double-blind, crossover study. Clin Nutr. 2009;28:636–41.
74. Liang X, Ying H, Wang H, Xu H, Yu H, Cai L, et al. Enhanced recovery program versus traditional care in laparoscopic hepatectomy. Medicine (United States). Lippincott Williams and Wilkins. 2016;95:e2835.
75. Sánchez-Pérez B. Fast-track program in laparoscopic liver surgery: theory or fact? World J Gastrointest Surg. Baishideng Publishing Group Inc. 2012;4:246.
76. Hevesi ZG, Lapukhin SY, Mezrich JD, Andrel AC, Lee M. Designated liver transplant anesthesia team reduces blood transfusion, need for mechanical ventilation, and duration of intensive care. Liver Transpl. 2009;15:460–5.
77. Chin A, Heywood L, Iu P, Pelecanos AM, Barrington MJ. The effectiveness of regional anaesthesia before and after the introduction of a dedicated regional anaesthesia service incorporating a block room. Anaesth Intensive Care. 2017;45:714–9.
78. Markin NW. Navigating to the goal: the importance of shared mental models in complex environments. J Cardiothorac Vasc Anesth. W.B. Saunders. 2018;32:2618–9.
79. Gardner AK, Scott DJ, AbdelFattah KR. Do great teams think alike? An examination of team mental models and their impact on team performance. Surgery (United States). Mosby Inc. 2017;161:1203–8.
80. McNally SJ, Revie EJ, Massie LJ, McKeown DW, Parks RW, Garden OJ, et al. Factors in perioperative care that determine blood loss in liver surgery. HPB. Blackwell Publishing Ltd. 2012;14:236–41.

Part III

Different Approaches in Robotic Liver Surgery

6 Different Approaches in Robotic Liver Surgery: Console Surgeon And Bedside Surgeon (The RoboLap Approach)

Francesca Ratti, Marco Catena, and Luca Aldrighetti

6.1 Introduction

In a recent European survey concerning the implementation of minimally invasive robotic and laparoscopic technologies in the field of liver surgery, it emerged that the most frequently used technique for parenchymal transection involves the combination of bipolar forceps (with the rationale of performing Kellyclasia) and monopolar scissors [1]. Interestingly, the majority of robotic liver surgeons declared themselves dissatisfied with the robotic instruments currently available for parenchymal transection, specifically complaining about the lack of an ultrasonic dissector [1]. In fact, laparoscopic ultrasonic-mediated parenchymal dissection is precise and effective: combined with the image magnification and the improvement in terms of spatial resolution provided by a mini-invasive approach, it allows effective reproduction of almost the whole portfolio of liver resections, thereby improving the feasibility and reproducibility of mini-invasiveness [2–4].

Each technique (open, laparoscopic, and robotic) described for liver surgery shows intrinsic limitations. In open surgery, the combination of instrument availability and human movements is favorable, but visibility of the surgical field is limited and the impact of surgery on patients is strong. In laparoscopic surgery, these problems are addressed (hence, leading to documented benefit for the patient) and adequate instruments for liver surgery are available but surgeons' movements are limited so that a significant risk of bleeding and conversion is still reported for procedures with a high degree of technical difficulty [3, 5–8]. The idea to combine robotic and laparoscopic approaches solves all criticisms regarding parenchymal transection, maximizing human performance while keeping all the advantages of minimal invasiveness [9]. In the RoboLap approach, the advantages of robotics and laparoscopy are exploited to the maximum, combining ultrasonic dissection performed by the bedside surgeon, with vessels coagulation, cutting, and vascular dissection performed by the surgeon at the console. In this perspective, liver resection is performed within a shared leadership [9].

6.2 Technical Aspects

6.2.1 Operating Room Setting and Trocars Configuration

The RoboLap approach can be performed irrespective of the characteristics and generation of the robotic platform. The first surgeon seats at the console and the first assistant is located between the patient's legs, which are apart. For all resec-

F. Ratti (✉) · M. Catena · L. Aldrighetti
Hepatobiliary Surgery Division, IRCCS San Raffaele Scientific Institute, Milan, Italy
e-mail: ratti.francesca@hsr.it; aldrighetti.luca@hsr.it

M. D'Hondt, I. Sucandy (eds.), *Textbook of Robotic Liver Surgery*,
https://doi.org/10.1007/978-3-031-76536-0_6

tions excluding those involving the left lateral section, a laparoscopic 10 mm trocar is positioned in a right pararectal position, along the umbilical line. Four robotic trocars are positioned in a standardized configuration, with one trocar in right flank, one along the mid-clavicular line, one along the midline, and one in the left hypochondrium (Fig. 6.1), and the robotic platform is docked to operating table coming from the head (Da Vinci X platform) or from the right side (Da Vinci Xi platform) of the patient, oriented in reverse-Trendelenburg position (Fig. 6.2). For resections involving the left lateral sector, the laparoscopic 10 mm trocar is positioned in a left pararectal position, while four robotic trocars are positioned as follows: one trocar along the right mid-clavicular line, one along the midline, one along the left mid-clavicular line, and one in the left flank. According to the complexity of the procedure, a second laparoscopic access may be required on the right flank (between robotic arms 1 and 2) or on the left (between robotic arms 3 and 4): the second laparoscopic access should be positioned only after docking of robotic arms, to exclude conflicts among laparoscopic and robotic instruments and to improve ergonomy.

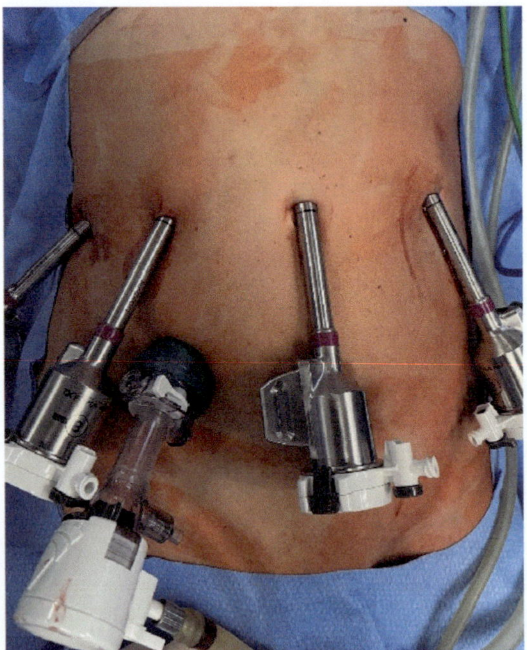

Fig. 6.1 Standard trocars configuration for all resections, excluding those for the lesions located in the left lateral section of the liver

It is important not to underestimate the space created due to distension from CO_2 of the abdominal cavity (especially in obese and elderly patients), which considerably distances the trocars from the target site. This phenomenon has specific negative consequences, especially on the accessibility of laparoscopic instruments to the surgical field. In particular, during the preparation phases, it is useful to take into consideration the length of the ultrasonic dissector and its possibility of reaching the furthest point of resection. If the abdominal cavity is wide or the resection area is cranial, the possibility of placing laparoscopic trocars on the same line of robotic trocars should be taken into consideration and should be balanced with the risk of scarce possibility of movement of laparoscopic instruments due to lateral external conflict with robotic arms. A camera is generally placed on arm 2 (on mid-clavicular line), aiming to have the line of transection in front of both the first surgeon's and the assistant's eyes.

6.2.2 Instrumentation

Standard robotic instrumentation includes prograsp forceps, Maryland bipolar forceps or long bipolar forceps, monopolar scissors, and robotic clip applier.

Laparoscopic instrumentation includes a suction system, small vascular clamps, titanium clips, hem-o-locks, vascular stapler, and ultrasonic dissector. For ultrasonic dissector, the choice may be based on the surgeon's preference, taking into consideration its length, flexibility, and efficacy.

6.2.3 Parenchymal Transection Technique

Ultrasonic dissector is used for parenchymal transection by the surgeon at the table, dissecting the liver parenchyma while preserving vessels

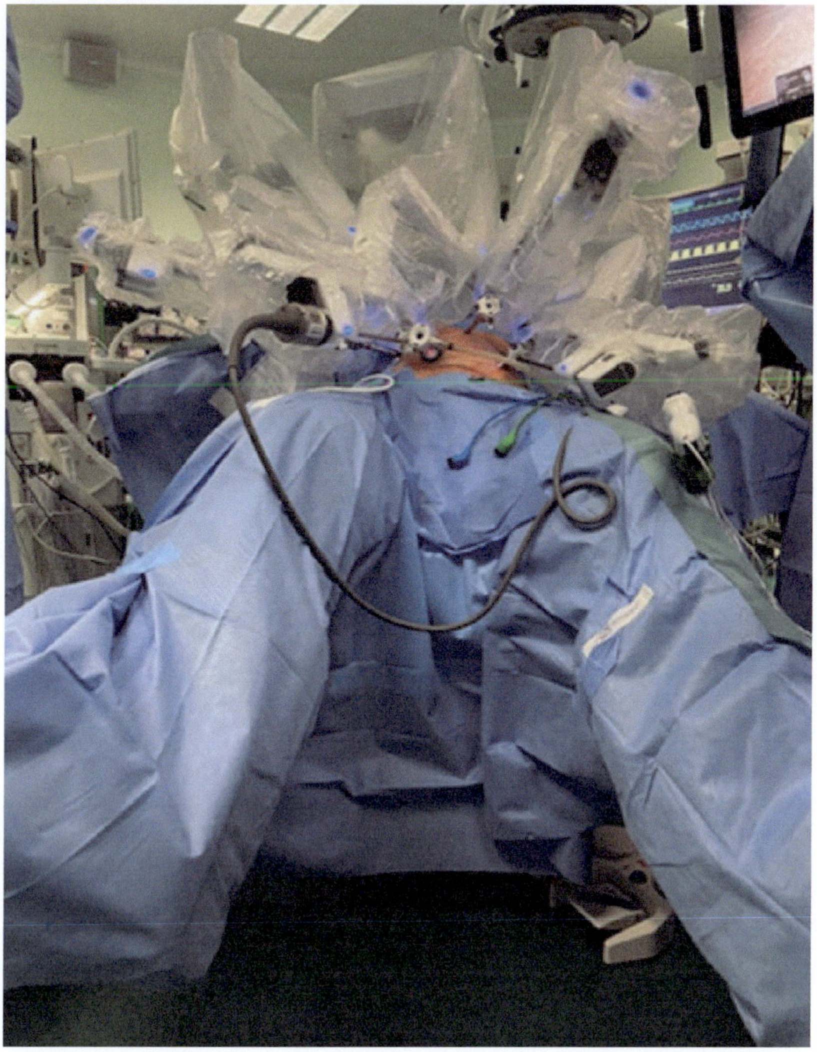

Fig. 6.2 Standard patient positioning

and biliary branches that were then coagulated or clipped according to their size (Fig. 6.3). Dissection technique should follow the same principles of ultrasonic-mediated transection in laparoscopic surgery. The direction of the tip of the ultrasonic dissector should be optimized in order to obtain an effective transection without being limited in its movement by robotic arms outside and robotic instruments inside. Maintaining the principle of having the transection area in the center of the visual field of both operators, it is possible to use the ultrasound dissector and the robotic bipolar for coagulation at the same time. By holding the robotic bipolar forceps in the console surgeon's left arm, the risk of conflicts with the ultrasound dissector is generally less. If the dissection field is particularly limited, a totally alternating use of the laparoscopic dissector and robotic instruments may be necessary. The surgeon at the console must be careful to bring the optics close enough to allow a good view to the bedside surgeon, without risking excessive fouling of the camera due to irrigation and dissection of the ultrasound instrument. Specific attention must be paid to phases of the procedure requiring the ultrasonic dissector working on the wall of hepatic veins: the direction of access of the ultrasonic dissector—from bottom to top—could in fact be less favorable, and consequently more dangerous, than in lapa-

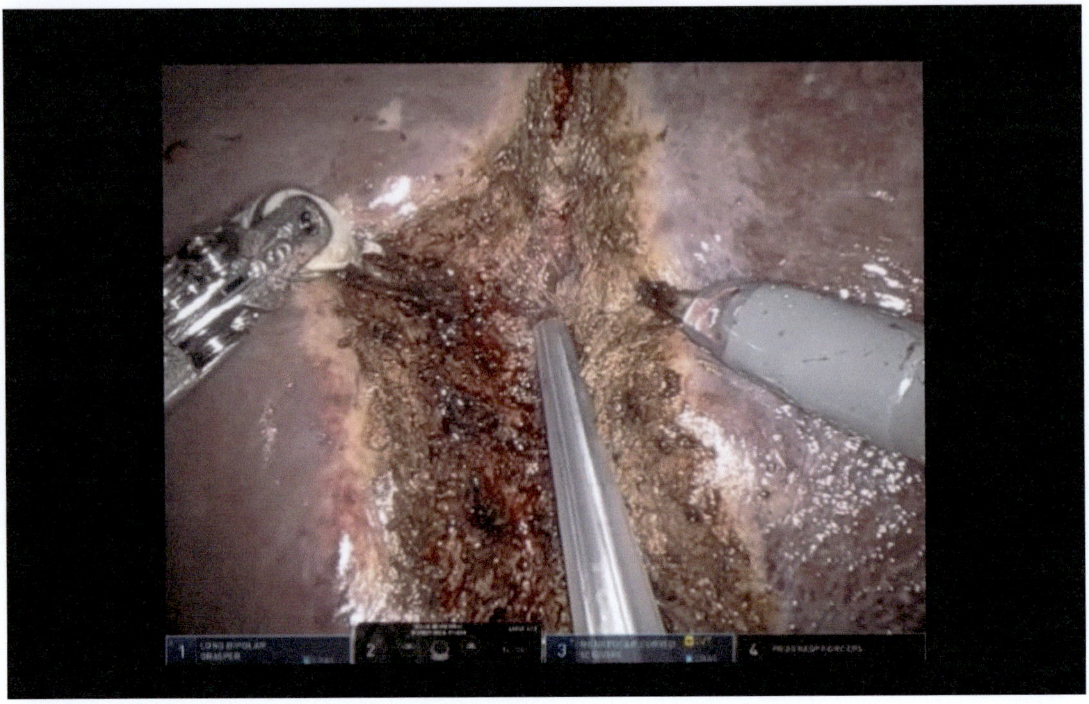

Fig. 6.3 Liver parenchymal transection in the RoboLap approach

roscopy. It is therefore advisable, in the proximity of hepatic vessels, to pay attention to avoid lesions of the lower and/or lateral wall of the vessel by first working on the anterior surface and eventually lowering the intensity of the ultrasound.

6.3 Literature Overview

The need to integrate the work of the console surgeon with the bedside surgeon was strongly felt in the setting of liver surgery, probably as the direct consequence of a still incomplete set of robotic instruments to fulfill all the tasks of hepatic resections. Perakkis et al. described a novel technique for the transection of liver parenchyma during robotic surgery using three devices (including a laparoscopic one) simultaneously: monopolar scissors and bipolar Maryland forceps of the robot and laparoscopic-guided waterjet [10]. They reported this technique in 28 cases, concluding that this technique of parenchyma dissection is a fundamental step forward to the standardization of robotic liver surgery for teaching purposes and the wider adoption of robotic hepatectomy into routine patient care. Camerlo et al. reported as well a two-surgeon technique using ultrasonic dissection and irrigated bipolar coagulation: in a mono-institutional series of 50 patients, the implementation of this approach appeared to be easy, safe, and effective [11]. With the same rationale of improving safety and efficacy by specifically targeting the issue of blood loss, Hawksworth et al. described the use of laparoscopic CUSA in robotic major hepatectomy and reported a mean blood loss of 300 mL and a transfusion rate of 10% in a series of 20 resections [12]. After the description of the RoboLap approach by our group [9], the technique was standardly implemented into a robotic resection program for resections with intermediate or high degree of complexity: within this perspective, the robotic approach demonstrated its superiority in terms of overall performance compared to both the laparoscopic and open approach, as reported in our series specifically analyzing the topic of lymphadenectomy in liver surgery: most patients in the robotic cohort were approached by the RoboLap technique, thus contributing to the large-scale diffusion of these procedures with a high profile of complexity [13–15]. Anyway, these results need to be validated on a larger scale to evaluate reproducibility and efficacy in terms of blood loss reduction (possibly resulting in a personalization of blood requirement according to the approach chosen [16] and reduction of in the risk of conversion) [17].

6.4 Discussion and Conclusions

The RoboLap approach is based on collaborative leadership between the surgeon at the console and the surgeon at the operating table. The synergy between the two members of the team in fact allows maximizing the benefit of the minimally invasive approach by combining the dexterity of the robot and the possibility of conducting an accurate dissection as in laparoscopy [9, 13].

Which criteria should be taken into consideration whether to use the RoboLap approach or the bipolar–scissors combination for parenchymal transection is an issue that needs to be clarified in the future. There are currently no data on the superiority of one technique over the other, and the choice is based on the preference of the surgeon and the availability of resources and devices. However, it could be speculated that selection criteria should be based on the technical difficulty of the resection: indeed, reserving the RoboLap approach to resections with a high profile of technical complexity allows maintaining a balance in costs and effectiveness. The other issue to be evaluated is the accessibility of the parenchymal transection plane by the ultrasonic dissector: in this perspective, anatomical resections (either anatomical or segmental) based on the intraoperative navigation through vascular landmarks (i.e., portobiliary pedicles and hepatic veins) constitute an excellent field of application as the transection line is generally in line with the eyes of surgeons and with the line of access of the ultrasonic dissector. Caution should be taken during laparoscopic ports positioning in order to optimize accessibility to the transection plane throughout the procedure.

The last topic concerns the background of the surgeon: it is possible—but no data are currently available—that surgeons proceeding from a minimally invasive laparoscopic program of liver resections to a robotic one have developed greater expertise with ultrasonic-mediated laparoscopic parenchymal dissection and therefore feel a greater need of this technique even in robotics. Instead, those who switch directly from an open to a robotic approach have a much more heterogeneous parenchymal dissection background (ultrasonic dissector vs. Kellyclasia) and therefore develop their own minimally invasive transection technique regardless of the ultrasonic dissector.

Among the whole portfolio of liver resections, right and left hepatectomy specifically have favorable characteristics to be the ideal field of application for this combined approach: on the one hand, indeed, the need for hepatic mobilization and accurate dissection of the hepatic hilum benefits from the joint movement of the robotic wrist, paving the way for greater feasibility of minimally invasive complex procedures and for a reduction in operating times; on the other hand, the transection line, extending from the free margin of the liver to the hepatocaval confluence and located in line with the eyes of both surgeons, allows an optimized use of the ultrasonic dissector by the surgeon at the table, increasing the accuracy of the dissection and the speed of the transection [18]. The presence—within the ultrasonic dissector—of an aspiration and irrigation systems overcomes the need to routinely use the aspirator.

This robotic–laparoscopic technique offers the possibility of improving the expertise and level of training in complex hepatectomies for the surgeon at the table, allowing them to deal with the challenges of the parenchymal transection phase in an "assisted" fashion. This may fasten the diffusion and the evolution of the robotic approach through the stages of surgical innovation [19]. However, the first assistant should already have a good level of expertise to maintain the procedure within the required level of safety. In particular, confidence with advanced laparoscopic techniques is required, as well as an adequate level of partnership between the two surgeons, to carry out the transection phases in a synergistic way. It is in fact mandatory to have an excellent level of understanding, communication, and collaboration between the two: unlike in the pure laparoscopic technique indeed, the first operator shares the surgical leadership in the parenchymal transection phase with the assistant. Furthermore, the assistant surgeon must have the expertise to timely and adequately manage any internal or external conflict among robotic instruments and must be able to manage an emergent conversion or other potentially life-threatening situations.

This technique is safe and reproducible in major hepatectomies and, in general, in robotic resections with a high profile of technical complexity of parenchymal transection, if the accessibility of the resection site by laparoscopic instruments is guaranteed.

References

1. MJW Z, Görgec B, Arabiyat A, CLM N, van der Poel MJ, Fichtinger RS, Berrevoet F, van Dam RM, Aldrighetti L, Fuks D, Hoti E, Edwin B, Besselink MG, Abu Hilal M, Hagendoorn J, Swijnenburg RJ, Dutch Liver Collaborative Group and E-AHPBA Innovation & Development Committee. Pan-European survey on the implementation of robotic and laparoscopic minimally invasive liver surgery. HPB (Oxford). 2022;24(3):322–31. https://doi.org/10.1016/j.hpb.2021.08.939. Epub 2021 Aug 21.
2. Ratti F, Cipriani F, Fiorentini G, Catena M, Paganelli M, Aldrighetti L. Have we really understood when the efforts of laparoscopic liver resection are justified?-a complexity-based appraisal of the differential benefit. Hepatobiliary Surg Nutr. 2022;11(3):363–74. https://doi.org/10.21037/hbsn-20-562. PMID: 35693407; PMCID: PMC9186211.
3. Cipriani F, Ratti F, Fiorentini G, Catena M, Paganelli M, Aldrighetti L. Pure laparoscopic right hepatectomy: a risk score for conversion for the paradigm of difficult laparoscopic liver resections. A single centre case series. Int J Surg. 2020;82:108–15. https://doi.org/10.1016/j.ijsu.2020.08.013. Epub 2020 Aug 27.
4. Ratti F, Fiorentini G, Cipriani F, Catena M, Paganelli M, Aldrighetti L. Technical insights on laparoscopic left and right hepatectomy for perihilar cholangiocarcinoma. Ann Surg Oncol. 2020;27(13):5191–2. https://doi.org/10.1245/s10434-020-08647-8. Epub 2020 May 26.
5. Masetti M, Fallani G, Ratti F, Ferrero A, Giuliante F, Cillo U, Guglielmi A, Ettorre GM, Torzilli G, Vincenti L, Ercolani G, Cipressi C, Lombardi R, Aldrighetti L, Jovine E. Minimally invasive treatment of colorectal liver metastases: does robotic surgery provide any technical advantages over laparoscopy? A multicenter analysis from the IGoMILS (Italian Group of Minimally Invasive Liver Surgery) registry. Updat Surg. 2022;74(2):535–45. https://doi.org/10.1007/s13304-022-01245-1. Epub 2022 Jan 31.
6. Cipriani F, Fiorentini G, Magistri P, Fontani A, Menonna F, Annecchiarico M, Lauterio A, De Carlis L, Coratti A, Boggi U, Ceccarelli G, Di Benedetto F, Aldrighetti L. Pure laparoscopic versus robotic liver resections: multicentric propensity score-based analysis with stratification according to difficulty scores. J Hepatobiliary Pancreat Sci. 2022;29(10):1108–23. https://doi.org/10.1002/jhbp.1022. Epub 2021 Aug 3.
7. Gavriilidis P, Roberts KJ, Aldrighetti L, Sutcliffe RP. A comparison between robotic, laparoscopic and open hepatectomy: a systematic review and network meta-analysis. Eur J Surg Oncol. 2020;46(7):1214–24. https://doi.org/10.1016/j.ejso.2020.03.227. Epub 2020 Apr 12.
8. Patriti A, Cipriani F, Ratti F, Bartoli A, Ceccarelli G, Casciola L, Aldrighetti L. Robot-assisted versus open liver resection in the right posterior section. JSLS. 2014;18(3):e2014.00040. https://doi.org/10.4293/JSLS.2014.00040. PMID: 25516700; PMCID: PMC4266223.
9. Aldrighetti L, Catena M, Ratti F. Maximizing performance in complex minimally invasive surgery of the liver: the RoboLap approach. J Gastrointest Surg. 2022;26(8):1811–3. https://doi.org/10.1007/s11605-022-05340-x. Epub 2022 May 9.
10. Perrakis A, Rahimli M, Gumbs AA, Negrini V, Andric M, Stockheim J, Wex C, Lorenz E, Arend J, Franz M, Croner RS. Three-device (3D) technique for liver parenchyma dissection in robotic liver surgery. J Clin Med. 2021;10(22):5265. https://doi.org/10.3390/jcm10225265. PMID: 34830547; PMCID: PMC8653962.
11. Camerlo A, Magallon C, Vanbrugghe C, Chiche L, Gaudon C, Rinaldi Y, Fara R. Robotic hepatic parenchymal transection: a two-surgeon technique using ultrasonic dissection and irrigated bipolar coagulation. J Robot Surg. 2021;15(4):539–46. https://doi.org/10.1007/s11701-020-01138-8. Epub 2020 Aug 10. PMID: 32779132.
12. Hawksworth J, Radkani P, Nguyen B, Belyayev L, Llore N, Holzner M, Mateo R, Meslar E, Winslow E, Fishbein T. Improving safety of robotic major hepatectomy with extrahepatic inflow control and laparoscopic CUSA parenchymal transection: technical description and initial experience. Surg Endosc. 2022;36(5):3270–6. https://doi.org/10.1007/s00464-021-08639-z. Epub 2021 Aug 9. PMID: 34370124.

13. Ratti F, Cipriani F, Ingallinella S, Tudisco A, Catena M, Aldrighetti L. Robotic approach for lymphadenectomy in biliary Tumours: the missing ring between the benefits of laparoscopic and reproducibility of open approach? Ann Surg. 2022;278:e780. https://doi.org/10.1097/SLA.0000000000005748. Epub ahead of print.
14. Zhang XF, Chakedis J, Bagante F, Chen Q, Beal EW, Lv Y, Weiss M, Popescu I, Marques HP, Aldrighetti L, Maithel SK, Pulitano C, Bauer TW, Shen F, Poultsides GA, Soubrane O, Martel G, Groot Koerkamp B, Guglielmi A, Itaru E, Pawlik TM. Trends in use of lymphadenectomy in surgery with curative intent for intrahepatic cholangiocarcinoma. Br J Surg. 2018;105(7):857–66. https://doi.org/10.1002/bjs.10827. Epub 2018 Apr 14.
15. Spolverato G, Yakoob MY, Kim Y, Alexandrescu S, Marques HP, Lamelas J, Aldrighetti L, Gamblin TC, Maithel SK, Pulitano C, Bauer TW, Shen F, Poultsides GA, Marsh JW, Pawlik TM. Impact of complications on long-term survival after resection of intrahepatic cholangiocarcinoma. Cancer. 2015;121(16):2730–9. https://doi.org/10.1002/cncr.29419. Epub 2015 Apr 22.
16. Pulitanò C, Arru M, Bellio L, Rossini S, Ferla G, Aldrighetti L. A risk score for predicting perioperative blood transfusion in liver surgery. Br J Surg. 2007;94(7):860–5. https://doi.org/10.1002/bjs.5731.
17. Aldrighetti L, Ratti F, Cillo U, Ferrero A, Ettorre GM, Guglielmi A, Giuliante F, Calise F. Italian Group of Minimally Invasive Liver Surgery (I GO MILS). Diffusion, outcomes and implementation of minimally invasive liver surgery: a snapshot from the I Go MILS (Italian Group of Minimally Invasive Liver Surgery) Registry. Updat Surg. 2017;69(3):271–83.
18. Ratti F, Cipriani F, Ariotti R, Giannone F, Paganelli M, Aldrighetti L. Laparoscopic major hepatectomies: current trends and indications. A comparison with the open technique. Updat Surg. 2015;67(2):157–67.
19. Berardi G, Van Cleven S, Fretland ÅA, Barkhatov L, Halls M, Cipriani F, Aldrighetti L, Abu Hilal M, Edwin B, Troisi RI. Evolution of laparoscopic liver surgery from innovation to implementation to mastery: perioperative and oncologic outcomes of 2,238 patients from 4 European specialized centers. J Am Coll Surg. 2017;225(5):639–49. https://doi.org/10.1016/j.jamcollsurg.2017.08.006. Epub 2017 Aug 31.

7 Different Approaches in Robotic Liver Surgery: Pure Robotic Approach

Mathieu D'Hondt and Edward Willems

7.1 Introduction

As discussed in previous chapters, the diffusion of robotic liver surgery (RLS) is slowly rising. However, compared to other surgical subspecialties, the introduction remains slow. Several factors contribute to the slow uptake of robotic liver surgery, including mainly technical aspects. For example, the technical complexity of minimally invasive liver surgery and the steep learning curve of these procedures were the cause of the slower uptake of laparoscopy in liver surgery and are also a factor in the slow rise of robotic liver surgery. Furthermore, the absence of robotic instruments for parenchyma transection such as CUSA® (Integra Lifesciences, Plainsboro, NJ, USA) requires important modifications in surgical techniques.

However, several surgeons have overcome these difficulties and have started a robotic liver surgery program without a fall in short-term postoperative results and several studies have reported on the safety and feasibility of robotic liver surgery [1–4]. The transition from laparoscopic to robotic liver surgery and the implementation of robotic liver surgery have been described in several studies [5–10]. A shorter learning curve for robotic hepatic resections has been suggested compared to laparoscopy.

In this chapter, we discuss two different approaches to robotic liver surgery: the pure robotic approach with a single console surgeon and the console surgeon and bedside surgeon (Robolap) approach. A comparison of both techniques is summarized in Table 5.1.

M. D'Hondt (✉) · E. Willems
Department of Digestive and Hepatobiliary/Pancreatic Surgery, Groeninge Hospital, Kortrijk, Belgium

M. D'Hondt, I. Sucandy (eds.), *Textbook of Robotic Liver Surgery*,
https://doi.org/10.1007/978-3-031-76536-0_7

Table 7.1 Advantages and disadvantages of pure robotic and Robolap approach

Pure robotic hepatectomy	
Advantages	*Disadvantages*
Operation leadership with console surgeon prevents conflicts during surgery and facilitates key steps Robotic system is used for operating, not for exposing Increase in independence of the console surgeon, resulting in a shorter learning curve Pure robotic surgery improves machine learning and the involvement of artificial intelligence Lower number of ports resulting in a lower total incision length and possible less pain No need for skilled bedside surgeon Cost-effective Lower surgical fatigue in an ergonomically seated console surgeon	Lack of haptic feedback No possibility of shared decision-making Higher threshold for implementation of robotic liver surgery
Robolap hepatectomy	
Advantages	*Disadvantages*
Possibility to use ultrasonic parenchyma transection devices as a familiar dissection tool Easier emergent conversion if needed Presence of the bedside surgeon in case of sudden need for laparoscopic assistance (bleeding control, …) Haptic feedback for bedside surgeon Shared decision-making and possibility to communicate in case of difficult decisions Lower threshold to transfer from laparoscopy to robotic surgery Faster learning curve in MI hepatectomy for bedside surgeon	Higher total costs Difficult assistant trocar placement for CUSA Uncomfortable position for bedside surgeon Decrease in independence for console surgeon Need for two experienced surgeons

7.2 Technical Aspects

Considerations regarding patient preparation and installation are extensively described in the previous chapter. Details on anesthesia in robotic liver surgery will be described in the next chapter.

For a pure robotic approach with a single console surgeon, the robotic system is docked from the right side of the patient. First, a 8 mm robotic trocar is placed at the level of the umbilicus, the remaining ports are placed at the same level. Depending on the type of transection, one or two 12 mm assistant ports can be placed. The bedside assistant is positioned between the patient's legs. In our experience, the bedside assistant can be either a surgical trainee or an experienced scrub nurse (Fig. 5.1).

When necessary, a fully intracorporeal Pringle maneuver can be performed using a 16Fr Foley catheter slung around the hepatoduodenal ligament (Huang's loop) [11]. This can be handled fully robotically without the need for additional incisions or trocars.

Depending on the indication and type of resection, mobilization of the liver and hilar dissection is carried out first. A fully robotic intraoperative ultrasound can be easily performed to detect occult lesions and estimate the size of the lesion(s) and their relationship to major vasculature. The Tilepro™ feature transfers the ultrasound images to the console, allowing a comfortable intraoperative ultrasound.

Parenchyma transection is traditionally started using a monopolar instrument, either a cautery hook or monopolar scissors. The superficial portion of the parenchyma can be safely dissected in this fashion as there are only small capillary vessels. In the absence of CUSA, deeper parenchyma transection needs to be performed carefully in order to prevent major hemorrhage or bile leak. Several techniques have been found to be safe and feasible.

In analogy with open liver surgery, Kelly clamping or Kellyclasia can be used to crush parenchyma. A fenestrated bipolar forceps or Maryland bipolar forceps is often used.

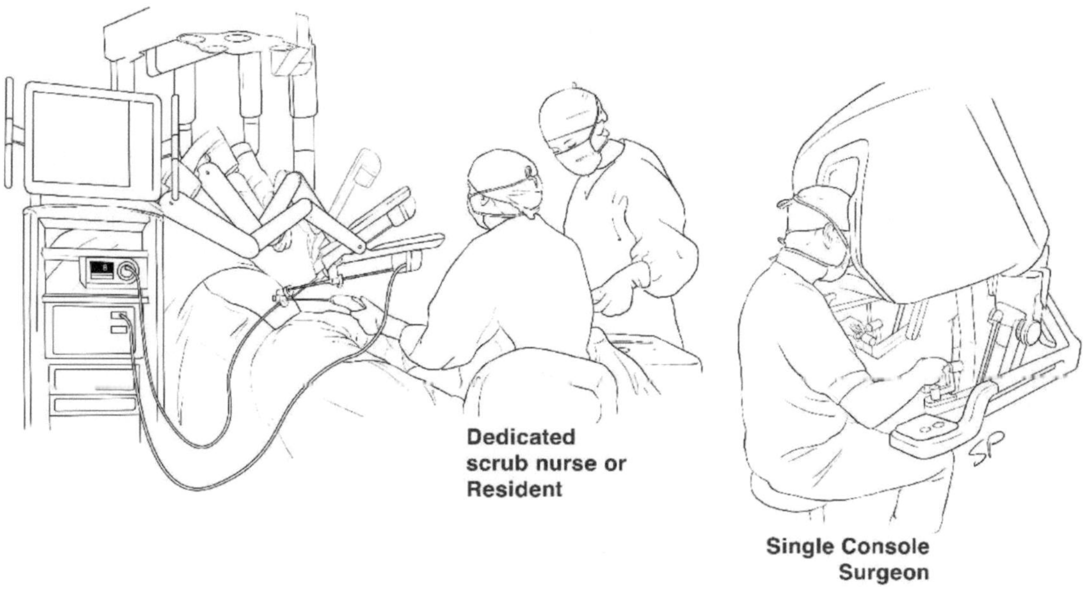

Fig. 7.1 Pure robotic approach or single console surgeon approach

Parenchyma is crushed by slowly closing the bipolar forceps and crushed parenchyma is washed away using a laparoscopic irrigation/aspiration device. Bipolar energy is cautiously applied to both sides of the transection line to obtain adequate hemostasis. Larger biliary or vascular structures are spared and can then be divided separately using a sealing device or clips. Some authors have described applying bipolar energy while crushing the parenchyma [12]. An alternative method of parenchyma transection following crushing with the bipolar forceps is to use the monopolar robotic scissors [12]. A comparable technique is to use a robotic sealing device for this Kellyclasia [13]. Either the robotic harmonic scalpel (Harmonic ACE®, Ethicon Endo-Surgery, Cincinnati, OH, USA), the Vessel Sealer Extend® or SynchroSeal® (Intuitive Surgical Inc., Sunnyvale, CA, USA). The sealing device is slowly closed to crush parenchyma and expose biliary and vascular structures within the transection plane. An additional advantage of this technique is that the device can be used to divide smaller vascular structures up to 7 mm diameter. Some surgeons activate the sealing device while slowly closing the jaws of the instrument, while others use it as a pure crush-clamp instrument and selectively activate the energy to treat vessels in the transection plane. Often, a bipolar forceps is used as a forceps for retraction purpose and allows for concurrent hemostasis of the transection plane. A laparoscopic irrigation/aspiration device is used to aspirate blood and crushed parenchyma and to rinse the transection plane when necessary, keeping the surgical field clean. As the transfer of robotic instruments is often somewhat time consuming, it is of vital importance to ensure adequate hemostasis and keep a clean surgical field.

A common complaint is that the Vessel Sealer Extend® is a large and blunt instrument that is too bulky for precise transection of the parenchyma [14]. The Synchroseal® provides a sufficient alternative. Compared to the Vessel Sealer, the SynchroSeal is a smaller instrument that can be used more precisely. However, the Vessel Sealer has a better ability to cut structures, which makes it more aggressive when used for Kellyclasia but more precise in dividing structures.

In general, the use of a bipolar forceps for Kellyclasia, along with a monopolar device, is preferred in patients with more stiff, fibrotic parenchyma.

The sealing devices can be used to divide vascular structures up to 7 mm. Larger vessels and biliary structures can be ligated, clipped using robotic WECK® Hem-o-lok® clips (Teleflex Medical, Research Triangle Park, NC, USA) or using laparoscopic clips. Rarely, larger pedicles need to be divided using a robotic or laparoscopic stapling device. However, several authors have described a decline in the use of staplers in robotic hepatectomy compared to laparoscopy [5].

The techniques described above allow for a purely robotic resection. Every step of the procedure, from the intraoperative ultrasound, over hilar dissection, application of a Pringle maneuver to parenchymal transection, is performed by the console surgeon. The role of the scrub nurse or resident at the bedside is purely to assist the console surgeon. For this purpose, one or two laparoscopic 12 mm trocars are inserted. The role of the bedside surgeon is to insert required materials such as gauzes and sutures, and to handle the irrigation/aspiration device: rinsing the transection surface and aspirating blood. Furthermore, the assistant port is necessary to insert the robotic ultrasound probe. Rarely, laparoscopic clips or a laparoscopic stapling device are necessary. Using this approach, there is an obvious "operating" surgeon in the console and an "assisting" surgeon bedside. The bedside surgeon is thus purely assisting the operating surgeon and is instructed by the console surgeon. This allows for a better cooperation between the two surgeons, where the bedside surgeon is often a (less experienced) surgical trainee or a trained scrub nurse. The fact that one person is leading the operation prevents the occurrence of conflicts during the operation [14, 15]. Furthermore, the limited role of the bedside surgeon allows for better ergonomics as the available space between the robotic arms is often very limited, preventing a comfortable position [16].

An important factor in the implementation of RLS is a stepwise implementation with dedicated training of the entire team [13, 15, 17]. Several authors have pointed out the importance of an optimal training of the entire surgical team. The complexity of the robotic system requires specific training, not only for the surgeon but for the entire team. This can be achieved by structured, dedicated training of the team as a whole. Different aspects are emphasized for surgeons, trainees, or scrub nurses. When these training are completed, a routine is obtained for the installation, docking, and exchange of robotic instruments.

7.3 Literature Overview

In recent years, several authors have described their experience with robotic liver surgery. Many of them are using the pure robotic approach. We described our own experience with the introduction of a pure robotic liver surgery program in 2022. The first 71 patients undergoing robotic liver surgery were compared to two historical cohorts of laparoscopic liver surgery, an "initial phase" and a "mastery phase." Parenchyma is transected using monopolar scissors, Maryland bipolar forceps, and a sealing device. Blood loss was lower in patients who underwent robotic surgery. All robotic resections were performed purely robotic. Sucandy et al. reported their institutional experience with robotic major hepatectomy in 2020 [10]. They performed 170 purely robotic liver resections, of which 100 were major hepatectomies, with excellent short-term outcomes. Parenchymal transection is performed using monopolar scissors and a bipolar vessel sealer. Machado et al. performed 50 purely robotic liver resections, including 32% major hepatectomies [12]. Their technique for parenchyma transection consists of using robotic bipolar forceps under continuous saline irrigation, followed by transection of the tissue with monopolar scissors. They reported one conversion and a postoperative complication rate of 8%. In their technique, the assistant surgeon is located on the left side of the patient and performs clipping and stapling if necessary. However, they do not describe the use of a laparoscopic parenchymal transection device. Magistri et al. used a combination of the robotic Harmonic ACE scalpel and Kellyclasia with the Maryland bipolar forceps in their series of 22 robotic major hepatectomies

[17]. Chen et al. reported their experience with robotic hepatectomies as they omitted the presence of the bedside surgeon [14]. Comparing procedures with and without bedside surgeon, including major hepatectomies, they report similar results. They even found shorter operative times in the group without bedside surgeon. A recent paper by Bianchi et al. compares the outcomes of robotic hepatectomies by the Da Vinci Si and Xi system in a total of 89 patients [15]. Several advantages were found in favor of the Xi system. However, all procedures were performed purely robotic using a combination of the Maryland bipolar forceps and the robotic Vessel Sealer. Di Benedetto et al. compared robotic and open liver resection for hepatocellular carcinoma (HCC) using propensity score matching [16]. After propensity score matching, 106 patients in both cohorts were compared. In the robotic group, they found longer operative times and higher blood loss, but a shorter hospital length of stay, a lower rate of posthepatectomy liver failure, and a lower number of ICU admissions. Their technique consists of a combination of Kellyclasia and either the Harmonic ACE scalpel or the Vessel Sealer. They clearly propagate a purely robotic approach and emphasize the agility of the flexible robotic instruments. Another propensity score-matched comparison between robotic and open major hepatectomies was performed by Sucandy and colleagues [18]. Forty-two patients in both groups were compared, showing favorable results for the robotic group, including lower blood loss, shorter ICU, and overall length of stay. The purely robotic technique, a so-called modified crush-clamp technique, is extensively described in their report. The robotic fenestrated bipolar forceps is used to crush the liver while using limited laparoscopic irrigation and applying bipolar energy to obtain hemostasis. Biliovascular structures are divided using the Vessel Sealer. The Pringle maneuver is not routinely used in this technique even for major hepatectomies. Choi et al. describe the implementation of robotic liver surgery for experienced laparoscopic surgeons [19]. A purely robotic approach is used in all procedures, and parenchyma transection is performed with bipolar forceps and the robotic harmonic scalpel. In the Robolap approach, two surgeons perform the procedure: a console surgeon and bedside surgeon assisting the console surgeon mostly with an ultrasonic aspirator during parenchymal transection [20] (Fig. 5.2).

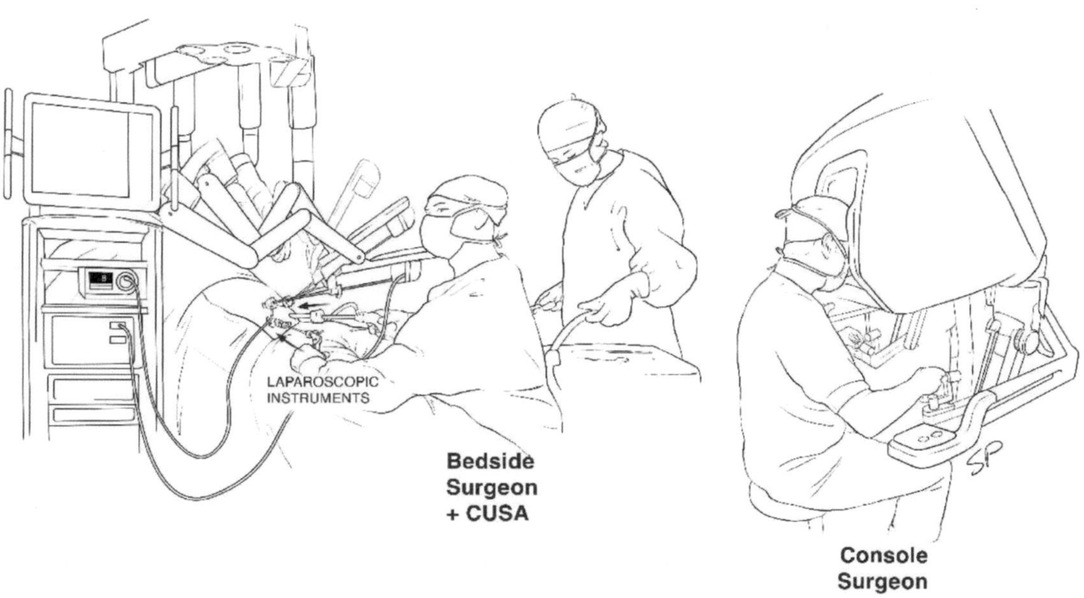

Fig. 7.2 The Robolap approach: a console "robotic" surgeon is assisted by a bedside "laparoscopic" surgeon

7.4 Conclusions

The implementation of robotic liver surgery is slowly rising, yet some concerns remain. One of these concerns is the absence of a robotic device for parenchyma transection, such as the CUSA device. Several authors have described their experience with a purely robotic approach for parenchyma transection with good results. A combination of Kellyclasia with a bipolar forceps and a robotic sealing device is most often used for parenchyma transection, although many variations of this technique are described. The purely robotic approach omits the necessity of an experienced bedside surgeon, placing full leadership of the operation with the console surgeon, seated comfortably at the robotic console. Future investigations need to point out the situations that may benefit from a combined robotic–laparoscopic approach.

References

1. Ciria R, Berardi G, Alconchel F, et al. The impact of robotics in liver surgery: a worldwide systematic review and short-term outcomes meta-analysis on 2,728 cases. J Hepatobiliary Pancreat Sci. 2020;29:181. https://doi.org/10.1002/jhbp.869.
2. Liu R, Wakabayashi G, Kim H, et al. International consensus statement on robotic hepatectomy surgery in 2018. World J Gastroenterol. 2019;25(12):1432–44. https://doi.org/10.3748/wjg.v25.i12.1432.
3. Montalti R, Berardi G, Patriti A, Vivarelli M, Troisi RI. Outcomes of robotic vs laparoscopic hepatectomy: a systematic review and meta-analysis. World J Gastroenterol. 2015;21(27):8441–51. https://doi.org/10.3748/wjg.v21.i27.8441.
4. Guan R, Chen Y, Yang K, et al. Clinical efficacy of robotassisted versus laparoscopic liver resection: a meta-analysis. Asian J Surg. 2019;42(1):19–31. https://doi.org/10.1016/j.asjsur.2018.05.008.
5. D'Hondt M, Devooght A, Willems E, Wicherts D, De Meyere C, Parmentier I, et al. Transition from laparoscopic to robotic liver surgery: clinical outcomes, learning curve effect, and cost-effectiveness. J Robot Surg. 2022;17(1):79–88. https://doi.org/10.1007/s11701-022-01405-w.
6. Lorenz E, Arend J, Franz M, et al. Robotic and laparoscopic liver resection-comparative experiences at a high-volume German academic center. Langenbeck's Arch Surg. 2021;406(3):753–61. https://doi.org/10.1007/s00423-021-02152-6.
7. Chen PD, Wu CY, Hu RH, et al. Robotic major hepatectomy: is there a learning curve? Surgery. 2017;161(3):642–9. https://doi.org/10.1016/j.surg.2016.09.025.
8. O'Connor VV, Vuong B, Yang ST, Difronzo A. Robotic minor hepatectomy offers a favorable learning curve and may result in superior perioperative outcomes compared with laparoscopic approach. Am Surg. 2017;83(10):1085–8. https://doi.org/10.1177/000313481708301014.
9. Salibi PN, Vrochides D, Martinie JB, Baker EH, Iannitti DA, Ocuin LM. Building up an effective robotic liver surgery team: who do we need? Laparosc Surg. 2020;4:28.
10. Sucandy I, Luberice K, Lippert T, Castro M, Krill E, Ross S, Rosemurgy A. Robotic major hepatectomy: an institutional experience and clinical outcomes. Ann Surg Oncol. 2020;27:4970–9.
11. Huang JW, Su WL, Wang SN. Alternative laparoscopic intracorporeal Pringle maneuver by Huang's loop. World J Surg. 2018;42(10):3312–5.
12. Machado MAC, Lobo-Filho MM, Mattos BH, Ardengh AO, Makdissi FF. Robotic liver resection. Report of the first 50 cases. Arq Gastroenterol. 2021;58(4):514–9.
13. Ayabe RI, Azimuddin A, Tran Cao HS. Robot-assisted liver resection: the real benefit so far. Langenbeck's Arch Surg. 2022;407(5):1779–87.
14. Chen P-D, Hu R-H, Liang J-T, Huang C-S, Wu Y-M. Toward a fully robotic surgery: performing robotic major liver resection with no table-side surgeon. Int J Med Compu Assist Surg. 2019;15:e1985.
15. Bianchi G, D'Angelis N, Musa N, Beghdadi N, Hentati H, Ammendola M, et al. Short-term outcomes of da Vinci Xi versus Si robotic systems for minor hepatectomies. Acta Biomed. 2022;93(5):e2022223. https://doi.org/10.23750/abm.v93i5.12851.
16. Di Benedetto F, Magistri P, Di Sandro S, Sposito C, Oberkofler C, Brandon E, et al. Safety and efficacy of robotic vs open liver resection for hepatocellular carcinoma. JAMA Surg. 2022;158:e225697. https://doi.org/10.1001/jamasurg.2022.5697.
17. Magistri P, Assirati G, Ballarin R, Di Sandro S, Di Benedetto F. Major robotic hepatectomies: technical considerations. Updat Surg. 2021;73:989–97.
18. Sucandy I, Shapera E, Syblis CC, Crespo K, Przetocki VA, Ross SB, Rosemurgy AS. Propensity score matched comparison of robotic and open major hepatectomy for malignant liver tumors. Surg Endosc. 2022;36:6724–32.
19. Choi SH, Han DH, Lee JH, Choi YR, Lee JH, Choi GH. Safety and feasibility of robotic major hepatectomy for novice surgeons in robotic liver surgery: a prospective multicenter pilot study. Surg Oncol. 2020;35:39–46.
20. Hawksworth J, Radkani P, Nguyen B, Belyayev L, Llore N, Holzner M, et al. Improving safety of robotic major hepatectomy with extrahepatic inflow control and laparoscopic CUSA parenchymal transection: technical description and initial experience. Surg Endosc. 2022;36:3270–6.

Part IV

Learning Curves and Training in Robotic Liver Surgery

Learning Curve in Robotic Liver Surgery

8

Michelle M. Dugan, Sharona Ross, and Iswanto Sucandy

8.1 Introduction

The field of surgery is continuously evolving with the introduction of new technological advancements and operative techniques. Minimally invasive techniques have revolutionized hepatobiliary surgery with consistently superior patient outcomes when compared with a traditional open approach, without compromising oncological outcomes and mid-/long-term survival [1, 2]. Some well-established advantages of robotic hepatectomy when compared with the open approach include shorter hospitalizations, improved postoperative pain control, lower operative blood loss, and better cosmesis [1–3].

Furthermore, the robotic platform offers significant technical advantages over the conventional laparoscopy, including the three-dimensional working environment, better visual magnification, "wristed" instrument technology, superior ergonomics, improved manual dexterity, tremor filtering, advanced energy devices for better hemostasis during parenchymal transection, and ease of intracorporeal suturing in challenging anatomical regions. Finally, the robotic platform also facilitates delicate suturing for vascular and biliary reconstruction [2, 4].

To implement these new advancements into daily practice, the surgeon will be required to overcome a learning curve. The time it takes to become proficient in any new skill will depend on many individual factors. To develop the skills necessary for robotic liver resection, contributing factors may include years of experience using the open approach, proficiency in laparoscopic technique, level of training, individual surgeon aptitude, institutional resources, and level of complexity of the procedure itself.

One widely accepted technique for measuring this learning curve is with the use of a Cumulative Sum (CUSUM) model. This method quantifies the number of cases, in chronological order, required to result in a significant improvement in a predefined dependent variable(s), which will be indicated by an inflection point in the plot representing the line between the "learning" phase and "competent" phase, sometimes further broken down into a third "mastery" phase. Continuous variables, such as operative time, estimated blood loss, and hospital length of stay, are generally more informative than binary variables such as rate of conversion to open procedure [5]. Other variables may then be analyzed (via univariate or

M. M. Dugan
Florida Atlantic University Schmidt College of Medicine, Boca Raton, FL, USA

Digestive Health Institute, AdventHealth Tampa, Tampa, FL, USA

S. Ross · I. Sucandy (✉)
Digestive Health Institute, AdventHealth Tampa, Tampa, FL, USA
e-mail: Iswanto.sucandy@adventhealth.com; Laura.Rosales@adventhealth.com

M. D'Hondt, I. Sucandy (eds.), *Textbook of Robotic Liver Surgery*,
https://doi.org/10.1007/978-3-031-76536-0_8

multivariate analysis) using the resulting case number as a set point to compare outcomes prior to versus after reaching the learning curve.

8.2 Overview of Published Literature

Published literature using various methods to measure a minimum number of cases required to be proficient in robotic liver resection is generally limited at this time, although studies have increased in recent years. An exact minimum number of cases to assure safe proficiency cannot be established at this time due to the heterogeneity of studies. The type of learning curve has been measured as a CUSUM plot, difficulty index score, rate of resection of posterosuperior segments, comparison with laparoscopic resection parameters, or other arbitrary methods [3, 6–16]. Even after narrowing down to studies using the CUSUM model, the parameters used for analysis vary greatly between studies. The most frequent CUSUM parameters used were operative duration, estimated blood loss (EBL), hospital length of stay, and open conversion rate. Furthermore, studies have varying methods of defining procedure type to include major hepatectomy, minor hepatectomy, or a combination of both [3, 7, 9, 12, 13, 16]. Relevant studies discussing the learning curve for robotic liver surgery are summarized in Fig. 8.1.

A study by Ahmad, Freeman, and Corn at the University of Kansas retrospectively analyzed 148 robotic hepatectomies (58 major, 90 minor) performed by a single surgeon between 2017 and 2020. Using the CUSUM analysis method for operative time and EBL, they found that proficiency was achieved after 22 cases of major hepatectomy and 34 cases of minor hepatectomy. The authors explained that there were multiple reasons for having a longer learning curve for minor versus major hepatectomy. Firstly, it was likely that more minor hepatectomy procedures were done early in their experience and subsequently, with increased experience, major hepatectomy was incorporated into the practice. Secondly, minor hepatectomy was a diverse group of procedures ranging from anterior resections (segments 4B and 5) to significantly more

Study (Year)	Country of study	n	Study period (Years)	Number of surgeons	Type of hepatectomy	Learning curve analysis metod(s)	Learning curve parameter(s)	Cases to overcome learning curve
Ahmad et al. (2022)	United States of America	148	4	Single surgeon	Separated: Major: n=58 Minor: n=90	$CUSUM_{OT/EBL}$	Operative time and EBL	Major: 22 Minor: 34
Görgec et al. (2022)	Netherlands	400	8	Institutional (multi-center)	Combined	$CUSUM_{EBL}$	EBL	33
Chua et al. (2021)	Multiple, Meta-analysis	Varied between studies	29	Varied between studies	Combined	CUSUM	Varied between studies	25
Magistri et al. (2019)	Italy	60	4	Institutional (multi-center)	Combined	$CUSUM_{OT}$	Operative time	30
Zhu et al. (2019)	China	140	3	2	Combined	$CUSUM_{Conversion}$	Open conversion rate	30
Chen et al. (2017)	Taiwan	92	4	Institutional (multi-center)	Major only	$CUSUM_{OT}$	Operative time	Phase 1: 15 Phase 2: 25 Phase 3: 52
Efanov et al. (2017)	Russia	40	7	2	Combined	Difficulty Index	Difficulty index Operative time	16
O'Connor et al. (2017)	United States of America	42	3	2	Minor only	Improved outcomes compared with laparoscopy	EBL LOS Complications	25

Fig. 8.1 Comparative studies measuring the learning curve necessary for proficiency in robotic liver resection. *CUSUM* Cumulative Sum model, *OT* operative time, *EBL* estimated blood loss, *LOS* length of stay

complex posterior-superior segmental resections. For major hepatectomy, proficiency was associated with an improvement in postoperative complication rate and hospital length of stay. For minor hepatectomy, proficiency was associated with an improvement in hospital length of stay. There were no differences found in conversion rate (to open), overall mortality rate, or 30-day readmission [3].

In a study by Chen et al., an analysis was conducted of 92 robotic major hepatectomies between 2012 and 2015. A CUSUM analysis for operative time revealed three consecutive phases of proficiency. The initial phase (1) was met after 15 cases, the intermediate phase (2) was met after 25 cases, and the mature phase (3) was met after 52 cases. More complex procedures were introduced during phase 2, such as trisegmentectomy, 8-5-4 trisegmentectomy, and donor hepatectomy, which were fully established by phase 3. On further analysis, both operative time and hospital length of stay were reduced after phase 1 (15 cases), and EBL was reduced after phase 2 (25 cases) [7].

A meta-analysis by Chua et al. included seven studies relating to the learning curve for robotic hepatectomy. The cases that were analyzed using the CUSUM model (in three studies) had a mean case number of 25 to surmount the learning curve. The parameters analyzed included operative time, EBL, hospital LOS, conversion to open, and complications. Six studies combined both major and minor hepatectomy and one study included only minor hepatectomy [9].

In a large multi-institute nationwide analysis by Görgec et al. of 400 robotic liver resections in the Netherlands (207 minor resections, 141 major resections) between 2014 and 2021, a CUSUM analysis for EBL revealed a learning curve of 33 cases [12]. Similar results were reported by Magistri et al., who analyzed 60 robotic hepatectomies from 2014 to 2017. A CUSUM analysis for operative time revealed a learning curve of 30 cases. On further analysis, a significant decrease in EBL and postoperative complications was also seen after the learning curve of 30 cases [13]. Another study by Zhu et al. on 140 robotic hepatectomies performed by two surgeons between 2015 and 2017 similarly found a learning curve of 30 cases, using the CUSUM model for conversion rate to open resection [16].

Several studies also incorporated a comparison between the learning curves for robotic versus laparoscopic liver resection. These studies have found the learning curve for robotic hepatectomy to be shorter than the learning curve for laparoscopic hepatectomy [4, 9, 10]. In a large meta-analysis by Chua et al. of 33 studies on laparoscopic hepatectomy, the authors found the mean learning curve to be 50 cases. After adjusting for the year of adoption of the robotic technique, the analysis found that the learning curve for robotic hepatectomy was 47.1 fewer cases than the learning curve for laparoscopic hepatectomy [9]. Lai et al. reviewed multiple studies using CUSUM models and found that the learning curve for laparoscopic hepatectomy was between 22 and 64 for minor cases and between 45 and 75 for major cases [4].

Finally, in a study by O'Connor et al. comparing robotic ($n = 42$) versus laparoscopic ($n = 51$) minor hepatectomies, they found that after gaining the experience of 25 robotic hepatectomy cases, there was a significantly lower EBL, hospital length of stay, and 30-day complication rate compared with the laparoscopic approach. Therefore, they concluded that after the learning curve of 25 robotic minor hepatectomies, patient outcomes may be superior to the laparoscopic approach [14]. Overall, studies have generally found that the learning curve for robotic hepatectomy decreases steadily over time, and that the learning curve for robotic hepatectomy is shorter than the learning curve for laparoscopic hepatectomy [3, 8–10, 15–17].

The author (IS) established the robotic liver surgery program at AdventHealth Tampa in November 2016 and has performed 430 robotic liver resections to date. Approximately 45% of those resections were major liver resections. Despite the seemingly low number of minimum cases required to achieve proficiency in robotic liver resections, the authors suggest to the learners that great caution must be exercised in the interpretation of these learning curve papers. Based on our center experience, we feel that the learning curve for robotic liver resection is about 100 cases since the difficulty of each individual case is highly variable. Various difficulty scoring systems have attempted to address this issue for laparoscopic liver resection but none for robotic liver resection. Liver factors such as cirrhotic or steatotic parenchyma often significantly change the operative complexity for operations that are expected to be technically simple based on the current difficulty scoring systems and the operating surgeon's expectation. We believe that performing cases beyond the published learning curve minimum number is very important to ensure operative safety and optimal outcomes for patients. The learners should not lower his/her vigilance for each individual case after meeting the learning curve minimum number. In fact, our first unplanned emergent conversion due to major bleeding from a technical error occurred in case #107, which is well beyond the required numbers to achieve robotic liver proficiency published by all the papers mentioned previously. Even after completing 200 robotic liver resections, our estimated blood loss and operative time continue to improve, which reflects that our learning curve has not plateaued even beyond 200 cases. This is especially important for high-complexity liver resections such as right posterior sectionectomy, extended hepatectomy, total caudate lobe resection, and central bisectionectomy.

8.3 Conclusions

Currently, there is no established standard of training for robotic hepatectomy. Based on the most current generalizable literature available at this time, the learning curve for robotic liver resection proficiency seems to fall between the range of 15–34 cases [3, 6, 7, 9, 10, 12, 13, 16]. The learning curve for robotic hepatectomy has been found to be shorter than the learning curve for laparoscopic hepatectomy [4, 9, 10]. Furthermore, many studies found that after the proficiency learning curve was achieved for robotic hepatectomy, outcomes continued to improve and the overall learning curve decreased steadily over time [6, 9, 16]. The author suggested the learners of robotic hepatectomy must be cautious and generous in the interpretation of the minimum case numbers needed to achieve proficiency.

Successful establishment of robotic hepatectomy ultimately requires a team approach, including an experienced assistant at the patient's bedside to appropriately manage instruments and positioning adjustments (Table 8.1). These necessary resources and support are most likely to be found at a high-volume specialized center [4].

Table 8.1 Key overall conclusions

Key points
The learning curve based on published reports for robotic hepatectomy is 15–34 cases.
The learning curve for robotic hepatectomy is shorter than the learning curve for laparoscopic hepatectomy.
It is important for surgeons to use the minimum required number to achieve proficiency in robotic liver surgery as an estimate only since individual case complexity is highly variable.

References

1. Chen PD, Wu CY, Hu RH, et al. Robotic versus open hepatectomy for hepatocellular carcinoma: a matched comparison. Ann Surg Oncol. 2017;24(4):1021–8.
2. Sucandy I, Giovannetti A, Ross S, et al. Institutional first 100 case experience and outcomes of robotic hepatectomy for liver tumors. Am Surg. 2020;86(3):200–7.
3. Ahmad A, Freeman HD, Corn SD. Robotic major and minor hepatectomy: critical appraisal of learning curve and its impact on outcomes. Surg Endosc. 2022;37(4):2915–22.
4. Lai ECH, Tang CN. Training robotic hepatectomy: the Hong Kong experience and perspective. Hepatobiliary Surg Nutr. 2017;6(4):222–9.
5. Woodall WH, Rakovich G, Steiner SH. An overview and critique of the use of cumulative sum methods with surgical learning curve data. Stat Med. 2021;40(6):1400–13.
6. Ceccarelli G, Andolfi E, Fontani A, et al. Robot-assisted liver surgery in a general surgery unit with a "Referral Centre hub&Spoke Learning Program". Early outcomes after our first 70 consecutive patients. Minerva Chir. 2018;73(5):460–8.
7. Chen PD, Wu CY, Hu RH, et al. Robotic major hepatectomy: is there a learning curve? Surgery. 2017;161(3):642–9.

8. Chong CC, Fuks D, Lee KF, et al. Propensity score-matched analysis comparing robotic and laparoscopic right and extended right hepatectomy. JAMA Surg. 2022;157(5):436–44.
9. Chua D, Syn N, Koh YX, et al. Learning curves in minimally invasive hepatectomy: systematic review and meta-regression analysis. Br J Surg. 2021;108(4):351–8.
10. Efanov M, Alikhanov R, Tsvirkun V, et al. Comparative analysis of learning curve in complex robot-assisted and laparoscopic liver resection. HPB (Oxford). 2017;19(9):818–24.
11. Goh BKP, Prieto M, Syn N, et al. Critical appraisal of the learning curve of minimally invasive hepatectomy: experience with the first 200 cases of a southeast Asian early adopter. ANZ J Surg. 2020;90(6):1092–8.
12. Gorgec B, Zwart M, Nota CL, et al. Implementation and outcome of robotic liver surgery in The Netherlands: a nationwide analysis. Ann Surg. 2022;277(6):e1269–77.
13. Magistri P, Guerrini GP, Ballarin R, et al. Improving outcomes defending patient safety: the learning journey in robotic liver resections. Biomed Res Int. 2019;2019:1835085.
14. O'Connor VV, Vuong B, Yang ST, et al. Robotic minor hepatectomy offers a favorable learning curve and may result in superior perioperative outcomes compared with laparoscopic approach. Am Surg. 2017;83(10):1085–8.
15. Patriti A, Marano L, Casciola L. MILS in a general surgery unit: learning curve, indications, and limitations. Updat Surg. 2015;67(2):207–13.
16. Zhu P, Liao W, Ding ZY, et al. Learning curve in robot-assisted laparoscopic liver resection. J Gastrointest Surg. 2019;23(9):1778–87.
17. Gravetz A, Sucandy I, Wilfong C, et al. Single-institution early experience and learning curve with robotic liver resections. Am Surg. 2019;85(1):115–9.

9 Training in Robotic Liver Surgery

Sara Shahrestani, Yuman Fong,
and Tamara M. H. Gall

9.1 Introduction

Just as the development of laparoscopic surgery in the 1980s revolutionised general surgery, robotics promises to redefine the minimally invasive surgical approach [39]. However, with the development of new techniques in surgery come questions of how best to train and privilege surgeons [31]. A proposed solution to the concern regarding morbidity associated with the learning curve is using simulation to negate the risk to patients [6].

While the learning curve for robotic surgery varies with case type, it is apparent that time is a uniform factor that improves performance across procedure types [31]. Factors such as trainee working hours, increased subspecialisation, medicolegal considerations, and ethical considerations have contributed to the development of simulators for learning surgery [7]. A cadaveric study examining the learning curve in robotic surgery in surgical trainees demonstrates that acquiring robotic surgery skills may be more rapid than laparoscopic surgery [16].

With the expansion of robotic surgery comes the potential to integrate systematic and competence-based training programs [35]. Training programs across residency and post-fellowship training curriculums exist [4] and may include didactic teaching, skills lab, virtual reality, cadaveric, live case observation, and proctoring [35]. While many simulation platforms are currently in use (RoSS, SEP, ProMIS, MdVT, and Da Vinci), studies validating their use in teaching and learning are still limited to small numbers [1]. Feedback methods for robotic training have also expanded with proctoring via crowdsourcing via open-access online platforms (i.e. C-SATS), machine learning, and artificial intelligence [4].

9.2 Simulation and Artificial Intelligence in Learning Robotic Surgery

Familiarity with robotic operating may occur through patient-side training; however, advanced skills, including excision and the use of instruments, require mentored development in either a virtual reality, dry lab, or wet lab environment [36]. Virtual reality can aid the development of

S. Shahrestani
Department of GI Surgery, Royal North Shore Hospital, Sydney, NSW, Australia

Y. Fong
Department of Surgery, City of Hope Medical Center, Duarte, CA, USA

T. M. H. Gall (✉)
Department of GI Surgery, Royal North Shore Hospital, Sydney, NSW, Australia

Department of Surgery and Cancer, Imperial College London, London, UK
e-mail: tamara.gall1@nhs.net

M. D'Hondt, I. Sucandy (eds.), *Textbook of Robotic Liver Surgery*,
https://doi.org/10.1007/978-3-031-76536-0_9

technical skills, but animate or cadaveric training better facilitates more complex skills development [23]. Hands-on side-console training with systematic evaluation of how closely the trainee's movements mimic the operating surgeons during a procedure has been developed [36].

Several simulators are in current use (Fig. 9.1).

Artificial intelligence incorporating neural networking analysis of endoscopic video, space-time information, and instrument motion is also burgeoning [8]. These systems show good internal correlation with established robotic skills assessment systems, including the Objective Structured Assessment of Technical Skills

SUMMARY OF ROBOTIC SIMULATION SKILLS DEVICES

Robot	System	Release Year	Key Design	Additional training Facilities
Intuitive Da Vinci	Da Vinci Skills Simulator	2011	• Needle control and dissection	• Dry and wet lab training facilities
	dV-Trainer	2007	• Xperience team • Maestro specific procedures (right partial nephrectomy, hysterectomy, inguinal hernia repair, and radical prostatectomy)	• Dry and wet lab training facilities
	ProMIS	2003	• Bodyform device with digital monitoring of instruments	
	SEP Robot	2005	• PATRIOT system monitors location of instruments	
	RobotiX Mentor	2016	• Procedures include prostatectomy, hysterectomy, vaginal cuff closure, lobectomy, inguinal hernia, hemicolectomy, and cholecystectomy	
	RoSS	2010	• Procedures include hysterectomy, radical prostatectomy, radical cystectomy, and extended lymph node dissection	
CMR Versius	Versius Virtual Reality headset	2022	• Wireless virtual reality headset that mimics the real world environment	• Surgical's Professional Education programme via ORSI academy
Medtronic Hugo	Hugo RAS	2021	• Instrument control and dissection modules	• Hugo Ascend training pathway via ORSI academy

Fig. 9.1 Summary of robotic simulation skills devices [27]

Table 9.1 Video recording software by company

Company	System	Features
Medtronic	Touch Surgery Enterprise	AI-powered surgical video recording and analysis. Annotations, cloud sharing, and team review
Intuitive	Da Vinci integrated recording	Direct to USB integrated recording system
	dVLogger	Under development. Allows recording and review of both video and motion log, with cloud sharing
CMR	Versius Connect	Video upload and analysis of cases, including instrument usage, operative phase timelines, and annotations

(OSATs) and Global Evaluative Assessment of Robotic skills (GEARs) [22]. Some centres are utilising crowdsourcing models via open access, allowing proctoring and feedback through online platforms (C-SATS) [4]. Biometric components, including computer vision, eye motion tracking, and muscle contraction analysis, demonstrate a strong correlation with performance [8]. One study has demonstrated a training program utilising adaptive machine logic, actively evaluating and focussing training based on performance in components of a task, results in more rapid skills acquisition [25].

There is also evidence that case review of operative recordings (self-performed and other surgeon) can improve the speed of skill acquisition in robotic surgery [23]. Video sharing via open-source networks can provide access to review and feedback without geographic limitation [4]. Several platforms exist for video review and sharing, as outlined in Table 9.1. A paper from the University of Pittsburgh outlines a methodology for developing a video library education, including integration into a fellowship teaching course and cost analysis [2].

9.3 Developing a Systematic Robotic Surgery Curricula

In a survey of robotic training programs in the United States, 105 training programs were identified, with around 68% having a formal training curriculum [37]. The primary curricula currently in use include the online (FRS and Da Vinci), onsite (sages robotics master class, robotics training network curriculum, fundamental skills of robotic-assisted training program), and simulation (robotic surgery simulator, dV trainer, da Vinci skills simulator, Robotix mentor) [10]. The FRS curriculum was validated in an international study across 33 centres with 122 participants, demonstrating faster performance with fewer errors than those trained without the systematic curriculum [34]. Even a 2-week training program in instrument mastery, simulation curriculum, suturing, and inanimate drills has demonstrated superior performance to non-systematic training [32].

A pilot validation study of a urological training curriculum consisting of a twelve-week training program improved the GEARS score, and 80% of trainees could safely perform robotic prostatectomy at the end of the training [38]. Phased approaches based on stepwise certification for resident training, with programmed skills refresher components, have been demonstrated [28]. Post-fellowship training also has an essential role in the demonstration of a significantly higher rate of adoption of pancreaticoduodenectomy at the consultant level following a systematic training program [26].

9.4 Application of Robotics to Liver Surgery

An increased interest in robotic liver surgery has developed, given the two-dimensional view and four-degree movement restriction in laparoscopic liver surgery [19]. Other benefits include a more rapid learning curve than laparoscopic, a lack of hand tremors, less postoperative pain, and a lower rate of vascular injury [14]. Integrated fluorescence guidance in the robotic approach provides advantages, while a paucity of parenchymal dissection devices and a lack of haptic feedback may be limiting factors [3].

An international consensus guideline recommends robotic liver resection for both minor and major resections, with advantages that include maintaining the abdominal wall and lower rates of postoperative ascites and postoperative complications (wound and respiratory) [3]. Long-term oncological outcomes, including survival and recurrence, are similar to open and laparoscopic methods in a 10-year multinational study [19]. Resection type moderate outcomes, with posterosuperior resections and minor resections (less than four segments) demonstrating a significantly shorter postoperative stay (4 days, compared with 8 days) [29].

While a few robots are in active use, each has varying advantages and disadvantages, specifically for liver surgery [12]. The Intuitive Da Vinci is the most commonly used robot in the liver surgery literature owing to its early development and availability [3]. Advantages of this system include the availability of a high-energy bipolar vessel sealer and robotic harmonic scalpel; however, this device does not have wristed manipulation. While there is an integrated stapling device, it requires length and may not be effective in thick tissues, and hence many centres combine the laparoscopic and robotic approaches. The bedside assistant also has the option of utilising energy devices, ligaclips, hem-o-loks, and the endoGIA; in our experience, this can prevent frequent instrument changes. In contrast, the CMR Versius does not have an associated high-energy device but monopolar and bipolar devices. The Medtronic Hugo, similar to Versius, has separate arms and, for this reason, can be more manoeuvrable; hence, a dynamic selection of target tissues may be easier. In our experience, presentation of the liver for dissection and control of bleeding is where the robot provides significant advantages over other minimally invasive approaches.

9.5 Learning Curves in Robotic Liver Surgery

Early experience with robotic liver surgery demonstrated a safe approach in three successive cohorts of 11 patients and decreased blood loss in these groups [18]. A meta-analysis shows the learning curve for robotic left liver resections is only 25 compared with 50 laparoscopic cases [11]. However, a primary study suggests case selection is an important factor, with the learning curve reached at 93 for robotic resections and 117 for laparoscopic when complex resections are included [20]. A difficulty index was developed based on tumour location, anatomic resection, proximity to vessels, and tumour size, with a minimum of sixteen robotic liver resections required to progress to a difficult resection, compared with 29 in the laparoscopic group [13]. Major resection has also been studied and shown to have three phases with decreased operation time and hospital stay after the first phase (15 patients) and less blood loss after the second phase (25 patients) [9].

9.6 Teaching Robotic Liver Surgery

The council of the International Laparoscopic Liver Society (ILLS) describes a 22-step process for performing laparoscopic liver resection. A German group has developed and validated a stepwise curriculum based on these steps [21]. A strong consensus among experts in hepatobiliary robotic surgery concedes that a systematic training program for liver resection is necessary [21]. The Carolinas Medical Centre has described a 12-month robotic hepatobiliary fellowship curriculum with eight core modules and six elective modules, of which two are liver-focused [5].

A Delphi consensus of international experts in liver surgery concludes training should commence with basic robotic surgical skills, a baseline evaluation of expertise, an index case of liver

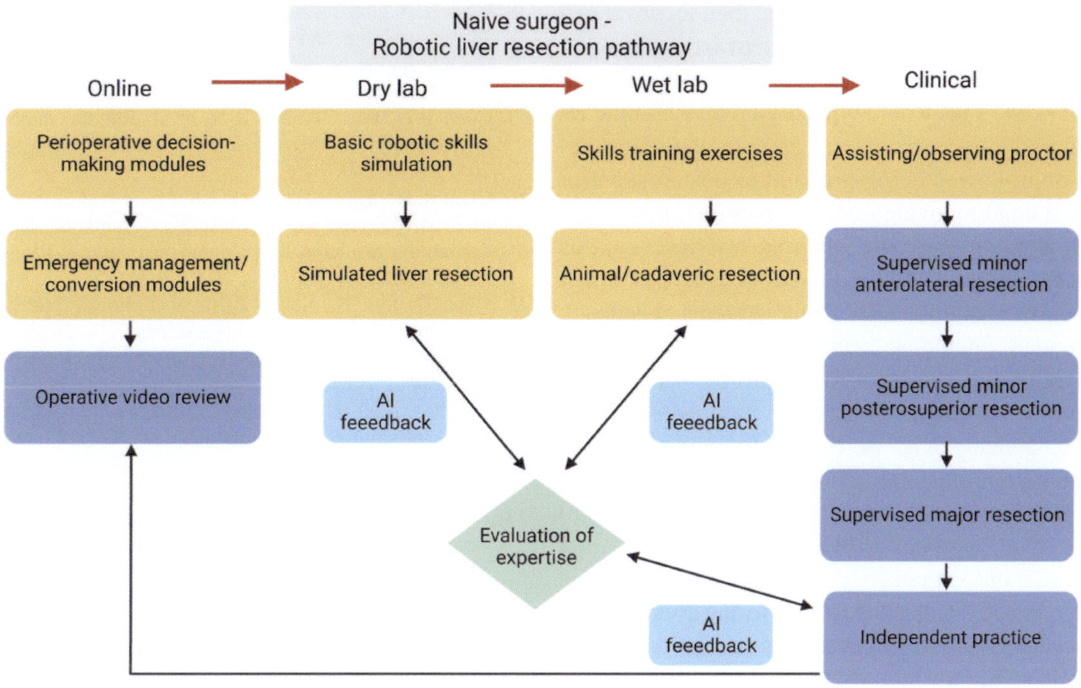

Fig. 9.2 Proposed training pathway in robotic liver resection for a naive surgeon

resection, cognitive skills development, stepwise video demonstration, port placement, troubleshooting, and emergency scenario management [15]. Departmental implementation of learning robotic liver resection has also been described with a stepwise model involving successive proctoring of five surgeons from tableside surgeon to minor anterolateral resection, minor posterosuperior resection, and finally, major liver resection [30]. This proposed pathway is outlined in Fig. 9.2.

9.7 Augmented Reality, Simulation, and Learning Robotic Liver Surgery

Augmented reality plays a unique role in the development of robotic liver resection and may have a role in learning safe practices. Three-dimensional visualisation with dynamic endoscopic overlay and red/green can alert the operator of proximity to the tumour, including lesions pre-existent to chemotherapy [17]. Simulated meta-sectionectomy utilising a patient-derived data model, two functioning instruments, and ultrasonic shears could successfully differentiate between novice and expert learners [24]. Successful simulation in an augmented reality environment and three-dimensional modelling for perioperative assessment, planning, and operative decision-making has been demonstrated [33], paving the way for integration into the robotic field. The application of further artificial intelligence, such as through neural networking analysis of performance, computer vision, motion tracking of eyes, and muscle contraction analysis as described in other areas of robotic surgery, could aid in the development of skills and targeting progression.

9.8 Conclusion

An internationally recognised and standardised robotic liver resection curriculum is necessary for effective training and safe expansion of robotic resection programs. Baseline performance evaluation, followed by basic skills, then an advanced robotic skills program with stepwise assessment and progression ensures systematic development of skills. This would typically involve commencing with an index case of an anterolateral resection, followed by progression to posterosuperior and finally moving to major liver resection.

Proctoring plays a vital role, and access to this can be improved with the development of open-source web-based feedback systems. Feedback models, including neural networking programs and artificial intelligence, could be utilised, and machine logic can guide skills training to focused components of a case leading to more rapid progression. Simulation has demonstrated success in other areas of surgery, and future works to develop a robotic liver resection simulation model are warranted. At the crossroads between technology and medicine, learning robotic liver resection demonstrates great potential for expansion and innovation in future.

References

1. Abboudi H, Khan MS, Aboumarzouk O, et al. Current status of validation for robotic surgery simulators – a systematic review. BJU Int. 2013;111:194–205.
2. Al Abbas AI, Jung JP, Rice MK, et al. Methodology for developing an educational and research video library in minimally invasive surgery. J Surg Educ. 2019;76:745–55.
3. Ayabe RI, Azimuddin A, Tran Cao HS. Robot-assisted liver resection: the real benefit so far. Langenbeck's Arch Surg. 2022;407:1779–87.
4. Azadi S, Green IC, Arnold A, et al. Robotic surgery: the impact of simulation and other innovative platforms on performance and training. J Minim Invasive Gynecol. 2021;28:490–5.
5. Baimas-George M, Watson M, Martinie J, et al. Curriculum matrix development for a hepato-pancreato-biliary robotic surgery fellowship. Can J Surg. 2021;64:E657–e662.
6. Bareeq RA, Jayaraman S, Kiaii B, et al. The role of surgical simulation and the learning curve in robot-assisted surgery. J Robot Surg. 2008a;2:11–5.
7. Bareeq RA, Jayaraman S, Kiaii B, et al. The role of surgical simulation and the learning curve in robot-assisted surgery. J Robot Surg. 2008b;2:11–5.
8. Chen IHA, Ghazi A, Sridhar A, et al. Evolving robotic surgery training and improving patient safety, with the integration of novel technologies. World J Urol. 2021;39:2883–93.
9. Chen P-D, Wu C-Y, Hu R-H, et al. Robotic major hepatectomy: is there a learning curve? Surgery. 2017;161:642–9.
10. Chen R, Rodrigues Armijo P, Krause C, et al. A comprehensive review of robotic surgery curriculum and training for residents, fellows, and postgraduate surgical education. Surg Endosc. 2020;34:361–7.
11. Chua D, Syn N, Koh Y-X, et al. Learning curves in minimally invasive hepatectomy: systematic review and meta-regression analysis. Br J Surg. 2021;108:351–8.
12. Costello DM, Huntington I, Burke G, et al. A review of simulation training and new 3D computer-generated synthetic organs for robotic surgery education. J Robot Surg. 2022;16:749–63.
13. Efanov M, Alikhanov R, Tsvirkun V, et al. Comparative analysis of learning curve in complex robot-assisted and laparoscopic liver resection. HPB (Oxford). 2017;19:818–24.
14. Finotti M, D'amico F, Mulligan D, et al. A review of the current and future role of robotic surgery in liver surgery and transplantation. Hepatobiliary Surg Nutr. 2021;12(1):56–68.
15. Fong Y, Buell JF, Collins J, et al. Applying the Delphi process for development of a hepatopancreaticobiliary robotic surgery training curriculum. Surg Endosc. 2020;34:4233–44.
16. Gall TMH, Alrawashdeh W, Soomro N, et al. Shortening surgical training through robotics: randomized clinical trial of laparoscopic versus robotic surgical learning curves. BJS Open. 2020;4:1100–8.
17. Giannone F, Felli E, Cherkaoui Z, et al. Augmented reality and image-guided robotic liver surgery. Cancers. 2021;13:6268.
18. Gravetz A, Sucandy I, Wilfong C, et al. Single-institution early experience and learning curve with robotic liver resections. Am Surg. 2019;85:115–9.
19. Khan S, Beard RE, Kingham PT, et al. Long-term oncologic outcomes following robotic liver resections for primary hepatobiliary malignancies: a multicenter study. Ann Surg Oncol. 2018;25:2652–60.
20. Krenzien F, Benzing C, Feldbrügge L, et al. Complexity-adjusted learning curves for robotic and laparoscopic liver resection: A word of caution. Annals of Surgery Open. 2022;3:e131.
21. Krenzien F, Schöning W, Brunnbauer P, et al. The ILLS laparoscopic liver surgery fellow skills curriculum. Ann Surg. 2020;272:786–92.
22. Lee D, Yu HW, Kwon H, et al. Evaluation of surgical skills during robotic surgery by deep learning-based multiple surgical instrument tracking in training and actual operations. J Clin Med. 2020;9:1964.

23. Lee JY, Mucksavage P, Sundaram CP, et al. Best practices for robotic surgery training and credentialing. J Urol. 2011;185:1191–7.
24. Lefor AK, Heredia Perez SA, Shimizu A, et al. Development and validation of a virtual reality simulator for robot-assisted minimally invasive liver surgery training. J Clin Med. 2022;11:4145.
25. Mariani A, Pellegrini E, Enayati N, et al. Design and Evaluation of a Performance-based Adaptive Curriculum for Robotic Surgical Training: a Pilot Study. 2018 40th Annual International Conference of the IEEE Engineering in Medicine and Biology Society (EMBC), 18–21 July 2018; 2018. pp. 2162–65.
26. Mark Knab L, Zenati MS, Khodakov A, et al. Evolution of a novel robotic training curriculum in a complex general surgical oncology fellowship. Ann Surg Oncol. 2018;25:3445–52.
27. Mazzone E, Dell'oglio P, Puliatti S. Basic skills training in robotic surgery: dry and wet-lab models and their application in robotic training pathways. In: Biyani CS, Van Cleynenbreugel B, Mottrie A, editors. Practical simulation in urology. Cham: Springer International Publishing; 2022.
28. Moit H, Dwyer A, De Sutter M, et al. A standardized robotic training curriculum in a general surgery program. JSLS. 2019;23:e2019.00045.
29. Nota CL, Hagendoorn J, Fong Y. ASO author reflections: the role of robotic surgery in liver resection. Ann Surg Oncol. 2019;26:591–2.
30. Nota CL, Molenaar IQ, Te Riele WW, et al. Stepwise implementation of robotic surgery in a high volume HPB practice in The Netherlands. HPB. 2020;22:1596–603.
31. Pernar LIM, Robertson FC, Tavakkoli A, et al. An appraisal of the learning curve in robotic general surgery. Surg Endosc. 2017;31:4583–96.
32. Ramirez Barriga M, Rojas A, Roggin KK, et al. Development of a two-week dedicated robotic surgery curriculum for general surgery residents. J Surg Educ. 2022;79:861–6.
33. Rashidian N, Giglio M, Van Herzeele I et al. HPB (Oxford). Effectiveness of an immersive virtual reality environment on curricular training for complex cognitive skills in liver surgery: a multicentric crossover randomized trial. 2022;24(12):2086–95.
34. Satava RM, Stefanidis D, Levy JS, et al. Proving the effectiveness of the Fundamentals of Robotic Surgery (FRS) skills curriculum: a single-blinded, multispecialty, multi-institutional randomized control trial. Ann Surg. 2020;272:384–92.
35. Schreuder HW, Wolswijk R, Zweemer RP, et al. Training and learning robotic surgery, time for a more structured approach: a systematic review. BJOG. 2012;119:137–49.
36. Sridhar AN, Briggs TP, Kelly JD, et al. Training in robotic surgery: an overview. Curr Urol Rep. 2017;18:58.
37. Tom CM, Maciel JD, Korn A, et al. A survey of robotic surgery training curricula in general surgery residency programs: how close are we to a standardized curriculum? Am J Surg. 2019;217:256–60.
38. Volpe A, Ahmed K, Dasgupta P, et al. Pilot validation study of the European Association of Urology robotic training curriculum. Eur Urol. 2014;68:292–9.
39. Wilson EB. The evolution of robotic general surgery. Scand J Surg. 2009;98:125–9.

10 Training Programs for Robotic Liver Surgery

Gabriela Pilz da Cunha, Marc G. Besselink, and Rutger-Jan Swijnenburg

10.1 Introduction

Liver surgery is complex, requiring a high level of surgical skill. Minimally invasive liver surgery (MILS) techniques such as laparoscopy and robotic surgery add to the technical complexity of surgical procedures but offer patients improved postoperative clinical outcomes such as less pain, quicker recovery, and shorter hospital stays [1]. In order to offer their patients the best available care, surgeons are required to master these techniques. There is currently much heterogeneity in robotic surgical training. To optimize the surgeons' learning process and protect patients from the surgeons' learning curve, it is important to establish systematic and standardized training programs for robotic liver surgery (RLS). The present lack of high-quality, formal training programs forms a hurdle in the widespread implementation of the robotic technique for liver surgery. Many hepato-pancreato-biliary (HPB) operating teams have a robotic system available and would like to implement it in their surgical practice. However, these surgeons are in need of guidelines and support to kickstart this at their respective centres [2]. This chapter outlines the current knowledge and important considerations in the implementation of a training program for RLS.

G. Pilz da Cunha · M. G. Besselink
R.-J. Swijnenburg (✉)
Department of Surgery, Amsterdam UMC, University of Amsterdam, Amsterdam, The Netherlands

Cancer Center Amsterdam,
Amsterdam, The Netherlands
e-mail: r.j.swijnenburg@amsterdamumc.nl

10.2 Surgical Skill and Patient Outcomes

It is crucial that surgeons perform to the highest of their capabilities as their skills can be reflected in the clinical outcomes of the patients on which they operate. In bariatric surgery, it has been demonstrated that higher levels of surgical proficiency correlate with improved patient outcomes [2, 3]. Moreover, increased surgical experience has also been shown to lead to decreased complication rates in pancreatic surgery [4]. This highlights the importance of adequate training in high-complexity surgery. A training program's design should promote quick progression of the surgeon's skills while also protecting the patient from mistakes during the learning curve caused by a paucity of surgical expertise. Improved clinical outcomes can be expected with increased RLS experience [5].

M. D'Hondt, I. Sucandy (eds.), *Textbook of Robotic Liver Surgery*,
https://doi.org/10.1007/978-3-031-76536-0_10

10.3 Current Experience with Large-Scale Training Programs

Since the introduction of robotic surgery, there has been much discussion on the topic of robotic surgical training. Experts agree that standardized curricula are necessary for the safe dissemination of robotic surgery [6, 7]. In the development of a robotic training program for liver surgery, we can potentially build on the experience of previous training programs for other minimally invasive surgical procedures.

In a series of consensus conferences, attended by 14 international surgical societies, the 'Fundamentals of Robotic Surgery' were developed through a Delphi method. The participants devised recommendations for teaching robotic surgery. Among the advices were calls to incorporate a varied program as a means of developing psychomotor as well as cognitive skills [8]. Additionally, a mastery-based curriculum was put forward as the most effective and ethical approach to learning. This is also referred to as 'proficiency-based progression' training. This style of training has also been endorsed by others as the best method for teaching surgical procedures [7, 9]. Trainees may only advance to the sequential step when demonstrating sufficient skill at their present level. This ensures that skills are honed before moving on to more complex stages. Important in this teaching method is that trainees are assessed on performance metrics. Validated tools have been developed to aid in the evaluation of surgical skill. Some examples are the objective structured assessment of technical skills (OSATS) (Fig. 10.1) or the global evaluative assessment of robotic skills (GEARS) (Fig. 10.2) [10, 11]. This, and other parameters such as operative time, allows for objective benchmarking of surgical skill. This type of structured training and assessment curriculum has been shown to be superior in improving operative performance when compared to conventional residency training [12, 13]. Simulation-based training plays a central role in proficiency-based progression training. Its principal advantage is the possibility for surgeons to practice in a low-risk environment before applying their skills to patients. Aviation

Grading Definition OSATS

RATING AND INTERPRETATION

1. Deficient / Traumatic
2. Lacking / Lacks finesse
3. Average
4. Skilled
5. Master / Flawless

GRADING ASPECTS AND ELUCIDATION

Gentleness	Gentle tissue handling that does not result in injury
Time and Motion	Economy of motion, maximum efficiency
Instrument Handling	Fluid use of instruments without awkwardness
Flow of Operation	Smooth transitions from one part of the operation to another
Tissue Exposure	Retraction that allows for good visualization and proper tissue alignment
Overall technical skill	Overall assessment of technical skill

Fig. 10.1 Modified OSATS grading [10]

DEPTH PERCEPTION

1	2	3	4	5
Constantly overshoots target, wide swings, slow to correct		Some overshooting or missing of target, but quick to correct		Accurately directs instruments in the correct plane to target

BIMANUEL DEXTERITY

1	2	3	4	5
Uses only one hand, ignores nondominant hand, poor coördination		Uses both hands, but does not optimize interaction between hands		Expertly uses both hands in a complementary way to provide best exposure

EFFICIENCY

1	2	3	4	5
Inefficient efforts; many uncertain movements; constantly changing focus or persisting without progress		Slow, but planned movements are reasonably organized		Confident, efficient and safe conduct, maintains focus on task, fluid progression

FORCE SENSITIVITY

1	2	3	4	5
Rough moves, tears tissue, injures nearby structures, poor control, frequent suture breakage		Handles tissues reasonably well, minor trauma to adjacent tissue, rare suture breakage		Applies appropriate tension, negligible injury to adjacent structures, no suture breakage

ROBOTIC CONTROL

1	2	3	4	5
Consistently does not optimize view, hand position, or repeated collisions even with guidance		View is sometimes not optimal. Occasionally needs to relocate arms. Occasional collisions and obstruction of assistant		Controls camera and hand position optimally and independently. Minimal collisions or obstruction of assistant

Fig. 10.2 Global evaluative assessment of robotic skills (GEARS) [11]

programs have successfully been implementing similar teaching methods to train pilots for decades [14].

Urologists have the most experience with implementing robot-assisted procedures. Important steps have been made in this field with regards to structured robotic surgical training. In 2015, the European Association of Urology published results from their validated training curriculum for robot-assisted radical prostatectomy [15]. The curriculum included e-learning, simulation-based training, and supervised operative training. Trainees demonstrated significant improvements in their skills after completing the program.

The University of Pittsburgh Medical Center (UPMC) also pioneered a curriculum for robotic pancreatic surgical training and evaluated its effects [16, 17]. This program consists of five stages: (1) virtual reality simulation training, (2) inanimate bio tissue curriculum, (3) procedure video library, (4) intraoperative evaluation, and (5) skills maintenance and audits. Each stage is respectively linked to the following competencies: instrument mastery, tissue handling mastery, mastery of the procedure, mastery of psychomotor performance, decision-making, and judgement. Outcomes of the training program have shown excellent potential. Surgical fellows demonstrated improvement in surgical skills as a result of program participation [16, 18–20].

In the Netherlands, the 'Dutch Liver Collaborative Group' and the 'Dutch Pancreatic Cancer Group' have demonstrated the feasibility and effectiveness of structured training programs for minimally invasive liver and pancreatic surgery with the LAELIVE (laparoscopic liver surgery) and LAELAPS (−1, −2, and − 3) nationwide efforts, respectively [21–24]. The LEARNBOT program has since also been initiated to train surgeons in robotic pancreatoduodenectomy across Europe [Netherlands Trial Register00NL8898] [25]. The LAELAPS-3 and LEARNBOT robotic training programs are modelled on UPMC's curriculum. Collectively, these initiatives have trained a large number of surgeons. It can be inferred that the availability of these programs is partly responsible for the increased proportion of these minimally invasive operations over the years [21–23, 26].

10.4 LIVEROBOT: A Pan-European Training Program for RLS

Probably owing to the high complexity of liver surgery, the implementation of robotics in this field has been slower than in other surgical specialities. Some studies have reported on single-centre experiences with the introduction of RLS training programs [27–29]. However, there is currently no validated structured training curriculum for RLS. A variety of RLS courses are offered each year, but their programs vary considerably. No consensus exists with regards to which program must be followed, resulting in a lack of uniformity. Centres looking to enrol an RLS program at their own facilities lack the structured guidance necessary to achieve this safely and effectively. A structured training program in RLS could facilitate knowledge sharing between centres wishing to embark on their robotic surgery journey and those with advanced experience. To date, a surgeon who is embarking on RLS can only rely on the existing observer and proctorship experience of Intuitive Surgical-certified proctors for robotic liver surgery. Currently, there are only six certified proctors for RLS in Europe, and a formal training program for RLS is lacking.

The LIVEROBOT project, endorsed by the European-African Hepato-Pancreato-Biliary Association (E-AHPBA), recently emerged as a response to the demand for a formal RLS training program. It will be the first of its kind for RLS. The program aims to facilitate the safe implementation of RLS. The first inclusions are expected at the beginning of 2023.

The valuable lessons learned from previous experiences with surgical training programs have provided the blueprint for this curriculum. A fundamental difference of hepatectomy when compared to other operations is the varying levels of complexity within liver resections. This is dependent on factors such as the size and location of the resection, often defined by the IWATE score [30]. The curriculum has been adapted to accommodate this unique characteristic of liver surgery. Utilizing this as an advantage by forming intermediary steps in the build-up to high complexity cases. In liver surgery, the following order of incremental difficulty is usually recommended: anterior resections, left lateral sectionectomy, right postero-superior resection, left hemihepatectomy, and right hemihepatectomy [31].

Specific inclusion criteria apply for enrolment in LIVEROBOT. Surgeons should demonstrate sufficient laparoscopic proficiency. Additionally, a minimum annual procedure volume has been set to safeguard the quality and safety of the operations performed in the context of the program as high-volume centres have improved postoperative outcomes [26, 32, 33].

The LIVEROBOT training is structured in four phases: preparation, minor RLS, major RLS, and mastery. As there may be much heterogeneity in the experience and skill level of the surgeons when enrolling in the curriculum, surgeons are allowed to begin at any step using the incremental inclusion criteria. The framework of the curriculum is outlined in Fig. 10.3. Throughout the curriculum, participants acquire skills from a variety of validated teaching tools. The training program makes use of simulation exercises, a video library, and off-site and on-site proctoring.

Data generated as a result of the training program will be prospectively registered in the European Minimally Invasive Liver Surgery (E-MILS) registry. This will allow for analysis of learning curves and clinical outcomes giving better insight into the effects in RLS. This registry will also allow for monitoring of quality following introduction.

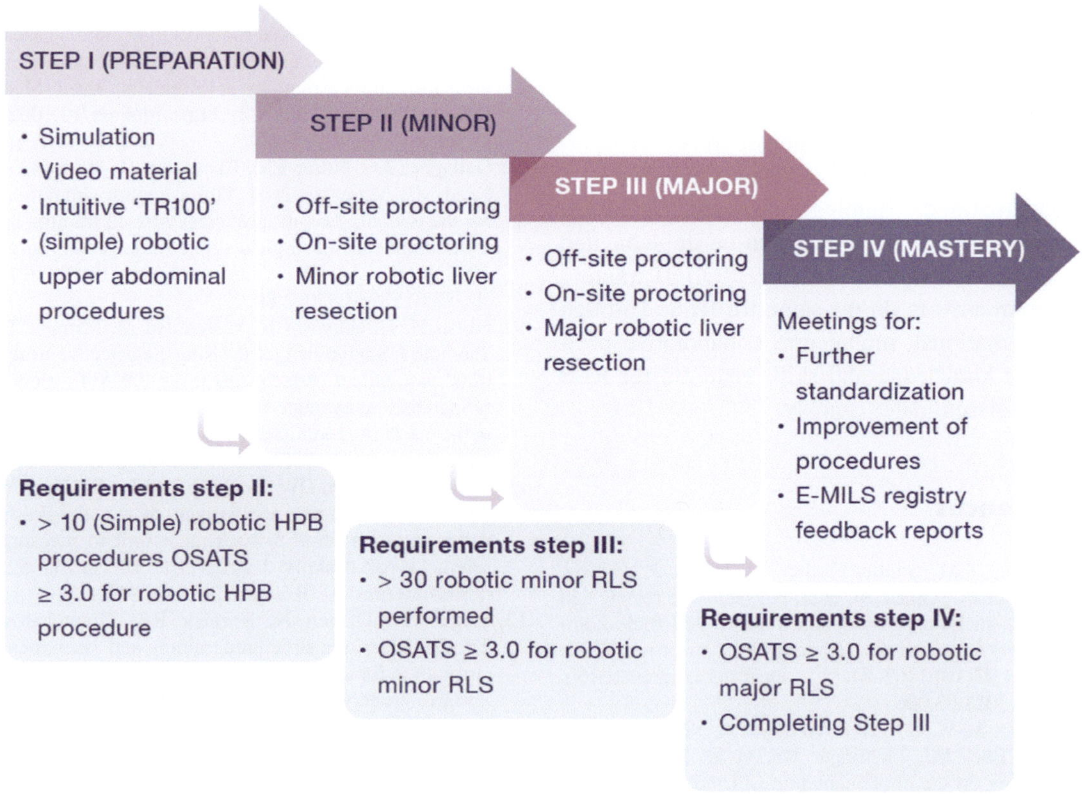

Fig. 10.3 LIVEROBOT training program structure

10.5 Teamwork Optimization

The introduction of the robot in the operating room presents surgical teams with new challenges [34]. Robotic operations are accompanied by additional steps such as the docking and undocking of the robot. At these moments, disruptions in workflow may arise [35, 36]. Additionally, the setup of the operation room is altered, with the surgeon physically distanced from the operating field. The location of the surgeon at the console, physically distanced from the operating field, makes the communication between the surgeon and other team members more difficult. Barriers in communication are one of the principal reasons for flow disruptions in robot-assisted surgeries [37]. To operate successfully and effectively, it is crucial that the whole operating team is well versed in dealing with the robotic platform. A team training protocol applied to gynaecological surgery involving theoretical training, hands-on training, and team communication skills enhancement was successful in optimizing performance and decreasing docking time and operating time, without compromising on safety [38]. Comparable methods should be considered by centres looking to optimize their robotic surgery workflow.

10.6 Conclusions and Future Perspectives

Mastery-based curricula provide the scaffolding for training complex surgical techniques. Training programs aim to mitigate the surgeon's learning curve, thereby protecting the patient. Technological innovations such as simulation-based training are central to this approach. Recently, training programs have been successful in integrating such tools into their curriculums and have demonstrated the feasibility of a proficiency-based progression approach for minimally invasive operations. When supported by a well-structured training program, complex robotic surgical techniques can be implemented safely and effectively. The LIVEROBOT training program aims to do the same for RLS. Through an international, multicentre collaborative project, the knowledge and skills required for RLS can be disseminated quickly.

References

1. Haney CM, Studier-Fischer A, Probst P, Fan C, Müller PC, Golriz M, et al. A systematic review and meta-analysis of randomized controlled trials comparing laparoscopic and open liver resection. HPB. 2021;23(10):1467–81. https://doi.org/10.1016/j.hpb.2021.03.006.
2. Zwart MJW, Görgec B, Arabiyat A, Nota CLM, van der Poel MJ, Fichtinger RS, et al. Pan-European survey on the implementation of robotic and laparoscopic minimally invasive liver surgery. HPB : the official journal of the International Hepato Pancreato Biliary Association. 2022;24(3):322–31. https://doi.org/10.1016/j.hpb.2021.08.939.
3. Varban OA, Thumma JR, Finks JF, Carlin AM, Ghaferi AA, Dimick JB. Evaluating the effect of surgical skill on outcomes for laparoscopic sleeve gastrectomy: a video-based study. Ann Surg. 2021;273(4):766–71. https://doi.org/10.1097/sla.0000000000003385.
4. Krautz C, Haase E, Elshafei M, Saeger HD, Distler M, Grützmann R, et al. The impact of surgical experience and frequency of practice on perioperative outcomes in pancreatic surgery. BMC Surg. 2019;19(1):108. https://doi.org/10.1186/s12893-019-0577-6.
5. Tsung A, Geller DA, Sukato DC, Sabbaghian S, Tohme S, Steel J, et al. Robotic versus laparoscopic hepatectomy: a matched comparison. Ann Surg. 2014;259(3):549–55. https://doi.org/10.1097/sla.0000000000000250.
6. Stefanidis D, Huffman EM, Collins JW, Martino MA, Satava RM, Levy JS. Expert consensus recommendations for robotic surgery credentialing. Ann Surg. 2022;276(1):88–93. https://doi.org/10.1097/sla.0000000000004531.
7. Petz W, Spinoglio G, Choi GS, Parvaiz A, Santiago C, Marecik S, et al. Structured training and competence assessment in colorectal robotic surgery. Results of a consensus experts round table. Int J Med Robot. 2016;12(4):634–41. https://doi.org/10.1002/rcs.1731.
8. Smith R, Patel V, Satava R. Fundamentals of robotic surgery: a course of basic robotic surgery skills based upon a 14-society consensus template of outcomes measures and curriculum development. Int J Med Robot. 2014;10(3):379–84. https://doi.org/10.1002/rcs.1559.
9. Gallagher AG, Ritter EM, Champion H, Higgins G, Fried MP, Moses G, et al. Virtual reality simulation for the operating room: proficiency-based training as a paradigm shift in surgical skills training. Ann Surg. 2005;241(2):364–72. https://doi.org/10.1097/01.sla.0000151982.85062.80.
10. Niitsu H, Hirabayashi N, Yoshimitsu M, Mimura T, Taomoto J, Sugiyama Y, et al. Using the objective structured assessment of technical skills (OSATS) global rating scale to evaluate the skills of surgical trainees in the operating room. Surg Today. 2013;43(3):271–5. https://doi.org/10.1007/s00595-012-0313-7.
11. Goh AC, Goldfarb DW, Sander JC, Miles BJ, Dunkin BJ. Global evaluative assessment of robotic skills: validation of a clinical assessment tool to measure robotic surgical skills. J Urol. 2012;187(1):247–52. https://doi.org/10.1016/j.juro.2011.09.032.
12. Palter VN, Orzech N, Reznick RK, Grantcharov TP. Validation of a structured training and assessment curriculum for technical skill acquisition in minimally invasive surgery: a randomized controlled trial. Ann Surg. 2013;257(2):224–30. https://doi.org/10.1097/SLA.0b013e31827051cd.
13. De Groote R, Puliatti S, Amato M, Mazzone E, Rosiello G, Farinha R, et al. Proficiency-based progression training for robotic surgery skills training: a randomized clinical trial. BJU International. 2022;130(4):528–35. https://doi.org/10.1111/bju.15811.
14. Collins JW, Wisz P. Training in robotic surgery, replicating the airline industry. How far have we come? World J Urol. 2020;38(7):1645–51. https://doi.org/10.1007/s00345-019-02976-4.
15. Volpe A, Ahmed K, Dasgupta P, Ficarra V, Novara G, van der Poel H, et al. Pilot validation study of the European Association of Urology robotic training curriculum. Eur Urol. 2015;68(2):292–9. https://doi.org/10.1016/j.eururo.2014.10.025.
16. Knab LM, Zureikat AH, Zeh HJ, Hogg ME. Towards standardized robotic surgery in gastrointestinal oncology. Langenbeck's Arch Surg. 2017;402(7):1003–14. https://doi.org/10.1007/s00423-017-1623-4.
17. Vining CC, Hogg ME. How to train and evaluate minimally invasive pancreas surgery. J Surg Oncol. 2020;122(1):41–8. https://doi.org/10.1002/jso.25912.
18. Hogg ME, Tam V, Zenati M, Novak S, Miller J, Zureikat AH, et al. Mastery-based virtual reality robotic simulation curriculum: the first step toward operative robotic proficiency. J Surg Edu. 2017;74(3):477–85. https://doi.org/10.1016/j.jsurg.2016.10.015.
19. Tam V, Zenati M, Novak S, Chen Y, Zureikat AH, Zeh HJ, et al. Robotic pancreatoduodenectomy biotissue curriculum has validity and improves technical

performance for surgical oncology fellows. J Surg Edu. 2017;74(6):1057–65. https://doi.org/10.1016/j.jsurg.2017.05.016.
20. Al Abbas A, Hogg M. Robotic biotissue curriculum for teaching the robotic pancreatoduodenectomy. APC. 2018;1:9. https://doi.org/10.21037/apc.2018.01.06.
21. de Rooij T, van Hilst J, Boerma D, Bonsing BA, Daams F, van Dam RM, et al. Impact of a Nationwide training program in minimally invasive distal Pancreatectomy (LAELAPS). Ann Surg. 2016;264(5):754–62. https://doi.org/10.1097/sla.0000000000001888.
22. de Rooij T, van Hilst J, Topal B, Bosscha K, Brinkman DJ, Gerhards MF, et al. Outcomes of a multicenter training program in laparoscopic Pancreatoduodenectomy (LAELAPS-2). Ann Surg. 2019;269(2):344–50. https://doi.org/10.1097/sla.0000000000002563.
23. Zwart MJW, Nota CLM, de Rooij T, van Hilst J, te Riele WW, van Santvoort HC, et al. Outcomes of a multicenter training program in robotic Pancreatoduodenectomy (LAELAPS-3). Ann Surg. 2022;276(6):e886–e95. https://doi.org/10.1097/sla.0000000000004783.
24. van der Poel MJ, Fichtinger RS, van Dam RM, Besselink MG. Outcomes of laparoscopic minor and major liver surgery in The Netherlands (LAELIVE): Nationwide retrospective cohort. HPB. 2018;20:S263. https://doi.org/10.1016/j.hpb.2018.06.210.
25. Impact of a European training program for robot pancreatoduodenectomy using a video databank, da Vinci® simulator and robot biotissue anastomoses on clinical outcomes (LEARNBOT): a pan-European prospective study. https://trialsearch.who.int/Trial2.aspx?TrialID=NL8898; 2020. Accessed 14 Dec 2021.
26. van der Poel MJ, Fichtinger RS, Bemelmans M, Bosscha K, Braat AE, de Boer MT, et al. Implementation and outcome of minor and major minimally invasive liver surgery in The Netherlands. HPB : the official journal of the International Hepato Pancreato Biliary Association. 2019;21(12):1734–43. https://doi.org/10.1016/j.hpb.2019.05.002.
27. Lai ECH, Tang CN. Training robotic hepatectomy: the Hong Kong experience and perspective. Hepatobiliary Surg Nutr. 2017;6(4):222–9. https://doi.org/10.21037/hbsn.2017.01.21.
28. Madureira FAV, Varela JLS, Madureira DF, D'Almeida LAV, Madureira FAV, Duarte AM, et al. Model of a training program in robotic surgery and its initial results. Revista do Colegio Brasileiro de Cirurgioes. 2017;44(3):302–7. https://doi.org/10.1590/0100-69912017003013.
29. King JC, Zeh HJ 3rd, Zureikat AH, Celebrezze J, Holtzman MP, Stang ML, et al. Safety in numbers: progressive implementation of a robotics program in an academic surgical oncology practice. Surg Innov. 2016;23(4):407–14. https://doi.org/10.1177/1553350616646479.
30. Tanaka S, Kawaguchi Y, Kubo S, Kanazawa A, Takeda Y, Hirokawa F, et al. Validation of index-based IWATE criteria as an improved difficulty scoring system for laparoscopic liver resection. Surgery. 2019;165(4):731–40. https://doi.org/10.1016/j.surg.2018.10.012.
31. Abu Hilal M, Aldrighetti L, Dagher I, Edwin B, Troisi RI, Alikhanov R, et al. The Southampton consensus guidelines for laparoscopic liver surgery: from indication to implementation. Ann Surg. 2018;268(1):11–8. https://doi.org/10.1097/sla.0000000000002524.
32. Eppsteiner RW, Csikesz NG, Simons JP, Tseng JF, Shah SA. High volume and outcome after liver resection: surgeon or center? Journal of gastrointestinal surgery : official journal of the Society for Surgery of the Alimentary Tract. 2008;12(10):1709–16; discussion 16. https://doi.org/10.1007/s11605-008-0627-3.
33. Fong Y, Gonen M, Rubin D, Radzyner M, Brennan MF. Long-term survival is superior after resection for cancer in high-volume centers. Ann Surg. 2005;242(4):540–4.; discussion 4-7. https://doi.org/10.1097/01.sla.0000184190.20289.4b.
34. Gillespie BM, Gillespie J, Boorman RJ, Granqvist K, Stranne J, Erichsen-Andersson A. The impact of robotic-assisted surgery on team performance: a systematic mixed studies review. Hum Factors. 2021;63(8):1352–79. https://doi.org/10.1177/0018720820928624.
35. Weber J, Catchpole K, Becker AJ, Schlenker B, Weigl M. Effects of flow disruptions on mental workload and surgical performance in robotic-assisted surgery. World J Surg. 2018;42(11):3599–607. https://doi.org/10.1007/s00268-018-4689-4.
36. Weigl M, Weber J, Hallett E, Pfandler M, Schlenker B, Becker A, et al. Associations of intraoperative flow disruptions and operating room teamwork during robotic-assisted radical prostatectomy. Urology. 2018;114:105–13. https://doi.org/10.1016/j.urology.2017.11.060.
37. Poulsen JL, Bruun B, Oestergaard D, Spanager L. Factors affecting workflow in robot-assisted surgery: a scoping review. Surg Endosc. 2022;36(12):8713–25. https://doi.org/10.1007/s00464-022-09373-w.
38. Vigo F, Egg R, Schoetzau A, Montavon C, Brezak M, Heinzelmann-Schwarz V, et al. An interdisciplinary team-training protocol for robotic gynecologic surgery improves operating time and costs: analysis of a 4-year experience in a university hospital setting. J Robot Surg. 2022;16(1):89–96. https://doi.org/10.1007/s11701-021-01209-4.

11 Proficiency-Based Progression Training

Metric-Based Simulation Training to Achieve Proficiency

Marie-Hélène Vinckier, Alexandre Mottrie, Anthony G. Gallagher, and Ruben De Groote

Abbreviations

PBP	Proficiency-based progression
RCT	Randomized clinical trial
VR	Virtual reality
IRR	Inter-rater reliability

11.1 Introduction

Today, novice apprentices still perform their first surgical steps on patients. However, the emergence of scientific data linking perioperative morbidity and mortality to surgical skill has raised some ethical concerns [1, 2]. Furthermore, reduced work hours and by default training hours for surgical trainees [3], as well as the rapid expansion of advanced technologies to support and perform procedures, challenge the traditional Halstedian approach to training and are forcing the surgical community to rethink new teaching modalities [4].

Simulation-based training has a strong foothold in safety-conscious industries such as aviation and nuclear power, and has been used in anesthesia to give individuals and teams the experience of emergency situations before encountering them in real life. A simulation-based, standardized surgical training system with objective and fair feedback in an out-of-hospital setting could improve patient safety [5].

Proficiency-based progression (PBP) is a training method in which an operative procedure is characterized in detail to identify objective performance metrics. After defining these unambiguous metric-based performance characterization, these metrics can be used to establish a benchmark (i.e., a level of proficiency) that trainees must demonstrate before they can progress in their training [6]. PBP has demonstrated its value in different surgical specialties [6, 7].

11.2 Literature Overview

Simulation-based training is a learning method that was first introduced by Satava et al. in 1993, who set up a blinded randomized clinical trial

M.-H. Vinckier
Department of Urology, OLV, Aalst, Belgium

Faculty of Medicine, KU Leuven, Leuven, Belgium

A. Mottrie · R. De Groote (✉)
Department of Urology, OLV, Aalst, Belgium

ORSI Academy, Melle, Belgium

A. G. Gallagher
Faculty of Medicine, KU Leuven, Leuven, Belgium

ORSI Academy, Melle, Belgium

School of Medicine, Faculty of Life and Health Sciences, Ulster University, Northern Ireland, UK

M. D'Hondt, I. Sucandy (eds.), *Textbook of Robotic Liver Surgery*,
https://doi.org/10.1007/978-3-031-76536-0_11

(RCT) demonstrating that trainees who performed simulation training using virtual reality (VR) performed better than surgeons who were trained traditionally [8]. This was the first study to introduce the initial thoughts of the PBP training method using the metrics system.

The requirement to demonstrate a predefined proficiency benchmark in training, combined with simulation-based practice, meant that PBP training was particularly effective and demonstrated performance improvements in over 40% in objectively assessed errors in comparison to traditional training in the areas of laparoscopy surgery [9–11].

A prospective, blinded RCT by Angelo et al. has also shown that implementation of a PBP curriculum, coupled with the use of a shoulder model simulator, produced a superior arthroscopic surgical skillset compared with traditional and simulator-enhanced training methods [12–14]. Breen et al. also showed in an RCT that the use of PBP for the training of the ISBAR (Identification, Situation, Background, Assessment, Recommendation) communication skillset in the context of a clinically deteriorating patient led to a significantly better communication performance [15]. Studies by Cates et al. and Srinivasan et al. showed better results while using PBP during respectively endovascular and anesthetic procedures [16, 17].

A recent meta-analysis by Mazzone et al. validates that PBP reduces procedural errors by 60% compared to conventional learning methods [18]. The PROVESA trial is an RCT by De Groote et al. and shows that the implementation of PBP resulted in superior surgical performance for robotic suturing and knot tying compared to conventional methods [19]. In this study, PBP participants were 10 times more likely to achieve the proficiency benchmark compared to the traditional training group.

As a conclusion from these studies, PBP can be considered a more effective approach to skills training.

PBP is thus proven to be an effective training method to improve skill; however, there is an urgent need to verify to what extent skills training impacts patient outcomes. Birkenmeyer et al. and Curtis et al. [1, 2] showed that peer-reviewed variation in surgical skill during laparoscopic bariatric and colorectal surgery, which can even be seen among experienced surgeons, is significantly associated with differences in postoperative patient outcomes. However, these studies use Likert-type scales, which have been proven to be unreliable [20]. A multicenter, prospective RCT is needed to verify to what extent skills training impacts patient outcomes.

11.3 Metrics

PBP is based on metrics, which are units of performance that define the configuration of any simulation that is developed to train skilled task performance. These metrics must capture the essence of the procedure in question and might include steps that should be performed, what instruments that should be used, and how they should be used [19].

Metrics must also describe what *not* to do during a specific procedure and in this way characterize a performance that deviates from the idea (an error) (Fig. 11.1). Errors could be defined as performance actions that are potentially unsafe, and critical errors are most certainly unsafe but do not necessarily lead to a bad outcome. This is very crucial since the aim is not only to concentrate on what should be done and the order in which it should be done, but also target performance errors in order to reduce or eliminate them. The crucial "error-defining" characteristic of metrics derives from the idea that bad outcomes are not accidental but usually caused by deviations from optimal procedure protocol.

Metric units should be unambiguously "defined" rather than "described" so that they can be scored as occurring or not occurring. This binary approach facilitates the reliable scoring of the metrics system in comparison to the more vague and subjective Likert scale [20].

Reaching an agreement between different groups of physicians and surgeons on performance metrics seems to be impossible. However, the majority of experienced and proficient doctors in the performance of a specific task can

Metric errors/criteria of injury assessment	Operational Definition
Procedure **START**	first contact of diathermy with tissue
Procedure **END**	last attachment is divided
FAILURE TO PROGRESS	No progress made in excising the gallbladder for an intire minute of the dissection. Dealing with the consequences of a predefined error represents lack of progress if no progress is made in excising the gallbladder during this period.
GALLBLADDER INJURY	There is gallbladder wall perforation with or without leakage of bile. Injury may be incurred with either hand.
LIVER INJURY	Necessitates capsule penetration and may have bleeding.
BURN NONTARGET TISSUE	Any application of electrocautery to non-target tissue with the exception of the final part of the fundic dissection where some current transmission may occur.
TEARING TISSUE	Uncontrolled tearing of tissue with the dissecting or retracting instrument.
INCORRECT PLANE OF DISSECTION	The dissection is conducted outside the recognized plane between the gallbladder and the liver (i.e. in the sub-mucosal plane on the gallbladder, or sub-capsular plane on the liver).
INSTRUMENT OUT OF VIEW	The dissecting instrument is placed outside the field of view of the telescope such that its tip is un-viewable and can potentially be in contact with tissue. No error will be attributed to an incident of a dissecting instrument out of view as the result of a sudden telescope movement.
ATTENDING TAKEOVER	The supervising attending surgeon takes the dissecting instrument (right hand), or retracting instrument (left hand) from the resident and performs a component of the procedure.

Fig. 11.1 Example of metric errors during laparoscopic cholecystectomy [7]

agree on what should be done, how it should be done, and what shouldn't be done for most parts of the performance. In the hands of proficient surgeons, these performance characteristics are automated and doctors rarely think about the performed procedure in that level of detail. It is thus a difficult primary process to identify these very detailed performance characteristics during the task analysis. An uncomplicated and straightforward reference procedure should be used, rather than a deviant task, to seek consensus on the units of performance. The task analysis stage of developing a training program is crucial since the metrics are the fundamentals of a qualitative simulation.

The procedural steps, errors, and critical errors should be validated at a Delphi consensus meeting.

If the metric identification and definition process is well performed, these performance units should be able to distinguish novices from experts (construct validity) and predict future performances (predictive validity) [21]. This construct validity will define which metrics best differentiate novice and expert performances and will help "benchmark" the process or "proficiency level" a trainee must acquire before they progress to the following training level.

11.4 Proficiency-Based Progression Training

PBP skills training is a scientific approach to the learning of surgical skills that is standardized, objective, transparent, and provides fair feedback. The performance metrics, which are the basis of PBP, should be constructed and validated by experienced and proficient surgeons (Fig. 11.2). These metrics are developed during a task analysis involving 3–5 experienced surgeons [6, 22].

Fig. 11.2 The proficiency-based progression method [7]

The metrics should identify the observable performance of an entire procedure. They are validated at a Delphi consensus and then via construct validity [21]. This construct validity requires that the metrics can be scored by independent instructors, with an inter-rater reliability (IRR) of more than 0.8. Secondly, the performance assessment has to be able to reliably discriminate between the objectively rated experienced and beginner surgeons. When these criteria for construct validity are met, a proficiency benchmark can be quantitatively defined on the basis of the mean performances of the experienced surgeons [23]. In addition, the performance metrics should use binary scores and not Likert-scale assessments. Furthermore, PBP-validated metrics should give trainees continuous constructive and objective performance feedback during training and as a consequence accelerate their learning using deliberate practice [24] rather than simple repeated practice. Trainees must always demonstrate the proficiency benchmarks before training progression, but they can repeat the training as often as necessary.

11.5 Conclusions

The valid and reliable assessment of surgical skill is imperative to effective and efficient training. Medicine should move away from a process-driven approach (e.g., time in training, number of procedures done) for the evaluation of skills and should be replaced by an outcome-based approach, where the skills of the trainee are verified according to the Institute of Medicine [25].

It is shown that better surgical skills lead to better outcomes [1, 2] and that is why the importance of surgical training to develop skill among (future) surgeons cannot be underestimated.

PBP is a new training methodology that uses the incorporation of surgical simulation, validated, binary performance metrics, and standardized procedure templates as cornerstones of training [7].

PBP is proven to be superior in comparison to high-quality simulation-based programs. Notably, PBP even decreases the number of procedural errors compared to traditional training [18].

References

1. Birkmeyer JD, Finks JF, O'Reilly A, et al. Surgical skill and complication rates after bariatric surgery. N Engl J Med. 2013;369:1434–42.
2. Curtis NJ, Foster JD, Miskovic D, et al. Association of surgical skill assessment with clinical outcomes in cancer surgery. JAMA Surg. 2020;155:590–8.
3. Chikwe J, De Souza AC, Pepper JR. No time to train the surgeons [Internet]. Vol. 328, British Medical Journal. BMJ Publishing Group; 2004 [cited 2020 Nov 11]. p. 418–9. Available from: https://pubmed.ncbi.nlm.nih.gov/14976074/.
4. Pellegrini CA. Surgical education in the United States: Navigating the white waters. In: Annals of Surgery [Internet]. Ann Surg; 2006 [cited 2021 Jan 3]. p. 335–42. Available from: https://pubmed.ncbi.nlm.nih.gov/16926559/.
5. Logan M. Why hospitals should fly: the ultimate flight plan to patient safety and quality care. Biomed Instrum Technol. 2010;44(2):110.
6. Gallagher AG, Ritter EM, Champion H, Higgins G, Fried MP, Moses G, et al. Virtual reality simulation for the operating room: proficiency-based training as a paradigm shift in surgical skills training. Ann Surg. 2005;241:364–72.
7. Gallagher AG. Metric-based simulation training to proficiency in medical education:– what it is and how to do it. Ulster Med J. 2012;81:107–13.
8. Satava RM. Virtual reality surgical simulator: the first steps. Surg Endosc. 1993;7:203–5.
9. Seymour N, Gallagher A, Roman S, et al. Virtual reality training improves operating room performance: results of a randomized, double-blinded study. Ann Surg. 2002;236:458–64.
10. Ahlberg G, Enochsson L, Gallagher AG, et al. Proficiency-based virtual reality training significantly reduces the error rate for residents during their first 10 laparoscopic cholecystectomies. Am J Surg. 2007;193:797–804.
11. Van Sickle KR, Ritter EM, Baghai M, et al. Prospective, randomized, double-blind trial of curriculum-based training for intracorporeal suturing and knot tying. J Am Coll Surg. 2008;207:560–8.
12. Angelo RL, Ryu RKN, Pedowitz RA, Beach W, Burns J, Dodds J, et al. A proficiency-based progression training curriculum coupled with a model simulator results in the acquisition of a superior arthroscopic Bankart skill set. Arthrosc – J Arthrosc Relat Surg. 2015;31(10):1854–71.
13. Angelo RL, Tauro J, St. Pierre P, Ross G, Voloshin I, Shafer B, et al. Arthroscopic rotator cuff repair metrics: establishing face, content, and construct validity in a cadaveric model. Arthrosc – J Arthrosc Relat Surg. 2020;36(1):71–79.e1.
14. Angelo RL, Pedowitz RA, Ryu RKN, Gallagher AG. The Bankart performance metrics combined with a shoulder model simulator create a precise and accurate training tool for measuring surgeon skill. Arthrosc – J Arthrosc Relat Surg. 2015;31(9):1639–54.
15. Breen D, O'Brien S, McCarthy N, Gallagher A, Walshe N. Effect of a proficiency-based progression simulation programme on clinical communication for the deteriorating patient: a randomised controlled trial. BMJ Open [Internet] 2019 Jul 1 [cited 2020 Nov 8];9(7). Available from: https://pubmed.ncbi.nlm.nih.gov/31289064/.
16. Cates CU, Lönn L, Gallagher AG. Prospective, randomised and blinded comparison of proficiency-based progression full-physics virtual reality simulator training versus invasive vascular experience for learning carotid artery angiography by very experienced operators. BMJ Simul Technol Enhanc Learn. 2016;2:1–5.
17. Srinivasan KK, Gallagher A, O'Brien N, et al. Proficiency-based progression training: an "end to end" model for decreasing error applied to achievement of effective epidural analgesia during labour: a randomised control study. BMJ Open. 2018;8:e020099.
18. Mazzone E, Puliatti S, Amato M, et al. A systematic review and meta-analysis on the impact of proficiency-based progression simulation training on performance outcomes. Ann Surg. 2021;274:281–9.
19. De Groote R, Puliatti S, Amato M, et al. Proficiency-based progression training for robotic surgery skills training: a randomized clinical trial. BJUI Int. 2022;130(4):528–35.
20. Satava RM, Stefanidis D, Levy JS, et al. Proving the effectiveness of the fundamentals of robotic surgery (FRS) skills curriculum: a single-blinded, multispecialty, multi-institutional randomized control trial. Ann Surg. 2020;272:384–92.
21. Puliatti S, Mazzone E, Amato M, De Groote R, Mottrie A, Gallagher AG. Development and validation of the objective assessment of robotic suturing and knot tying skills for chicken anastomotic model. Surg Endosc [Internet]. 2020 [cited 2021 Jan 3]; Available from: https://pubmed.ncbi.nlm.nih.gov/32857241/.
22. Gallagher AG, O'Sullivan GC, Gallagher AG, O'Sullivan GC, O'Sullivan GC. Simulations for procedural training. In: Fundamentals of surgical simulation. London: Springer; 2011. p. 39–66.

23. Gallagher AG, De Groote R, Paciotti M, et al. Proficiency-based progression training: a scientific approach to learning surgical skills. Eur Urol. 2022;81(4):394–5. In press.
24. Ericsson KA, Krampe RT, Tesch-Römer C. The role of deliberate practice in the acquisition of expert performance. Psychol Rev. 1993;100:363–406.
25. Asch DA, Weinstein DF. Innovation in Medical Education. N Engl J Med [Internet]. 2014 Aug 28 [cited 2021 Feb 8];371(9):794–5. Available from: https://pubmed.ncbi.nlm.nih.gov/25075620/.

Part V

Robotic Liver Surgery for Malignant Tumors

12 Robotic Surgery Trends in General Oncological Surgery

Mark Hall and Ali Ahmad

12.1 Introduction

Minimally invasive surgery (MIS) has evolved through the early phases of laparoscopy to the current more sophisticated robotic platforms, gradually overcoming the overt criticism from stalwarts of traditional open surgery. Application of minimally invasive surgery in surgical oncology has been challenging secondary to limitations of technology, specifically in maintaining principles of oncological resections. The adoption of laparoscopy in the late 1990s paved the path for technology in medicine, which subsequently led to the birth of the robotic surgical platform.

The robotic platform offered promise in overcoming conventional limitations of laparoscopy, primarily from advanced dexterity and visuals. In early phases, theoretical benefits of the robotic platform were overshadowed by cumbersome machine setup issues and steep surgeon-specific learning curves. However, with time, evolution of newer robotic systems led to increased efficiency and ease, which translated into a more streamlined adoption pathway.

M. Hall · A. Ahmad (✉)
Department of Surgery, University of Kansas, School of Medicine-Wichita, Wichita, KS, USA

12.2 Adoption in Oncological Surgery

12.2.1 Colorectal Surgery

Over seven million procedures have been performed via the Intuitive Surgical da Vinci system since its inception, varying from general surgery procedures to thoracic, urological, and gynecological surgery. Benefits of improved three-dimensional visualization, multimodal dexterity of the instruments, and enhanced ergonomics have allowed for increased interest in various aspects of surgical oncology. Over the past 10 years, there has been more than a fivefold increase in the adoption of robotic surgery in oncological surgeries [1].

The adoption of robotic surgery in colorectal surgery has gained substantial traction. From 2013 to 2018, the rate of robotic approach to colorectal surgery in the United States increased from 2.8% to 11.4% annually [2]. However, there is a lack of concrete data supporting the benefit of robotic platform over traditional laparoscopy. Results from the ROLARR trial, a multicenter randomized controlled trial (RCT), failed to show benefit of the robotic approach over laparoscopy in rectal cancer resections [3]. Results of such trials have faced criticism such as the 8% conversion rate to open seen in ROLARR, which is significantly higher than most recent studies [4, 5]. Multiple other studies have shown substantial

M. D'Hondt, I. Sucandy (eds.), *Textbook of Robotic Liver Surgery*,
https://doi.org/10.1007/978-3-031-76536-0_12

benefits associated with the robotic platform compared with the laparoscopy in colorectal surgery, which include decreased morbidity and hospital length of stay [6, 7].

12.2.2 Pancreatico-Biliary Surgery

Pancreatico-biliary surgery is often considered to be a technically demanding field with high morbidity, which in turn has led to a limited adoption of minimally invasive approach. Results of the LEOPARD-1 RCT paved the path for MIS distal pancreatectomy procedures [8]. Expansion of robotic surgery into the left-sided/distal pancreatectomy procedures is widespread primarily due to the success of the laparoscopic approach. Results from the LEOPARD-2 RCT, however, demonstrated inferior outcomes with laparoscopic pancreatico-duodenectomy (PD) compared to the traditional open approach [9]. But despite such dismal results with laparoscopy, the technical advantages of robotic surgical systems have allowed surgeons to re-explore the option of MIS approach to PD, with excellent results in early studies [10, 11]. Even though robotic PD procedure is still in its infancy, it appears to be gaining increased acceptance throughout the globe. There remains a debate regarding the true benefit of the MIS approach to PD but based on the bright prospects of modern era technology, it is likely that the MIS approach to PD will grow noticeably in the future.

12.2.3 Hepatectomy

Interest in the field of minimally invasive hepatectomy has seen substantial growth. Introduction of the robotic platform has allowed for a gradual and sustained increase in the incorporation of minimally invasive hepatectomy at high-volume institutions. Despite excellent outcomes with laparoscopic hepatectomy, technical limitations have hindered the rate of widespread adoption compared to other subspecialties [12, 13].

The safety and feasibility of robotic hepatectomy has been proven across multiple previously published studies [14, 15]. Due to the complexity of the procedure, the learning curve of such procedures is still being investigated. However, proficiency leads to improvements in outcomes as demonstrated by Hawksworth et al. [16]. Additionally, the learning curve of robotic hepatectomy appears to be shorter than conventional laparoscopic approach [17]. In our practice, we have recognized substantial improvements in morbidity and length of hospital stay with surgeon proficiency after overcoming the learning curve.

In the field of surgical oncology, apart from better operative techniques, the advancements in imaging modalities, multidisciplinary approach, and a better understanding of cancer biology have all been translated into substantial improvements in cancer-related outcomes. In the era of parenchyma-preserving hepatectomy paired with the ability to detect smaller lesions due to enhanced imaging and surveillance, the morbidity of hepatectomy procedures has decreased. Procedures such as left lateral hepatectomy and anterior segment wedge resections have been performed via laparoscopic approach for a long time; however, robotics has allowed for a sustained expansion of the MIS approach to formal major hepatectomy procedures in addition to non-anatomic posterior–superior segmental resections [16, 18].

In the management of liver metastases, MIS approaches have seen increasing interest primarily owing to the trend toward parenchymal-sparing techniques. In a recent national database study by Carpenter et al. [19], rates of robotic hepatectomy are increasing; however, they are mostly still confined to less technically demanding resections. At our institution, since the inception of the robotic hepatobiliary program in 2017, we have observed a considerable increase in the rate of robotic hepatectomies for both anatomic and minor resections (Fig. 12.1).

Multiple factors can influence the complexity of robotic hepatectomy. In our experience, we observed factors such as liver cirrhosis, obesity (body mass index $\geq$30 kg/m^2), and larger tumors ($\geq$ 10 cm) can bear an impact on outcomes. Chen et al. demonstrated excellent outcomes with

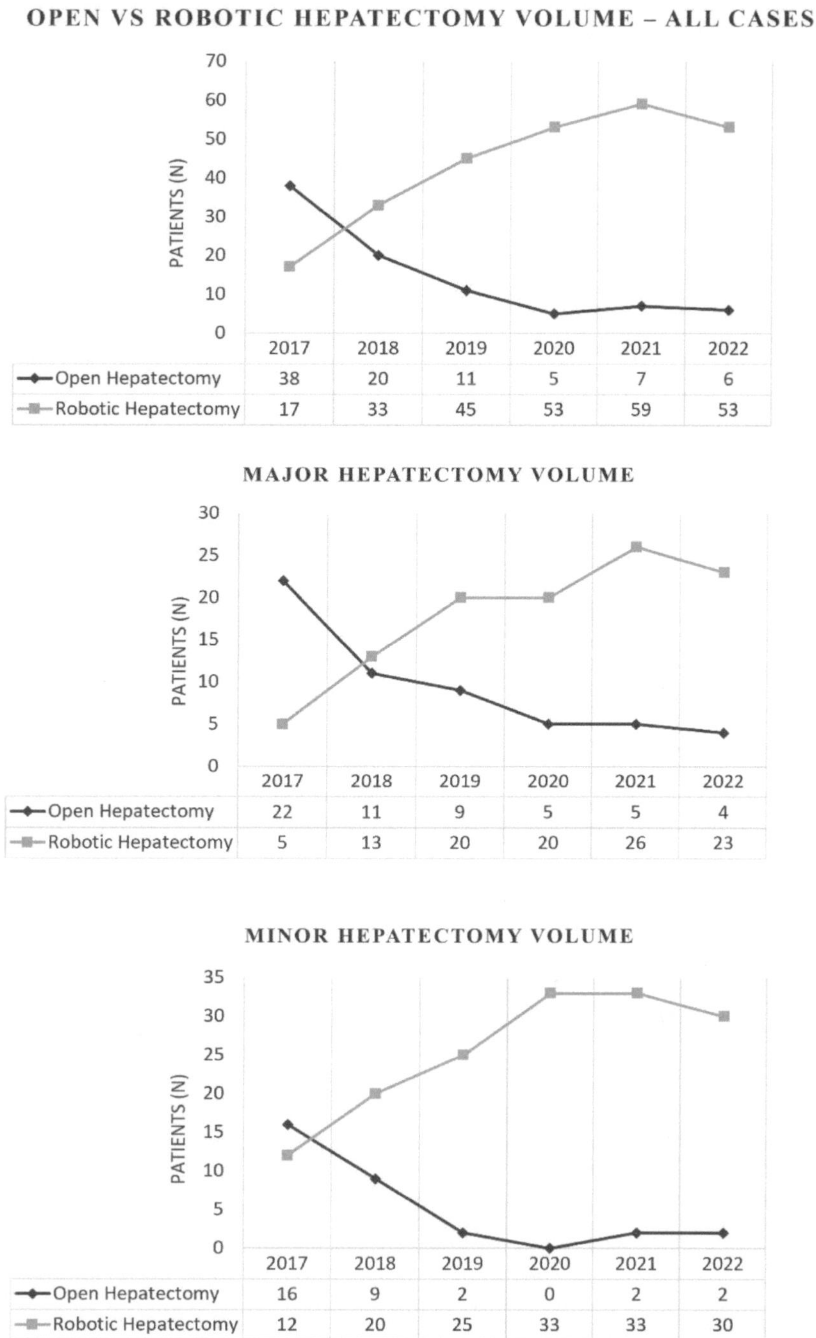

Fig. 12.1 Robotic surgery trends at our institution

robotic hepatectomy for hepatocellular carcinoma in the setting of underlying cirrhosis [20]. Similarly, recent studies have shown safety of the robotic approach in obese patients as well as the elderly [21, 22]. Such promising results in various technically demanding scenarios have allowed for expansion of robotic hepatectomy beyond the realm of favorably located nonanatomic resections.

Robotic hepatectomy, without a doubt, is experiencing rapid growth across the globe. The benefits offered by the robotic platform have allowed for safe expansion of the MIS approach to hepatectomy procedures. Future robotic surgical systems and advances in technology will likely further the adoption of robotic hepatectomies.

References

1. Stewart CL, Ituarte PH, Melstrom KA, Warner SG, Melstrom LG, Lai LL, Fong Y, Woo Y. Robotic surgery trends in general surgical oncology from the National Inpatient Sample. Surg Endosc. 2019;33(8):2591–601.
2. Hajirawala LN, Leonardi C, Orangio GR, Davis KG, Barton JS. Trends in open, laparoscopic, and robotic approaches to colorectal operations. Am Surg. 2021;28:00031348211034754.
3. Jayne D, Pigazzi A, Marshall H, Croft J, Corrigan N, Copeland J, Quirke P, West N, Rautio T, Thomassen N, Tilney H. Effect of robotic-assisted vs conventional laparoscopic surgery on risk of conversion to open laparotomy among patients undergoing resection for rectal cancer: the ROLARR randomized clinical trial. JAMA. 2017;318(16):1569–80.
4. Myrseth E, Nymo LS, Gjessing PF, Kørner H, Kvaløy JT, Norderval S. Lower conversion rate with robotic assisted rectal resections compared with conventional laparoscopy; a national cohort study. Surg Endosc. 2022;36(5):3574–84.
5. Phan K, Kahlaee HR, Kim SH, Toh JW. Laparoscopic vs. robotic rectal cancer surgery and the effect on conversion rates: a meta-analysis of randomized controlled trials and propensity-score-matched studies. Tech Coloproctol. 2019;23(3):221–30.
6. Sun Y, Xu H, Li Z, Han J, Song W, Wang J, Xu Z. Robotic versus laparoscopic low anterior resection for rectal cancer: a meta-analysis. World J Surg Oncol. 2016;14(1):1–8.
7. Chang YS, Wang JX, Chang DW. A meta-analysis of robotic versus laparoscopic colectomy. J Surg Res. 2015;195(2):465–74.
8. de Rooij T, van Hilst J, van Santvoort H, Boerma D, van den Boezem P, Daams F, van Dam R, Dejong C, van Duyn E, Dijkgraaf M, van Eijck C. Minimally invasive versus open distal pancreatectomy (LEOPARD): a multicenter patient-blinded randomized controlled trial. Ann Surg. 2019;269(1):2–9.
9. Van Hilst J, de Rooij T, Bosscha K, Brinkman DJ, van Dieren S, Dijkgraaf MG, Gerhards MF, de Hingh IH, Karsten TM, Lips DJ, Luyer MD. Laparoscopic versus open pancreatoduodenectomy for pancreatic or periampullary tumours (LEOPARD-2): a multicentre, patient-blinded, randomised controlled phase 2/3 trial. Lancet Gastroenterol Hepatol. 2019;4(3):199–207.
10. Rosemurgy A, Ross S, Bourdeau T, Craigg D, Spence J, Alvior J, Sucandy I. Robotic pancreaticoduodenectomy is the future: here and now. J Am Coll Surg. 2019;228(4):613–24.
11. Baimas-George M, Watson M, Murphy KJ, Iannitti D, Baker E, Ocuin L, Vrochides D, Martinie JB. Robotic pancreaticoduodenectomy may offer improved oncologic outcomes over open surgery: a propensity-matched single-institution study. Surg Endosc. 2020;34(8):3644–9.
12. Tsung A, Geller DA, Sukato DC, Sabbaghian S, Tohme S, Steel J, Marsh W, Reddy SK, Bartlett DL. Robotic versus laparoscopic hepatectomy: a matched comparison. Ann Surg. 2014;259(3):549–55.
13. Chong CC, Lok HT, Fung AK, Fong AK, Cheung YS, Wong J, Lee KF, Lai P. Robotic versus laparoscopic hepatectomy: application of the difficulty scoring system. Surg Endosc. 2020;34(5):2000–6.
14. Kingham TP, Leung U, Kuk D, Gönen M, D'Angelica MI, Allen PJ, DeMatteo RP, Laudone VP, Jarnagin WR, Fong Y. Robotic liver resection: a case-matched comparison. World J Surg. 2016;40(6):1422–8.
15. Qiu J, Chen S, Chengyou D. A systematic review of robotic-assisted liver resection and meta-analysis of robotic versus laparoscopic hepatectomy for hepatic neoplasms. Surg Endosc. 2016;30(3):862–75.
16. Hawksworth J, Radkani P, Llore N, Holzner M, Winslow E, Satoskar R, Jha R, Terry C, Meslar E, Haddad N, Fishbein T. Medium and high complexity robotic hepatectomy: learning curve and outcomes. HPB. 2021;1(23):S606–7.
17. Efanov M, Alikhanov R, Tsvirkun V, Kazakov I, Melekhina O, Kim P, Vankovich A, Grendal K, Berelavichus S, Khatkov I. Comparative analysis of learning curve in complex robot-assisted and laparoscopic liver resection. HPB. 2017;19(9):818–24.
18. Sucandy I, Luberice K, Lippert T, Castro M, Krill E, Ross S, Rosemurgy A. Robotic major hepatectomy: an institutional experience and clinical outcomes. Ann Surg Oncol. 2020;27(13):4970–9.
19. Carpenter EL, Thomas KK, Adams AM, Valdera FA, Chick RC, Kemp Bohan PM, Spitzer HV, Clifton GT, Bader JO, Nelson DW, Vreeland TJ. Modern trends in minimally invasive versus open hepatectomy for colorectal liver metastasis: an analysis of ACS-NSQIP. Surg Endosc. 2022;7:1–2.
20. Chen PD, Wu CY, Hu RH, Chou WH, Lai HS, Liang JT, Lee PH, Wu YM. Robotic versus open hepatectomy for hepatocellular carcinoma: a matched comparison. Ann Surg Oncol. 2017;24(4):1021–8.
21. Tee MC, Chen L, Peightal D, Franko J, Kim PT, Brahmbhatt RD, Raman S, Scudamore CH, Chung SW, Segedi M. Minimally invasive hepatectomy is associated with decreased morbidity and resource utilization in the elderly. Surg Endosc. 2020;34(11):5030–40.
22. Sucandy I, Attili A, Spence J, Bordeau T, Ross S, Rosemurgy A. The impact of body mass index on perioperative outcomes after robotic liver resection. J Robot Surg. 2020;14(1):41–6.

Robotic Liver Surgery for Hepatocellular Carcinoma

13

Ning Qi Pang, Paolo Magistri, Stefano Di Sandro, and Fabrizio Di Benedetto

13.1 Introduction

Minimally invasive liver surgery has demonstrated several benefits over the classical open approach, including lower estimated blood loss and postoperative pain, shorter length of hospital stay, and improved cosmesis [1–4]. Prior to the advent of robotic surgery, laparoscopy was considered the standard of care when adopting minimally invasive approaches for liver resection [5–8]. Initial concerns about oncological outcomes proved to be unfounded as laparoscopic liver surgery has been proven to provide equivalent survival compared to traditional open surgery [9–12]. The Louisville and Morioka Consensus Conferences in 2008 and 2014, respectively, and the Southampton Guidelines Meeting in 2017 further established the minimally invasive approach as the gold standard in specific liver resection procedures [8, 13, 14].

In recent years, the use of the robotic platform as an alternative to laparoscopic liver surgery has gained traction. The excitement in its adoption stems from its advantages over laparoscopy: the stability it provides, three-dimensional and magnified high-definition views, increased dexterity and degrees of freedom of instruments, as well as filtering of tremors [15]. These characteristics of the robotic platform allow for very precise dissection and suturing even in narrow operative fields and awkward operative angles. In liver resections, this may mean an enhanced experience when dissecting the hepatic hilum, performing hilar lymphadenectomy and biliary reconstructions [16, 17]. In addition, the robot reduces the conversion rate compared to laparoscopy in the setting of high-difficulty procedures [18] and has been shown to require a shorter learning curve compared to laparoscopy for specific procedures [19].

Therefore, adopting a robotic approach may increase the opportunity for patients to be treated with the advantages of minimally invasive surgery, even when a complex procedure is needed [20, 21].

In this chapter, we will review some key points for a better understanding of the robotic approach to the liver and its application in the field of hepatocellular carcinoma (HCC).

N. Qi Pang
Division of Hepatobiliary & Pancreatic Surgery, Department of Surgery, National University Hospital, Singapore, Singapore

P. Magistri · S. Di Sandro · F. Di Benedetto (✉)
Hepato-Pancreato-Biliary Surgery and Liver Transplantation Unit, University of Modena and Reggio Emilia, Modena, Italy
e-mail: fabrizio.dibenedetto@unimore.it

M. D'Hondt, I. Sucandy (eds.), *Textbook of Robotic Liver Surgery*,
https://doi.org/10.1007/978-3-031-76536-0_13

13.2 Patient Selection and Indications

The indication for surgical resection in a modern hepatobiliary center should be always discussed in a multidisciplinary meeting involving surgeons, radiologists, anesthesiologists, hepatologists, and oncologists to ensure the accuracy and appropriateness of the therapeutic strategy [22, 23]. Indocyanine green (ICG) clearance and evaluation of portal hypertension measuring the hepatic venous pressure gradient (HVPG) may help refine the indication and prevent post-hepatectomy liver failure [24]. In selected cases, hepatobiliary scintigraphy may be an option, especially in cases of two-stage resections to test the function of the future liver remnant. Up to now, there are no formal contraindications to performing a robotic liver resection, except the general contraindication for pneumoperitoneum. The presence of macrovascular tumor infiltration, large exophytic tumors, and the need for major vascular reconstructions can be considered relative contraindications for the robotic approach, and the decision to include those patients relies on the experience of the center. Cirrhotic patients with HVPG >10 mmHg should be evaluated case by case, according to the extent of the resection needed. In the case of repeat resections, minimally invasive approaches and in particular the use of the robot have been proven to be safe and effective [25, 26].

13.3 Technical Aspects

13.3.1 Patient Positioning and System Setup

Even if other robotic systems are available on the market, most of the evidence produced in the literature so far was obtained using DaVinci robotic platforms (Intuitive Surgical Inc., Sunnyvale, CA, USA), and therefore we will discuss system setup details of the most recent DaVinci model. The DaVinci platform includes a surgeon's console, a patient-side cart, and a vision system. Patients are usually positioned supine, 10–15° anti-Trendelenburg, and can be rotated to the left to allow easier access to the right and posterior segments. The legs of the patient are parted and the assistant surgeon is positioned here to obtain comfortable access to the operative field. The position of the trocars varies according to each patient's particular body habitus and lesion localization, and to the robotic platform in use [15].

Using the Xi platform, the first step, according to the manufacturer, is to determine the target anatomy, which is not the disease location but the area where the midline of the surgical workspace intersects the far edge of the surgical workspace boundary. The initial endoscope port should be placed 10–20 cm back from the target anatomy, on the opposite edge of the surgical workspace boundary. The other ports are usually placed on a straight line perpendicular to the target anatomy: for liver procedures, two arms are controlled with the right hand (patient's left side) and one with the left hand (patient's right side). The trocars should be placed 6–10 cm apart according to patient body habitus and at least 2 cm away from bony structures. Assistant ports can be inserted at least 7 cm from robotic ports, with adequate triangulation to enable the table-side surgeon to reach the desired anatomy and ensure physical access to the port.

13.3.2 Use of Indocyanine Green-Based Fluorescence

Another novel feature of the DaVinci robotic platform is the integrated Firefly, an indocyanine green (ICG)-based fluorescence imaging system that allows for a seamless transition between normal light and fluorescence-guided imaging. There are several ways this technology can be utilized to reach the goal of a personalized and tailored surgery. In liver surgery, ICG fluorescence can be used for tumor identification, surface mapping, and real-time cholangiography [27–29]. Tumor cells can be either hypo-fluorescent or hyper-fluorescent according to their histological features: well-differentiated HCCs show a homogeneous fluorescence pattern, while poorly differentiated HCCs demonstrate an inhomogeneous or even a rim-type fluorescence pattern [27]. ICG can therefore also aid in tumor localization.

For this kind of visualization, an intravenous injection of 0.5 mg/kg bodyweight of ICG is recommended 2–14 days prior to surgery. This recommendation is based on studies and observations in patients who underwent surgery after an ICG clearance test, which explains the dose and the timing. In our experience, a dose of 0.25 mg/kg bodyweight 12 h before surgery is sufficiently effective for HCC visualization in noncirrhotic patients (Fig. 12.1).

Arteries and veins light up around 5–60 seconds after intravenous injection of ICG, and the biliary tree is illuminated 45–60 minutes after ICG administration as ICG is excreted in bile [15]. This allows delineation of segmental and biliary anatomy. Two different techniques can be used for the definition of segmental anatomy: positive and negative staining. In the negative staining method, the inflow vessels of the hepatic parenchyma to be removed must be temporarily clamped to confirm the demarcation line, followed by systemic injection of 2.5 mg of ICG, resulting in fluorescent enhancement of the hepatic parenchyma to be preserved [30]. The positive staining technique requires direct injection of 2.5 mg ICG diluted in 10 mL of saline solution into the tumor-bearing portal branch through a 16-gauge needle. This way, the segment to be removed becomes fluorescence-enhanced on the liver surface and can be marked with monopolar energy to guide the transection between fluorescing and nonfluorescing parenchyma [30] (Fig. 13.1).

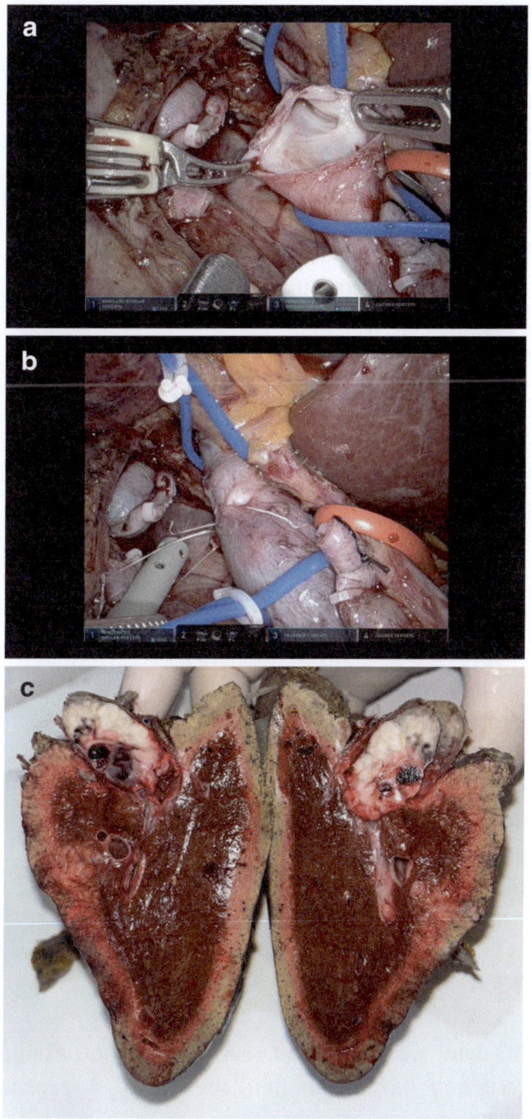

Fig. 13.1 (**a**) Venoplasty of the portal vein bifurcation during right hepatectomy due to right portal vein tumor thrombosis. (**b**) Final aspect of the venoplasty suture. (**c**) Specimen with evident macrovascular tumor thrombosis

13.3.3 Parenchymal Transection

For parenchymal transection, we adopt a Kelly crush technique with the use of bipolar forceps, combined with monopolar energy and use of the DaVinci Harmonic ACE (Ethicon, Somerville, NJ, USA) for deeper layers. The robotic platform does not support liver-specific devices for parenchyma dissection such as the Cavitron Ultrasonic Surgical Aspirator (CUSA); however, it can be used by the table-side surgeon even though this approach is not comfortable and requires placing the assistant port very high. Moreover, the transection line is determined by the surgeon using the CUSA rather than the console surgeon. From this perspective, the use of the CUSA from the table-side dispels most of the advantages of the robotic approach and, for those reasons, it does not represent our first choice.

It is well known that nonanatomical liver resections are more difficult to approach with minimally invasive strategies compared to the straight transection plane of major hepatectomies. In this perspective, the robot adds several benefits thanks to instrument flexibility and 3-D visualization, which help reach deep and narrow spaces such as posterior segments, with safe control of vascular structures.

Similar considerations for laparoscopic surgery apply to robotic liver resection. The Pringle maneuver may be used according to surgeon preference. The central venous pressure should be less than 5 mmHg to reduce intraoperative blood loss [31]. Transection should be performed layer by layer to avoid going into a deep hole within the liver parenchyma with limited visualization and access. Minor vessels can be secured with bipolar cautery or ultrasonic harmonic shears. Large vessels and pedicles should be ligated, with options ranging from Hem-o-lok® clips, metal clips, silk ties, prolene suture ligation to vascular staplers. After transection, the raw area of the liver should be inspected for bleeding and bile leak. We routinely place the specimen in an endoscopic plastic bag and remove it via a Pfannenstiel incision.

13.4 Literature Overview

To date, well over 150 reports on robotic liver resection have been published [32]. A recent comprehensive systematic review by Ciria et al. showed that the number of reported cases worldwide has clearly taken off over the past decade, with nearly 3000 cases in the literature, ranging from minor wedge resection to complex liver surgery. More than half (53.4%) of these reported robotic resections are for HCC [32]. It is beyond the aims of this chapter to discuss every review of the literature on this topic, and several systematic reviews have been published for this purpose [32–39]. The following discussion will therefore focus mainly on perioperative safety and oncological safety of robotic liver resection for HCC.

13.4.1 Perioperative Safety

Several studies have shown that robotic surgery is as safe as open and laparoscopic approaches for HCC in terms of operative complications and postoperative morbidity. For example, in an Eastern series of 183 patients with HCC treated with robotic liver resection compared to 275 open surgery controls, the robotic group showed longer operation times (343 vs. 220 min), shorter hospital stays (7.5 vs. 10.1 days), and lower dosages of postoperative patient-controlled analgesia (350 vs. 554 ng/kg) [40]. There was no statistically significant difference in terms of postoperative complication rates. Similar findings have been reported in studies comparing laparoscopic to robotic resections for HCC, with no increase in perioperative complication rates [41, 42]. Our recent Western multicenter series comparing robotic cases from high-volume HPB centers versus open cases from a center not performing robotic surgery not only confirmed these results but also showed that robotic approach reduced the incidence of post-hepatectomy liver failure (PHLF) and the length of stay in the intensive care unit [43]. The use of the robot proved to be safe and effective also in the field of advanced liver procedures such as associated liver partition and portal vein ligation for staged hepatectomy (ALPPS) for HCC [44, 45].

13.4.2 Oncological Safety

There has been a number of retrospective cohorts reporting on survival outcomes after robotic liver resection for HCC [17, 40, 46–53]. Choi et al., in their series, reported 13 HCC patients all with negative margins and a 92.3% rate of anatomic resection, without recurrences during the 11-month median follow-up [46]. Overall survival (OS) and disease-free survival (DFS) at 3 years after robotic liver resection for HCC are reported to range between 97.7–92.6% and 72.2–71.9%, respectively [54]. Magistri et al. performed a comprehensive review of the literature across 10 retrospective cohorts on robotic resection of HCC [34]. They found the reported mean margin to range from 5.85 mm to 21 mm. The reported DFS ranged from 77.8% to 91.5% at 1 year and 65.5% to 84.3% at 3 years. The OS was 100% at 1 year for all studies that reported it and ranged from 86% to 97.7% at 3 years. Similar survival results have been reported when comparing open, laparoscopic and robotic resections for HCC [42, 55]. When talking about oncological radicality, a positive oncological effect of the minimally invasive approach might be foreseen, particularly in patients at an earlier stage of disease. Moreover, by reducing the risk of postoperative liver decompensation, more patients may benefit from a radical surgical resection. However, in our multicenter study comparing open and robotic approach, a subgroup analysis of patients within Milan criteria did not show superiority of the robotic approach over the standard open. Additional prospective studies on this issue are warranted.

13.5 Role of Robotics in Transplant Oncology

Given the abovementioned characteristics of the technology, the robot is a valuable tool for bridging and down-staging strategies [56–58] and an alternative to locoregional treatments like percutaneous ablation (PA) [59, 60]. The cumulative incidence of recurrence was found to be decreased in the robotic group versus the PA group in a cohort of very early and early newly diagnosed HCCs [61]. The underlying reason for this different recurrence pattern can be found in the 20% incidence of satellitosis, which is more efficiently treated with a surgical approach. Besides, the robotic approach may ultimately reduce the formation of abdominal adhesions for future liver transplantation [58].

The robotic approach may also be used in the transplant setting for living-donor liver procurement, which today probably represents one of the highest technical challenge of robotic liver surgery. In a study from Korea reporting on 52 cases, the authors applied selection criteria at the beginning of the series including less complex cases before approaching advanced vascular variants or larger grafts [62]. The results were outstanding, with low blood loss (109.8 ± 101.5 mL), acceptable operative time (493.6 ± 91.5 min), and short in-hospital stay (9 ± 2.1 days), with a statistically significant advantage over the open approach. The incidence of postoperative morbidity was 23.1%, with only one case of a complication >3a, according to Clavien-Dindo. Also, the results from the King Faisal Hospital in Riyadh show excellent results for both donors and recipients [63]. Moreover, the first series of full robotic approach for whole graft liver transplantation was recently reported, with HCC being the leading indication so far [64].

It is clear now that robotic and laparoscopic resections for HCC are as safe as the standard open approach when performed in expert and high-volume centers, where surgeons completed the learning curve of both open and minimally invasive procedures. The community of HPB surgeons should contribute to a collective effort to offer patients the best treatment, improving surgical education programs and referring patients to high-volume centers. Given the equipoise of the technical approaches, it remains the issues of indication, timing, and benefit between liver resection and liver transplantation under the light of modern transplant oncology conception [65]. A tumor biology-guided approach should be pursued including a multidimensional evaluation of each case, balancing all the available strategies, including advanced downstaging with combined surgery, radiology and checkpoint inhibitors agents, and considering living donation as a valuable tool to match the perfect timing for liver transplantation.

13.5.1 Limitations of Robotic Liver Resections

Minimally invasive liver surgery is still considered an advanced resection technique. Specialized training and proctorship are required to ensure patient safety. While the learning curve for some robotic procedures has been shown to be shorter than their laparoscopic counterparts [19], training has to be provided for not just the console surgeon, but also the assistant surgeon and OR nurses. The robotic dual console was designed to accelerate the learning of the surgical team [66]. Specifically for liver surgery, the array of instruments available for parenchyma transection remains limited. Although the DaVinci Harmonic ACE is efficient in cutting and coagulating, currently it still lacks the EndoWrist technology that other robotic instruments provide, hence limiting the degrees of movement available for increased precision.

13.6 Conclusions

The robotic approach to the liver is a safe, effective, and reproducible strategy for the treatment of HCC, from nonanatomical resections to more advanced procedures, including ALPPS and living-donor hepatectomy. Perioperative and long-term outcomes are at least comparable to the traditional open and laparoscopic approaches. It is an additional tool for surgeons that may aid in expanding the indications for surgical resections for HCC.

References

1. Ciria R, Cherqui D, Geller DA, Briceno J, Wakabayashi G. Comparative short-term benefits of laparoscopic liver resection: 9000 cases and climbing. Ann Surg. 2016;263(4):761–77. https://doi.org/10.1097/SLA.0000000000001413.
2. Bhojani FD, Fox A, Pitzul K, et al. Clinical and economic comparison of laparoscopic to open liver resections using a 2-to-1 matched pair analysis: an institutional experience. J Am Coll Surg. 2012;214(2):184–95. https://doi.org/10.1016/j.jamcollsurg.2011.10.020.
3. Jackson NR, Hauch A, Hu T, Buell JF, Slakey DP, Kandil E. The safety and efficacy of approaches to liver resection: a meta-analysis. JSLS. 2015;19(1):e2014.00186. https://doi.org/10.4293/JSLS.2014.00186.
4. Gavriilidis P, Roberts KJ, Aldrighetti L, Sutcliffe RP. A comparison between robotic, laparoscopic and open hepatectomy: a systematic review and network meta-analysis. Eur J Surg Oncol. 2020;46(7):1214–24. https://doi.org/10.1016/j.ejso.2020.03.227.
5. Nguyen KT, Marsh JW, Tsung A, Steel JJL, Gamblin TC, Geller DA. Comparative benefits of laparoscopic vs open hepatic resection: a critical appraisal. Arch Surg. 2011;146(3):348–56. https://doi.org/10.1001/archsurg.2010.248.
6. Nguyen KT, Gamblin TC, Geller DA. World review of laparoscopic liver resection-2,804 patients. Ann Surg. 2009;250(5):831–41. https://doi.org/10.1097/SLA.0b013e3181b0c4df.
7. Soubrane O, Goumard C, Laurent A, et al. Laparoscopic resection of hepatocellular carcinoma: a French survey in 351 patients. HPB (Oxford). 2014;16(4):357–65. https://doi.org/10.1111/hpb.12142.
8. Buell JF, Cherqui D, Geller DA, et al. The international position on laparoscopic liver surgery: the Louisville Statement, 2008. Ann Surg. 2009;250(5):825–30. https://doi.org/10.1097/sla.0b013e3181b3b2d8.
9. Memeo R, de'Angelis N, Compagnon P, et al. Laparoscopic vs. open liver resection for hepatocellular carcinoma of cirrhotic liver: a case-control study. World J Surg. 2014;38(11):2919–26. https://doi.org/10.1007/s00268-014-2659-z.
10. Dagher I, Belli G, Fantini C, et al. Laparoscopic hepatectomy for hepatocellular carcinoma: a European experience. J Am Coll Surg. 2010;211(1):16–23. https://doi.org/10.1016/j.jamcollsurg.2010.03.012.
11. Vibert E, Perniceni T, Levard H, Denet C, Shahri NK, Gayet B. Laparoscopic liver resection. Br J Surg. 2006;93(1):67–72. https://doi.org/10.1002/bjs.5150.
12. Gobardhan PD, Subar D, Gayet B. Laparoscopic liver surgery: an overview of the literature and experiences of a single centre. Best Pract Res Clin Gastroenterol. 2014;28(1):111–21. https://doi.org/10.1016/j.bpg.2013.11.012.
13. Wakabayashi G, Cherqui D, Geller DA, et al. Recommendations for laparoscopic liver resection: a report from the second international consensus conference held in Morioka. Ann Surg. 2015;261(4):619–29. https://doi.org/10.1097/SLA.0000000000001184.
14. Abu Hilal M, Aldrighetti L, Dagher I, et al. The Southampton consensus guidelines for laparoscopic liver surgery: from indication to implementation. Ann Surg. 2018;268(1):11–8. https://doi.org/10.1097/SLA.0000000000002524.
15. Giulianotti PC, Bianco FM, Daskalaki D, Gonzalez-Ciccarelli LF, Kim J, Benedetti E. Robotic liver surgery: technical aspects and review of the literature. Hepatobiliary Surg Nutr. 2016;5(4):311–21. https://doi.org/10.21037/hbsn.2015.10.05.
16. Milone L, Daskalaki D, Fernandes E, Damoli I, Giulianotti PC. State of the art in robotic hepatobiliary surgery. World J Surg. 2013;37(12):2747–55. https://doi.org/10.1007/s00268-013-2276-2.
17. Giulianotti PC, Coratti A, Sbrana F, et al. Robotic liver surgery: results for 70 resections. Surgery. 2011;149(1):29–39. https://doi.org/10.1016/j.surg.2010.04.002.
18. Cipriani F, Fiorentini G, Magistri P, et al. Pure laparoscopic versus robotic liver resections: multicentric propensity score-based analysis with stratification according to difficulty scores. J Hepatobiliary Pancreat Sci. 2022;29(10):1108–23. https://doi.org/10.1002/jhbp.1022.

19. Ocuin LM, Tsung A. Robotic liver resection for malignancy: current status, oncologic outcomes, comparison to laparoscopy, and future applications. J Surg Oncol. 2015;112(3):295–301. https://doi.org/10.1002/jso.23901.
20. Boggi U, Caniglia F, Amorese G. Laparoscopic robot-assisted major hepatectomy. J Hepatobiliary Pancreat Sci. 2014;21(1):3–10. https://doi.org/10.1002/jhbp.34.
21. Magistri P, Assirati G, Ballarin R, Di Sandro S, Di Benedetto F. Major robotic hepatectomies: technical considerations. Updat Surg. 2021;73(3):989–97. https://doi.org/10.1007/s13304-020-00940-1.
22. European Association for the Study of the Liver. Electronic address: easloffice@easloffice.eu, European Association for the Study of the Liver. EASL Clinical Practice Guidelines: management of hepatocellular carcinoma. J Hepatol. 2018;69(1):182–236. https://doi.org/10.1016/j.jhep.2018.03.019.
23. Elsayes KM, Hooker JC, Agrons MM, et al. 2017 version of LI-RADS for CT and MR imaging: an update. Radiographics. 2017;37(7):1994–2017. https://doi.org/10.1148/rg.2017170098.
24. Ruzzenente A, Bagante F, Ratti F, et al. Minimally invasive versus open liver resection for hepatocellular carcinoma in the setting of portal vein hypertension: results of an international multi-institutional analysis. Ann Surg Oncol. 2020;27(9):3360–71. https://doi.org/10.1245/s10434-020-08444-3.
25. Machado MA, Surjan RC, Basseres T, Makdissi F. Robotic repeat right hepatectomy for recurrent colorectal liver metastasis. Ann Surg Oncol. 2019;26(1):292–5. https://doi.org/10.1245/s10434-018-6996-2.
26. Levi Sandri GB, Colasanti M, Aldrighetti L, et al. Correction to: is minimally invasive liver surgery a reasonable option in recurrent HCC? A snapshot from the I Go MILS registry. Updat Surg. 2022;74(2):799. https://doi.org/10.1007/s13304-021-01200-6.
27. Franz M, Arend J, Wolff S, et al. Tumor visualization and fluorescence angiography with indocyanine green (ICG) in laparoscopic and robotic hepatobiliary surgery—valuation of early adopters from Germany. Innov Surg Sci. 2021;6(2):59–66. https://doi.org/10.1515/iss-2020-0019.
28. Gonzalez-Ciccarelli LF, Quadri P, Daskalaki D, Milone L, Gangemi A, Giulianotti PC. Robotic approach to hepatobiliary surgery. Chirurg. 2017;88(Suppl 1):19–28. https://doi.org/10.1007/s00104-016-0223-0.
29. Kinoshita M, Kawaguchi T, Tanaka S, et al. Application of indocyanine green fluorescence imaging for tumor localization during robot-assisted hepatectomy. Cancers (Basel). 2023;15(17):4205. https://doi.org/10.3390/cancers15174205.
30. Marino MV, Di Saverio S, Podda M, Gomez Ruiz M, Gomez FM. The application of indocyanine green fluorescence imaging during robotic liver resection: a case-matched study. World J Surg. 2019;43(10):2595–606. https://doi.org/10.1007/s00268-019-05055-2.
31. Li Z, Sun YM, Wu FX, Yang LQ, Lu ZJ, Yu WF. Controlled low central venous pressure reduces blood loss and transfusion requirements in hepatectomy. World J Gastroenterol. 2014;20(1):303–9. https://doi.org/10.3748/wjg.v20.i1.303.
32. Ciria R, Berardi G, Alconchel F, et al. The impact of robotics in liver surgery: a worldwide systematic review and short-term outcomes meta-analysis on 2,728 cases. J Hepatobiliary Pancreat Sci. 2022;29(2):181–97. https://doi.org/10.1002/jhbp.869.
33. Hu Y, Guo K, Xu J, et al. Robotic versus laparoscopic hepatectomy for malignancy: a systematic review and meta-analysis. Asian J Surg. 2021;44(4):615–28. https://doi.org/10.1016/j.asjsur.2020.12.016.
34. Magistri P, Tarantino G, Assirati G, et al. Robotic liver resection for hepatocellular carcinoma: a systematic review. The International Journal of Medical Robotics and Computer Assisted Surgery. 2019;15(4):e2004. https://doi.org/10.1002/rcs.2004.
35. Aboudou T, Li M, Zhang Z, et al. Laparoscopic versus robotic hepatectomy: a systematic review and meta-analysis. J Clin Med. 2022;11(19):5831. https://doi.org/10.3390/jcm11195831.
36. Guan R, Chen Y, Yang K, et al. Clinical efficacy of robot-assisted versus laparoscopic liver resection: a meta analysis. Asian J Surg. 2019;42(1):19–31. https://doi.org/10.1016/j.asjsur.2018.05.008.
37. Zhang L, Yuan Q, Xu Y, Wang W. Comparative clinical outcomes of robot-assisted liver resection versus laparoscopic liver resection: a meta-analysis. PLoS One. 2020;15(10):e0240593. https://doi.org/10.1371/journal.pone.0240593.
38. Wang JM, Li JF, Yuan GD, He SQ. Robot-assisted versus laparoscopic minor hepatectomy: a systematic review and meta-analysis. Medicine (Baltimore). 2021;100(17):e25648. https://doi.org/10.1097/MD.0000000000025648.
39. Ziogas IA, Giannis D, Esagian SM, Economopoulos KP, Tohme S, Geller DA. Laparoscopic versus robotic major hepatectomy: a systematic review and meta-analysis. Surg Endosc. 2021;35(2):524–35. https://doi.org/10.1007/s00464-020-08008-2.
40. Chen PD, Wu CY, Hu RH, et al. Robotic versus open hepatectomy for hepatocellular carcinoma: a matched comparison. Ann Surg Oncol. 2017;24(4):1021–8. https://doi.org/10.1245/s10434-016-5638-9.
41. O'Connell RM, Bucheeri M, Quidwai O, Bourke M, Gallagher T, Hoti E. Robotic, laparoscopic, and open liver resection for hepatocellular carcinoma: a propensity score matched analysis of perioperative outcomes. Surgery in Practice and Science. 2023;14:100196. https://doi.org/10.1016/j.sipas.2023.100196.
42. Duong LM, Cai H, Shrubsole MJ, Bailey C, Idrees K, Shu XO. Outcomes of robotic-assisted liver surgery versus laparoscopic liver surgery for treatment of stage I hepatocellular carcinoma. Cancer. 2022;128(4):762. https://doi.org/10.1002/cncr.33979.

43. Di Benedetto F, Magistri P, Di Sandro S, et al. Safety and efficacy of robotic vs open liver resection for hepatocellular carcinoma. JAMA Surg. 2023;158(1):46–54. https://doi.org/10.1001/jamasurg.2022.5697.
44. Di Benedetto F, Assirati G, Magistri P. Full robotic ALPPS for HCC with intrahepatic portal vein thrombosis. Int J Med Robot. 2020;16(2):e2087. https://doi.org/10.1002/rcs.2087.
45. de Fernandes ESM, de Barros F, Magistri P, et al. Total robotic ALPPS approach for hepatocellular carcinoma in cirrhotic liver. Int J Med Robot. 2021;17(3):e2238. https://doi.org/10.1002/rcs.2238.
46. Choi GH, Choi SH, Kim SH, et al. Robotic liver resection: technique and results of 30 consecutive procedures. Surg Endosc. 2012;26(8):2247–58. https://doi.org/10.1007/s00464-012-2168-9.
47. Lai ECH, Yang GPC, Tang CN. Robot-assisted laparoscopic liver resection for hepatocellular carcinoma: short-term outcome. Am J Surg. 2013;205(6):697–702. https://doi.org/10.1016/j.amjsurg.2012.08.015.
48. Wu YM, Hu RH, Lai HS, Lee PH. Robotic-assisted minimally invasive liver resection. Asian J Surg. 2014;37(2):53–7. https://doi.org/10.1016/j.asjsur.2014.01.015.
49. Sandro S, Lauterio A, Giacomoni A, et al. Totally robotic liver resection for hepatocellular carcinoma in cirrhotic patients: safety and feasibility. J Robot Surg. 2014;8:357–64. https://doi.org/10.1007/s11701-014-0479-7.
50. Morel P, Jung M, Cornateanu S, et al. Robotic versus open liver resections: a case-matched comparison. The International Journal of Medical Robotics and Computer Assisted Surgery. 2017;13(3):e1800. https://doi.org/10.1002/rcs.1800.
51. Marino MV, Gulotta G, Komorowski AL. Fully robotic left hepatectomy for malignant tumor: technique and initial results. Updat Surg. 2019;71(1):129–35. https://doi.org/10.1007/s13304-018-0560-2.
52. Wang WH, Kuo KK, Wang SN, Lee KT. Oncological and surgical result of hepatoma after robot surgery. Surg Endosc. 2018;32(9):3918–24. https://doi.org/10.1007/s00464-018-6131-2.
53. Magistri P, Tarantino G, Guidetti C, et al. Laparoscopic versus robotic surgery for hepatocellular carcinoma: the first 46 consecutive cases. J Surg Res. 2017;217:92–9. https://doi.org/10.1016/j.jss.2017.05.005.
54. Di Benedetto F, Petrowsky H, Magistri P, Halazun KJ. Robotic liver resection: hurdles and beyond. Int J Surg. 2020;82S:155–62. https://doi.org/10.1016/j.ijsu.2020.05.070.
55. Lai ECH, Tang CN. Long-term survival analysis of robotic versus conventional laparoscopic hepatectomy for hepatocellular carcinoma: a comparative study. Surg Laparosc Endosc Percutan Tech. 2016;26(2):162–6. https://doi.org/10.1097/SLE.0000000000000254.
56. Vitali GC, Laurent A, Terraz S, et al. Minimally invasive surgery versus percutaneous radio frequency ablation for the treatment of single small (≤3 cm) hepatocellular carcinoma: a case-control study. Surg Endosc. 2016;30(6):2301–7. https://doi.org/10.1007/s00464-015-4295-6.
57. Magistri P, Tarantino G, Ballarin R, Coratti A, Di Benedetto F. Robotic liver surgery is the optimal approach as bridge to transplantation. World J Hepatol. 2017;9(4):224–6. https://doi.org/10.4254/wjh.v9.i4.224.
58. Magistri P, Olivieri T, Assirati G, et al. Robotic liver resection expands the opportunities of bridging before liver transplantation. Liver Transpl. 2019;25(7):1110–2. https://doi.org/10.1002/lt.25477.
59. Inchingolo R, Posa A, Mariappan M, Spiliopoulos S. Locoregional treatments for hepatocellular carcinoma: current evidence and future directions. World J Gastroenterol. 2019;25(32):4614–28. https://doi.org/10.3748/wjg.v25.i32.4614.
60. Conticchio M, Inchingolo R, Delvecchio A, et al. Radiofrequency ablation vs surgical resection in elderly patients with hepatocellular carcinoma in Milan criteria. World J Gastroenterol. 2021;27(18):2205–18. https://doi.org/10.3748/wjg.v27.i18.2205.
61. Magistri P, Catellani B, Frassoni S, et al. Robotic liver resection versus percutaneous ablation for early HCC: short- and long-term results. Cancers (Basel). 2020;12(12):3578. https://doi.org/10.3390/cancers12123578.
62. Rho SY, Lee JG, Joo DJ, et al. Outcomes of robotic living donor right hepatectomy from 52 consecutive cases: comparison with open and laparoscopy-assisted donor hepatectomy. Ann Surg. 2022;275(2):e433–42. https://doi.org/10.1097/SLA.0000000000004067.
63. Broering D, Sturdevant ML, Zidan A. Robotic donor hepatectomy: a major breakthrough in living donor liver transplantation. Am J Transplant. 2022;22(1):14–23. https://doi.org/10.1111/ajt.16889.
64. Pinto-Marques H, Sobral M, Magistri P, Gomes da Silva S, Guerrini GP, Mega R, Guidetti C, Coelho JS, Di Sandro S, Di Benedetto F. Full Robotic Whole Graft Liver Transplantation: A Step Into The Future. Ann Surg. 2024. https://doi.org/10.1097/SLA.0000000000006420. Epub ahead of print. PMID: 38953528.
65. Di Sandro S, Sposito C, Ravaioli M, et al. Surgical treatment of hepatocellular carcinoma: multicenter competing-risk analysis of tumor-related death following liver resection and transplantation under an intention-to-treat perspective. Transplantation. 2023;107(9):1965–75. https://doi.org/10.1097/TP.0000000000004593.
66. Fernandes E, Elli E, Giulianotti P. The role of the dual console in robotic surgical training. Surgery. 2014;155(1):1–4. https://doi.org/10.1016/j.surg.2013.06.023.

14 Robotic Liver Surgery for Colorectal Metastases

Marcel Autran Machado
and Fabio Ferrari Makdissi

14.1 Introduction

Liver resection is the standard treatment for colorectal liver metastases (CRLM). Metastases of colorectal origin, unlike other primary and metastatic liver diseases, have several features that distinguish them from other liver resections. First, most patients undergo neoadjuvant chemotherapy, which affects the quality of the liver because we almost always approach the liver with some degree of hepatocellular underlying diseases, whether steatosis, steatohepatitis, or sinusoidal obstruction syndrome (blue liver). The most common chemotherapy-associated liver injuries are steatohepatitis and sinusoidal obstruction syndrome [1–3]. Second, in many cases, the metastases are multiple and can affect both sides of the liver, which affects the resection strategy [4]. Therefore, resection is often not standardized and the technique to be used can only be decided intraoperatively. The type of resection can vary and may include trisectionectomies, hemihepatectomies, segmentectomies, nonanatomic resections, and sometimes a combination of one or more techniques [5–7]. Third, recurrence is a common event and repeat hepatectomy is very common. Therefore, many of the liver resections for metastases of colorectal origin are rehepatectomies [8–10]. Therefore, many adhesions can be expected during the procedure. Even in the first hepatectomy, we can find adhesions from the resection of the primary colorectal tumor. Fourth, many of the metastases are synchronous, and in certain circumstances, resection of the primary tumor may be performed before, during, or after liver resection (liver first) [11]. Unlike hepatocellular carcinoma, portal vein thrombus is rare but can occur, and the robotic platform is extremely useful to treat this complication [12]. For these reasons, the management of colorectal metastases has its own nuances that may affect the form and planning of surgical treatment. In this chapter, we discuss the various aspects of robotic liver surgery in patients with colorectal metastases.

14.2 Chemotherapy-Associated Liver Injury (CALI)

Patients with colorectal liver metastases often receive chemotherapy before liver resection. However, chemotherapy causes liver damage, which may increase operative risk and decrease tolerance to further chemotherapy [1–3]. Some authors analyzed the reversibility of liver damage after the interruption of chemotherapy [13]. Liver

M. A. Machado (✉)
Sirio Libanes Hospital, São Paulo, Brazil

Department of Hepatobiliary Surgery, Nove de Julho Hospital, São Paulo, Brazil
e-mail: dr@drmarcel.com.br

F. F. Makdissi
Sirio Libanes Hospital, São Paulo, Brazil

M. D'Hondt, I. Sucandy (eds.), *Textbook of Robotic Liver Surgery*,
https://doi.org/10.1007/978-3-031-76536-0_14

damage persists for a long time after chemotherapy. Sinusoidal dilatation and nodular regenerative hyperplasia (Fig. 14.1) do not regress until 9 months after the end of chemotherapy, while steatosis and steatohepatitis persist even after this long interval [13].

The most common chemotherapy-associated liver injuries are steatohepatitis and sinusoidal obstruction syndrome (Fig. 14.1). CALI appears to be therapy-specific and generally includes two main types of liver damage, vascular changes, and fatty changes, primarily associated with the development of sinusoidal obstruction syndrome, which leads to cellular damage and activates apoptosis pathways. The prevalence of CALI increases with the duration of chemotherapy, and currently there are no convincing data on the reversibility of CALI [13].

Therefore, preoperative assessment of liver quality is of paramount importance as it may influence the outcome of liver resection [1, 2]. Postoperative liver failure may be more frequent, so a larger volume may be required as future liver remnant volume. Intraoperative blood loss is also affected by a damaged liver, especially in patients with sinusoidal obstruction syndrome. Liver rupture and hemorrhage may occur more easily in patients with massive steatosis than in normal livers (Fig. 14.1). Liver mobilization may be difficult in patients with enlarged, friable, or hard livers. However, a recent multicenter study has shown that neoadjuvant chemotherapy had mini-

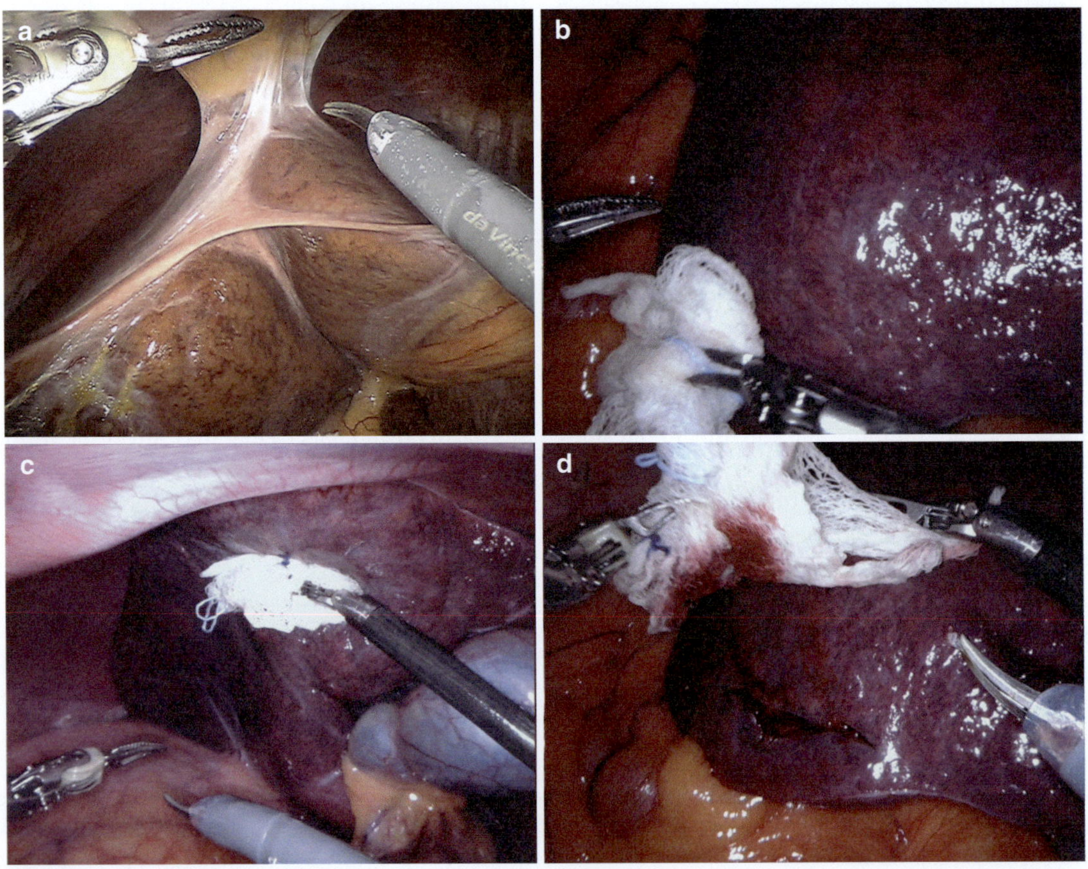

Fig. 14.1 Chemotherapy-associated liver injury. (**a**) Intraoperative photograph showing a yellow liver with steatohepatitis. (**b**) Intraoperative photograph showing a blue liver with sinusoidal obstruction syndrome. (**c**) Intraoperative photograph showing a blue liver with sinusoidal dilatation and nodular regenerative hyperplasia. (**d**) Intraoperative image showing liver rupture during mobilization

mal impact on the difficulty and outcomes of minimally invasive major liver resection CRLM. The main reason is that this study included high-volume centers that have learned how to deal with these patients.

14.3 Multiple and/or Bilateral Liver Metastasis

With the improvement in chemotherapeutic agents, more and more patients are undergoing surgical treatment of colorectal metastases. Patients with multiple, bilateral metastases (Figs. 14.2 and 14.3), who until recently were considered unsuitable for surgical treatment, are now undergoing surgery [4]. However, this has complicated the performance of hepatectomies as many of these patients have lesions with significant response and are sometimes difficult to find. For this reason, the use of intraoperative ultrasound is essential. The liver surgeon must master the screening technique and use this tool intelligently to locate and plan the resection of liver metastases. The robotic platform with the integration of the ultrasound image into the console helps the surgeon to plan the hepatectomy (Fig. 14.4). Our technique for resection of liver metastasis is published elsewhere [5, 6]. Briefly, after demarcation of the lateral and deep margins with the aid of the intraoperative ultrasound (Fig. 14.4), we proceed with division of the liver

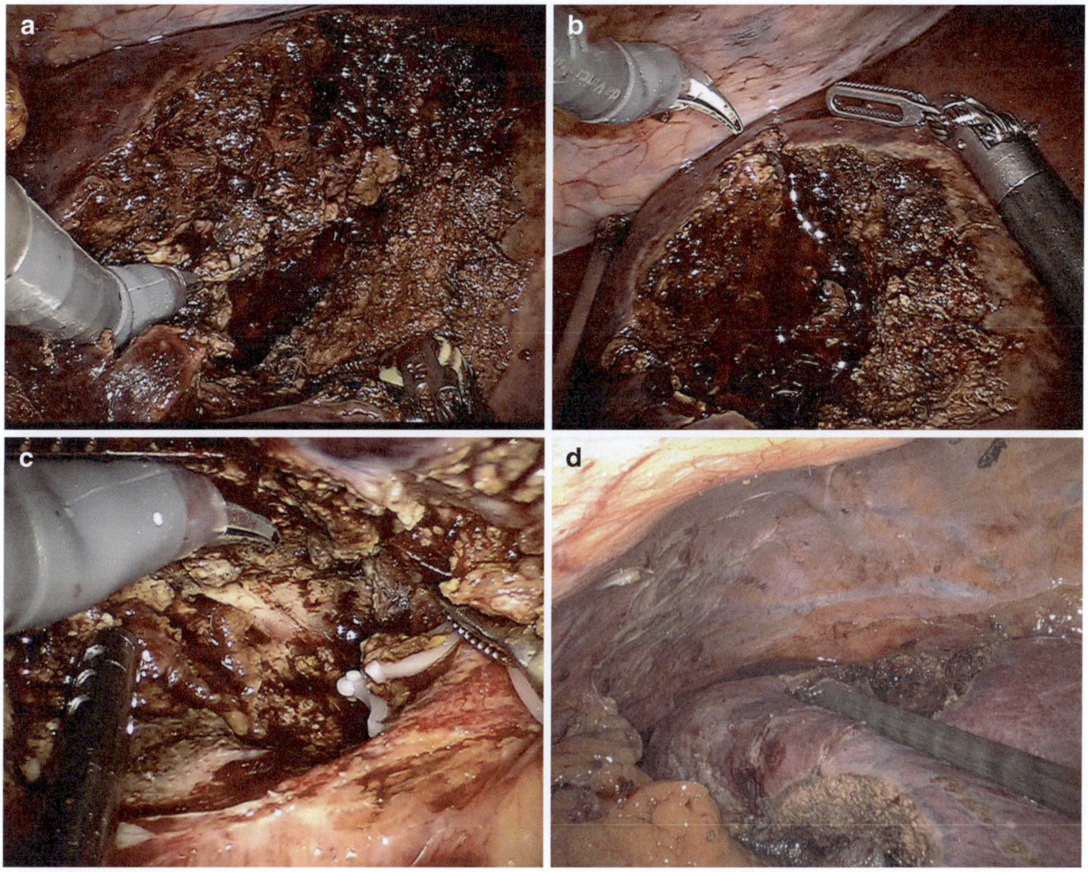

Fig. 14.2 Robotic resection of multiple and bilateral liver metastases. (**a**) Intraoperative photograph after resection of metastases in segment 3. (**b**) Same patient as (**a**). Intraoperative photograph after resection of metastases in segment 6. (**c**) Same patient as (**a**). Intraoperative photograph after resection of metastases in segment 4b near the hepatic hilum. (**d**) Intraoperative photograph after resection of two metastases in the right liver

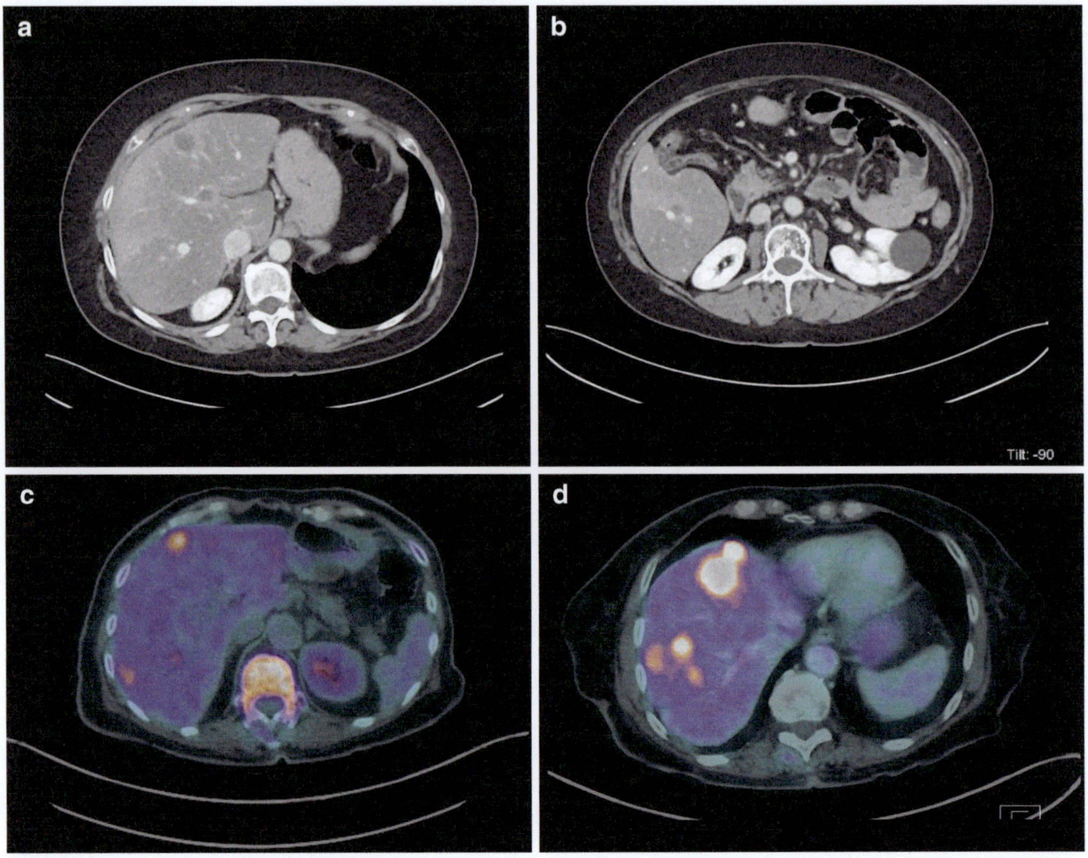

Fig. 14.3 Robotic resection of multiple and bilateral liver metastases. Preoperative imaging. (**a**) CT scan shows two metastases in segments 3 and 4. (**b**) The same patient as (**a**). CT scan shows a metastasis in segment 6. (**c**) PET-CT shows multiple and bilateral metastases in favorable location. (**d**) PET- CT shows multiple and bilateral metastases not suitable for MI approach

parenchyma using a combination of robotic Maryland bipolar forceps and robotic scissors with continuous saline irrigation (Fig. 14.4c) under intermittent Pringle maneuver (Fig. 14.5) using Huang loop [14].

Recently, a tumor burden score was developed to help understand and plan treatment in patients with multiple liver metastases. It appears to be an accurate tool to evaluate the impact of the number and size of metastases on long-term survival [15]. Prognosis is also different for bilateral and unilateral liver metastases when the same tumor burden score is used [16].

A multidisciplinary team of oncologists, radiologists, and liver surgeons is essential in this process. Whenever possible, resection of all metastases should be performed with maximal preservation of the liver parenchyma to reduce the risk of postoperative liver failure and allow for repeat hepatectomy if disease recurs. Preservation of adequate future liver remnant is essential in planning any major liver resection, especially in patients with bilateral colorectal liver metastases. Procedures such as portal vein embolization and hepatic venous deprivation for one- or two-stage hepatectomy, and the ALPPS procedure (see Chap. 26), have been developed to allow curative hepatectomy for colorectal liver metastases in patients with an initially inadequate future liver remnant [17, 18].

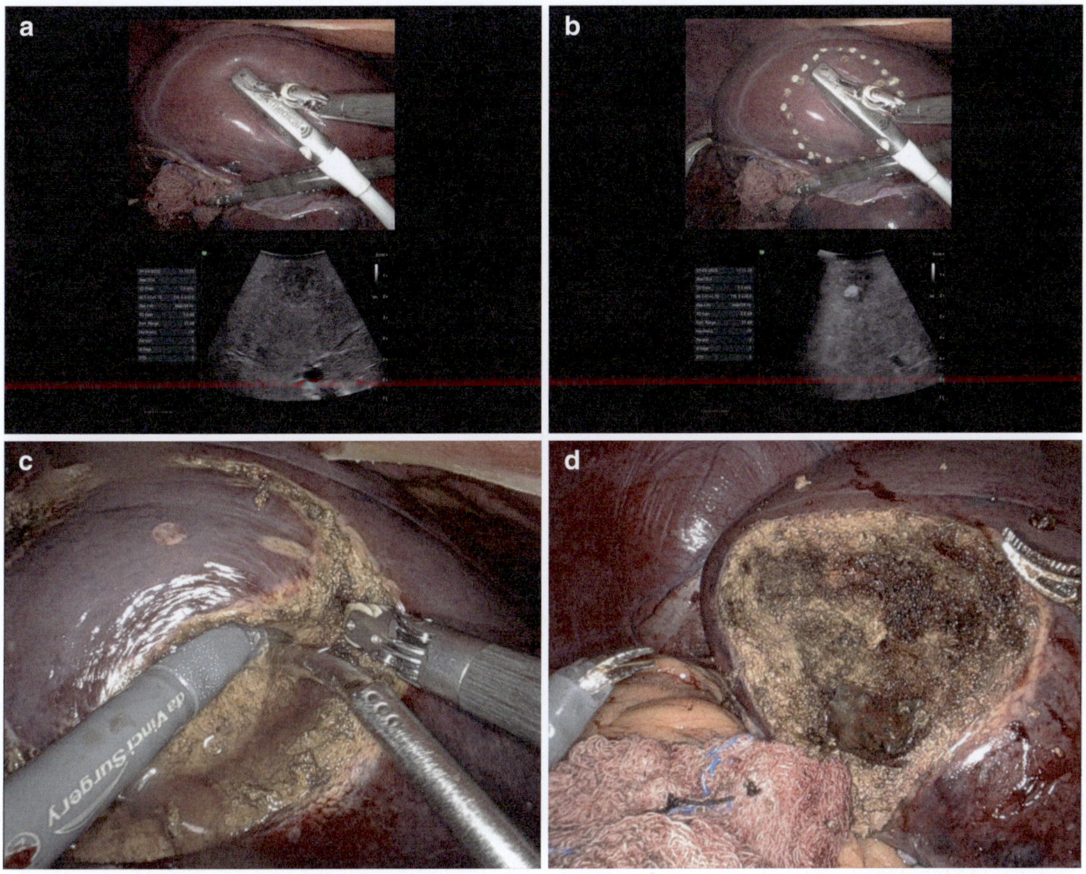

Fig. 14.4 Intraoperative ultrasound during robotic liver resection. (**a**) Intraoperative ultrasound is helpful in identifying a deep-seated metastasis. (**b**) Intraoperative ultrasound is used to delineate the lateral and deep margins. (**c**) Resection of the liver metastasis is performed using intermittent Pringle maneuver and bipolar forceps, scissors, and saline irrigation. (**d**) Intraoperative image shows the final aspect after resection of the metastasis

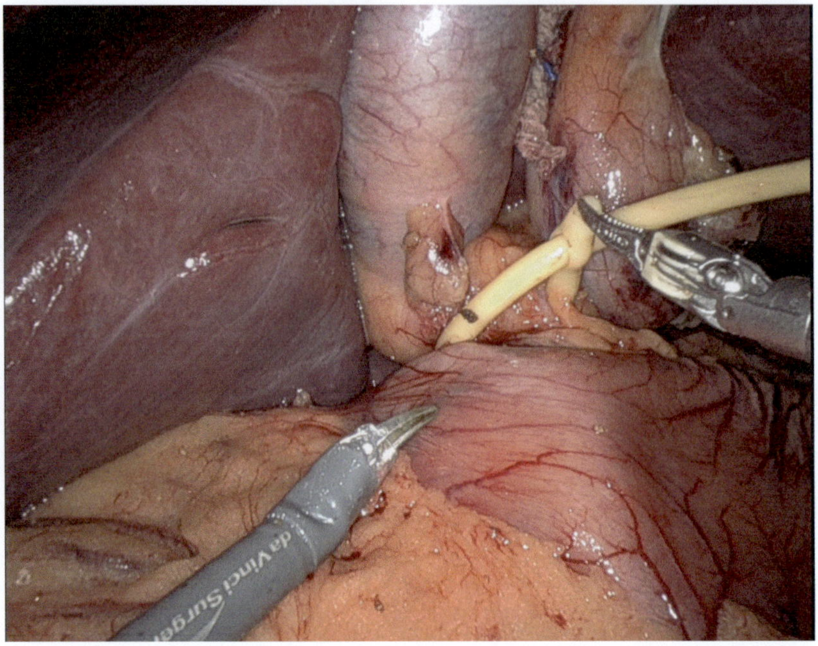

Fig. 14.5 Huang loop for intracorporeal Pringle maneuver [14]

14.4 Repeat Hepatectomy

Recurrence of colorectal liver metastases is a persistent problem, occurring in up to 70% of patients [19]. Repeat hepatectomy is a viable treatment and may offer favorable survival. Repeat hepatectomy is technically challenging, so minimally invasive repeat hepatectomy has been used in a limited number of patients. Adhesions from previous surgeries may increase operative time (Fig. 14.6), complications, and conversion, especially in patients with previous open surgeries [8]. Laparoscopic repeat liver resections can be performed safely, especially in patients with previous laparoscopic resections.

Open abdominal surgeries usually result in postoperative adhesions, which may prolong the operative time of subsequent surgeries because adhesiolysis is required (Fig. 14.7). In addition, these patients are at increased risk for bowel injury, complications, and the need for laparotomy. In modern oncological treatment of colorectal liver metastases, resectable recurrent metastatic disease is common. In this situation, repeat liver resection is indicated and has a positive effect on survival [19].

Open repeat hepatectomy is the standard of care, but more and more minimally invasive (MI) liver resections are being performed worldwide. Therefore, it is now common for a patient who has previously undergone MI surgery to require a repeat liver resection. In this situation, a repeat MI liver resection appears attractive. The first study reporting the benefits of the laparoscopic approach was published by Belli et al. [20] and focused on recurrent hepatocellular carcinoma in cirrhotic patients. Further, larger studies have shown that laparoscopic repeat liver resections can be performed safely, especially in patients with previous laparoscopic resections. However, the indication criteria for using a minimally invasive approach for repeat hepatectomy are not clear and are likely to be used only in selected cases. We no longer avoid the MI approach in patients who have previously undergone open liver resection. The decision to perform MI liver resection is the same regardless of whether prior surgery has been performed in the liver or upper abdomen. Adhesiolysis is greatly facilitated by the robotic approach and the use of wristed instruments [8–10]. This could reduce the risk of bowel injury and other complications. Liver

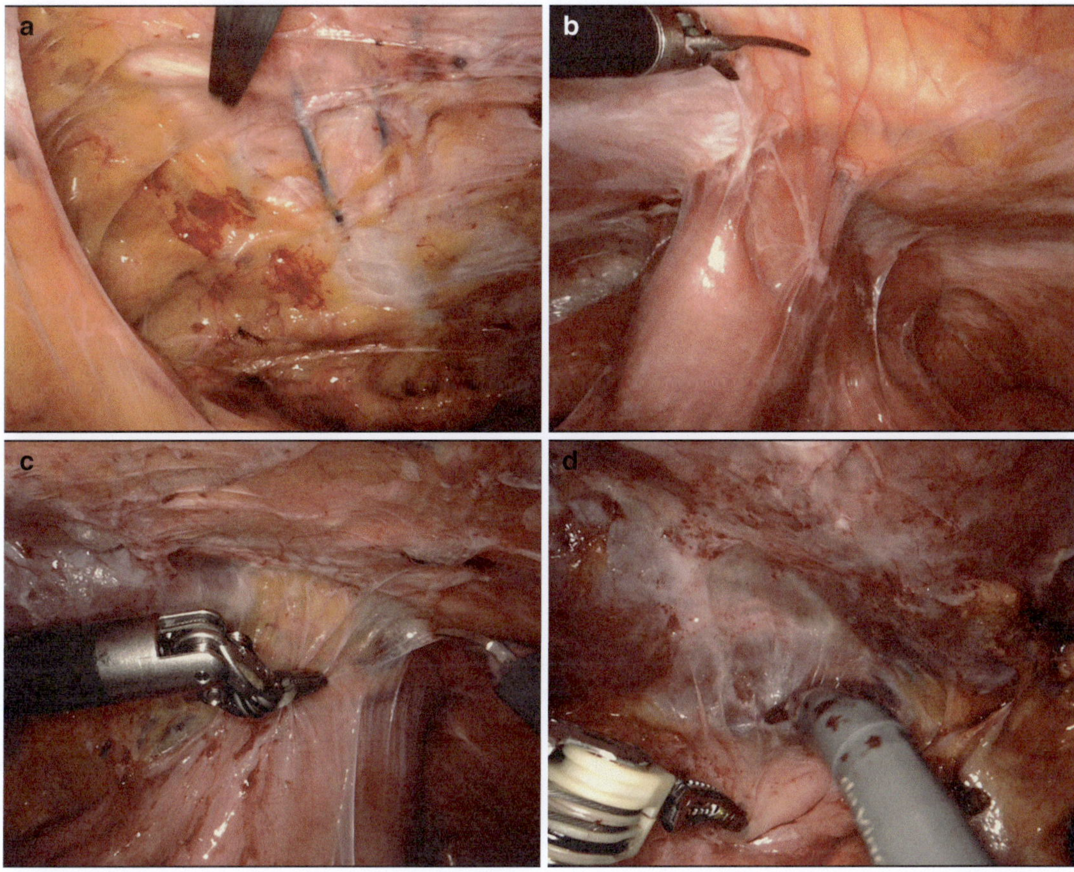

Fig. 14.6 Robotic repeat liver resection. (**a**) Intraoperative photograph shows the camera pointed at the abdominal wall and the suture from the previous open liver resection. (**b**) The small bowel is attached to the abdominal wall along the previous laparotomy. (**c**) The small bowel is carefully removed. (**d**) Intraoperative image shows sutures in the liver surface from the previous open hepatectomy

resection can be performed after adhesiolysis and mobilization of the affected liver (Fig. 14.8). However, hard and fibrotic adhesions may hinder visualization and dissection of the hepatic hilum, which may be necessary to perform liver resection. The liver capsule may bleed during mobilization and division of adhesions, increasing blood loss and blurring the surgical field.

We have used the laparoscopic approach for repeat hepatectomy since 2013 with good results and a low conversion rate (data not published). The use of a robotic system may improve certain steps of minimally invasive repeat liver resection. The magnified three-dimensional view allows better definition of the surgical field, which has been altered by previous operations, and thus better recognition of the structures of the liver hilum, allowing selective inflow control. In a limited working area, common in repeated liver resection, the use of wristed instruments allows precise dissection and suturing whenever needed.

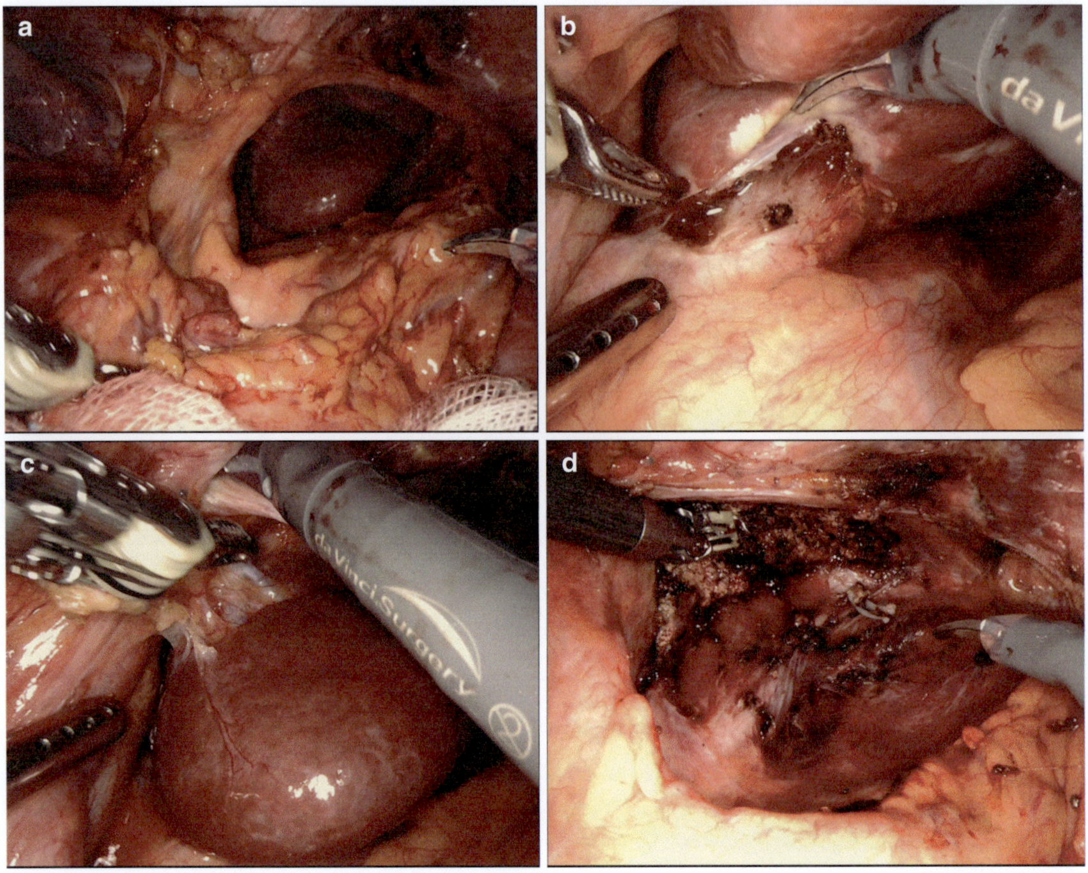

Fig. 14.7 Robotic repeat liver resection. Resection of S1 after open left hepatectomy. (**a**) Intraoperative photograph shows segment 1 after removal of adhesions from previous left hepatectomy. (**b**) Segment 1 is detached from the inferior vena cava. (**c**) The S1 Glissonian pedicle is identified and transected. (**d**) Intraoperative photograph after resection of segment 1

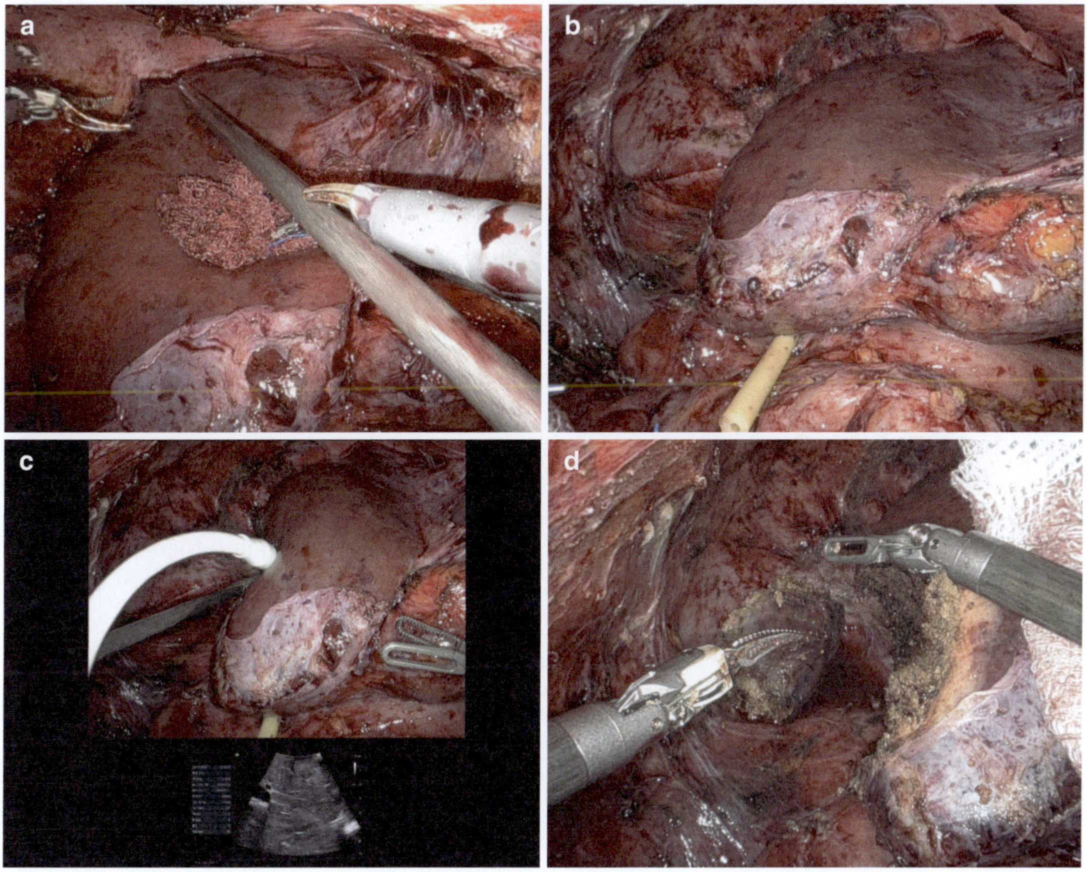

Fig. 14.8 Robotic repeat liver resection. Resection of metastases in the remnant liver (segment 4) after open right hepatectomy. (**a**) Intraoperative photograph showing mobilization of the liver. (**b**) Intraoperative photograph after liver mobilization and adhesiolysis. (**c**) Intraoperative ultrasound is used to identify and delineate the lateral and deep margins. (**d**) Intraoperative image after resection of recurrent liver metastasis

14.5 Synchronous Metastases

Synchronous colorectal liver metastases (CRLM) are present in 14–18% of patients at diagnosis [11]. Numerous studies have been published demonstrating the safety and feasibility of simultaneous resection [21]. One randomized controlled trial demonstrated that simultaneous resection of a primary colorectal tumor and CRLM resulted in similar morbidity rates and better overall survival compared with delayed resection [22].

Da Vinci Xi is the newer and most widely used platform worldwide. It can be used to perform multivisceral resections with a high success rate. The advantage of this system is most evident in the resection of pathologies in opposing quadrants, such as the combined resection of the colon and CLRM. The versatility and flexibility of the Xi system have allowed simplification of port configuration and easier multiquadrant access for multiple organ resection. In addition, the development of an operating table with integrated table movement has further revolutionized the role of robotic surgery. This allows patients to be repositioned without undocking the robots and instruments, making the simultaneous performance of complex organ resections more feasible and less time consuming.

Whenever indicated, we prefer to perform both colorectal and liver surgery simultaneously. In doing so, we use the newer Xi robotic platform, which allows for better planning when using opposing quadrants. Even for low rectal tumors, we can perform both surgeries simultaneously if a major hepatectomy is not programmed [7]. However, many patients with synchronous disease are referred after surgery for the primary tumor.

14.6 Worldwide Experience with Robotic Surgery for Colorectal Liver Metastases

Robotic liver resection is gaining popularity because it allows articulation of instruments, elimination of tremor, and improved three-dimensional visualization for fine dissections [23]. However, there are limited data on the outcomes of this method in colorectal liver metastases. A multicenter study compared robotic and laparoscopic liver resection and found that outcomes were generally similar between groups [24]. The median tumor size was 2.5 and 2.4 cm for robotic and laparoscopic resection, respectively; most patients had one or two tumors and similar resection types. There were no differences in postoperative complications, length of stay, ICU admission, or resection margin, and the two groups had equivalent 5-year overall survival rates. Another study compared minimally invasive techniques with open liver resection for CRLM [25] and found that robotic techniques were comparable to laparoscopy in terms of postoperative complication rate, R0 resection, and operative time. The results of this study suggest that the robotic technique may be associated with less blood loss than laparoscopy or the open approach, but the authors acknowledge that the number of robotic cases included was small and should be interpreted with caution.

Since May 2018, all of our minimally invasive liver surgeries have been performed using the robotic platform [23]. Our experience with this new system has grown during this period, and so have the indications for more complex cases [26, 27]. The robotic approach, with its additional degrees of freedom and robotic platform stability, can provide options for minimally invasive performance of complex liver resections that were a relative contraindication for laparoscopy [23, 28, 29]. Procedures that require excellent accuracy and dexterity are the best candidates for robotic surgery. Vascular structures such as the portal vein [12], hepatic artery, and hepatic veins are magnified by the robotic vision. The magnified vision camera can expose the anatomical structure of the hilum. The Da Vinci robot provides 20-fold magnified 3D vision, improving the precision of hilar dissection so that vascular sutures, vein reconstructions [12], and bile duct anastomoses can be performed at difficult angles with the nondominant hand when necessary. Excellent visualization is key to controlling intraoperative bleeding during mobilization and transection of the liver.

The proportion of robot-assisted liver resections is 56.8% of all our liver resections [23]. In our experience, laparoscopy was completely replaced by the robotic approach. The open approach was reserved for patients in whom there were contraindications to the minimally invasive technique [30]. In short, any patient with a straightforward liver resection can and should be operated on with a minimally invasive approach. There is no consensus on a contraindication to the minimally invasive approach. Patients who until recently had a contraindication to the minimally invasive approach are now being operated on with the robotic approach. Therefore, the indications for the use of the robotic technique are changing rapidly, and the proportion of minimally invasive liver resections will certainly increase as experience with this new technology increases.

According to Luberice et al. [26], this impact of robotic technology on outcome is independent of the degree of difficulty of liver resection. This has been noted by us since the beginning of our robotic program and has been applied to robotic hepatectomy by these authors with their analysis of the Iwate criteria for difficulty of laparoscopic liver resection.

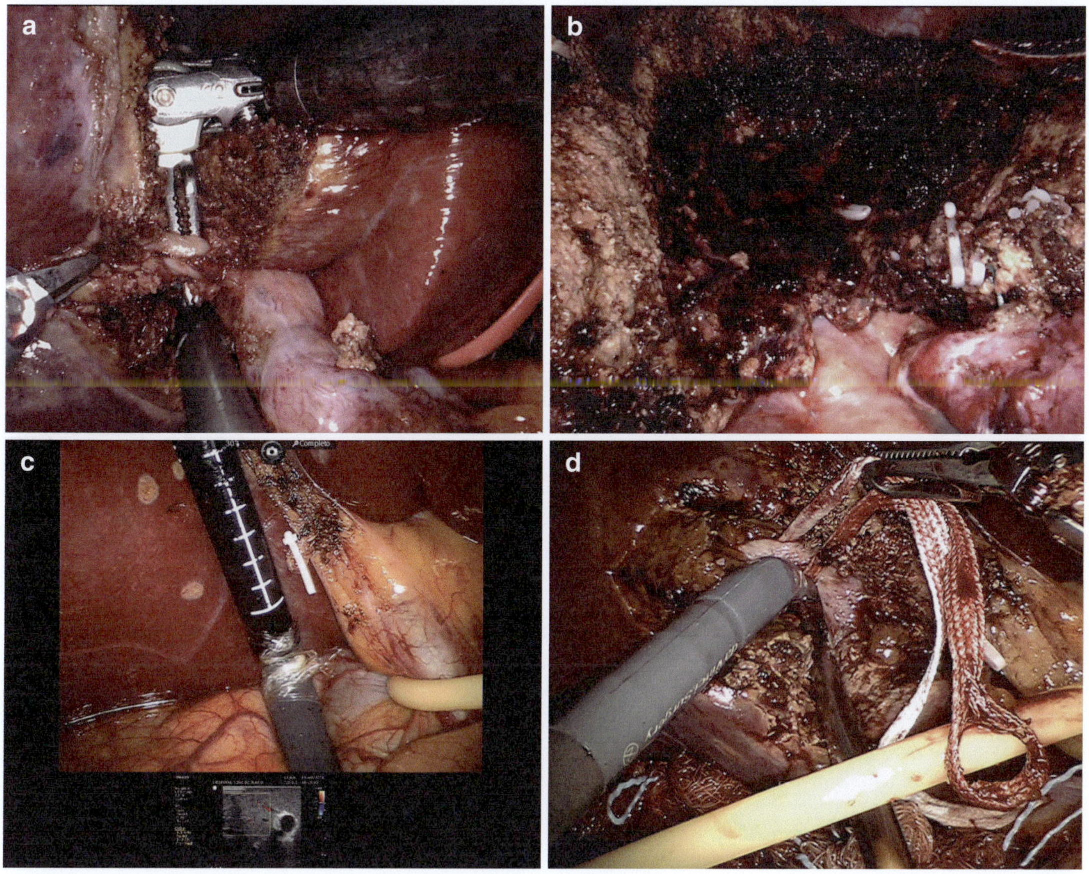

Fig. 14.9 Glissonian approach during robotic liver resection. (**a**) The pedicle of segment 4b is encircled during resection of segment 4. (**b**) Intraoperative photograph after resection of segment 4. (**c**) Intraoperative ultrasound is used to identify the Glissonian pedicle of segment 6. (**d**) Pedicle of segment 6 is circled prior to anatomic resection of segment 6

The robotic approach is useful for precise dissection of the hepatic hilum. Individual dissection and identification of the portal triad appears to be faster and easier than with laparoscopy [31]. The Glissonian approach is our preferred method of inflow control for anatomic resections because it allows rapid control of portal pedicles while allowing segmental liver resections [32, 33]. The use of this approach in laparoscopic liver resection required an adaptation of the technique [34, 35]. Instead of circling the Glissonian pedicle for individual control of the appropriate portal pedicle (as in the open intrahepatic Glissonian technique), the laparoscopic Glissonian approach was achieved with blind insertion of a vascular clamp around the target pedicle using specific anatomic landmarks [34, 35]. The use of the robotic platform allowed safe encirclement of the Glissonian pedicle in the same manner as originally described for open liver resection (Figs. 14.9 and 14.10), precluding the hilar dissection, even though the robotic technique facilitates such dissection [5, 6].

Fluorescence imaging after indocyanine green injection is an important tool to define the limits of liver resection and to check liver perfusion and bile leakage after resection [36]. Our initial experience with robotic liver surgery has raised several issues that may be important for any surgeon who wishes to explore this new technology. Previous experience in both open and laparoscopic liver surgery is essential for better out-

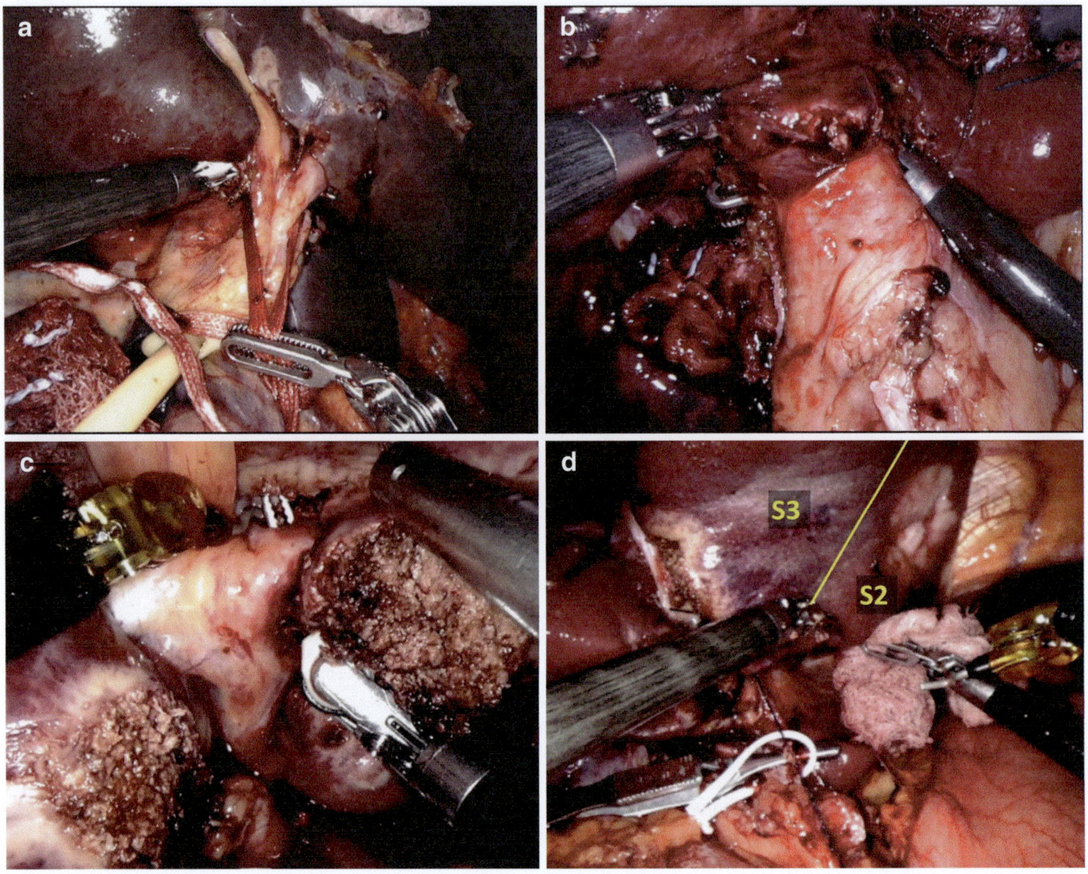

Fig. 14.10 Glissonian approach during robotic liver resection. (**a**) The left main Glison pedicle is encircled before the left hepatectomy. (**b**) The right Glison pedicle is encircled before the right hepatectomy. (**c**) The pedicle of segment 3 is circled before anatomic left lateral resection. (**d**) Ischemic delineation of segment 3 (S3). The Glissonian pedicle of segment 2 (S2) is not yet clamped, and we can see a well-perfused S2 just before the S2-S3 bi-segmentectomy

comes. Trocar placement may vary depending on the patient's biotype and differs from laparoscopic hepatectomy. Proper trocar placement is one of the most important steps for successful robotic surgery. Another important point is that there are some surgical instruments commonly used in liver surgery that are not yet available in robotic systems, such as CUSA, the Cavitron ultrasound aspirator (used for liver transection). However, these instruments can be used by the bedside surgeon, so sufficient experience in advanced liver surgery is also a requirement for the bedside surgeon. Conversion to an open approach is less common than with laparoscopy because bleeding is easier to resolve with the use of wristed instruments that allow precise suturing when needed. Indeed, several patients with major bleeding were easily controlled by suture during our experience [23].

Five different retrospective cohort studies compared the robotic versus laparoscopic approach for resection of CRLM in a total of 1869 patients [24, 37–42]. Reported outcomes included perioperative safety, length of stay, and survival. There were no differences between the two groups in estimated blood loss, transfusion rates, or perioperative morbidities. No differences were noted in 5-year disease-free rate (38 vs. 44%, ns) or overall survival (61 vs. 60%, $p > 0.05$). There were conflicting reports regarding operative times. Rahimli and colleagues [39] found in their series that robotics ($n = 12$) was associated with a significantly longer operative time (342 vs. 200 min) but a higher tendency to R0 resection (100% vs. 77%, $p > 0.05$) compared to laparoscopy ($n = 12$) [37]. In their multicenter analysis, Masetti and colleagues found no differences in operative times between groups, and robotics was associated with lower rates of R1 resection (16.9 vs. 28.8%, ns) with greater surgical margin than laparoscopy [41]. Beard and colleagues examined the collective experience at six high-volume tertiary referral centers in the United States and Belgium [24]. Propensity matching was performed to minimize differences between laparoscopic and robotic patients. Most procedures were parenchymal-sparing wedge resections. After matching, there were no differences in reoperation rates (0.9% vs. 3.5%, ns), perioperative complications (27.8% vs. 31.3%, ns), perioperative mortality rates, or margin status. After a median follow-up of 2.8–3.1 years, there were no differences in 5-year overall survival or disease-free survival. A separate meta-analysis evaluated seven retrospective cohort studies that evaluated laparoscopic and robotic liver resection in a cohort of 525 patients [40]. There were no differences in perioperative complication rates, perioperative mortality rates, rates of conversion to an open procedure, R1 resections, blood transfusions, operative times, or length of hospital stay. No survival data could be extrapolated from the cohort studies.

References

1. Donati F, Cioni D, Guarino S, Mazzeo ML, Neri E, Boraschi P. Chemotherapy-induced liver injury in patients with colorectal liver metastases: findings from MR imaging. Diagnostics (Basel). 2022;12(4):867.
2. Calistri L, Rastrelli V, Nardi C, Maraghelli D, Vidali S, Pietragalla M, Colagrande S. Imaging of the chemotherapy-induced hepatic damage: yellow liver, blue liver, and pseudocirrhosis. World J Gastroenterol. 2021;27(46):7866–93.
3. White MA, Fong Y, Singh G. Chemotherapy-associated hepatotoxicities. Surg Clin North Am. 2016;96:207–17.
4. Schadde E, Grunhagen DJ, Verhoef C, Krzywon L, Metrakos P. Limitations in resectability of colorectal liver metastases 2020 – a systematic approach for clinicians and patients. Semin Cancer Biol. 2021;71:10–20.
5. Machado MA, Mattos BH, Lobo Filho M, Makdissi F. Intrahepatic Glissonian approach for robotic left hepatectomy. Surg Oncol. 2021;38:101601.
6. Machado MA, Mattos BH, Filho ML, Makdissi F. Intrahepatic Glissonian approach for robotic right hepatectomy. Surg Oncol. 2021;38:101579.
7. Machado MA, Mattos BH, Lobo Filho MM, Makdissi FF. Robotic resection of postero-superior liver segments (7,8) (with video). J Gastrointest Surg. 2021;25(2):574–5.
8. Machado MA, Surjan RC, Basseres T, Makdissi F. Robotic repeat right hepatectomy for recurrent colorectal liver metastasis. Ann Surg Oncol. 2019;26(1):292–5.
9. Machado MAC, Mattos BV, Lobo Filho MM, Makdissi F. Glissonian approach during robotic mesohepatectomy for recurrent colorectal liver metastasis. Ann Surg Oncol. 2022;29(13):8449–51.
10. Machado MA, Mattos BV, Lobo Filho MM, Makdissi F. Glissonian approach during robotic Mesohepatectomy for recurrent colorectal liver metastasis. Ann Surg Oncol. 2022;29(13):8449–51.
11. Jegatheeswaran S, Mason JM, Hancock HC, Siriwardena AK. The liver-first approach to the management of colorectal cancer with synchronous hepatic metastases: a systematic review. JAMA Surg. 2013;148(4):385–91.
12. Machado MA, Mattos BH, Lobo Filho MM, Makdissi FF. Robotic right hepatectomy with portal vein thrombectomy for colorectal liver metastasis (with video). J Gastrointest Surg. 2021;25(7):1932–5.
13. Vigano L, De Rosa G, Toso C, Andres A, Ferrero A, Roth A, Sperti E, Majno P, Rubbia-Brandt L. Reversibility of chemotherapy-related liver injury. J Hepatol. 2017;67(1):84–91.
14. Huang JW, Su WL, Wang SN. Alternative laparoscopic intracorporeal Pringle maneuver by Huang's loop. World J Surg. 2018;42(10):3312–5.
15. Sasaki K, Morioka D, Conci S, Margonis GA, Sawada Y, Ruzzenente A, Kumamoto T, Iacono C, Andreatos N, Guglielmi A, Endo I, Pawlik TM. The Tumor Burden Score: a new "Metro-ticket" prognostic tool for colorectal liver metastases based on tumor size and number of tumors. Ann Surg. 2018;267(1):132–41.
16. Moazzam Z, Shaikh CF, Tsilimigras DI, Pawlik TM. Resection of colorectal liver metastases: prognostic impact of tumor burden score varies with unilateral versus bilateral metastatic spread. J Gastrointest Surg. 2023;27(2):404–6.
17. Ayabe RI, Vauthey JN, Newhook TE. Optimizing the future liver remnant: Portal vein embolization, hepatic venous deprivation, and associating liver partition and portal vein ligation for staged hepatectomy. Surgery. 2023;174:116–8.

18. Petrowsky H, Linecker M, Raptis DA, Kuemmerli C, Fritsch R, Kirimker OE, Balci D, Ratti F, Aldrighetti L, Voskanyan S, Tomassini F, Troisi RI, Bednarsch J, Lurje G, Fard-Aghaie MH, Reese T, Oldhafer KJ, Ghamarnejad O, Mehrabi A, Abraham MET, Truant S, Pruvot FR, Hoti E, Kambakamba P, Capobianco I, Nadalin S, Fernandes ESM, Kron P, Lodge P, Olthof PB, van Gulik T, Castro-Benitez C, Adam R, Machado MA, Teutsch M, Li J, Scherer MN, Schlitt HJ, Ardiles V, de Santibañes E, Brusadin R, Lopez-Lopez V, Robles-Campos R, Malagó M, Hernandez-Alejandro R, Clavien PA. First long-term oncologic results of the ALPPS procedure in a large cohort of patients with colorectal liver metastases. Ann Surg. 2020;272(5):793–800.
19. Butte JM, Gönen M, Allen PJ, Peter Kingham T, Sofocleous CT, DeMatteo RP, Fong Y, Kemeny NE, Jarnagin WR, D'Angelica MI. Recurrence after partial hepatectomy for metastatic colorectal cancer: potentially curative role of salvage repeat resection. Ann Surg Oncol. 2015;22(8):2761–71.
20. Belli G, Cioffi L, Fantini C, D'Agostino A, Russo G, Limongelli P, Belli A. Laparoscopic redo surgery for recurrent hepatocellular carcinoma in cirrhotic patients: feasibility, safety, and results. Surg Endosc. 2009;23(8):1807–11.
21. Xian SH, Tan HJ, Marino MV, Thiruchelvam N, Heng AK. Review of robotic simultaneous resection of colorectal cancer with synchronous liver metastases using Da Vinci Xi: technical considerations and outcomes. Chirurgia (Bucur). 2023;118(1):20–6.
22. Boudjema K, Locher C, Sabbagh C, Ortega-Deballon P, Heyd B, Bachellier P, Métairie S, Paye F, Bourlier P, Adam R, Merdrignac A, Tual C, Le Pabic E, Sulpice L, Meunier B, Regimbeau JM, Bellissant E, METASYNC Study Group. Simultaneous versus delayed resection for initially resectable synchronous colorectal cancer liver metastases: a prospective, open-label, randomized, controlled trial. Ann Surg. 2021;273(1):49–56.
23. Machado MA, Lobo-Filho MM, Mattos BH, Ardengh AO, Makdissi FF. Robotic liver resection. Report of the first 50 cases. Arq Gastroenterol. 2021;58(4):514–9.
24. Beard RE, Khan S, Troisi RI, Montalti R, Vanlander A, Fong Y, Kingham TP, Boerner T, Berber E, Kahramangil B, Buell JF, Martinie JB, Vrochides D, Shen C, Molinari M, Geller DA, Tsung A. Long-term and oncologic outcomes of robotic versus laparoscopic liver resection for metastatic colorectal cancer: a multicenter, propensity score matching analysis. World J Surg. 2020;44(3):887–95.
25. Gumbs AA, Croner R, Lorenz E, Cacciaguerra AB, Tsai TJ, Starker L, Flanagan J, Yu NJ, Chouillard E, Abu HM. Survival study: international multicentric minimally invasive liver resection for colorectal liver metastases (SIMMILR-2). Cancers (Basel). 2022;14(17):4190.
26. Luberice K, Sucandy I, Modasi A, Castro M, Krill E, Ross S, Rosemurgy A. Applying IWATE criteria to robotic hepatectomy: is there a "robotic effect"? HPB (Oxford). 2021;23(6):899–906.
27. Machado MA, Makdissi FF, Surjan RC, Abdalla RZ. First robotic-assisted laparoscopic liver resection in Latin America. Arq Gastroenterol. 2009;46(1):78–80.
28. Wong DJ, Wong MJ, Choi GH, Wu YM, Lai PB, Goh BKP. Systematic review and meta-analysis of robotic versus open hepatectomy. ANZ J Surg. 2019;89(3):165–70.
29. Tsung A, Geller DA, Sukato DC, Sabbaghian S, Tohme S, Steel J, Marsh W, Reddy SK, Bartlett DL. Robotic versus laparoscopic hepatectomy: a matched comparison. Ann Surg. 2014;259(3):549–55.
30. Wakabayashi G, Cherqui D, Geller DA, Buell JF, Kaneko H, Han HS, Asbun H, O'Rourke N, Tanabe M, Koffron AJ, Tsung A, Soubrane O, Machado MA, Gayet B, Troisi RI, Pessaux P, Van Dam RM, Scatton O, Abu Hilal M, Belli G, Kwon CH, Edwin B, Choi GH, Aldrighetti LA, Cai X, Cleary S, Chen KH, Schön MR, Sugioka A, Tang CN, Herman P, Pekolj J, Chen XP, Dagher I, Jarnagin W, Yamamoto M, Strong R, Jagannath P, Lo CM, Clavien PA, Kokudo N, Barkun J, Strasberg SM. Recommendations for laparoscopic liver resection: a report from the second international consensus conference held in Morioka. Ann Surg. 2015;261(4):619–29.
31. Giulianotti PC, Sbrana F, Coratti A, Bianco FM, Addeo P, Buchs NC, Ayloo SM, Benedetti E. Totally robotic right hepatectomy: surgical technique and outcomes. Arch Surg. 2011;146(7):844–50.
32. Machado MA, Herman P, Machado MC. A standardized technique for right segmental liver resections. Arch Surg. 2003;138(8):918–20.
33. Machado MA, Herman P, Machado MC. Anatomical resection of left liver segments. Arch Surg. 2004;139(12):1346–9.
34. Machado MA, Makdissi FF, Galvão FH, Machado MC. Intrahepatic Glissonian approach for laparoscopic right segmental liver resections. Am J Surg. 2008;196(4):e38–42.
35. Machado MA, Makdissi FF, Surjan RC, Herman P, Teixeira AR, Machado C, MC. Laparoscopic resection of left liver segments using the intrahepatic Glissonian approach. Surg Endosc. 2009;23(11):2615–9.
36. Marino MV, Di Saverio S, Podda M, Gomez Ruiz M, Gomez FM. The application of indocyanine green fluorescence imaging during robotic liver resection: a case-matched study. World J Surg. 2019;43(10):2595–606.
37. Lo WM, Tohme ST, Geller DA. Recent advances in minimally invasive liver resection for colorectal cancer liver metastases-a review. Cancers (Basel). 2022;15(1):142.
38. Guerra F, Guadagni S, Pesi B, Furbetta N, Di Franco G, Palmeri M, Annecchiarico M, Eugeni E, Coratti A, Patriti A, Morelli L. Outcomes of robotic liver resections for colorectal liver metastases. A multi-institutional analysis of minimally invasive ultrasound-guided robotic surgery. Surg Oncol. 2019;28:14–8.
39. Rahimli M, Perrakis A, Schellerer V, Gumbs A, Lorenz E, Franz M, Arend J, Negrini VR, Croner RS. Robotic

and laparoscopic liver surgery for colorectal liver metastases: an experience from a German Academic Center. World J Surg Oncol. 2020;18(1):333.

40. Ziogas IA, Giannis D, Esagian SM, Economopoulos KP, Tohme S, Geller DA. Laparoscopic versus robotic major hepatectomy: a systematic review and meta-analysis. Surg Endosc. 2021;35(2):524–35.
41. Masetti M, Fallani G, Ratti F, Ferrero A, Giuliante F, Cillo U, Guglielmi A, Ettorre GM, Torzilli G, Vincenti L, Ercolani G, Cipressi C, Lombardi R, Aldrighetti L, Jovine E. Minimally invasive treatment of colorectal liver metastases: does robotic surgery provide any technical advantages over laparoscopy? A multicenter analysis from the IGoMILS (Italian Group of Minimally Invasive Liver Surgery) registry. Updat Surg. 2022;74(2):535–45.
42. Ghotbi J, Aghayan D, Fretland Å, Edwin B, Syn NL, Cipriani F, Alzoubi M, Lim C, Scatton O, Long TCD, Herman P, Coelho FF, Marino MV, Mazzaferro V, Chiow AKH, Sucandy I, Ivanecz A, Choi SH, Lee JH, Prieto M, Vivarelli M, Giuliante F, Ruzzenente A, Yong CC, Yin M, Fondevila C, Efanov M, Morise Z, Di Benedetto F, Brustia R, Dalla Valle R, Boggi U, Geller D, Belli A, Memeo R, Mejia A, Park JO, Rotellar F, Choi GH, Robles-Campos R, Wang X, Sutcliffe RP, Pratschke J, Tang CN, Chong CCN, D'Hondt M, Monden K, Lopez-Ben S, Kingham TP, Ferrero A, Ettorre GM, Levi Sandri GB, Pascual F, Cherqui D, Liang X, Mazzotta A, Wakabayashi G, Giglio M, Troisi RI, Han HS, Cheung TT, Sugioka A, Chen KH, Liu R, Soubrane O, Fuks D, Aldrighetti L, Abu Hilal M, Goh BKP. International robotic and laparoscopic liver resection study group investigators. Impact of neoadjuvant chemotherapy on the difficulty and outcomes of laparoscopic and robotic major liver resections for colorectal liver metastases: a propensity-score and coarsened exact-matched controlled study. Eur J Surg Oncol. 2023;49(7):1209–16.

Robotic Livery Surgery for Malignant Tumors of the Biliary Tract

15

Ismaël Chaoui and Mathieu D'Hondt

15.1 Introduction

Biliary tract cancer (BTC) represents less than 1% of all cancers and generally carries a poor prognosis [1]. It encompasses intrahepatic (IHCCA), perihilar (PHCCA), and distal (DCCA) cholangiocarcinoma (CCA), as well as gallbladder carcinoma (GBC) [2]. Its grim prognosis is owed to the fact that most patients present with locally advanced or metastatic disease and are deemed ineligible for surgery upon diagnosis. Nonsurgical treatments still lack efficacy. In case of resectable disease, surgery is the only treatment modality associated with increased long-term survival [3, 4].

Arising from the biliary epithelium, CCA is the second most common primary liver cancer after hepatocellular carcinoma. PHCCA (60%) originates between the second-order bile ducts and the cystic duct insertion into the common bile duct (CBD) and can further be classified into four types based on biliary involvement, according to the Bismuth–Corlette classification [5]. Each type carries different therapeutic and prognostic consequences. In addition to lymph node dissection, extrahepatic biliary resection generally suffices for type 1 and 2 tumors. Type 3 and 4 tumors may require a combined hepatectomy, bile duct resection and bilio-enteric reconstruction, while sometimes also necessitating vascular reconstruction. These procedures are highly challenging, and mortality rates of up to 18% have been reported, even with an open approach [6]. Five-year survival rates following resection range from 25 to 50%, with locoregional recurrence limiting long-term survival [7]. DCCA (30%) typically requires a pancreaticoduodenectomy (Whipple procedure), yielding a median 5-year survival of 20–40%, contingent upon the extent of disease [1]. Finally, localized IHCCA (6–10%), arising from the second-order bile ducts, may be amenable to complete surgical resection if diagnosed early (20–40%), with 5-year survival ranging from 10 to 49% [8].

GBC is the most common biliary tract malignancy (80–95%) [9]. It originates in the gallbladder or cystic duct and is often an incidental finding, with an estimated incidence of 0.2–2% after routine cholecystectomy [10]. A lack of specific symptomatology leads to frequent presence of advanced or metastatic disease upon diagnosis, making only 20% of tumors amenable to resection [11]. Five-year survival rates range between 7 and 20% because of a shorter time to recurrence and survival after recurrence in comparison to PHCCA [12]. Disease stage deter-

I. Chaoui · M. D'Hondt (✉)
Department of Digestive and Hepatobiliary/Pancreatic Surgery, Groeninge Hospital, Kortrijk, Belgium
e-mail: ismael.chaoui@azgroeninge.be

M. D'Hondt, I. Sucandy (eds.), *Textbook of Robotic Liver Surgery*,
https://doi.org/10.1007/978-3-031-76536-0_15

mines the need for additional resection. While a simple cholecystectomy is sufficient for localized T1a tumors with 5-year survival rates as high as 95–100%. Higher T-stages may require more extended (repeat) resections [13] ranging from cholecystectomy with a wedge resection of the gallbladder bed or anatomical IVb/V segmentectomy and regional lymphadenectomy, to potential extrahepatic bile duct resection and adjacent organ resection. If R0 resection is not feasible or tolerated, palliative surgery can be considered to improve quality of life.

Minimally invasive liver surgery (MILS) has found its way into current surgical practice. In fact, a European survey of 103 centers found that a minimally invasive approach is used in about one-third of liver resections, with laparoscopy being the most common approach [14]. Compared with open resections, laparoscopic liver surgery has been demonstrated to reduce intraoperative blood loss, postoperative complications, and length of stay (LOS), with similar oncological outcomes [15–17]. However, for highly complex major resections with biliary and potentially vascular reconstructions, wide implementation of laparoscopic surgery has been hampered by its technical limitations with regard to fine dissection during resection and lymph node dissection (LND), as well as advanced suturing during reconstruction. Therefore, an open approach is often used as a recourse for (parts of) these procedures. The advent of robotic surgery, with its enhanced 3D vision, tremor filtering, ICG guidance, and improved dexterity, has the potential to overcome this barrier and subsequently cause a further shift toward MILS.

15.2 Literature Overview

15.2.1 Safety and Feasibility: The First Robotic Experiences

The robotic platform is now widely used in a variety of surgical subspecialties and has had a positive effect on short-term patient outcomes compared with open and laparoscopic approaches for select procedures. Notably for hepatic resections, a recent multicenter propensity score-matched analysis demonstrated a lower conversion rate and reduced blood loss in the robotic group compared to its laparoscopic counterpart [18, 19]. A positive impact of the robotic platform on postoperative complications and length of stay has also been reported, with similar oncological outcomes [20]. Thus, the platform is considered suitable for oncological liver surgery.

MIS in the management of CCA and GBC, however complex, has been deemed safe and feasible in expert hands [21–27]. Nevertheless, reports describing a pure laparoscopic approach for PHCCA are scarce, and conversion rates for major liver resections range between 9 and 42% in laparoscopic liver surgery [28]. In 2010, Giulianotti et al. first described the robotic management of a PHCCA [29]. They performed a right hepatectomy with lymphadenectomy and left hepaticojejunostomy, with a previous robotic experience consisting of about 70 hepatic resections and considerable laparoscopic experience. Later in 2012, Shen et al. published their first experience with extended radical robotic resection of GBC and lymphadenectomy, with a previous robotic experience of 20 hepatobiliary/pancreatic surgeries [30]. Both groups relayed the general and short-term oncological safety and feasibility of the approach. Since the publishing of these reports, several high-volume expert centers have expanded their MIS practice, proposing robotic management for increasingly complex cases involving biliary malignancies.

15.2.2 Robotic Management of Biliary Cancer

15.2.2.1 Perihilar Cholangiocarcinoma

Beyond small single-center case series, robust outcome data regarding robotic treatment of biliary tumors is still limited. Owing to the rarity of the disease, large cohorts are equally scarce, and a lack of prospective and randomized data is evident. However, the literature reflects a growing interest in the robotic platform for radical surgical excision of PHCCA in recent years, irrespective of Bismuth–Corlette type, as shown in Table 15.1.

Still, the only available comparative study is that of Xu et al., dating back to 2016 [32]. This single-center retrospective study included 10 robotic and 32 open resections of PHCCA. The authors noted a significantly longer operative time, a higher major morbidity rate, and poor oncological outcomes in the robotic group. Consequently, the group recommended against the robotic approach for resection of PHCCA. These inferior outcomes were likely linked to insufficient patient selection (40% Bismuth–Corlette IV rate) too early in the robotic learning curve. In addition, several technical limitations were cited, such as the bulky robotic instruments hindering liver mobilization, retraction, and dissection. The platform itself has since been refined, and conflicts inside and outside the abdomen have largely been addressed in the newer robotic systems. When reported, the used robotic versions are detailed in Table 15.1. PHCCA is now exclusively managed with the latest systems.

More recently, Sucandy et al. reported on the feasibility and clinical and oncological outcome experience of three Western expert centers in the management of type I–IV PHCAA [44]. In 38 consecutive resections, with an 84% concomitant major hepatectomy rate and Bismuth–Corlette ≥III in 66% of patients, one conversion was needed. Average blood loss was 200 cc and major morbidity was reported in 16% of patients with a median LOS of 6 days. R0 resection was achieved in 82% of patients (87% perioperative frozen-section R0 margins) with a median of seven harvested lymph nodes. After a mean follow-up period of 15 months, an overall survival of 81% and disease-free survival of 68% were observed.

Sucandy et al.'s experience aligns with that of the largest study to date by Li et al. [34], affirming that complex robotic hepatobiliary surgery can be performed safely and effectively in high-volume expert centers, without compromising clinical and oncological outcomes. In fact, a recent systematic review by Guadagni et al. confirmed these findings specifically for type III and IV PHCCA, reporting acceptable oncological results in this aggressive disease subgroup with a 50% recurrence rate after a mean follow-up period of 10.1 months [6]. In comparison, a recent study involving 24 expert centers defined surgical and oncological benchmark values for PHCCA and reported an acceptable R0 resection rate of ≥57% and 5-year overall survival rate of ≥39.7% [46]. Dugan et al. are the latest to report long-term oncological outcomes since Xu et al.'s experience in 2016, with an 85% robotic R0 resection rate resulting in a 5-year OS of 55%, once again highlighting the importance of margin status in increasing OS [45]. The authors attribute their consistently high R0 resection rate to the superior visualization and dissection facilitated by the robotic platform. These reports encourage the pursuit of wider, multicenter investigations in order to confirm the long-term results of robotic management of PHCCA.

15.2.2.2 Gallbladder Carcinoma

Similarly to PHCCA, the literature concerning robotic management of GBC is limited to low-evidence single-center small series, though steadily increasing in recent years with the adoption of robotic surgery in the hepatobiliary field (Table 15.2).

Radical surgical resection is equally paramount in the management of GBC. Given the indistinct disease symptomatology, GBC is mostly identified incidentally, in about 2% of cholecystectomy specimens [9, 10]. As T-stage determines the best surgical approach, this often results in a need to revisit a recently operated field with obliterated tissue planes, inflammatory adhesions, and a short cystic duct stump. Even for de novo GBC, advanced disease could make a minimally invasive approach challenging, certainly considering the demanding dissection during central hepatectomy and lymphadenectomy, and the potential need for advanced suturing for bilio-enteric anastomosis. In addition to the increased dexterity, 3D vision with intraoperative ICG fluorescence imaging can facilitate identification of crucial hilar structures during lymph

Table 15.1 Reports of robotic resection of PHCCA: demographic and preoperative clinical data, surgical interventions, and oncological and surgical outcomes

Study	Year	Study type	Da Vinci version	N	M/F	Bismuth–Corlette type (I/II/IIIA/IIIB/IV)	Age	BMI	Neoadjuvant therapy	Preoperative biliary drainage	PVE/ALPSS
Giulianotti et al. [29]	2010	Case report	–	1	1/0	IIIA	66	23	0	1	1 (PVE)
Liu et al. [31]	2012	Case series	–	39	–	–	–	–	–	–	–
Xu et al. [32]	2016	Case–control (robotic/open)	S	10 32	8/2 20/12	0/1/4/1/4 0/4/17/11	54 (36–77) 59 (37–77)	–	0 0	6 19	1 (PVE) 4 (PVE)
Quijano et al. [33]	2016	Case report	S	1	–	IIIB	–	–	–	–	–
Li et al. [34]	2020	Case series	Si	48	28/20	20/6/5/17/0	62.9 (±9)	23.7 (±3.1)	0	20	–
Machado et al. [35]	2020	Case report	Si	1	0/1	IIIB	76	30	1	1	–
Marino et al. [36]	2020	Case report	Xi	1	1/0	IIIB	57	–	–	–	–
Cillo et al. [37]	2021	Case series	Xi	4	1/3	IIIB	59.5 (44–79)	–	1	4	0
Sucandy et al. [38]	2021	Case report	Xi	1	0/1	IIIB	77	–	0	1	–
Sucandy et al. [39]	2021	Case series	Xi	15	9/6	2/5/6/1	73 (±8.9)	27 (±6.3)	-	14	–
Di Benedetto et al. [40]	2022	Case report	Si	1	0/1	IIIA	74	NR	0	1	1 (ALPSS/PVE)
Tee et al. [41]	2022	Case report	Xi	1	1/0	I	58	49	0	1	–
Jacoby et al. [42]	2022	Case series	Xi	21	14/7	4/5/10/2	72 (55–90)	25 (19–40)	3	21	–
D'Hondt et al. [43]	2023	Case series	Xi	5	1/4	0/0/2/2/1	75 (46–77)	32 (±17.5)	0	1/5	0
Sucandy et al. [44]	2024	Case series	Xi	38	25/23	6/7/4/16/5	69 (±10,2)	26 (±5.8)	5	31	–
Dugan et al. [45]	2024	Case series	–	27	19/8	5/5/15/2	71 (±9.6)	29 (±7.1)	5	–	–

Study	Operation	Lymph node dissection	Lymph nodes retrieved	No. of ducts in biliary anastomosis	Operative time (minutes)	Blood loss (cc)	Transfusion (N)	Length of stay (days)	Conversion	Clavien-Dindo ≥ 3	R0 resection	30-day mortality	FU (months)	Recurrence rate (%)	Deaths during FU
Giulianotti et al. [29]	RH + S1 + BR	Yes	–	1	540	800	1	11	0	0	1	0	8	0	0
Liu et al. [31]	3 LH – 15 BR – 21 other	–	–	–	–	–	–	–	1	–	–	1	–	–	–
Xu et al. [32]	6 RH + S1 + BR 4 LH + S1 + BR	Yes	–	–	**703 (±62)** **475 (±121)**	1360 (±809) 1014 (±811)	6 17	16 (9–58) 14 (4–54)	0 -	**3** **5**	7	1 2	60 52	90 56	9 –
Quijano et al. [33]	LH + S1 + BR	Yes	13	–	510	1000	–	16	1	0	1	0	3	–	–
Li et al. [34]	RH/LH + S1 + BR	Yes	–	–	276 (170–500)	150 (20–1500)	13	9 (4–52)	0	5	35	0	–	–	–
Machado et al. [35]	LH + S1 + BR	Yes	–	1	480	740	1	–	0	0	1	0	5	0	0
Marino et al. [36]	LH + S1 + BR	Yes	9	1	295	280	–	6	0	0	1	0	–	–	–
Cillo et al. [37]	4 LH + S1 + BR	Yes	10 (mean)	1 – 2 – 3 – 5	840 (750–950)	700 (600–800)	0	9 (7–11)	1	0	3	0	7.5 (6–12)	25	0
Sucandy et al. [38]	LH + S1 + BR	Yes	–	1	360	–	–	6	0	–	1	0	12	0	0
Sucandy et al. [39]	–	Yes	3.4 (±2.2)	–	443 (±85)	182 (±138.4)	–	6 (±3.2)	0	1	11	0	12 (±11.8)	13	3
Di Benedetto et al. [40]	RH + S1 + BR	Yes	21	2	370	450	0	19	0	0	0	0	13	0	0
Tee et al. [41]	BR	Yes	12	1	540	100	0	5	0	0	1	0	12	0	0
Jacoby et al. [42]	1 RH + BR 7 LH + BR 8 CH + BR 5 BR	Yes	4 (1–12)	–	458 (279–573)	150 (10–450)	0	5 (3–14)	0	2	19	0	21 (1–63)	–	6
D'Hondt et al. [43]	3 LH + S1 + BR 2 RH + S1 + BR	Yes	9 (8–16)	2 (1) – 1 (4)	380 (290–460)	100 (50–350)	0	8 (6–16)	0	0	5	0	1	0	0
Sucandy et al. [44]	4 RH + S1 + BR 13 LH + S1 + BR 11 CH – 2 ERH – 2 ELH	Yes	8 (±6.6)	1 (31) – 2 (6) – 3 (1)	485 (±157)	252 (±234)	–	9 (±8.8)	1	6	31	1	19 (±16)	13	7
Dugan et al. [45]	13 major resections	Yes	–	–	439 (±111.2)	150 (±118.1)	–	4 (±5)	0	2	23	0	23 (±17.16)	–	–

Statistics reported as mean (± standard deviation) or median (range). Bold values indicate statistical significance ($P < 0.05$)

Abbreviations: *N* Patient population, *M* male, *F* female, *BMI* body mass index, *PVE* portal vein embolization, *ALPPS* Associating Liver Partition and Portal vein Ligation for Staged hepatectomy, *RH* right hepatectomy, *ERH* extended right hepatectomy, *ELH* extended left hepatectomy, *LH* left hepatectomy, *CH* central hepatectomy, *BR* (biliary resection and reconstruction), *S1* segment 1 resection, and *FU* follow-up

Table 15.2 Reports of robotic resection of GBC: demographic and preoperative clinical data, surgical interventions, and oncological and surgical outcomes

Study	Year	Study type	Da Vinci version	N	M/F	T-stage (</≥ T2)	Age	BMI	Neoadjuvant therapy	Adjuvant therapy	PVE/ ALPSS
Shen et al. [30]	2012	Case series	–	5	2/3	–	60 (46–63)	–	–	–	1 (PVE)
Zeng et al. [47]	2018	Case series	Si	3	1/2	0/3	61 (57–69)	–	–	–	–
Sinagra et al. [48]	2018	Case series	–	3	3/3	3/0	–	–	–	–	1 (PVE)
Khan et al. [49]	2018	Case series	–	11	5/6	–	63 (57–85)	25 (20–44)	–	3	–
Goel et al. [50]	2019	Case–control (robotic/open)	Xi	27 70	9/18 25/45	5/22 12/58	54 (28–75) 54.5 (-)	24.4 (18.1–33.2) -	1 –	22	–
Araujo et al. [51]	2020	Case series	–	3	1/2	3/0	49 (33–53)	30.9 (29.9–31.8)	1	–	–
Ahmad et al. [52]	2020	Case series	–	10	4/6	0/10	69.5 (49–84)	–	–	–	–
Byun et al. [53]	2020	Case–control (robotic/open)	–	16 34	10/6 14/20	5/11 5/29	64.3 (±10.4) 67.4 (±9.7)	24.8 (±2.6) 24.1 (±3.9)	– –	– –	– –
Tschuor et al. [54]	2021	Case–control (robotic/open)	Si/Xi	20 23	6/14 7/16	2/18 3/19	68 (39–82) 63 (40–82)	– –	4 2	– –	– –
Sucandy et al. [55]	2022	Case series	–	15	3/12	–	70 (56–90)	25.7 (19.6–32.3)	–	–	–
Cho et al. [56]	2023	Case–control (robotic/open)		41 41	25/16 24/17	10/31 8/33	66.8 (±9.0) 68.8 (±8.0)	23.8 (±2.7) 23.9 (±2.8)	– –	23 26	– –
D'Hondt et al. [43]	2023	Case series	Xi	2	0/2	0/2	72.5 (68–77)	29 (27–31)	–	–	1 (PVE)
Dugan et al. [45]	2024	Case series	–	19	7/12	–	69 (±13)	26.7 (±9.1)	3	3	–

Study	Operation	Primary/ completion surgery	Lymph node dissection	Lymph nodes retrieved	Positive nodal status	Operative time (minutes)	Blood loss (cc)	Transfusion (N)	Length of stay (days)	Conversion	Clavien-Dindo ≥ 3	R0 resection	30-day mortality	FU (months)	Recurrence rate (%)	Deaths during FU
Shen et al. [30]	5 rCCE	3/2	5	10 (3–11)	1	160 (120–300)	200 (50–400)	–	7 (7–8)	0	0	–	0	11 (1–17)	1 (20%)	1
Zeng et al. [47]	1 rCCE – 2 CH/LND	1/2	3	7 (1–11)	1	360 (220–530)	200 (50–700)	0	3 (3–4)	0	0	3 (100%)	0	–	–	–
Sinagra et al. [48]	3 CH/LND	0/3	3	17.5 (14–22)	1	290 (250–310)	175 (100–350)	–	6 (5–10)	0	0	3 (100%)	0	21 (6–32)	0	0
Khan et al. [49]	4 CCE/ CH – 7CH (7 LND)	4/7	7	5 (0–9)	1	219 (99–790)	50 (10–200)	0	4 (2–9)	0	1	9 (81.8%)	0	41.6 (±4.9)	3 (27%)	3
Goel et al. [50]	27 rCCE 70 rCCE	25/2 –	27 70	10 (2–21) 9 (2–25)	– –	**295** **(200–710)** **200 (85–400)**	**200 (20–70)** **600 (50–3000)**	– –	**4 (2–12)** **5 (3–15)**	4 –	**1** **15**	27 (100%) –	– –	9 (1–46) 9 (1–46)	2 (7%) 12 (17%)	0 0
Araujo et al. [51]	3 CH/LND	0/3	3	4 (3–6)	0	385 (380–410)	200 (50–300)	0	3 (3–3)	0	0	3 (100%)	0	3	0 (0%)	0
Ahmad et al. [52]	10 CH/LND	3/7	10	5.5 (2–8)	8	175 (95–240)	67.5 (25–250)	–	3 (2–6)	0	0	10 (100%)	0	5.8 (2–16)*	–	–
Byun et al. [53]	16 rCCE 34 rCCE	16/0 34/0	16 34	7.2 (±3.3) 7.4 (±4.5)	3 11	198.3 (±36.9) 200.4 (±73.4)	295 (50–1500) 400 (240–850)	– –	**7 (6–8)** **8 (8–10)**	0 –	1 5	16 (100%) –	0 1	8.8 (±5.1) 8.9 (±6.8)	2 (12.5%) 2 (5.9%)	0 –
Tschuor et al. [54]	9 rCCE – 11 CH/LND 9 rCCE – 14 CH/LND	9/11 9/14	20 23	5 (2–15) 3 (2–11)	– –	193 (112–447) 208 (50–498)	**150 (5–1200)** **350 (75–1500)**	– –	**2.5 (0–6)** **6 (2–22)**	0 –	2 4	16 (100%) 19 (82.6%)	0 0	12.8 (1–62) 14.2 (1–131)	– –	8 14
Sucandy et al. [55]	15 rCCE	11/4	15	–	–	222 (116–453)	200 (10–500)	-	3 (1–16)	0	1	13 (86.7%)	0	30*	2 (13%)	0
Cho et al. [56]	– –	– –	41 41	7.1 (±4.1) 8.0 (±4.8)	– –	202.4 (±35.2) 195.2 (±54.4)	**382.7** **(±386.2)** **717.2** **(±809.1)**	1 5	**6.9 (±2.1)** **8.5 (±4.5)**	0 –	5 4	41 (100%) 41 (100%)	0 0	60 60	4 (9.6%) 16 (39%)	– –
D'Hondt et al. [57]	1 RH-BR-LND-HJ 1 RH-SIVb-BR-LND-HJ	1/1	2	8 (7–9)	2	342.5 (285–400)	95 (40–150)	0	11 (9–13)	0	0	2 (100%)	0	–	–	–
Dugan et al. [58]	19 CH/LND	–	19	6 (±2.7)*	-	282.6 (±117.4)	200.3 (±128.9)	–	4.3 (±3.4)	0	2	17 (89.5%)	0	33.2 (±19.9)	–	9

Statistics reported as mean (± standard deviation) or median (range). Bold values indicate statistical significance ($P < 0.05$)

Abbreviations: *N* Patient population, *M* male, *F* female, *BMI* body mass index, *PVE* portal vein embolization, *ALPPS* Associating Liver Partition and Portal vein Ligation for Staged hepatectomy, *eCCE* extended cholecystectomy/cholecystectomy with lymph node dissection, *rCCE* radical cholecystectomy/cholecystectomy with lymph node dissection and IVb/V wedge resection, *CH* central liver resection, *LND* lymph node dissection, *RH* right hepatectomy, *BR* biliary resection and reconstruction, and *HJ* hepaticojejunostomy node dissection. In contrast to reported laparoscopic conversion rates, ranging from 9 to 20%, only one robotic study to date reported a need for conversion in 4 out of 27 patients with near-perfect R0 resection rates across the literature [57–59]. Tschuor et al. reported no conversions despite a 55% reintervention rate after prior cholecystectomy [54]. Thus, the argument can be made for the superiority of the robotic approach over laparoscopy in adjusting to a hostile operative field and impromptu complexities, without compromising a correct oncological resection.

Nonetheless, compared to the classical open approach, laparoscopic management of GBC has previously been associated with less intraoperative bleeding and shortened hospital stay, with comparable long-term survival outcomes [59, 60]. Four case–control studies are available comparing robotic and open management of GBC [50, 53, 54, 56]. A meta-analysis of three of these retrospective cohorts by Kone et al. revealed a convincing trend toward less perioperative blood loss, a significantly shorter hospital stay, and no significant difference in operative time [61]. The most recent comparative study by Cho et al. matched 41 patients undergoing open versus robotic extended cholecystectomy for T2 or higher stage GBC, and found the robotic group performed better in terms of hospital stay, blood loss, and postoperative pain, with comparable oncological (lymph node yield and R0 resection rate) and long-term survival outcomes [56]. Consequently, the robotic approach is put forward as a valid alternative to its conventional open and laparoscopic counterparts in the management of advanced GBC.

15.2.2.3 Intrahepatic Cholangiocarcinoma

The laparoscopic approach for IHCC has previously been proven safe and feasible [15] and is associated with lower morbidity, perioperative blood loss, and length of stay in comparison to open surgery, with similar survival rates [62, 63]. In recent years, the robotic approach is being increasingly utilized in the management of IHCC [64]. though literature around the topic has yet to catch up. The current evidence is limited and should interpreted carefully (Table 15.3).

Only two comparative studies are available to date, pitting robotic surgery against the open approach for IHCC [64, 65]. Compared with open surgery, a national cohort study by Hamad et al. reported similar oncological outcomes after robotic surgery (lymph node yield, R0 resection, and long-term overall survival), with a shorter hospital stay. However, a larger proportion of patients underwent minor hepatectomy in the robotic group. Shapera et al. found reduced perioperative blood loss during robotic resection, with similar median OS after R0 resection [65]. Compared to recently published perioperative and oncological benchmark values, the available robotic cohorts perform equivalent or better than the reported values [67]. Therefore, the limited evidence provisionally suggests improved short-term outcomes after robotic surgery for IHCC without compromising oncological safety.

15.2.3 Lymphadenectomy for Biliary Cancer

15.2.3.1 Towards a Minimally Invasive Lymphadenectomy

Irrespective of the surgical approach, two key oncological tenets should be abided by during curative-intent surgery for malignant disease: obtaining negative resection margins and performing adequate lymphadenectomy in the right patients [1, 68]. These elements cannot be compromised in favor of short-term outcomes such as length of stay and morbidity. The available literature shows that both open and minimally invasive lymph node dissection for biliary tumors is still widely underperformed [64, 69**], and when performed, rarely meets the standards set forward in the current guidelines. In fact, LND has been identified as the oncological factor with the largest performance difference between high- and low-volume minimally invasive centers [70, 71]. Reluctance toward LND is likely rooted in an uncertainty of the impact of this demanding additional procedure and dissent on the scope of the required nodal clearance. Both subjects are addressed below.

Table 15.3 Reports of robotic resection of IHCCA: demographic and preoperative clinical data, surgical interventions, and oncological and surgical outcomes

Study	Year	Study type	Da Vinci version	N	M/F	Differentiation grade (well/moderate/poor/undifferentiated/ND)	Age	BMI	Tumor size	Tumor multifocality	Neoadjuvant therapy	Adjuvant therapy
Hamad et al. [64]	2022	Case–control (robotic/open)	–	72 *R* 1876 *O*	30/42 899/977	5/44/14/1/8 186/940/465/19/266	68 (±12) 64 (±11)	–	5.1 (±2.3) 6.8 (±8.4)	– –	7 254	21 258
Shapera et al. [65]	2022	Case–control (robotic/open)	Xi	15 *R* 19 *O*	10/5 13/6	11/-/5/-/- 10/-/9/-/-	68.4 (±12.1) 60.7 (±11.4)	31.6 (±7.8) 27 (± 6.2)	5.5 (±2.8) 5.5 (±4.3)	– –	**1** **7**	4 9
Willems et al. [66]	2023	Case report	Xi	1	0/1	–	45	–	–	0	–	–

Study	Operation (major/minor hepatectomy)	Lymph node dissection	Lymph nodes retrieved	Positive nodal status	Operative time (minutes)	Blood loss (cc)	Trans-fusion (N)	Length of stay (days)	Conver-sion	Clavien-Dindo ≥ 3	R0 resection	30-day mortality	FU (months)	Recurrence rate (%)	Deaths during FU
Hamad et al. [64]	**30/36** **1185/561**	34 1027	3.5 (2–7) 4.0 (2–7)	1.0 (1.0–1.5) 1.0 (1.0–2.0)	– –	– –	– –	**5.78 (±4.63)** **8.92 (±10.24)**	-	– –	54 1418	1 44	11 (1–17)	1 (20%)	1
Shapera et al. [65]	13/2 18/1	15 19	6.47 (±1.37) 5.75 (±2.98)	– –	316.5 (±121.1) 332.5 (±117.6)	**258 (±394.1)** **727 (±731.4)**	0 1	6.3 (±6.1) 8.2 (±5.0)	0	2 5	13 12	1 1	–	–	–
Willems et al. [66]	0/1	1	10	0	270	20	0	3	0	0	1	0	21 (6–32)	0	0

Statistics reported as mean (± standard deviation) or median (range). Bold values indicate statistical significance $P < 0.05$

Abbreviations: *N* Patient population, *R* robotic, *O* open, *M* male, *F* female, *BMI* body mass index, and *FU* follow-up

Beyond these factors, the need for a thorough lymphadenectomy in a deep and perilous space has historically been one of the primary obstacles in the implementation of minimally invasive surgery in the hepatobiliary field. Though laparoscopic lymphadenectomy for biliary tumors has been shown to be feasible, with lower blood loss and shorter length of stay, the technical complexities of the procedure have hampered widespread dissemination beyond high-volume centers [72]. The requirement for complex vascular and biliary dissection and control in a deep and narrow space is addressed by the robotic-enhanced and stable 3D/ICG camera, increased range of instrument motion, and superior ergonomics during these often long and demanding procedures. These factors are already known to contribute to a lower conversion rate in robotic surgery and offer the possibility to expand the indications of minimally invasive surgery. Moreover, an NCDB study by Kim et al. showed high-volume centers outperformed low-volume centers in terms of R0-resection rate, LND rate, and lymph node yield >6 with respect to laparoscopic surgery for biliary cancer, highlighting the complexity of these procedures [71]. In the robotic cohort, no significant difference was reported; however, the group was considerably smaller. The robotic platform has the potential to overcome technical difficulties and long learning curves, thus leveling the playing field and making minimally invasive lymphadenectomy more accessible.

Similarly to robotic surgery for gastric cancer, some studies have translated these advantages into a higher lymph node yield using a robotic approach for biliary tumors [43, 73, 74]. Ratti et al. matched patients treated by way of open, laparoscopic and robotic surgery, and demonstrated the oncological noninferiority of the robotic approach for biliary tumors requiring LND [74]. While preserving the short-term benefits of minimally invasive surgery (blood loss, functional recovery, length of stay), admittedly with longer operative times, the robotic approach also proved superior to its laparoscopic and open counterparts with regard to retrieval of >6 lymph nodes (resp. 96%, 86.6%, 79.6%). The duration of robotic LND was significantly reduced in comparison to the laparoscopic group and draws near to the duration for open LND. Interestingly, the authors recorded no learning curve, theorizing that minimally invasive skills need only be acquired once and are fully transferable to the robotic platform, further encouraging the broadening indications of robotic surgery.

15.2.3.2 Indications and Extent

Perihilar Cholangiocarcinoma

Survival after curative-intent resection of PHCCA is determined by a variety of prognostic factors, such as R0 resection, perineural invasion, tumor differentiation, size, and staging. Positive resection margins and presence of lymph node metastasis represent the most important prognostic factors for poorer overall and DFS in PHCCA [4]. Even so, adequate LND remains ill-defined and consequently underperformed, leading to disease understaging and suboptimal patient stratification for adjuvant treatment.

The incidence of lymph node metastasis ranges around 30–60% for PHCCA and increases with disease stage. There is a lack of robust data regarding the ideal total lymph node yield during LND for PHCC. Several retrospective cohorts found a significantly improved survival with a higher lymph node count [75–78]. While each identified different cutoffs, these reports suggest an important role of adequate lymph node sampling in accurate staging and prognostication. Similarly to breast cancer, some authors have advocated for the use of the positive lymph node ratio to identify the impact of the number of involved nodes to the total collected lymph nodes [79–82]. However, total lymph node yield in PHCCA is generally limited and small differences offset LNR greatly, challenging its clinical value pending further investigation.

Metastatic nodes are mostly found along the bile duct, portal vein, and common hepatic artery [4]. In accordance with these findings, the current AJCC guidelines suggest performing lymphadenectomy along the hepatoduodenal ligament. Lymph nodes beyond this region are considered a hallmark of distant metastatic disease. Some reports have suggested that extended

LND (including the para-aortic, celiac, and superior mesenteric nodes) provides a survival benefit for patients with distant lymph node metastasis. Mantel et al. later confirmed the association between extended LND and higher lymph node yield, better staging, and survival. However, the survival benefit was limited to patients with M0 disease after R0 resection [83, 84]. Consequently, there is insufficient evidence to infer a therapeutic effect of extended LND for distant lymph node metastasis. Instead, these patients may be better served with adjuvant chemotherapy [85].

Gallbladder Carcinoma

The lymphatic system represents the primary metastatic route for GBC, making LND fundamental for proper oncological resection, accurate staging, and patient selection for adjuvant therapy [86]. Whilst the evidence supporting LND for T2 and T3 GBC regardless of nodal status is overwhelming, LND is still widely underperformed.

There has been some discussion in the past on the role of LND for early-stage GBC. When confined to the mucosal layer, early GBC is classified as T1a and is routinely considered to be localized disease. Lymph node involvement is described in around 2% of patients. Therefore, contingent upon negative resection margins, a simple cholecystectomy yields a near-perfect 5-year survival rate of 95–100%. Extended cholecystectomy (involving lymph node dissection) is not associated with better survival and consequently not advised [87, 88].

In T1b tumors, invading the muscularis, lymph node metastasis is more common, described in 11–21% of all cases. Consequently, in addition to cholecystectomy with en-bloc liver resection, nodal clearance is required, and results in better disease staging and survival [87–92]. A recent meta-analysis confirmed significantly improved overall survival benefit in patients undergoing LND for T1b to T3 tumors, with insufficient data on T4 tumors to complete a meta-analysis [92].

While generally considered a contraindication for resection, some surgeons have suggested that aggressive surgery for advanced GBC, even with para-aortic lymph node metastasis, may increase long-term survival [93, 94]. Conversely, extended lymphadenectomy may cause unnecessary complications due to the complex nature of this advanced procedure. While it may contribute to staging and prognostication, a wider LND in the para-aortic, celiac, or SMA region is not linked to improved outcomes as the involvement of these nodes represents distant metastatic disease [95]. Similarly, a 2020 study by Sahara et al. found the survival benefit of LND for T4 tumors and patients with CA19.9 $\geq$ 200 UI/mL to be limited, linking both factors to systemic spreading [12]. To date, there is insufficient evidence to support curative intent resection in these patients, leading the AJCC to leave the decision to the surgeon's discretion [13].

The scope of the lymphadenectomy, as well as the number of lymph nodes required for accurate staging, has long been subject to debate. In fact, the definition of the N-stage was changed in every edition of the American Joint Committee on Cancer (AJCC) staging manual between 1997 and 2017 [92]. For accurate staging and risk stratification, the eighth edition AJCC Manual currently recommends collecting at least six lymph nodes. This direction is supported by a number of retrospective studies reporting a significantly longer DFS at this cutoff for N0 disease, while presence of metastatic nodes leads to poorer survival [12, 66, 96, 97]. Besides the minimum required nodes, current guidelines do not specify the recommended nodal stations to include in a qualitative lymphadenectomy specimen. The HDL nodes are most frequently implicated, followed by the postero-superior pancreaticoduodenal nodes. Thus, for sufficient nodal yield, dissection should at least include the N1 nodes, from the cystic duct node to the CBD and HDL nodes and may extend to the postero-superior pancreaticoduodenal region. Positive nodes outside this scope (e.g., para-aortic, celiac, and SMA) are associated with poorer prognosis and are now considered distant metastatic disease [98, 99]. Finally, routine CBD resection has not been found to increase lymph node yield or improve survival [12, 68, 100].

Intrahepatic Cholangiocarcinoma

Curative-intent treatment of IHCC centers around radical surgical resection [101, 102]. Lymph node metastasis is found in approximately 20–50% of patients with IHCC [103] and is independently associated with poor prognosis [104, 105]. The therapeutic role of routine lymphadenectomy remains equivocal, as current recommendations are based on expert opinion [70].

Some studies have reported a lack of survival advantage compared to standalone liver resection [105, 106], deeming LND a purely diagnostic, but complex procedure, with a significant risk of vascular and biliary injury [107]. Increased postoperative morbidity has been reported after LND, in particular among cirrhotic patients [108]. Consequently, many surgeons opt out of performing LND. Nevertheless, assessment of the nodal status is critical for accurate staging and defines a population that may benefit from adjuvant therapy [86, 109]. The current NCCN and AJCC guidelines recommend harvesting at least six lymph nodes for accurate staging. Since its publishing, this recommendation has been endorsed by several large international reports, reporting poorer survival when harvesting <6 nodes in pN0 patients [110, 111].

Some studies have shown an association between an adequate LND during hepatectomy for IHCC and a trend toward improved local disease control and survival [103, 107, 112, 113], though literature on long-term oncological benefit remains conflicting. A recent meta-analysis of 17 studies could not demonstrate a significant difference in DFS or OS in patients undergoing LND in comparison to those who did not [114]. This conclusion is nuanced by the unmatched comparisons and heterogeneity in approach in the included retrospective series. Conversely, a recent retrospective study by Sposito et al. reported better DFS and OS after LND in cN0 patients with pathological positive nodal status, demonstrating a potentially therapeutic effect of LND [113].

Historically, most publications have not accurately outlined the scope of their lymph node dissection, nor have they acknowledged the fact that the pattern of lymphatic spread may be informed by the anatomic location of the tumor. While the NCCN guidelines recommend LND along the porta hepatis, the AJCC endorses more extensive lymphadenectomy based on tumor location. Standard portal LND involves clearing the HDL (station 12) and the nodes along the common hepatic artery (station 8) and is recommended for all patients undergoing IHCCA resection. Based on the laterality of lymphatic drainage patterns, additional lymphadenectomy may be needed. For IHCCA of the right liver, additional clearing of the peri-duodenal and peripancreatic nodes is recommended, while LND along the lesser curve and paracardial region is recommended for left-sided tumors. Celiac, caval, and para-aortic nodal involvement is considered distant metastatic disease [115, 116].

15.3 Conclusion

Biliary tract cancer is rare and carries a dismal prognosis. Radical surgical resection is the only treatment modality associated with long-term survival benefits. While MILS has found its way into current practice, the open approach endures as the mainstay in the treatment of biliary tumors. Since the first accounts on robotic management of PHCCA and GBC, more reports are gradually emerging, indicating a shift toward minimally invasive surgery for even the most complex hepatobiliary procedures. In view of the disease's uncommon nature, high mortality, and low resectability, there is an explicit lack of prospective and randomized data. Still, the available literature indicates that a robotic approach to complex liver and bile duct resections is safe and feasible and, in experienced hands, presents an oncologically sound alternative to its open and laparoscopic counterpart. Moreover, the short-term benefits of the minimally invasive approach are preserved, resulting in decreased blood loss and shorter hospital stay, with infrequent need for conversion.

Lymphadenectomy, with a requisite for fine dissection in a narrow and perilous space, has constituted one of the primary challenges in the implementation and wider propagation of laparoscopy for biliary tumors. The robotic system, with its stable high-quality 3D camera, tremor filtering, and increased range of motion, has unmistakably addressed the limited maneuverability of its laparoscopic counterpart. Robust outcome data regarding robotic LND for biliary tumors is limited, but oncological noninferiority has been displayed, with some reports even describing superior results with respect to lymph node yield. From an oncological perspective, qualitative LND remains the most accurate staging procedure. While some studies report a small fraction of long-term survivors with lymph node metastasis after extended LND, evidence regarding the therapeutic effect of LND for biliary tumors is conflicting. Prognosis remains largely driven by surgical margin and nodal status, making these factors essential for patient stratification and further treatment allocation. Although adjuvant therapy is not yet standardized, future advancements may rely on adequate staging to convincingly demonstrate treatment benefits. Still, perioperative risks should be carefully weighed, especially in cirrhotic patients. Ultimately there is a need for standardized reporting to specify the role of LND in the management of biliary cancer.

References

1. Vogel A, Bridgewater J, Edeline J, Kelley RK, Klümpen HJ, Malka D, Primrose JN, Rimassa L, Stenzinger A, Valle JW, Ducreux M, ESMO Guidelines Committee. Biliary tract cancer: ESMO Clinical Practice Guideline for diagnosis, treatment and follow-up. Ann Oncol. 2023;34(2):127–40. https://doi.org/10.1016/j.annonc.2022.10.506.
2. de Jong MC, Marques H, Clary BM, Bauer TW, Marsh JW, Ribero D, Majno P, Hatzaras I, Walters DM, Barbas AS, Mega R, Schulick RD, Choti MA, Geller DA, Barroso E, Mentha G, Capussotti L, Pawlik TM. The impact of portal vein resection on outcomes for hilar cholangiocarcinoma: a multi-institutional analysis of 305 cases. Cancer. 2012;118(19):4737–47. https://doi.org/10.1002/cncr.27492.
3. Rea DJ, Munoz-Juarez M, Farnell MB, Donohue JH, Que FG, Crownhart B, Larson D, Nagorney DM. Major hepatic resection for hilar cholangiocarcinoma: analysis of 46 patients. Arch Surg (Chicago, IL: 1960). 2004;139(5):514–25. https://doi.org/10.1001/archsurg.139.5.514.
4. Liang L, Li C, Jia HD, Diao YK, Xing H, Pawlik TM, Lau WY, Shen F, Huang DS, Zhang CW, Yang T. Prognostic factors of resectable perihilar cholangiocarcinoma: a systematic review and meta-analysis of high-quality studies. Ther Adv Gastrointest Endosc. 2021;14:2631774521993065. https://doi.org/10.1177/2631774521993065.
5. Bismuth H, Corlette MB. Intrahepatic cholangioenteric anastomosis in carcinoma of the hilus of the liver. Surg Gynecol Obstet. 1975;140(2):170–8.
6. Guadagni S, Comandatore A, Furbetta N, Di Franco G, Carpenito C, Bechini B, Vagelli F, Ramacciotti N, Palmeri M, Di Candio G, Morelli L. Robotic hepatectomy plus biliary reconstruction for Bismuth type III and type IV hilar cholangiocarcinoma: state of the art and literature review. J Pers Med. 2023;14(1):12. https://doi.org/10.3390/jpm14010012.
7. Kobayashi A, Miwa S, Nakata T, Miyagawa S. Disease recurrence patterns after R0 resection of hilar cholangiocarcinoma. Br J Surg. 2010;97(1):56–64. https://doi.org/10.1002/bjs.6788.
8. de Jong MC, Nathan H, Sotiropoulos GC, Paul A, Alexandrescu S, Marques H, Pulitano C, Barroso E, Clary BM, Aldrighetti L, Ferrone CR, Zhu AX, Bauer TW, Walters DM, Gamblin TC, Nguyen KT, Turley R, Popescu I, Hubert C, Meyer S, et al. Intrahepatic cholangiocarcinoma: an international multi-institutional analysis of prognostic factors and lymph node assessment. J Clin Oncol. 2011;29(23):3140–5. https://doi.org/10.1200/JCO.2011.35.6519.
9. Stinton LM, Shaffer EA. Epidemiology of gallbladder disease: cholelithiasis and cancer. Gut and liver. 2012;6(2):172–87. https://doi.org/10.5009/gnl.2012.6.2.172.
10. Konstantinidis IT, Deshpande V, Genevay M, Berger D, Fernandez-del Castillo C, Tanabe KK, Zheng H, Lauwers GY, Ferrone CR. Trends in presentation and survival for gallbladder cancer during a period of more than 4 decades: a single-institution experience. Arch Surg (Chicago, IL: 1960). 2009;144(5):441–7. https://doi.org/10.1001/archsurg.2009.46.
11. Edeline J, Benabdelghani M, Bertaut A, Watelet J, Hammel P, Joly JP, Boudjema K, Fartoux L, Bouhier-Leporrier K, Jouve JL, Faroux R, Guerin-Meyer V, Kurtz JE, Assénat E, Seitz JF, Baumgaertner I, Tougeron D, de la Fouchardière C, Lombard-Bohas C, Boucher E, et al. Gemcitabine and Oxaliplatin Chemotherapy or Surveillance in Resected Biliary Tract Cancer (PRODIGE 12-ACCORD 18-UNICANCER GI): A Randomized Phase III Study. J Clin Oncol. 2019;37(8):658–67. https://doi.org/10.1200/JCO.18.00050.
12. Sahara K, Tsilimigras DI, Maithel SK, Abbott DE, Poultsides GA, Hatzaras I, Fields RC, Weiss M, Scoggins C, Isom CA, Idrees K, Shen P, Endo I, Pawlik TM, The US Extrahepatic Biliary Malignancy Consortium. Survival benefit of lymph-

adenectomy for gallbladder cancer based on the therapeutic index: an analysis of the US extrahepatic biliary malignancy consortium. J Surg Oncol. 2020;121(3):503–10. https://doi.org/10.1002/jso.25825.

13. Lee SE, Kim KS, Kim WB, Kim IG, Nah YW, Ryu DH, Park JS, Yoon MH, Cho JY, Hong TH, Hwang DW, Choi DW, Korean Association of Hepato-Biliary and Pancreas Surgery. Practical guidelines for the surgical treatment of gallbladder cancer. J Korean Med Sci. 2014;29(10):1333–40. https://doi.org/10.3346/jkms.2014.29.10.1333.
14. Zwart MJW, Görgec B, Arabiyat A, Nota CLM, van der Poel MJ, Fichtinger RS, Berrevoet F, van Dam RM, Aldrighetti L, Fuks D, Hoti E, Edwin B, Besselink MG, Abu Hilal M, Hagendoorn J, Swijnenburg RJ, Dutch Liver Collaborative Group and E-AHPBA Innovation & Development Committee. Pan-European survey on the implementation of robotic and laparoscopic minimally invasive liver surgery. HPB. 2022;24(3):322–31. https://doi.org/10.1016/j.hpb.2021.08.939.
15. Wakabayashi G, Cherqui D, Geller DA, Buell JF, Kaneko H, Han HS, Asbun H, O'Rourke N, Tanabe M, Koffron AJ, Tsung A, Soubrane O, Machado MA, Gayet B, Troisi RI, Pessaux P, Van Dam RM, Scatton O, Abu Hilal M, Belli G, et al. Recommendations for laparoscopic liver resection: a report from the second international consensus conference held in Morioka. Ann Surg. 2015;261(4):619–29. https://doi.org/10.1097/SLA.0000000000001184.
16. Fretland ÅA, Dagenborg VJ, Bjørnelv GMW, Kazaryan AM, Kristiansen R, Fagerland MW, Hausken J, Tønnessen TI, Abildgaard A, Barkhatov L, Yaqub S, Røsok BI, Bjørnbeth BA, Andersen MH, Flatmark K, Aas E, Edwin B. Laparoscopic versus open resection for colorectal liver metastases: the OSLO-COMET randomized controlled trial. Ann Surg. 2018;267(2):199–207. https://doi.org/10.1097/SLA.0000000000002353.
17. Qin T, Wang M, Zhang H, Li J, Deng X, Zhang Y, Zhao W, Fan Y, Li D, Chen X, Feng Y, Zhu S, Xing Z, Yu G, Xu J, Xie J, Dou C, Ma H, Liu G, Shao Y, et al. The long-term outcome of laparoscopic resection for Perihilar Cholangiocarcinoma compared with the open approach: a real-world multicentric analysis. Ann Surg Oncol. 2023;30(3):1366–78. https://doi.org/10.1245/s10434-022-12647-1.
18. Krenzien F, Schmelzle M, Pratschke J, Feldbrügge L, Liu R, Liu Q, Zhang W, Zhao JJ, Tan HL, Cipriani F, Hoogteijling TJ, Aghayan DL, Fretland ÅA, Siow TF, Lim C, Scatton O, Herman P, Coelho FF, Marino MV, Mazzaferro V, et al. Propensity score-matching analysis comparing robotic versus laparoscopic limited liver resections of the posterosuperior segments: an international multicenter study. Ann Surg. 2024;279(2):297–305. https://doi.org/10.1097/SLA.0000000000006027.
19. Chong Y, Prieto M, Gastaca M, Choi SH, Sucandy I, Chiow AKH, Marino MV, Wang X, Efanov M, Schotte H, D'Hondt M, Choi GH, Krenzien F, Schmelzle M, Pratschke J, Kingham TP, Giglio M, Troisi RI, Lee JH, Lai EC, et al. An international multicentre propensity score matched analysis comparing between robotic versus laparoscopic left lateral sectionectomy. Surg Endosc. 2023;37(5):3439–48. https://doi.org/10.1007/s00464-022-09790-x.
20. Liu L, Lewis N, Mhaskar R, Sujka J, DuCoin C. Robotic-assisted foregut surgery is associated with lower rates of complication and shorter post-operative length of stay. Surg Endosc. 2023;37(4):2800–5. https://doi.org/10.1007/s00464-022-09814-6.
21. Liu S, Liu X, Li X, Li O, Yi W, Khan J, Yang P, Guo C, Peng C, Jiang B. Application of laparoscopic radical resection for type III and IV hilar Cholangiocarcinoma treatment. Gastroenterol Res Pract. 2020;2020:1506275. https://doi.org/10.1155/2020/1506275.
22. Zhang Y, Dou C, Wu W, Liu J, Jin L, Hu Z, Zhang C. Total laparoscopic versus open radical resection for hilar cholangiocarcinoma. Surg Endosc. 2020;34(10):4382–7. https://doi.org/10.1007/s00464-019-07211-0.
23. Ratti F, Fiorentini G, Cipriani F, Catena M, Paganelli M, Aldrighetti L. Perihilar cholangiocarcinoma: are we ready to step towards minimally invasiveness? Updat Surg. 2020;72(2):423–33. https://doi.org/10.1007/s13304-020-00752-3.
24. Franken LC, van der Poel MJ, Latenstein AEJ, Zwart MJ, Roos E, Busch OR, Besselink MG, van Gulik TM. Minimally invasive surgery for perihilar cholangiocarcinoma: a systematic review. J Robot Surg. 2019;13(6):717–27. https://doi.org/10.1007/s11701-019-00964-9.
25. Tang W, Qiu JG, Deng X, Liu SS, Cheng L, Liu JR, Du CY. Minimally invasive versus open radical resection surgery for hilar cholangiocarcinoma: comparable outcomes associated with advantages of minimal invasiveness. PLoS One. 2021;16(3):e0248534. https://doi.org/10.1371/journal.pone.0248534.
26. Agarwal AK, Javed A, Kalayarasan R, Sakhuja P. Minimally invasive versus the conventional open surgical approach of a radical cholecystectomy for gallbladder cancer: a retrospective comparative study. HPB. 2015;17(6):536–41. https://doi.org/10.1111/hpb.12406.
27. Itano O, Oshima G, Minagawa T, Shinoda M, Kitago M, Abe Y, Hibi T, Yagi H, Ikoma N, Aiko S, Kawaida M, Masugi Y, Kameyama K, Sakamoto M, Kitagawa Y. Novel strategy for laparoscopic treatment of pT2 gallbladder carcinoma. Surg Endosc. 2015;29(12):3600–7. https://doi.org/10.1007/s00464-015-4116-y.
28. Kasai M, Cipriani F, Gayet B, Aldrighetti L, Ratti F, Sarmiento JM, Scatton O, Kim KH, Dagher I, Topal B, Primrose J, Nomi T, Fuks D, Abu Hilal M. Laparoscopic versus open major hepatectomy: a systematic review and meta-analysis of individual

patient data. Surgery. 2018;163(5):985–95. https://doi.org/10.1016/j.surg.2018.01.020.
29. Giulianotti PC, Sbrana F, Bianco FM, Addeo P. Robot-assisted laparoscopic extended right hepatectomy with biliary reconstruction. Journal of laparoendoscopic & advanced surgical techniques. Part A. 2010;20(2):159–63. https://doi.org/10.1089/lap.2009.0383.
30. Shen BY, Zhan Q, Deng XX, Bo H, Liu Q, Peng CH, Li HW. Radical resection of gallbladder cancer: could it be robotic? Surg Endosc. 2012;26(11):3245–50. https://doi.org/10.1007/s00464-012-2330-4.
31. Liu QD, Chen JZ, Xu XY, Zhang T, Zhou NX. Incidence of port-site metastasis after undergoing robotic surgery for biliary malignancies. World J Gastroenterol. 2012;18(40):5695–701. https://doi.org/10.3748/wjg.v18.i40.5695.
32. Xu Y, Wang H, Ji W, Tang M, Li H, Leng J, Meng X, Dong J. Robotic radical resection for hilar cholangiocarcinoma: perioperative and long-term outcomes of an initial series. Surg Endosc. 2016;30(7):3060–70. https://doi.org/10.1007/s00464-016-4925-7.
33. Quijano Y, Vicente E, Ielpo B, Duran H, Diaz E, Fabra I, Olivares S, Ferri V, Ortega I, Malavé L, Ferronetti A, Piccinni G, Caruso R. Robotic liver surgery: early experience from a single surgical center. Surg Laparosc Endosc Percutan Tech. 2016;26(1):66–71. https://doi.org/10.1097/SLE.0000000000000227.
34. Li J, Tan X, Zhang X, Zhao G, Hu M, Zhao Z, Liu R. Robotic radical surgery for hilar cholangiocarcinoma: A single-centre case series. The international journal of medical robotics + computer assisted surgery. MRCAS. 2020;16(2):e2076. https://doi.org/10.1002/rcs.2076.
35. Machado MA, Mattos BV, Lobo Filho MM, Makdissi F. Robotic resection of hilar cholangiocarcinoma. Ann Surg Oncol. 2020;27(11):4166–70. https://doi.org/10.1245/s10434-020-08514-6.
36. Marino MV, Pellino G, Ahmad A. The robotic-assisted approach for left-side predominance hilar cholangiocarcinoma: a video technique. Updat Surg. 2020;72(3):911–2. https://doi.org/10.1007/s13304-020-00777-8.
37. Cillo U, D'Amico FE, Furlanetto A, Perin L, Gringeri E. Robotic hepatectomy and biliary reconstruction for perihilar cholangiocarcinoma: a pioneer western case series. Updat Surg. 2021;73(3):999–1006. https://doi.org/10.1007/s13304-021-01041-3.
38. Sucandy I, Ross S, Rosemurgy A. Robotic resection of a type IIIB Klatskin tumor. J Gastrointest Surg. 2021;25(7):1939–40. https://doi.org/10.1007/s11605-021-04968-5.
39. Sucandy I, Shapera E, Jacob K, Luberice K, Crespo K, Syblis C, Ross SB, Rosemurgy AS. Robotic resection of extrahepatic cholangiocarcinoma: institutional outcomes of bile duct cancer surgery using a minimally invasive technique. J Surg Oncol. 2022;125(2):161–7. https://doi.org/10.1002/jso.26674.
40. Di Benedetto F, Magistri P, Guerrini GP, Di Sandro S. Robotic liver partition and portal vein embolization for staged hepatectomy for perihilar cholangiocarcinoma. Updat Surg. 2022;74(2):773–7. https://doi.org/10.1007/s13304-021-01209-x.
41. Tee MC, Brahmbhatt RD, Franko J. Robotic resection of type I hilar Cholangiocarcinoma with Intrapancreatic bile duct dissection. Ann Surg Oncol. 2022;29(2):964–9. https://doi.org/10.1245/s10434-021-10811-7.
42. Jacoby H, Rayman S, Ross S, Crespo K, Syblis C, Rosemurgy A, Sucandy I. Robotic resection of hilar cholangiocarcinoma: a single institution experience. Mini-invasive Surgery. 2022;6:58. https://doi.org/10.20517/2574-1225.2022.58.
43. D'Hondt M, Wicherts DA. Pure robotic major hepatectomy with biliary reconstruction for hepatobiliary malignancies: first European results. Surg Endosc. 2023;37(6):4396–402. https://doi.org/10.1007/s00464-023-09863-5.
44. Sucandy I, Marques HP, Lippert T, Magistri P, Coelho JS, Ross SB, Chumbinho B, Di Sandro S, DiBenedetto F. Clinical outcomes of robotic resection for perihilar cholangiocarcinoma: A first, multicenter, trans-Atlantic, expert-center, collaborative study. Ann Surg Oncol. 2024;31(1):81–9. https://doi.org/10.1245/s10434-023-14307-4.
45. Dugan MM, Sucandy I, Ross SB, Crespo K, Syblis C, Alogaidi M, Rosemurgy A. Analysis of survival outcomes following robotic hepatectomy for malignant liver diseases. Am J Surg. 2024;228:252–7. https://doi.org/10.1016/j.amjsurg.2023.10.034.
46. Mueller M, Breuer E, Mizuno T, Bartsch F, Ratti F, Benzing C, Ammar-Khodja N, Sugiura T, Takayashiki T, Hessheimer A, Kim HS, Ruzzenente A, Ahn KS, Wong T, Bednarsch J, D'Silva M, Koerkamp BG, Jeddou H, López-López V, de Ponthaud C, et al. Perihilar cholangiocarcinoma – novel benchmark values for surgical and oncological outcomes from 24 expert centers. Ann Surg. 2021;274(5):780–8. https://doi.org/10.1097/SLA.0000000000005103.
47. Zeng G, Teo NZ, Goh BKP. Short-term outcomes of minimally invasive surgery for patients presenting with suspected gallbladder cancer: report of 8 cases. J Min Access Surg. 2019;15(2):109–14. https://doi.org/10.4103/jmas.JMAS_229_17.
48. Sinagra E, Garritano S, Iacopinelli SM, Messina M, Raimondo D, Rossi F, Spada M, Martorana G, Spampinato MG. Minimally invasive surgical approach for radicalization of incidental post-cholecystectomy gallbladder carcinoma: safety, feasibility and outcomes. MITAT. 2018;27(4):217–20. https://doi.org/10.1080/13645706.2017.1410489.
49. Khan S, Beard RE, Kingham PT, Fong Y, Boerner T, Martinie JB, Vrochides D, Buell JF, Berber E, Kahramangil B, Troisi RI, Vanlander A, Molinari M, Tsung A. Long-term oncologic outcomes following robotic liver resections for primary hepatobiliary malignancies: A multicenter study. Ann Surg Oncol.

2018;25(9):2652–60. https://doi.org/10.1245/s10434-018-6629-9.
50. Goel M, Khobragade K, Patkar S, Kanetkar A, Kurunkar S. Robotic surgery for gallbladder cancer: operative technique and early outcomes. J Surg Oncol. 2019;119(7):958–63. https://doi.org/10.1002/jso.25422.
51. Araujo RLC, de Sanctis MA, Coelho TRV, Felippe FEC, Burgardt D, Wohnrath DR. Robotic surgery as an alternative approach for reoperation of incidental gallbladder cancer. J Gastrointest Cancer. 2020;51(1):332–4. https://doi.org/10.1007/s12029-019-00264-3.
52. Ahmad A. Use of indocyanine green (ICG) augmented near-infrared fluorescence imaging in robotic radical resection of gallbladder adenocarcinomas. Surg Endosc. 2020;34(6):2490–4. https://doi.org/10.1007/s00464-019-07053-w.
53. Byun Y, Choi YJ, Kang JS, Han Y, Kim H, Kwon W, Jang JY. Early outcomes of robotic extended cholecystectomy for the treatment of gallbladder cancer. J Hepatobiliary Pancreat Sci. 2020;27(6):324–30. https://doi.org/10.1002/jhbp.717.
54. Tschuor C, Pickens RC, Isenberg EE, Motz BM, Salibi PN, Robinson JN, Murphy KJ, Iannitti DA, Baker EH, Vrochides D, Martinie JB. Robotic resection of gallbladder cancer: A single-center retrospective comparative study to open resection. Am Surg. 2023;89(4):888–96. https://doi.org/10.1177/00031348211047491.
55. Sucandy I, Jabbar F, Syblis C, Crespo K, Ross S, Rosemurgy A. Robotic central hepatectomy for the treatment of gallbladder carcinoma. Outcomes of minimally invasive approach. Am Surg. 2022;88(3):348–51. https://doi.org/10.1177/00031348211047457.
56. Cho YJ, Yun WG, Jung HS, Lee M, Han Y, Kwon W, Jang JY. Oncologic safety of robotic extended cholecystectomy for gallbladder cancer. Surg Endosc. 2023;37(12):9089–97. https://doi.org/10.1007/s00464-023-10463-6.
57. Lee JW, Kwon JH, Lee JW. Oncologic and long-term outcomes of laparoscopic and open extended cholecystectomy for gallbladder cancer. J Clin Med. 2022;11(8):2132. https://doi.org/10.3390/jcm11082132.
58. Ong CT, Leung K, Nussbaum DP, Sun Z, Gloor B, Blazer DG 3rd, Worni M. Open versus laparoscopic portal lymphadenectomy in gallbladder cancer: is there a difference in lymph node yield? HPB. 2018;20(6):505–13. https://doi.org/10.1016/j.hpb.2017.10.015.
59. Zhang W, Ouyang DL, Che X. Short- and long-term outcomes of laparoscopic vs open surgery for T2 gallbladder cancer: A systematic review and meta-analysis. World J Gastrointest Surg. 2022;14(12):1387–96. https://doi.org/10.4240/wjgs.v14.i12.1387.
60. Dou C, Zhang C, Zhang C, Liu J. Propensity score analysis of outcomes following laparoscopic or open radical resection for gallbladder cancer in T2 and T3 stages. J Gastrointest Surg. 2022;26(7):1416–24. https://doi.org/10.1007/s11605-022-05288-y.
61. Kone LB, Bystrom PV, Maker AV. Robotic surgery for biliary tract cancer. Cancers. 2022;14(4):1046. https://doi.org/10.3390/cancers14041046.
62. Ratti F, Casadei-Gardini A, Cipriani F, Fiorentini G, Pedica F, Burgio V, Cascinu S, Aldrighetti L. Laparoscopic surgery for intrahepatic cholangiocarcinoma: A focus on oncological outcomes. J Clin Med. 2021;10(13):2828. https://doi.org/10.3390/jcm10132828.
63. Hobeika C, Cauchy F, Fuks D, Barbier L, Fabre JM, Boleslawski E, Regimbeau JM, Farges O, Pruvot FR, Pessaux P, Salamé E, Soubrane O, Vibert E, Scatton O, AFC-LLR-2018 study group. Laparoscopic versus open resection of intrahepatic cholangiocarcinoma: nationwide analysis. Br J Surg. 2021;108(4):419–26. https://doi.org/10.1093/bjs/znaa110.
64. Hamad A, Ansari A, Li Y, Shen C, Cloyd J, Pawlik TM, Ejaz A. Short- and long-term outcomes following robotic and open resection for intrahepatic cholangiocarcinoma: A national cohort study. Surg Oncol. 2022;43:101790. https://doi.org/10.1016/j.suronc.2022.101790.
65. Shapera EA, Ross S, Syblis C, Crespo K, Rosemurgy A, Sucandy I. Analysis of oncological outcomes after robotic liver resection for intrahepatic cholangiocarcinoma. Am Surg. 2023;89(6):2399–412. https://doi.org/10.1177/00031348221093933.
66. Willems E, D'Hondt M. Robotic left hemihepatectomy with lymph node dissection for intrahepatic cholangiocarcinoma (with video). J Visc Surg. 2023;160(2):151–2. https://doi.org/10.1016/j.jviscsurg.2023.02.005.
67. Alaimo L, Endo Y, Catalano G, Ruzzenente A, Aldrighetti L, Weiss M, Bauer TW, Alexandrescu S, Poultsides GA, Maithel SK, Marques HP, Martel G, Pulitano C, Shen F, Cauchy F, Koerkamp BG, Endo I, Kitago M, Pawlik TM. Benchmarks in liver resection for intrahepatic cholangiocarcinoma. Ann Surg Oncol. Advance online publication. 2024;31:3043. https://doi.org/10.1245/s10434-023-14880-8.
68. Benson AB 3rd, D'Angelica MI, Abbott DE, Abrams TA, Alberts SR, Saenz DA, Are C, Brown DB, Chang DT, Covey AM, Hawkins W, Iyer R, Jacob R, Karachristos A, Kelley RK, Kim R, Palta M, Park JO, Sahai V, Schefter T, et al. NCCN guidelines insights: hepatobiliary cancers, version 1.2017. JNCCN. 2017;15(5):563–73. https://doi.org/10.6004/jnccn.2017.0059.
69. Hamad A, Cloyd JM, Dillhoff M, Manilchuk A, Pawlik TM, Tsung A, Ejaz A. Comparison of lymph node evaluation and yield among patients undergoing open and minimally invasive surgery for gallbladder adenocarcinoma. Surg Endosc. 2021;35(5):2223–8. https://doi.org/10.1007/s00464-020-07635-z.
70. Edge SB, American Joint Committee on Cancer, A.C.S. AJCC cancer staging handbook: from the

AJCC cancer staging manual, vol. 19. New York: Springer; 2010. p. 718.
71. Kim BJ, Newhook TE, Tzeng CD, Ikoma N, Chiang YJ, Chun YS, Vauthey JN, Tran Cao HS. Lymphadenectomy and margin-negative resection for biliary tract cancer surgery in the United States-differential technical performance by approach. J Surg Oncol. 2022;126(4):658–66. https://doi.org/10.1002/jso.26924.
72. Ratti F, Cipriani F, Ariotti R, Gagliano A, Paganelli M, Catena M, Aldrighetti L. Safety and feasibility of laparoscopic liver resection with associated lymphadenectomy for intrahepatic cholangiocarcinoma: a propensity score-based case-matched analysis from a single institution. Surg Endosc. 2016;30(5):1999–2010. https://doi.org/10.1007/s00464-015-4430-4.
73. Guerrini GP, Esposito G, Magistri P, Serra V, Guidetti C, Olivieri T, Catellani B, Assirati G, Ballarin R, Di Sandro S, Di Benedetto F. Robotic versus laparoscopic gastrectomy for gastric cancer: the largest meta-analysis. Int J Surg (London, England). 2020;82:210–28. https://doi.org/10.1016/j.ijsu.2020.07.053.
74. Ratti F, Cipriani F, Ingallinella S, Tudisco A, Catena M, Aldrighetti L. Robotic approach for lymphadenectomy in biliary tumors: the missing ring between the benefits of laparoscopic and reproducibility of open approach? Ann Surg. 2023;278(4):e780–8. https://doi.org/10.1097/SLA.0000000000005748.
75. Bagante F, Tran T, Spolverato G, Ruzzenente A, Buttner S, Ethun CG, Groot Koerkamp B, Conci S, Idrees K, Isom CA, Fields RC, Krasnick B, Weber SM, Salem A, Martin RC, Scoggins C, Shen P, Mogal HD, Schmidt C, Beal E, et al. Perihilar cholangiocarcinoma: number of nodes examined and optimal lymph node prognostic scheme. J Am Coll Surg. 2016;222(5):750–759.e2. https://doi.org/10.1016/j.jamcollsurg.2016.02.012.
76. Schwarz RE, Smith DD. Lymph node dissection impact on staging and survival of extrahepatic cholangiocarcinomas, based on U.S. population data. J Gastrointest Surg. 2007;11(2):158–65. https://doi.org/10.1007/s11605-006-0018-6.
77. Ito K, Ito H, Allen PJ, Gonen M, Klimstra D, D'Angelica MI, Fong Y, DeMatteo RP, Brennan MF, Blumgart LH, Jarnagin WR. Adequate lymph node assessment for extrahepatic bile duct adenocarcinoma. Ann Surg. 2010;251(4):675–81. https://doi.org/10.1097/SLA.0b013e3181d3d2b2.
78. Kambakamba P, Linecker M, Slankamenac K, DeOliveira ML. Lymph node dissection in resectable perihilar cholangiocarcinoma: a systematic review. Am J Surg. 2015;210(4):694–701. https://doi.org/10.1016/j.amjsurg.2015.05.015.
79. Guglielmi A, Ruzzenente A, Campagnaro T, Pachera S, Conci S, Valdegamberi A, Sandri M, Iacono C. Prognostic significance of lymph node ratio after resection of peri-hilar cholangiocarcinoma. HPB. 2011;13(4):240–5. https://doi.org/10.1111/j.1477-2574.2010.00277.x.
80. Li J, Zhou MH, Ma WJ, Li FY, Deng YL. Extended lymphadenectomy in hilar cholangiocarcinoma: what it will bring? World J Gastroenterol. 2020;26(24):3318–25. https://doi.org/10.3748/wjg.v26.i24.3318.
81. Oshiro Y, Sasaki R, Kobayashi A, Murata S, Fukunaga K, Kondo T, Oda T, Ohkohchi N. Prognostic relevance of the lymph node ratio in surgical patients with extrahepatic cholangiocarcinoma. Eur J Surg ncol. 2011;37(1):60–4. https://doi.org/10.1016/j.ejso.2010.10.011.
82. Aoba T, Ebata T, Yokoyama Y, Igami T, Sugawara G, Takahashi Y, Nimura Y, Nagino M. Assessment of nodal status for perihilar cholangiocarcinoma: location, number, or ratio of involved nodes. Ann Surg. 2013;257(4):718–25. https://doi.org/10.1097/SLA.0b013e3182822277.
83. Mantel HT, Wiggers JK, Verheij J, Doff JJ, Sieders E, van Gulik TM, Gouw AS, Porte RJ. Lymph node micrometastases are associated with worse survival in patients with otherwise node-negative hilar cholangiocarcinoma. Ann Surg Oncol. 2015;22(Suppl 3):S1107–15. https://doi.org/10.1245/s10434-015-4723-9.
84. Jena SS, Mehta NN, Nundy S. Surgical management of hilar cholangiocarcinoma: controversies and recommendations. Ann Hepatobiliary Pancreat Surg. 2023;27(3):227–40. https://doi.org/10.14701/ahbps.23-028.
85. Rangarajan K, Simmons G, Manas D, Malik H, Hamady ZZ. Systemic adjuvant chemotherapy for cholangiocarcinoma surgery: A systematic review and meta-analysis. Eur J Surg Oncol. 2020;46(4 Pt A):684–93. https://doi.org/10.1016/j.ejso.2019.11.499.
86. Primrose JN, Fox RP, Palmer DH, Malik HZ, Prasad R, Mirza D, Anthony A, Corrie P, Falk S, Finch-Jones M, Wasan H, Ross P, Wall L, Wadsley J, Evans JTR, Stocken D, Praseedom R, Ma YT, Davidson B, Neoptolemos JP, et al. Capecitabine compared with observation in resected biliary tract cancer (BILCAP): a randomised, controlled, multicentre, phase 3 study. Lancet Oncol. 2019;20(5):663–73. https://doi.org/10.1016/S1470-2045(18)30915-X.
87. Lee SE, Jang JY, Lim CS, Kang MJ, Kim SW. Systematic review on the surgical treatment for T1 gallbladder cancer. World J Gastroenterol. 2011;17(2):174–80. https://doi.org/10.3748/wjg.v17.i2.174.
88. Pilgrim C, Usatoff V, Evans PM. A review of the surgical strategies for the management of gallbladder carcinoma based on T stage and growth type of the tumour. Eur J Surg Oncol. 2009;35(9):903–7. https://doi.org/10.1016/j.ejso.2009.02.005.
89. Goel M, Pandrowala S, Patel P, Patkar S. Node positivity in T1b gallbladder cancer: A high volume centre experience. Eur J Surg Oncol. 2022;48(7):1585–9. https://doi.org/10.1016/j.ejso.2022.03.013.

90. Jensen EH, Abraham A, Jarosek S, Habermann EB, Al-Refaie WB, Vickers SA, Virnig BA, Tuttle TM. Lymph node evaluation is associated with improved survival after surgery for early stage gallbladder cancer. Surgery. 2009;146(4):706–13. https://doi.org/10.1016/j.surg.2009.06.056.
91. Tran TB, Nissen NN. Surgery for gallbladder cancer in the US: a need for greater lymph node clearance. J Gastrointest Oncol. 2015;6(5):452–8. https://doi.org/10.3978/j.issn.2078-6891.2015.062.
92. Widmann B, Warschkow R, Beutner U, Weitzendorfer M, Ukegjini K, Schmied BM, Tarantino I, Steffen T. Effect of lymphadenectomy in curative gallbladder cancer treatment: a systematic review and meta-analysis. Langenbeck's Arch Surg. 2020;405(5):573–84. https://doi.org/10.1007/s00423-020-01878-z.
93. Nishio H, Nagino M, Ebata T, Yokoyama Y, Igami T, Nimura Y. Aggressive surgery for stage IV gallbladder carcinoma; what are the contraindications? J Hepato-Biliary-Pancreat Surg. 2007;14(4):351–7. https://doi.org/10.1007/s00534-006-1187-0.
94. Kang MJ, Song Y, Jang JY, Han IW, Kim SW. Role of radical surgery in patients with stage IV gallbladder cancer. HPB. 2012;14(12):805–11. https://doi.org/10.1111/j.1477-2574.2012.00544.x.
95. Kondo S, Nimura Y, Hayakawa N, Kamiya J, Nagino M, Uesaka K. Regional and Para-aortic lymphadenectomy in radical surgery for advanced gallbladder carcinoma. Br J Surg. 2000;87(4):418–22. https://doi.org/10.1046/j.1365-2168.2000.01384.x.
96. Ito H, Ito K, D'Angelica M, Gonen M, Klimstra D, Allen P, DeMatteo RP, Fong Y, Blumgart LH, Jarnagin WR. Accurate staging for gallbladder cancer: implications for surgical therapy and pathological assessment. Ann Surg. 2011;254(2):320–5. https://doi.org/10.1097/SLA.0b013e31822238d8.
97. Negi SS, Singh A, Chaudhary A. Lymph nodal involvement as prognostic factor in gallbladder cancer: location, count or ratio? J Gastrointest Surg. 2011;15(6):1017–25. https://doi.org/10.1007/s11605-011-1528-4.
98. Kishi Y, Nara S, Esaki M, Hiraoka N, Shimada K. Extent of lymph node dissection in patients with gallbladder cancer. Br J Surg. 2018;105(12):1658–64. https://doi.org/10.1002/bjs.10913.
99. Leigh NL, Solomon D, Feingold D, Hiotis SP, Labow DM, Magge DR, Sarpel U, Golas BJ. Staging gallbladder cancer with lymphadenectomy: the practical application of new AHPBA and AJCC guidelines. HPB. 2019;21(11):1563–9. https://doi.org/10.1016/j.hpb.2019.03.372.
100. Gani F, Buettner S, Margonis GA, Ethun CG, Poultsides G, Tran T, Idrees K, Isom CA, Fields RC, Krasnick B, Weber SM, Salem A, Martin RC, Scoggins C, Shen P, Mogal HD, Schmidt C, Beal E, Hatzaras I, Shenoy R, et al. Assessing the impact of common bile duct resection in the surgical management of gallbladder cancer. J Surg Oncol. 2016;114(2):176–80. https://doi.org/10.1002/jso.24283.
101. Aljiffry M, Abdulelah A, Walsh M, Peltekian K, Alwayn I, Molinari M. Evidence-based approach to cholangiocarcinoma: a systematic review of the current literature. J Am Coll Surg. 2009;208(1):134–47. https://doi.org/10.1016/j.jamcollsurg.2008.09.007.
102. Si A, Li J, Xiang H, Zhang S, Bai S, Yang P, Zhang X, Xia Y, Wang K, Yan Z, Lau WY, Shi L, Shen F. Actual over 10-year survival after liver resection for patients with intrahepatic cholangiocarcinoma. Oncotarget. 2017;8(27):44521–32. https://doi.org/10.18632/oncotarget.17815.
103. Ohtsuka M, Ito H, Kimura F, Shimizu H, Togawa A, Yoshidome H, Miyazaki M. Results of surgical treatment for intrahepatic cholangiocarcinoma and clinicopathological factors influencing survival. Br J Surg. 2002;89(12):1525–31. https://doi.org/10.1046/j.1365-2168.2002.02268.x.
104. Uchiyama K, Yamamoto M, Yamaue H, Ariizumi S, Aoki T, Kokudo N, Ebata T, Nagino M, Ohtsuka M, Miyazaki M, Tanaka E, Kondo S, Uenishi T, Kubo S, Yoshida H, Unno M, Imura S, Shimada M, Ueno M, Takada T. Impact of nodal involvement on surgical outcomes of intrahepatic cholangiocarcinoma: a multicenter analysis by the Study Group for Hepatic Surgery of the Japanese Society of Hepato-Biliary-Pancreatic Surgery. J Hepatobiliary Pancreat Sci. 2011;18(3):443–52. https://doi.org/10.1007/s00534-010-0349-2.
105. Yamamoto M, Takasaki K, Yoshikawa T. Lymph node metastasis in intrahepatic cholangiocarcinoma. Jpn J Clin Oncol. 1999;29(3):147–50. https://doi.org/10.1093/jjco/29.3.147.
106. Shimada K, Sano T, Nara S, Esaki M, Sakamoto Y, Kosuge T, Ojima H. Therapeutic value of lymph node dissection during hepatectomy in patients with intrahepatic cholangiocellular carcinoma with negative lymph node involvement. Surgery. 2009;145(4):411–6. https://doi.org/10.1016/j.surg.2008.11.010.
107. Yoh T, Cauchy F, Le Roy B, Seo S, Taura K, Hobeika C, Dokmak S, Farges O, Gelli M, Sa Cunha A, Adam R, Uemoto S, Soubrane O. Prognostic value of lymphadenectomy for long-term outcomes in node-negative intrahepatic cholangiocarcinoma: A multicenter study. Surgery. 2019;166(6):975–82. https://doi.org/10.1016/j.surg.2019.06.025.
108. Bagante F, Spolverato G, Weiss M, Alexandrescu S, Marques HP, Aldrighetti L, Maithel SK, Pulitano C, Bauer TW, Shen F, Poultsides GA, Soubrane O, Martel G, Groot Koerkamp B, Guglielmi A, Itaru E, Ruzzenente A, Pawlik TM. Surgical management of intrahepatic cholangiocarcinoma in patients with cirrhosis: impact of lymphadenectomy on peri-operative outcomes. World J Surg. 2018;42(8):2551–60. https://doi.org/10.1007/s00268-017-4453-1.
109. Reames BN, Bagante F, Ejaz A, Spolverato G, Ruzzenente A, Weiss M, Alexandrescu S, Marques HP, Aldrighetti L, Maithel SK, Pulitano C, Bauer

TW, Shen F, Poultsides GA, Soubrane O, Martel G, Koerkamp BG, Guglielmi A, Itaru E, Pawlik TM. Impact of adjuvant chemotherapy on survival in patients with intrahepatic cholangiocarcinoma: a multi-institutional analysis. HPB. 2017;19(10):901–9. https://doi.org/10.1016/j.hpb.2017.06.008.

110. Zhang XF, Xue F, Dong DH, Weiss M, Popescu I, Marques HP, Aldrighetti L, Maithel SK, Pulitano C, Bauer TW, Shen F, Poultsides GA, Soubrane O, Martel G, Koerkamp BG, Itaru E, Lv Y, Pawlik TM. Number and station of lymph node metastasis after curative-intent resection of intrahepatic cholangiocarcinoma impact prognosis. Ann Surg. 2021;274(6):e1187–95. https://doi.org/10.1097/SLA.0000000000003788.

111. Kim SH, Han DH, Choi GH, Choi JS, Kim KS. Recommended minimal number of harvested lymph nodes for intrahepatic cholangiocarcinoma. J Gastrointest Surg. 2021;25(5):1164–71. https://doi.org/10.1007/s11605-020-04622-6.

112. Kim SH, Han DH, Choi GH, Choi JS, Kim KS. Oncologic impact of lymph node dissection for intrahepatic cholangiocarcinoma: a propensity score-matched study. J Gastrointest Surg. 2019;23(3):538–44. https://doi.org/10.1007/s11605-018-3899-2.

113. Sposito C, Ratti F, Cucchetti A, Ardito F, Ruzzenente A, Di Sandro S, Maspero M, Ercolani G, Di Benedetto F, Guglielmi A, Giuliante F, Aldrighetti L, Mazzaferro V. Survival benefit of adequate lymphadenectomy in patients undergoing liver resection for clinically node-negative intrahepatic cholangiocarcinoma. J Hepatol. 2023;78(2):356–63. https://doi.org/10.1016/j.jhep.2022.10.021.

114. Li F, Jiang Y, Jiang L, Li Q, Yan X, Huang S, Chen J, Yuan S, Fu Y, Liu J. Effect of lymph node resection on prognosis of resectable intrahepatic cholangiocarcinoma: A systematic review and meta-analysis. Front Oncol. 2022;12:957792. https://doi.org/10.3389/fonc.2022.957792.

115. Orimo T, Kamiyama T, Mitsuhashi T, Kamachi H, Yokoo H, Wakayama K, Shimada S, Nagatsu A, Taketomi A. Impact of tumor localization on the outcomes of surgery for an intrahepatic cholangiocarcinoma. J Gastroenterol. 2018;53(11):1206–15. https://doi.org/10.1007/s00535-018-1469-8.

116. Umeda Y, Takagi K, Matsuda T, Fuji T, Kojima T, Satoh D, Hioki M, Endo Y, Inagaki M, Oishi M, Yagi T, Fujiwara T. Clinical implications and optimal extent of lymphadenectomy for intrahepatic cholangiocarcinoma: a multicenter analysis of the therapeutic index. Ann Gastroenterol Surg. 2022;7(3):512–22. https://doi.org/10.1002/ags3.12642.

Part VI

Robotic Liver Surgery: Procedures and Technical Aspects

Parenchymal Transection in Robotic Liver Surgery

16

Anish J. Jain, Pejman Radkani, and Jason S. Hawksworth

16.1 Introduction

Parenchymal transection remains the most critical part of any hepatectomy, due to the complex intrahepatic biliary and vascular anatomy and risk of catastrophic bleeding. There are a variety of techniques to improve the safety of parenchymal transection and reduce the risk of vascular and biliary morbidity. These include the use of inflow control, parenchymal transection with energy or water-jet devices, and low central venous pressure maintenance during hepatectomy. Specifically, a variety of energy devices have been successfully employed in open hepatic resection (OHR) including ultrasonic dissectors, harmonic scalpels, and water jets, among others [1].

A. J. Jain
Department of Surgery, Stanford University, Stanford, CA, USA

P. Radkani
Georgetown University School of Medicine, Washington, DC, USA

MedStar Georgetown Transplant Institute, Washington, DC, USA

J. S. Hawksworth (✉)
Liver Transplant and Hepatobiliary Surgery, Columbia University Hospital, New York, NY, USA

Columbia University Irving Medical Center, New York, NY, USA
e-mail: jh4824@cumc.columbia.edu

OHR is the traditional approach for liver resections; however, minimally invasive surgical (MIS) hepatic resection has become increasingly accepted due to its reduced length of stay (LOS), improved pain control, and equivalent oncologic outcomes [2–8]. However, the laparoscopic hepatic resection experience worldwide has been limited to experienced, high-volume institutions due to steep learning curve and potential difficulty controlling hemorrhage during parenchymal transection. Advanced robotic technology may overcome some of the limitations of laparoscopy including improved optics and articulating instruments [2, 4, 7–10].

Currently there is no consensus on the preferred method for parenchymal transection in robotic hepatic resections. This chapter describes the predominant methods/instruments used in robotic hepatic surgery worldwide, highlighting the advantages, disadvantages, and the technical aspects of each method.

16.2 Technical Aspects

Currently there are three predominant techniques/instruments for transection of hepatic parenchyma in robotic hepatic resections: da Vinci Harmonic ACE (Ethicon, Somerville, NJ, USA), cavitron ultrasonic surgical aspirator device (Integra Lifesciences, Plainsboro, NJ, USA), and clamp-crush technique using robotic

M. D'Hondt, I. Sucandy (eds.), *Textbook of Robotic Liver Surgery*,
https://doi.org/10.1007/978-3-031-76536-0_16

TECHNIQUE/INSTRUMENT	ADVANTAGES	DISADVANTAGES
CUSA	• Mimics open surgical techniques • Facilitates precise dissection around hepatic vascular and biliary structures • Enables non-anatomic resections • Allows continued CUSA parenchyma division during robotic instrument changesresection	• Robotic CUSA NOT currently available, thereby requires second skilled bedside surgeon to operate laparoscopic CUSA • May require placement of additional 12mm assist port • CUSA port placement may influence robotic port alignment
Harmonic Scalpel	• Mimics open surgical techniques • Robotic-integrated instrument, controlled by primary surgeon at consoleEnables non-anatomic resections • Enables concurrent coagulation and parenchymal transection	• Instrument lacks articulation • Instrument length is limited, which may influence port placement • Potential increased risk of iatrogenic partial injury to vascular or biliary structures during blind parenchymal transection
Robotic Vessel Sealer	• Robotic instrument, controlled by primary surgeon at console • Potentially faster transection • Mimics traditional open Clamp-Crush technique	• Vessel Sealer has bulky instrument head, can be disadvantageous for precise parenchymal transection and/or deep parenchymal transection • Potential increased risk of iatrogenic partial injury to vascular or biliary structures during blind parenchymal transection • May be suboptimal for cirrhotic livers

Fig. 16.1 Methods of robotic hepatic parenchymal transection

vessel sealers. See Fig. 16.1 for summarization of advantages and disadvantages of each technique/instrument in robotic hepatectomy.

In each of these methods, monopolar and bipolar energy devices are commonly used to aid in hemostasis and dissection. These devices can vary from the standard bipolar to the curved bipolar variant and in most cases the Maryland bipolar grasper [10–15]. Robotic Maryland forceps can also be used to clamp and crush the parenchyma to help expose the vascular and biliary structures. These structures were then divided using a combination of the transection methods mentioned above with clips, scissors, and/or staplers [16].

16.2.1 Cavitron Ultrasonic Surgical Aspirator (CUSA) Device

The cavitron ultrasonic surgical aspirator (CUSA) device uses ultrasonic energy to fragment and aspirate parenchymal tissue, which allows for precise transection planes thus preserving normal hepatic tissue (Fig. 16.2). It also allows for exposure of biliary and vascular structures to be clipped, ligated, and/or reconstructed at the surgeon's discretion [1, 17].

While transection time with the CUSA has been reported to be slower than other methods, it provides a well-defined transection plane, performs fine dissection, and functions well even in cirrhotic livers, and the technology may be associated with decreased bile leak [18]. For many HPB centers, the CUSA is the preferred instrument/technique for parenchymal transection [1]. In robotic surgery, parenchymal transection with the CUSA can improve visualization of intrahepatic vascular and biliary structures, resulting in decreased hemorrhage and biliary morbidity in minor and major hepatectomies [19–22].

A major limitation for robotic hepatic surgery is that the CUSA is not integrated into the robotic platform, requiring an additional skilled bedside surgeon to operate a laparoscopic CUSA. This limitation may be one of the major factors preventing a broader application of RHR among surgeons who typically use the CUSA during open and laparoscopic hepatectomy. Conversely, in the appropriate setting the utilization of a bedside surgeon operating a bedside CUSA with a second surgeon controlling the robot at the console may

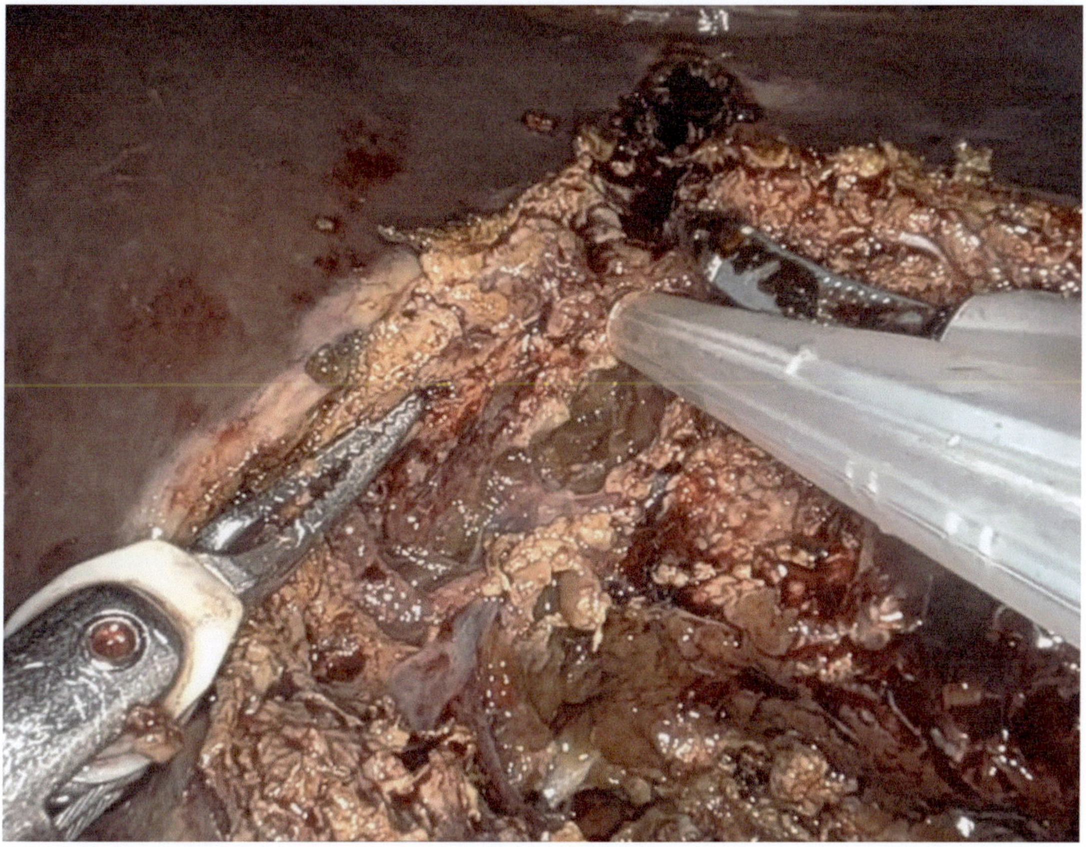

Fig. 16.2 Use of the cavitron ultrasonic surgical aspirator (CUSA) device by the bedside surgeon assisting the console surgeon during parenchymal transection of the liver

improve the choreography of the parenchymal transection [4, 23, 24]. Specifically, the bedside surgeon can continually dissect the liver parenchyma during robotic instrument exchanges, potentially reducing operative time.

Technical considerations regarding the use of the laparoscopic CUSA during robotic hepatic resection include placement of an extra 12-mm assistant port. The robotic ports are placed across the upper abdomen, in line with the CUSA port to prevent interference between the robotic instruments and the bedside surgeon operating the CUSA. This port can be placed between the camera port and the most lateral right-sided port (see Fig. 16.3). An additional consideration with placement of the assist port for the CUSA is that it should be in parallel with the planned parenchymal transection plane [19–22]).

16.2.2 da Vinci Harmonic ACE (Harmonic Scalpel)

The harmonic scalpel is another commonly used energy device for parenchymal transection during open or MIS hepatectomy. The harmonic uses ultrasonically activated shears, with blades vibrating at a frequency of 55.5 kHz, which can easily dissect through hepatic parenchyma. Additionally, it provides a coagulation effect upon contact with vasculature up to 2–3 mm in diameter [1, 18]. Unlike the CUSA, the harmonic scalpel has been integrated into the robotic platform and can be used by the console surgeon during parenchymal transection. This may be advantageous in situations where a bedside surgeon with the skillset required to operate a laparoscopic CUSA is unavailable. Surgeons have

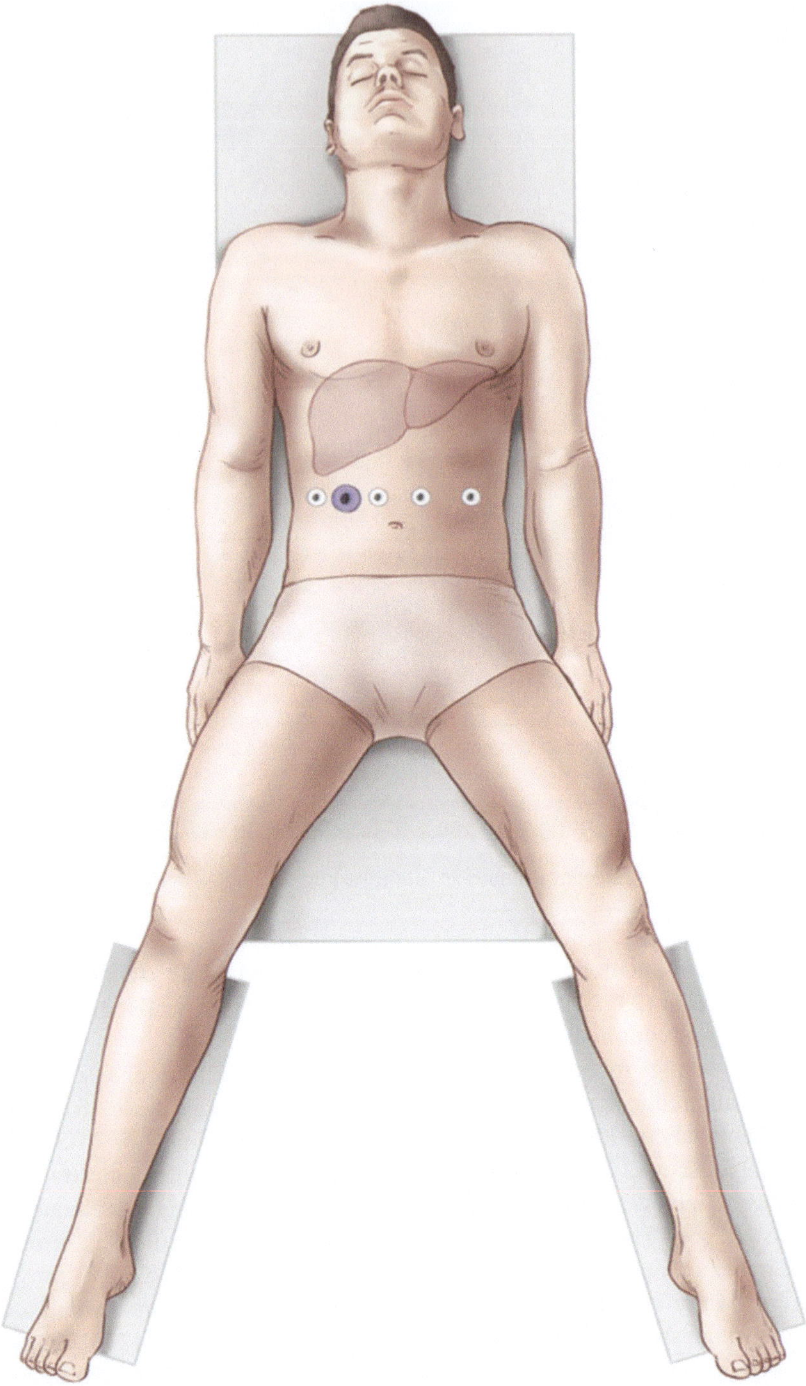

Fig. 16.3 Trocar placement for robotic hepatic resection using laparoscopic CUSA. Robotic trocars (white) and one 12-mm assistant port for CUSA device (purple)

also successfully employed the harmonic scalpel during transection of cirrhotic livers [1, 25].

Disadvantages of the harmonic scalpel include limited dissection around major vascular structures, which may limit complex hepatic resections in some robotic cases. This may also translate to a higher incidence of postoperative biliary fistula due to inadequate identification of biliary structures in the setting of incomplete hemostasis during parenchymal transection [1,

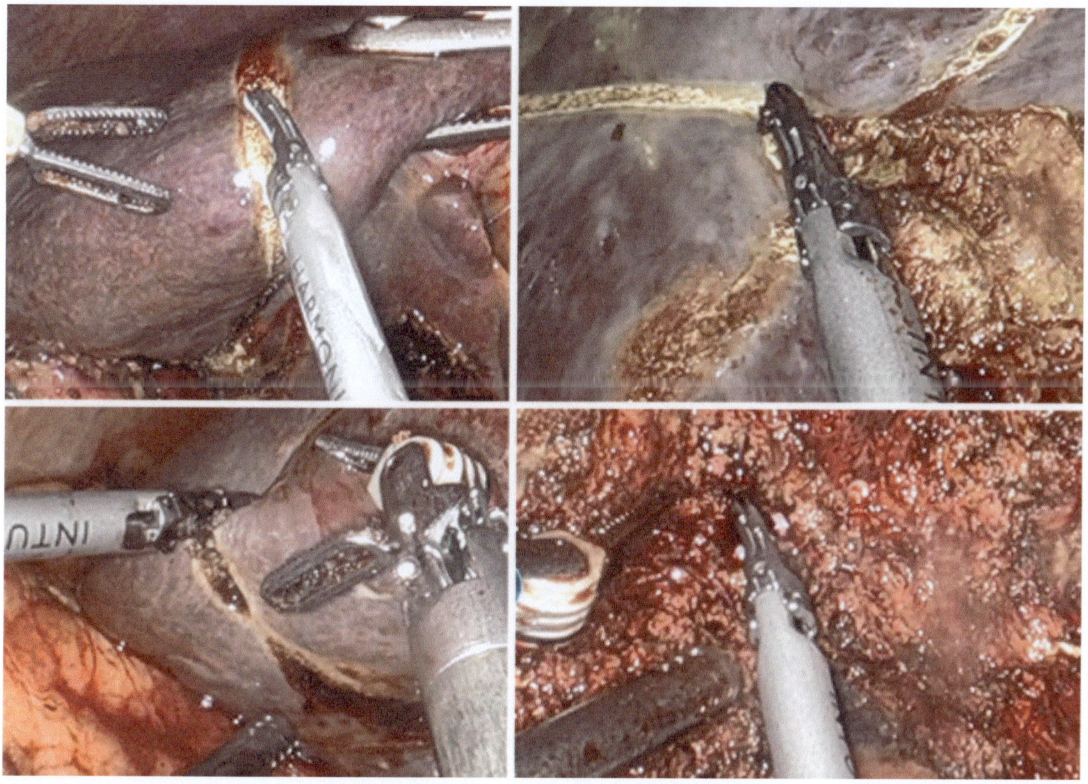

Fig. 16.4 Use of the robotic harmonic scalpel during parenchymal transection

25–27]. An additional limitation of the harmonic scalpel is the tip's inability to articulate, which requires port placement so that the instrument is in line with the planned parenchymal transection plane similar to utilization of the laparoscopic CUSA [2] (Fig. 16.4).

16.2.3 Clamp-Crushing Technique ("Clamp-Crush") Utilizing the Robotic Vessel Sealer

The clamp-crush technique was first reported in 1974 and, for decades, has remained the standard means of parenchymal transection for many HPB surgeons [17, 28]. A modification of the finger fracture technique, the fundamental concepts behind it involve clamping the liver parenchyma (often with forceps or a Kelly clamp) while ligating/dividing exposed vascular and biliary structures that are subsequently exposed [1, 18, 29]. Notably the clamp-crush technique may not be ideal in a cirrhotic liver [1].

Utilization of the robotic vessel sealer during parenchymal transection is a modified crush clamp technique. The vessel sealer jaws can be used to deliver parenchyma and expose underlying vascular and biliary radicals, similar to the open technique. However, the vessel sealer adds coagulation and ligation abilities and can be controlled by the console surgeon. In addition, the vessel sealer articulates, giving the surgeon additional degrees of freedom during parenchymal transection. This liberalizes port placement when using this technique in robotic hepatectomy. It should be noted that some surgeons prefer the use of this technique with the SynchroSeal device, which is a lower profile version of the vessel sealer (Figs. 16.5 and 16.6).

A disadvantage of the vessel sealer is that it is bulky and imprecise around major vascular and

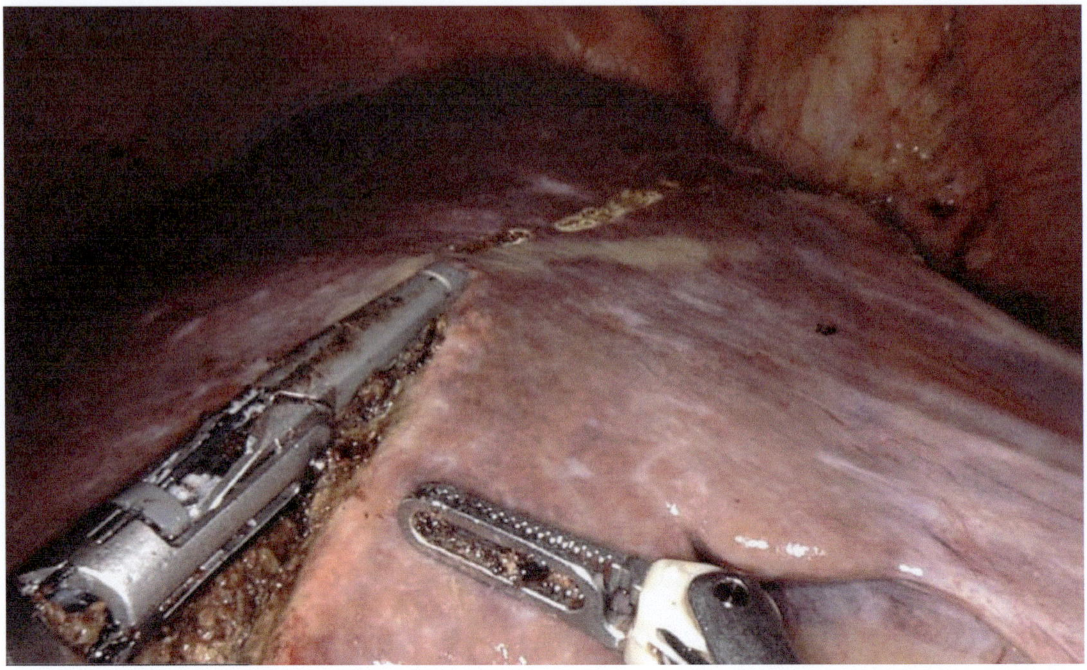

Fig. 16.5 Utilization of the robotic vessel sealer for parenchymal transection (left hemihepatectomy)

Fig. 16.6 Use of the SynchroSeal device for dissection of the pedicle of segment P8 ventral

biliary structures. In addition, this instrument may not be ideal for cirrhotic livers, similar to the traditional open crush clamp technique [30].

16.3 Literature Overview

In 2016, the first report of integration of the laparoscopic CUSA into robotic hepatectomy was published by Eric Lai based in Hong Kong [23, 24, 31]. In this study, they retrospectively reviewed 100 patients who underwent a total of 102 robotic hepatectomies at their institution between 2009 and 2015. The primary surgeon would use the robotic harmonic scalpel and electrocautery forceps for parenchymal transection, with an assistant surgeon using the CUSA for fine parenchymal transection [24, 31, 32]. The mean operative time was 207.4 min, with a mean estimated blood loss (EBL) of 334.6 mL. Four cases (4%) were converted to an open approach. The major morbidity rate was 14%, including one patient who required reoperation for bleeding and another patient who ultimately required liver transplant after developing postoperative liver failure. Mean length of stay (LOS) was 7.3 days with no reported perioperative mortalities [32].

In 2020, a French group led by Antoine Carmelo also reported the integration of laparoscopic CUSA into robotic hepatectomy. Transection of the superficial 2 cm of the parenchyma was done using bipolar coagulation in combination with a laparoscopic CUSA manipulated by the second surgeon. Deeper transection, occasionally with a pringle maneuver in cases of bleeding, was performed with a combination of the CUSA and the clamp-crush technique using the bipolar forceps. The CUSA was incorporated to provide a clean transection plane and delicate dissection in zones close to major vascular structures [33].

This series included 50 patients, with 23 undergoing major hepatectomy (46%) and 27 undergoing minor hepatectomies (54%). The median operative time was 251 min, with a median EBL of 238 mL and only two patients (5%) requiring intraoperative blood transfusion. Three cases (6%) were converted to open surgery due to arterial bleeding from left hepatic artery injury (in one of the earliest cases in the series), suspicion of right posterior bile duct injury, and air embolism leading to arrhythmia. Only four patients had Clavien-Dindo grade III complications, with only one requiring reoperation for biliary peritonitis, and the median LOS was 7 days with no 90-day mortality [33].

The largest North American experience of successful integration of the laparoscopic CUSA into robotic hepatectomy was reported by Hawksworth et al. It should be noted that this group also uses the robotic vessel sealer and bipolar electrocautery to complement the laparoscopic CUSA during parenchymal transection [19]. This case series included 54 patients who underwent robotic hepatectomies with the use of laparoscopic CUSA between September 2018 and December 2020. The series included a diverse case mix of hepatic resections such as major hepatectomies (39%), caudate resections (6%), and right posterior resections (26%). The median case time was 257 min for minor hepatectomies and 387 min for major hepatectomies. Median EBL was 285 mL with a low blood transfusion rate of only 9% (five patients). The major complication rate (Clavien-Dindo grade III or greater) was also low at 4% including one biloma and one pneumothorax. Additionally, only one case was converted to open due to failure to progress. Median length of stay (LOS) was 2 days, and there were no 90-day mortalities [19].

The most widely reported technique to date is use of the robotic harmonic scalpel for parenchymal transection in robotic hepatectomy.

Giulianotti et al. conducted a retrospective review of 70 RHRs performed between 2002 and 2009 at two different centers (based in the United States and Italy) by a single surgeon. The robotic harmonic scalpel and bipolar forceps were used to perform parenchymal transection in all cases. Forty-three cases (61.4%) were minor hepatectomies, with a median operative time of 198 min and EBL of 150 mL. The other 27 cases (38.6%) were major hepatectomies, with a median operative time of 313 min and EBL of 300 mL. Nine patients (20.9%) undergoing minor hepatectomy required intraoperative transfusions, five of whom were cirrhotic. Six patients (22.2%) undergoing major liver resections required transfusion. Conversion to open surgery was performed in four cases (5.7%), including one due to bleeding (from left hepatic vein during a left lobectomy in a cirrhotic patient) and the other three for oncologic reasons. No conversions occurred during the final 38 robotic hepatic resections in this series. Major morbidity occurred in eight patients (11.4%), none of whom required reoperation. The median LOS was 7 days, with no perioperative mortality [12].

In a subsequent study published by Giulianotti, his group retrospectively reviewed 68 robotic hepatic resections performed between 2009 and 2013 at a single center (based in the United States) using the same harmonic scalpel technique for parenchymal transection. Twenty-nine cases (42.65%) were considered major hepatectomies, with a mean operative time of 404 min and mean EBL of 570 mL. The other 39 cases (57.35%) were minor hepatectomies, with a mean operative time of 223.2 min and EBL of 354.7 mL. Nine patients (13.2%) required intraoperative transfusion, and six patients (8.8%) required conversion to open surgery. Three patients (4.4%) suffered Clavien-Dindo grade III–IV complications, all of which were biliary leaks, but none required reoperation. The Mean LOS was 6.8 days, with no 30-day mortalities [13].

A Belgian group, led by Roberto Troisi, retrospectively reviewed 36 patients between 2008 and 2014 who underwent robotic hepatectomies predominantly using the harmonic scalpel technique. Notably in cases that required angulated transection planes or dissection near major vessels, the clamp-crush technique using endowristed bipolar precise forceps was preferred. The mean operative time was 306 min, with a mean EBL of 415 mL. Five cases (13.9%) were converted to open laparotomies, with the reasons being bleeding (n = 2), oncological concerns (n = 2), and technical difficulty (n = 1). Four patients (11.1%) suffered major complications including postoperative liver failure requiring transplant, myocardial infarction, and inferior vena cava syndrome. The mean LOS was 6 days with one (2.8%) perioperative mortality [10].

A Korean group led by Jin Sub Choi published a review of the 69 robotic hepatectomies performed within in their institution from 2008 to 2016 also using the robotic harmonic scalpel. Similar to the Belgian experience, the Maryland bipolar forceps were used for hemostasis, dissection, and isolation of major vessels. The clamp-crush technique was used in deeper parenchymal transection. The review consisted of 54 (78%) major hepatectomies with a median operative time of 518 min and median EBL of 200 mL. Fifteen cases (22%) were considered minor hepatectomies with a median operative time of 360 min and EBL of 100 mL. Blood transfusions were given in six (8.7%) of cases. Additionally, six cases (8.7%) were converted to open, including two due to bleeding and/or bile duct injury. Seven patients (10.6%) suffered Clavien-Dindo grade III complications. The median length of stay was 8 days with no perioperative mortality [11].

Finally, a Chinese group led by Rong Liu and Wan Yee Lau retrospectively reviewed 92 patients who underwent robotic hemihepatectomies at a single center from 2011to 2017. Similar to the studies above, the robotic harmonic scalpel was used to transect the superficial 2 cm of hepatic parenchyma, while deeper parenchyma was transected via a combination of the harmonic and bipolar cautery. The mean operative time for robotic cases was 195.5 min, with an EBL of 243 mL. One patient (1.1%), who had a massive hemangioma in the right liver, was converted to laparotomy. Additionally, one patient (1.1%) suffered a Clavien-Dindo grade III complication that was managed nonoperatively. The mean LOS was 7.41 days, with no 30-day postoperative mortalities [14].

It should be noted that in many of the case series above, parenchymal transection with the harmonic scalpel was often augmented with use of Maryland bipolar cautery. The bipolar cautery was used in a clamp-crush technique, especially during deeper parenchymal transection, to aid in hemostasis of the cut surface [11–14, 34–36].

The crush-clamp technique using the robotic vessel sealer is also commonly reported [30, 37, 38]. Additionally, given the improved dissection of modern surgical devices, robotic parenchymal transection using the vessel sealer in a clamp-crush technique is often augmented with other devices, most notably the harmonic scalpel (especially for linear transection) and/or robotic electrocautery [10, 38–43].

An American group, led by Peter Kingham, at Memorial Sloan Kettering Cancer Center retrospectively reviewed 64 patients who underwent robotic hepatectomy at their institution between 2010 and 2014. The hepatic parenchyma was divided using a combination of hook cautery, the harmonic scalpel or vessel sealer, and the bipolar PK dissecting forceps. The median operative time was 163 min, with an EBL of 100 mL. Four cases (6%) required conversion to open due to difficulty exposing vascular anatomy, a stapler malfunction and resultant hemorrhage, discovery of additional hepatic tumors, and inability to visualize the tumor and resection planes. The median LOS was 4 days, with a major complication rate of 5% ($n = 3$) and 90-day mortality rate of 3% ($n = 2$) [30].

In 2018, Tsung et al. published a multicenter (all based in the United States), retrospective study of 61 patients who underwent robotic hepatic resections between 2006 and 2016. The robotic vessel sealer, harmonic scalpel, and endovascular GIA staplers were used for parenchymal transection and control of vasculature. Cirrhosis was actually present in 14 patients (22.8%) in this series. The median operative time was 240 min with an EBL of 100 mL. Seven cases (11.5%) required conversion to open, but only one was due to uncontrolled bleeding and one due to technical difficulties. Seven patients (11.5%) suffered Clavien-Dindo grade 3+ complications, with one patient who had HCC and alcohol cirrhosis developing liver failure requiring liver transplantation. The median LOS of the entire cohort was 5 days, with no perioperative mortalities [38].

In 2019, Martinie et al. published a retrospective case series of 57 robotic assisted major hepatectomies performed between 2011 and 2016 at a single center in the United States (Carolinas Medical Center). Parenchymal transection was performed predominantly with the robotic bipolar vessel-sealing device or in some cases the robotic ultrasonic shears. The median operative time was 194 min with a median EBL of 250 mL. The median LOS was 4 days with only four patients (7%) suffering major postoperative complications [42].

In 2019, Sucandy et al. published a retrospective case series of 80 patients who underwent robotic hepatectomies between 2013 and 2018 at a different HPB center in the United States (Digestive Health Institute, AdventHealth Tampa). The surgeons utilized robotic monopolar electrocautery for <2 cm superficial liver parenchymal transection, followed by a robotic vessel sealer for the deeper parenchymal transection. The median operative time was 233 min with a median EBL of 150 mL. One patient (1.25%) required conversion to an open approach due to difficulty accessing a tumor located in posteromedial segment 8. Ten (12.5%) patients suffered from postoperative complications including bile leaks (three patients), respiratory failure (two patients), wound complications (two patients), enterocutaneous fistula (one patient), urinary retention (one patient), and atrial fibrillation (one patient). The median LOS was 3 days with one mortality (1.25%) within 30 days, which was due to respiratory failure [43].

A group based in Hong Kong led by Paul Lai published a case series of 91 patients who underwent robotic hepatectomy between 2003 and 2017. All operations were performed by the same team of HPB surgeons who each used the robotic harmonic scalpel in combination with a robotic vessel sealer for parenchymal transection. Endovascular staplers were used to divide larger vascular pedicles, while bleeding from small vessels was controlled with metal clips. Seventy-two patients (79.1%) underwent minor hepatectomy, while the other 19 patients (20.9%) underwent major hepatectomy. The mean operative time was 259.3 min, with a mean EBL of 274.6 mL. Open conversion was only needed in seven cases (7.7%), specifically for uncontrolled bleeding ($n = 5$) and slow progress ($n = 2$). The mean LOS was 4.8 days, with only three patients (3.3%) suffering major complications [37].

Several of the above case series report concurrent use of electrocautery of small blood vessels, clipping/suture ligation of slightly larger vessels, and linear stapling of main hepatic veins to aid

Group (Location)	Year	Cases (n)	Method of Transection	Operative Time (min)*	EBL (mL)*	Conversion to Open (%)	Major Complications (%)**
Lai (Hong Kong) [32]	2016	100	CUSA Harmonic Bipolar Forceps	207.4	334.6	4	14
Camerlo (France)[33]	2020	50	CUSA Bipolar Forceps	251	238	6.0	8.0
Hawksworth (US)[19]	2021	54	CUSA Vessel Sealer Bipolar Forceps	257 (minor) 387 (major)	285	1.8	3.7
Giulianotti (US &Italy)[12]	2011	70	Harmonic Bipolar Forceps	198 (minor) 313 (major)	150 (minor) 300 (major)	5.7	11.4
Giulianotti (US)[13]	2017	68	Harmonic Bipolar Forceps	223.2 (minor) 404 (major)	354.7 (minor) 570 (major)	8.8	3.4
Troisi (Belgium)[10]	2016	36	Harmonic Bipolar Forceps	306	415	13.9	11.1
Choi (Korea)[11]	2017	69	Harmonic Bipolar Forceps	360 (minor) 518 (major)	100 (minor) 200 (major)	8.7	10.6
Liu & Lau (China)[14]	2019	92	Harmonic Bipolar Forceps	195.5	243	1.1	1.1
Kingham (US)[30]	2016	64	Harmonic Bipolar Forceps Vessel sealer	163	100	6.0	5.0
Tsung (US)[38]	2018	61	Harmonic Vessel sealer	240	100	11.5	11.5
Martinie (US)[42]	2019	57	Vessel sealer Harmonic	194	250	N/A	7
Sucandy (US)[43]	2020	80	Vessel sealer Robotic Monopolar	233	150	1.25	12.5
Lai (Honk Kong)[37]	2020	91	Vessel sealer Harmonic	259.3	274.6	7.7	3.3

Fig. 16.7 Hepatic parenchymal transection methods in robotic hepatectomy case series

with hemostasis in robotic hepatectomies performed with clamp-crush technique. There was also selective use of a pringle maneuver in some of these case series as well to further aid in hemostasis [10, 39–43].

It should be noted that all groups and studies, regardless of their parenchymal transection method, discussed the importance of intraoperative US and/or indocyanine green (ICG) fluoroscopy to assist in the demarcation of the transection plane. All robotic HPB operations should involve their use to obtain adequate resection margins and avoid damage to major vascular/biliary structures (Fig. 16.7).

16.4 Conclusion

Parenchymal transection is the most dangerous aspect of liver surgery and potentially a barrier to the wide adaptation of robotic hepatectomy (Table 16.1). As robotic parenchymal techniques

Table 16.1 Tips for robotic hepatic parenchymal transection

Administer ICG after inflow control → can delineate transection plane via negative staining
ICG cholangiogram can identify major intrahepatic biliary radicals during parenchymal transection
Intraoperative US → paramount in identification of hepatic vein and portal pedicle location
Maintain parenchymal transection plane tension using constant retraction with third robotic arm
Hem-o-lock clips are most reliable for control/ligation of vascular and biliary structures
Rapid hemostatic sutures (3–0 Monocryl on SH needle cut to 12-cm length) should always be available in event of hemorrhage during transection

continue to improve, the application of robotic minimally invasive liver surgery will increase. The current available techniques have a variety of advantages and disadvantages and ultimately reflect surgeon preference, their respective experience, and available technology at their institution.

References

1. Poon RTP. Current techniques of liver transection. HPB. 2007;9(3):166–73.
2. Giulianotti PC, Bianco FM, Daskalaki D, Gonzalez-Ciccarelli LF, Kim J, Benedetti E. Robotic liver surgery: technical aspects and review of the literature. Hepatobiliary Surg Nutr. 2016;5(4):311–21.
3. Guan R, Chen Y, Yang K, Ma D, Gong X, Shen B, et al. Clinical efficacy of robot-assisted versus laparoscopic liver resection: a meta analysis. Asian J Surg. 2019;42(1):19–31.
4. Di Benedetto F, Petrowsky H, Magistri P, Halazun KJ. Robotic liver resection: hurdles and beyond. Int J Surg. 2020;82:155–62.
5. Nguyen KT, Gamblin TC, Geller DA. World review of laparoscopic liver resection-2,804 patients. Ann Surg. 2009;250(5):831–41.
6. Nguyen KT, Marsh JW, Tsung A, Steel JJ, Gamblin TC, Geller DA. Comparative benefits of laparoscopic vs open hepatic resection: a critical appraisal. Arch Surg. 2011;146(3):348–56.
7. Tsilimigras DI, Moris D, Vagios S, Merath K, Pawlik TM. Safety and oncologic outcomes of robotic liver resections: a systematic review. J Surg Oncol. 2018;117(7):1517–30.
8. Wong DJ, Wong MJ, Choi GH, Wu YM, Lai PB, Goh BKP. Systematic review and meta-analysis of robotic versus open hepatectomy. ANZ J Surg. 2019;89(3):165–70.
9. Idrees K, Bartlett DL. Robotic liver surgery. Surg Clin North Am. 2010;90(4):761–74.
10. Montalti R, Scuderi V, Patriti A, Vivarelli M, Troisi RI. Robotic versus laparoscopic resections of posterosuperior segments of the liver: a propensity score-matched comparison. Surg Endosc. 2016;30(3):1004–13.
11. Choi GH, Chong JU, Han DH, Choi JS, Lee WJ. Robotic hepatectomy: the Korean experience and perspective. Hepatobiliary Surg Nutr. 2017;6(4):230–8.
12. Giulianotti PC, Coratti A, Sbrana F, Addeo P, Bianco FM, Buchs NC, et al. Robotic liver surgery: results for 70 resections. Surgery. 2011;149(1):29–39.
13. Daskalaki D, Gonzalez-Heredia R, Brown M, Bianco FM, Tzvetanov I, Davis M, et al. Financial impact of the robotic approach in liver surgery: a comparative study of clinical outcomes and costs between the robotic and open technique in a single institution. J Laparoendosc Adv Surg Tech. 2017;27(4):375–82.
14. Wang Z-Z, Tang W-B, Hu M-G, Zhao Z-M, Zhao G-D, Li C-G, et al. Robotic vs laparoscopic hemihepatectomy: a comparative study from a single center. J Surg Oncol. 2019;120(4):646–53.
15. Wikiel KJ, Robinson TN, Jones EL. Energy in robotic surgery. Ann Laparosc Endosc Surg. 2021;6:9.
16. Berber E, Akyildiz HY, Aucejo F, Gunasekaran G, Chalikonda S, Fung J. Robotic versus laparoscopic resection of liver tumours. HPB. 2010;12(8):583–6.
17. Bodzin AS, Leiby BE, Ramirez CG, Frank AM, Doria C. Liver resection using cavitron ultrasonic surgical aspirator (CUSA) versus harmonic scalpel: a retrospective cohort study. Int J Surg. 2014;12(5):500–3.
18. Aragon RJ, Solomon NL. Techniques of hepatic resection. J Gastrointest Oncol. 2012;3(1):28–40.
19. Hawksworth J, Radkani P, Holzner M, Llore N, Winslow E, Fishbein T. Parenchymal transection with the laparoscopic cusa during robotic hepatectomy. HPB. 2021;23:S541–S2.
20. Hawksworth J, Llore N, Holzner ML, Radkani P, Meslar E, Winslow E, et al. Robotic hepatectomy is a safe and cost-effective alternative to conventional open hepatectomy: a single-center preliminary experience. J Gastrointest Surg. 2021;25(3):825–8.
21. Hawksworth J, Radkani P, Llore N, Holzner M, Meslar E, Winslow E, et al. Minimizing blood loss during robotic hepatectomy: technical description and initial experience. HPB. 2020;22:S128.
22. Hawksworth J, Radkani P, Nguyen B, Belyayev L, Llore N, Holzner M, et al. Improving safety of robotic major hepatectomy with extrahepatic inflow control and laparoscopic CUSA parenchymal transection: technical description and initial experience. Surg Endosc. 2022;36(5):3270–6.
23. Lai ECH, Yang GPC, Tang CN. Robot-assisted laparoscopic liver resection for hepatocellular carcinoma: short-term outcome. Am J Surg. 2013;205(6):697–702.
24. Lai ECH, Tang CN, Yang GPC, Li MKW. Multimodality laparoscopic liver resection for hepatic malignancy—from conventional total

laparoscopic approach to robot-assisted laparoscopic approach. Int J Surg. 2011;9(4):324–8.
25. Sugo H, Mikami Y, Matsumoto F, Tsumura H, Watanabe Y, Kojima K, et al. Hepatic resection using the harmonic scalpel. Surg Today. 2000;30(10):959–62.
26. Kim J, Ahmad SA, Lowy AM, Buell JF, Pennington LJ, Soldano DA, et al. Increased biliary fistulas after liver resection with the harmonic scalpel. Am Surg. 2003;69(9):815–9.
27. Aldrighetti L, Pulitanò C, Arru M, Catena M, Finazzi R, Ferla G. "Technological" approach versus clamp crushing technique for hepatic parenchymal transection: a comparative study. J Gastrointest Surg. 2006;10(7):974–9.
28. Lin TY. A simplified technique for hepatic resection: the crush method. Ann Surg. 1974;180(3):285–90.
29. Scalzone R, Lopez-Ben S, Figueras J. How to transect the liver? A history lasting more than a century. Dig Surg. 2012;29(1):30–4.
30. Kingham TP, Leung U, Kuk D, Gönen M, D'Angelica MI, Allen PJ, et al. Robotic liver resection: a case-matched comparison. World J Surg. 2016;40(6):1422–8.
31. Chan OCY, Tang CN, Lai ECH, Yang GPC, Li MKW. Robotic hepatobiliary and pancreatic surgery: a cohort study. J Hepatobiliary Pancreat Sci. 2011;18(4):471–80.
32. Lai ECH, Tang CN. Long-term survival analysis of robotic versus conventional laparoscopic hepatectomy for hepatocellular carcinoma: a comparative study. Surg Laparosc Endosc Percutan Tech. 2016;26(2):162–6.
33. Camerlo A, Magallon C, Vanbrugghe C, Chiche L, Gaudon C, Rinaldi Y, et al. Robotic hepatic parenchymal transection: a two-surgeon technique using ultrasonic dissection and irrigated bipolar coagulation. J Robot Surg. 2021;15(4):539–46.
34. Boggi U, Caniglia F, Amorese G. Laparoscopic robot-assisted major hepatectomy. J Hepatobiliary Pancreat Sci. 2014;21(1):3–10.
35. Ji W-b, Wang H-g, Zhao Z-m, Duan W-d, Lu F, Dong J-h. Robotic-assisted laparoscopic anatomic hepatectomy in China: initial experience. Ann Surg. 2011;253(2):342.
36. Hu M, Liu Y, Li C, Wang G, Yin Z, Lau WY, et al. Robotic versus laparoscopic liver resection in complex cases of left lateral sectionectomy. Int J Surg. 2019;67:54–60.
37. Chong CCN, Lok HT, Fung AKY, Fong AKW, Cheung YS, Wong J, et al. Robotic versus laparoscopic hepatectomy: application of the difficulty scoring system. Surg Endosc. 2020;34(5):2000–6.
38. Khan S, Beard RE, Kingham PT, Fong Y, Boerner T, Martinie JB, et al. Long-term oncologic outcomes following robotic liver resections for primary hepatobiliary malignancies: a multicenter study. Ann Surg Oncol. 2018;25(9):2652–60.
39. Troisi RI, Patriti A, Montalti R, Casciola L. Robot assistance in liver surgery: a real advantage over a fully laparoscopic approach? Results of a comparative bi-institutional analysis. Int J Med Robot. 2013;9(2):160–6.
40. Patriti A, Cipriani F, Ratti F, Bartoli A, Ceccarelli G, Casciola L, et al. Robot-assisted versus open liver resection in the right posterior section. JSLS. 2014;18(3):e2014.00040.
41. Tranchart H, Ceribelli C, Ferretti S, Dagher I, Patriti A. Traditional versus robot-assisted full laparoscopic liver resection: a matched-pair comparative study. World J Surg. 2014;38(11):2904–9.
42. Fruscione M, Pickens R, Baker EH, Cochran A, Khan A, Ocuin L, et al. Robotic-assisted versus laparoscopic major liver resection: analysis of outcomes from a single center. HPB. 2019;21(7):906–11.
43. Sucandy I, Schlosser S, Bourdeau T, Spence J, Attili A, Ross S, et al. Robotic hepatectomy for benign and malignant liver tumors. J Robot Surg. 2020;14(1):75–80.

Approaching the Liver Hilum in Robotic Liver Surgery and Pringle Maneuver

17

Yee L. Cheah and Caroline J. Simon

17.1 Introduction: Robotic Dissection of the Liver Hilum

One of the main advantages of the robotic platform in complex liver surgery is the ability to perform meticulous dissection of the liver hilum. Comparable to the human wrist, the 7 degrees of freedom of the robotic instruments facilitates isolation of the various structures in the hilum, while the nonworking arms provides stable retraction, and the robotic optic provides magnified 3D camera view.

Indications for dissection of the liver hilum include:

1. Extra-Glissonian approach to major hepatectomy for inflow control.
2. Dissection and preservation of inflow vessels during live donor hepatectomy.
3. Hilar lymphadenectomy for liver, bile duct, gallbladder, and pancreatic malignancies.

Preoperative review of the cross-sectional imaging scans is useful to delineate the hilar anatomy. We will detail the technical aspects of the approach to the liver hilum and describe dissection the hepatic artery, portal vein, and main bile duct branches. The Pringle maneuver is feasible in robotic liver surgery, and our technique is similar to maneuvers performed by our laparoscopic colleagues.

Y. L. Cheah (✉) · C. J. Simon
Division of Transplantation, Houston Methodist Hospital, Houston, TX, USA
e-mail: ycheah@houstonmethodist.org

17.2 Technical Aspects

17.2.1 Port Placement

These are our usual port placements for right-sided versus left-sided liver resection (Fig. 17.1a and b). The placement of the camera provides adequate visualization of the right or left side of the liver hilum as needed. The third arm is used for retraction and placed as a right-handed instrument. Two laparoscopic assistant ports are placed in the lower abdomen, and the specimen is usually extracted via a Pfannenstiel incision.

17.2.2 Standard Robotic Instruments

1. Maryland bipolar (+ electrocautery)
2. Camera
3. Scissors (+ electrocautery)
4. Cadiere forceps

17.2.3 Exposure of the Hilum

The ligamentum teres and falciform ligamentum are divided with robotic scissors using electro-

M. D'Hondt, I. Sucandy (eds.), *Textbook of Robotic Liver Surgery*,
https://doi.org/10.1007/978-3-031-76536-0_17

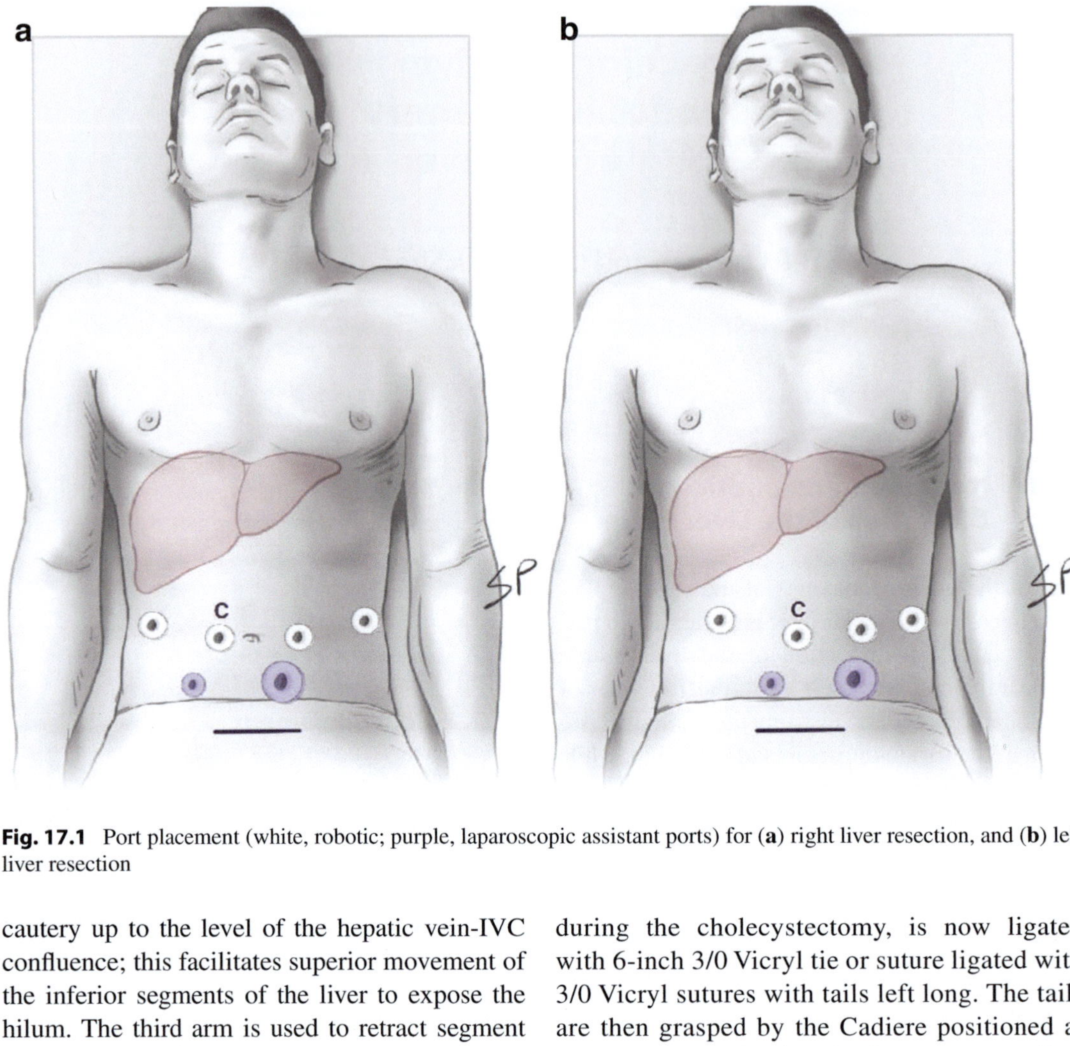

Fig. 17.1 Port placement (white, robotic; purple, laparoscopic assistant ports) for (**a**) right liver resection, and (**b**) left liver resection

cautery up to the level of the hepatic vein-IVC confluence; this facilitates superior movement of the inferior segments of the liver to expose the hilum. The third arm is used to retract segment 4B and gallbladder anterosuperiorly; a mini-lap pad is placed between the tip and elbow of the Cadiere to prevent injury to the liver surface (Fig. 17.2a). The laparoscopic assistant retracts the duodenum by gently pushing it inferiorly/dorsally with blunt-tipped instrument or suction device.

17.2.4 Dissection of Right Hepatic Artery

A standard cholecystectomy is performed, and the gallbladder specimen is placed in Morrison's pouch for later removal. The cystic duct stump, which should have been clipped during the cholecystectomy, is now ligated with 6-inch 3/0 Vicryl tie or suture ligated with 3/0 Vicryl sutures with tails left long. The tails are then grasped by the Cadiere positioned at segment 4B, and using the forceps as a pulley, the cystic duct is retracted anteriorly and slightly left, exposing the lateral and right posterior aspect of the common hepatic duct (CHD) (Fig. 17.2b).

The pulsation of a standard right hepatic artery (RHA) is usually visible behind the CHD at this stage, and the lymphatic and nerve tissues lateral and posterior to the CHD are slowly dissected until the lateral wall of the artery is visualized. Care is taken not to injure the CHD and common bile duct (CBD), either via traction or cautery burn. If the artery is not found in this manner, plan B is to follow the cystic artery to its origin, which is usually the right hepatic artery or a branch of it.

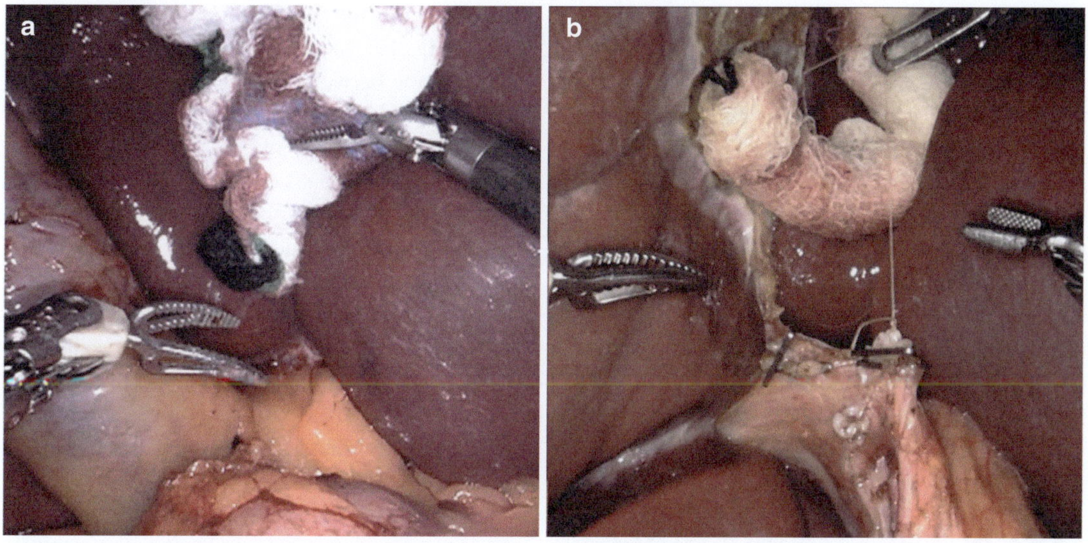

Fig. 17.2 Usage of the third arm (**a**) with a gauze to retract segment 4b to expose the hilum and (**b**) as a pulley to retract the ligature placed at the cystic duct stump, which will rotate the CBD and CHD anteriorly and toward the patient's left, in order to expose the area posterior to the duct where the right hepatic artery courses

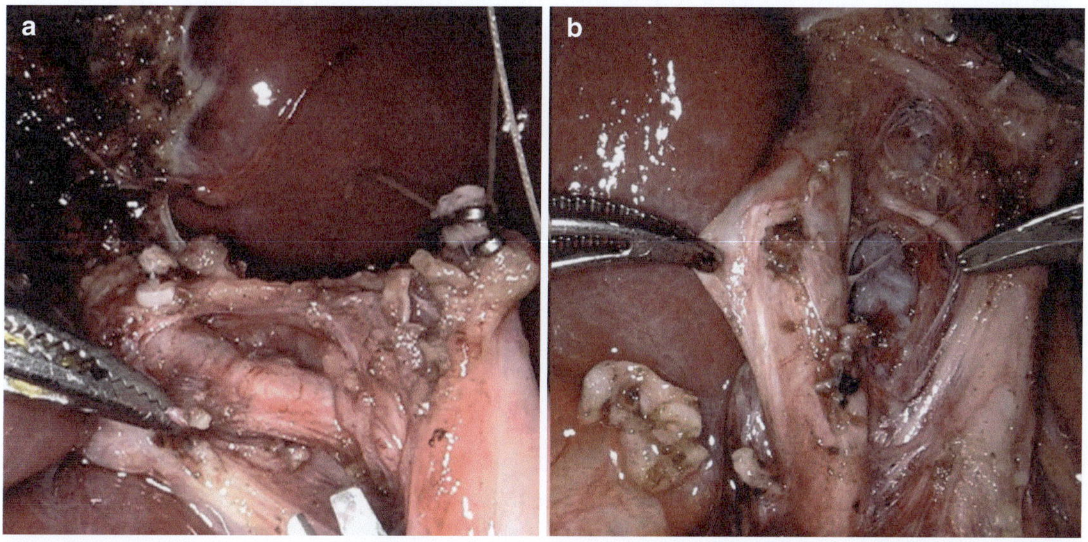

Fig. 17.3 Most common variations of RHA anatomy: (**a**) standard which courses posterior to the CHD and (**b**) replaced RHA from the superior mesenteric vein which courses along the lateral posterior aspect of the CBD and CHD

As more of the connective tissue is cleared between the duct and RHA, the space should open up between the two structures, and more retraction may be used on the cystic duct tie to expose the anterior surface of the RHA (Fig. 17.3a). The posterior aspect of the standard RHA may be adherent to the anterior surface of the portal vein (PV), and the artery will need to be gently dissected off of the PV. Once the anterior, lateral, and posterior walls of the artery are cleared, the tissues at the medial aspect of the RHA can be gently pushed away from the arterial wall and the artery encircled with a vessel loop.

If the preoperative imaging reveals a replaced RHA from the superior mesenteric artery, this artery is usually found lying posterior laterally along the course CBD and CHD. In this position, the artery is covered by a lymph node that commonly stretches the length of the bile duct. This node will need to be excised to gain access to the replaced RHA. The main difference in dissection of the replaced vs standard RHA is the close association with the portal vein posteriorly and CBD anterolaterally through the length of the artery (Fig. 17.3b).

In cases where we plan to take the inflow of the right lobe prior to parenchymal transection, the encircled RHA can be divided between clips now. This will facilitate the right portal vein (RPV) dissection as detailed below.

17.2.5 Dissection of Right Portal Vein

Division of the RHA will expose the anterior surface of the RPV enabling access to the portal vein bifurcation. In cases where the RHA is to be preserved, e.g., donor right hepatectomy, the path to the bifurcation can be reached by carefully retracting the vessel loop of the RHA with the third arm; this is similar to the retraction of the cystic duct stump ligature. If there is a replaced RHA, retraction of the vessel loop toward the patient's right side will help expose the PV bifurcation (Fig. 17.4a).

The plane between the CHD and anterior surface of the PV is gently developed by pushing the portal vein posteriorly off the attachments to the duct and RHA. Dissection proceeds cephalad until the bifurcation is encountered. Clear identification of the bifurcation is confirmed by visualization of the right, left, and main portal veins. The RPV anterosuperior surface is then carefully teased off the hilar plate by gently pressing the RPV posteriorly with the Maryland and sweeping the tissues off the vein with the scissors.

Attention is then moved to the dissection of the posterior RPV. The peritoneal lining on the posterior aspect of the RPV is divided and retracted posterior inferiorly by the laparoscopic assistant. The posterior wall of the portal vein is dissected free from the peritoneum surface until the caudate branches are visualized. The scissors is then switched to a large needle driver, and this instrument, in tandem with the Maryland forceps, helps encircle these short branches with 3/0 silk ties (Fig. 17.4b). The PV side of these branches is always ligated, while the caudate side may be clipped to avoid clips on the PV hampering future clipping or stapling of the RPV.

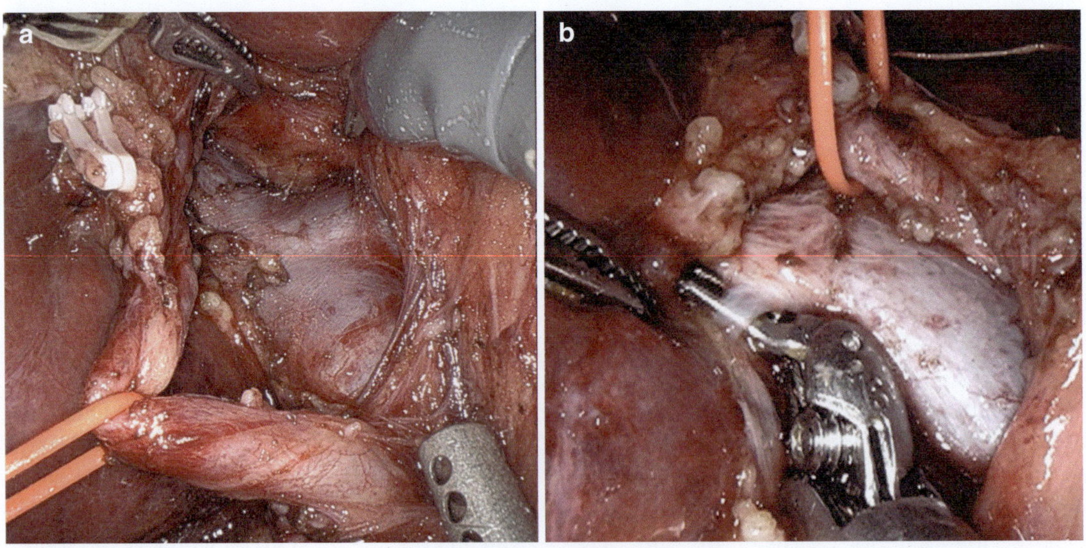

Fig. 17.4 Approaching the RPV: (**a**) anterior portal vein bifurcation exposed by retracting the replaced RHA toward the patient's right; (**b**) caudate branch on the posterior wall of RPV encircled using a large needle driver and Maryland forceps

Once all caudate branches are divided, the posterior RPV is then dissected off the hilar plate moving cephalad until we meet our dissection from the anterior side. Once this space is developed, the RPV is encircled with a vessel loop. In cases where we plan to take the inflow of the right lobe prior to parenchymal transection, the RPV may be divided between large plastic clips at this stage. If the RPV is particularly short, the proximal RPV inflow can be clipped without division and completion of this process performed later in the procedure (transection of the parenchyma and right hepatic duct increases the space around the RPV which will enable placement of additional clips and division).

17.2.6 Dissection of the Left Hepatic Artery

Since the left-sided hilar structures are not covered anteriorly by the CBD, cholecystectomy may be performed at this stage or later in the operation. The third arm is used, as described above, to retract segment 4B.

The peritoneum over the left side of the hilum is opened, and pulsation of the left hepatic artery (LHA) is located. This artery usually courses along the left side of the hilum from the base to the umbilical fissure, and it is common for the LHA to be short as it may divide early into the segment 4 and segment 2/3 branches. All lymphatic and nerve tissues are dissected away from the LHA including the station 12a node which is usually adherent to the left side of the both the LHA and left portal vein (LPV) (Fig. 17.5a). The posterior distal LHA or branches may need to be mobilized off the anterior LPV. The LHA and its branches, if necessary, are encircled with vessel loops (Fig. 17.5b).

In cases where we plan to take the inflow of the left lobe prior to parenchymal transection, the encircled LHA can be divided between clips or ties now. Leaving the tail long on the distal tie will enable retraction of the left distal artery anteriorly by the third arm. This will facilitate LPV dissection as detailed below.

17.2.7 Dissection of the Left Portal Vein

Division of the LHA will expose the anterior surface of the LPV. In cases where the LHA is to be preserved, e.g., donor left-sided hepatectomy, exposure of the LPV can be performed while the laparoscopic assistant retracts the vessel loop of the LHA laterally.

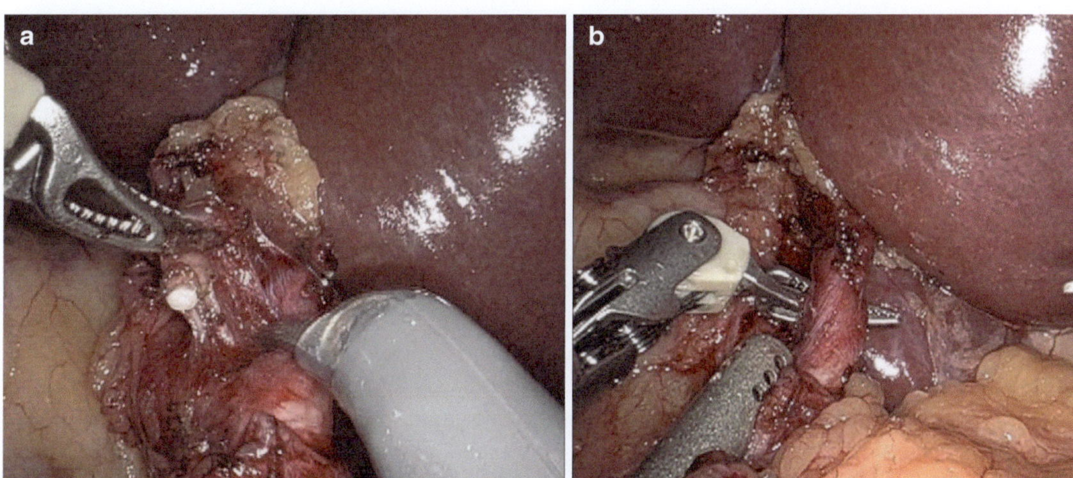

Fig. 17.5 LHA coursing from the left base of the hilum to the umbilical fissure: (**a**) lymphatic and nerve tissues dissected from the wall of the LHA; (**b**) LHA encircled with Maryland forceps

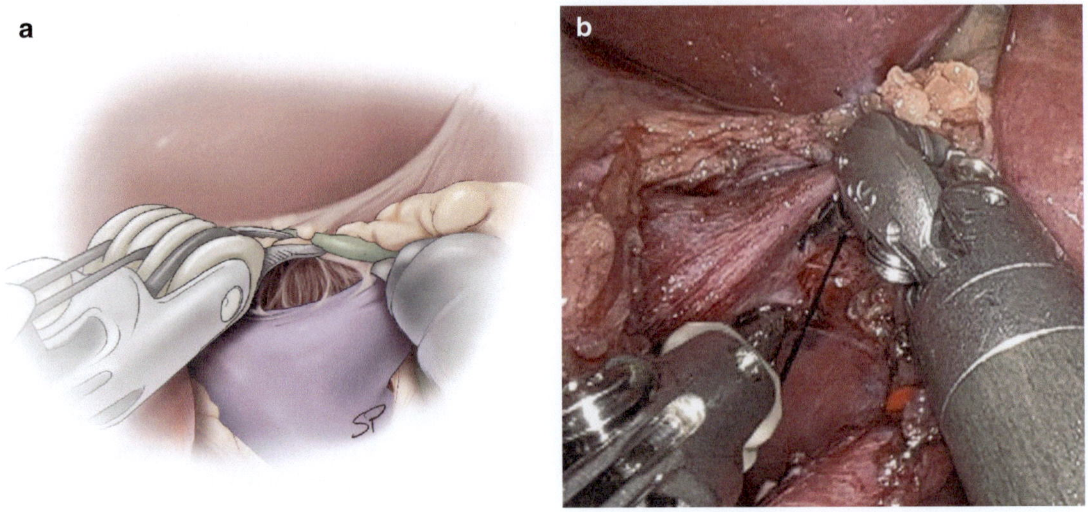

Fig. 17.6 Approaching the LPV: (**a**) the vein is freed from the attachments to the hilar plate and segment 4b; (**b**) ligature is about to be placed around a caudate branch on the posterior aspect of the LPV

The plane between the left hepatic duct or hilar plate and superior anterior surface of the LPV is gently developed by pushing the portal vein posteriorly and inferiorly off the attachments. Dissection proceeds transversely along the base of segment 4B from the portal vein bifurcation until the umbilical fissure (Fig. 17.6a). Several small segment 4B branches from the LPV are divided in a similar manner to the caudate branches.

Attention is then moved to the dissection of the posterior LPV. Retraction of the vessel loop, if used on the LHA, is released. The peritoneal lining the posterior aspect of the LPV is divided and retracted posterior inferiorly by the laparoscopic assistant. The posterior wall of the portal vein is dissected free from the peritoneum surface until the caudate branches are visualized. If our plan is not to resect the caudate lobe, these branches are divided after ligation in a similar manner as the right-sided caudate branches (Fig. 17.6b).

Once all caudate branches are divided, the posterior LPV is then dissected off the hilar plate moving cephalad until we meet our dissection from the anterior side. Once this space is developed, the LPV is encircled with a vessel loop. In cases where we plan to take the inflow of the left lobe prior to parenchymal transection, the LPV is usually long enough to be divided between large plastic clips or even a stapler at this stage.

17.2.8 Indocyanine Green Injection and Firefly™

Demarcation of right vs left liver lobes is enhanced by indocyanine green (ICG) dye injection (Fig. 17.7). We use a 5-mg dose of ICG given after inflow to the relevant lobe has been ligated or, in a donor hepatectomy, the inflow vessels have been clamped with laparoscopic bulldogs. Injecting the dye at this stage will also allow enough time for biliary excretion which will assist in identification of the relevant-sided hepatic duct during parenchymal transection.

17.2.9 Dissection of Hepatic Duct

We divide the relevant-sided hepatic duct after about 75% of the parenchyma has been transected. By this stage, all the parenchyma on a plane anterior to the hepatic duct has been divided, exposing most of the circumference of the duct.

Firefly is used to identify the bile duct bifurcation and the relevant-sided duct. The point of

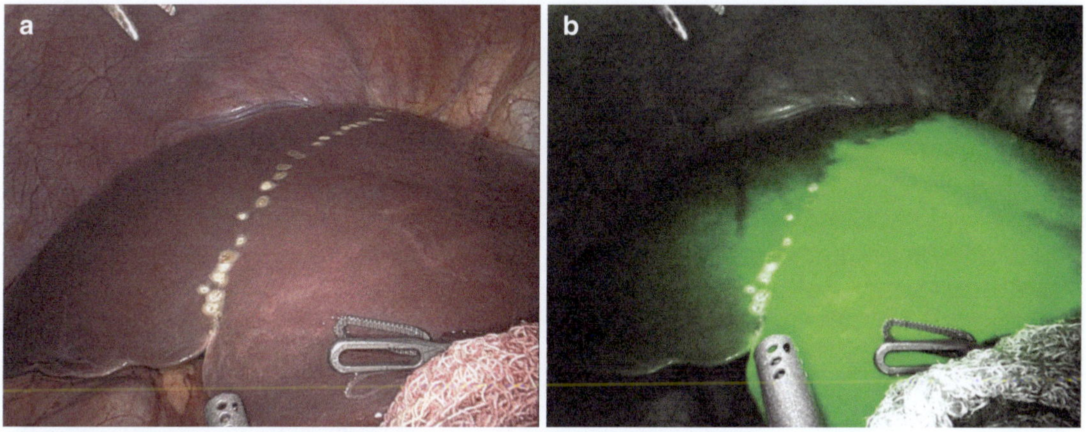

Fig. 17.7 After clamping of the RHA and RPV at the hilum, (**a**) demarcation of the right lobe is subtle on the normal view but (**b**) enhanced by ICG injection on the Firefly™ view

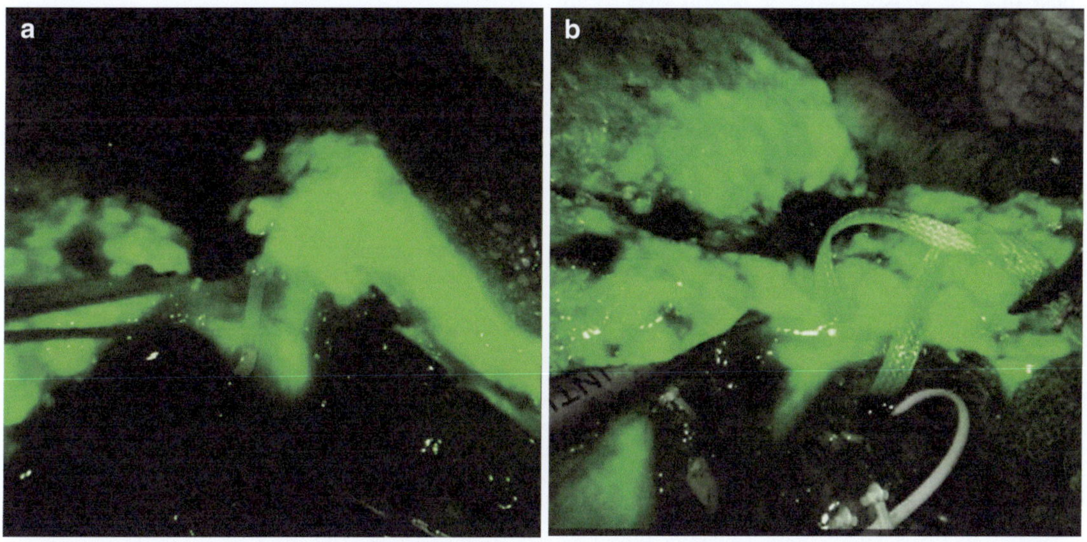

Fig. 17.8 Visualization of the bile duct bifurcation via excretion of the ICG in the biliary system: (**a**) a plastic clip has been placed on the distal right hepatic duct; (**b**) left hepatic duct encircled with an umbilical tape

hepatic duct division is chosen. There are two ways to transect this duct:

1. "Clip and cut": Encircling the duct and its surrounding hilar plate with an umbilical tape (Fig. 17.8a and b).

   The Maryland forceps is used to encircle an umbilical tape around the entire hepatic duct with its surrounding hilar plate tissue. A large plastic clip is then placed in the distal hepatic duct making sure that it does not encroach onto the bifurcation. The proximal hepatic duct on the specimen side can then be divided. The stump of the distal hepatic duct is suture ligated with a running 5/0 PDS.
2. "Cut and clip": Cutting the duct without initial clipping.

   The hepatic duct is cut at the chosen point without cutting the hilar tissues that are posterior to the duct (there is usually a vessel in this tissue). The hilar plate tissue is clipped with two small metal clips prior to division. The stump of the distal hepatic duct is suture ligated with a running 5/0 PDS.

17.2.10 Pringle Maneuver

Most reports of Pringle maneuver for minimally invasive liver surgery involve extracorporeal control of the tourniquet. In this technique, control of the Pringle maneuver in robotic surgery is in the hands of the bedside laparoscopic assistant surgeon. One end of a long moistened umbilical tape is inserted via a 5-mm laparoscopic port leaving the other end outside the abdomen. The position of this port is usually on the opposite side of the lobe to be resected, i.e., port placed in the left abdomen in a right hepatic resection and vice versa. The tape is slung around the liver hilum and the end externalized via the same port. The laparoscopic trocar is then removed, and a small-bore chest tube or silastic catheter is slid over both strings with the ends remaining extracorporeal. The bedside surgeon can then control the Pringle application in the style of a Rummel tourniquet.

Alternatively, intracorporeal Pringle maneuvers have been described including applying a laparoscopic Satinsky clamp across the hilum, cinching a trimmed urinary catheter around the hilum or to leave the ends of the above tapes intracorporeal and cinching the tourniquet by inserting a large occluding clip.

17.3 Tips and Tricks

- An aberrant RHA may course anterior to the CBD; in these cases, care must be taken to recognize the position of the CBD in relation to the RHA to avoid injury to the duct during dissection of the artery.
- After the RPV has been dissected and encircled, a 6/0 Prolene stitch may be placed superficially on the wall of the PV to mark the bifurcation; this stitch is useful to guide the clipping or stapling of the RPV to avoid impinging the bifurcation and causing stenosis.
- If the length or the angle of the RPV makes encirclement by a Maryland forceps difficult, an alternative to a Pringle maneuver for inflow control of the right lobe is to occlude the RPV with a laparoscopic bulldog clamp during parenchymal transection; once the parenchyma is transected, there is usually more room to accommodate encirclement.

17.4 Conclusion

The robotic platform facilitates meticulous dissection of the liver hilum where important structures are in close proximity to one another. This has enabled isolation of the vessels at the hilum for inflow control during major hepatectomy. ICG dye application is a useful adjunct to delineate biliary anatomy during hilar dissection.

Further Reading

1. Choi TW, Chung JW, Kim HC, et al. Anatomic variations of the hepatic artery in 5625 patients. Radiol Cardiothorac Imaging. 2021;3(4):e210007.
2. Sucandy I, Luberice K, Lippert T, et al. Robotic major hepatectomy: an institutional experience and clinical outcomes. Ann Surg Oncol. 2020;27:4970–9.
3. Sucandy I, Gravetz A, Ross S, et al. Technique of robotic left hepatectomy: how we approach it. J Robot Surg. 2019;13(2):201–7.
4. Laurenzi A, Cherqui D, Figueroa R, et al. Totally intra-corporeal Pringle maneuver during laparoscopic liver resection. HPB. 2018;20(2):128–31.
5. Choi GH, Chong JU, Han DH, et al. Robotic hepatectomy: the Korean experience and perspective. Hepatobiliary Surg Nutr. 2017;6(4):230–8.
6. Chen P-D, Chao-Yin W, Rey-Heng H, et al. Robotic major hepatectomy: is there a learning curve? Surgery. 2017;161(3):642–9.
7. Giulianotti PC, Bianco FM, Daskalaki D, et al. Robotic liver surgery: technical aspects and review of the literature. Hepatobiliary Surg Nutr. 2016 Aug;5(4):311–21.
8. Huang JW, Su WL, Wang SN. Alternative laparoscopic intracorporeal pringle maneuver by huang's loop. World J Surg. 2018;42(10):3312–5.

Intraoperative Liver Ultrasound: Technical Considerations During Robotic Surgery

18

Ellen J. Hagopian

18.1 Introduction

Ultrasound (US) is an essential adjunct during liver surgery. Given its lack of transparency and its three-dimensionality, it is impossible to visualize inside of the liver to identify anatomy, locate tumor, mark vasculature, and find other pathology. US provides this missing link in the operating room and is considered one of the pillars of liver surgery, in addition to the knowledge of anatomy and use of vascular control [1]. The use of intraoperative US becomes even more important when the hands of the surgeon are removed from direct contact of the organ, such as in minimally invasive procedures. The European consensus conference on laparoscopic liver resection asserts a strong recommendation that intraoperative US should be available for every laparoscopic liver resection [2]. This recommendation should be no different for robot-assisted liver resection.

The objective of this chapter is to provide a foundation for the use of intraoperative US during robot-assisted liver surgery, including its indications, evolution, and technical procedural steps.

E. J. Hagopian (✉)
University of Toledo College of Medicine and Life Sciences, Toledo, OH, USA
e-mail: Ellen.Hagopian@utoledo.edu

18.2 Uses and Indications

Intraoperative US is an essential component of *all* liver surgical procedures. The use of US during liver surgery can be conceptualized into confirming what is present on the preoperative imaging (known), finding what is not present on the preoperative imaging (unknown), and actively guiding the procedure [1]. During the diagnostic US procedure, liver anatomy can be defined and suspected lesions located or excluded. US is a highly accurate and sensitive tool to detect new lesions. Furthermore, vascular anatomy and any anomality can be clearly identified, and vascular and/or biliary involvement by tumor can be defined. In this aspect, US is imperative to *confirm* what is known and *find* what is unknown, thereby diagnosing the extent of disease.

Understanding the extent of the disease, including the number and location of tumors and proximity to vascular and biliary structures, the procedure can be planned. In this aspect, US is imperative to assist the surgeon to *decide* the strategy of the procedure.

Once the procedure has been planned, US can be used to *guide* the procedure. Important liver anatomic information can be identified, such as the location of the hepatic and portal veins to mark the limits of resection. For example, during right or left hepatectomy, the middle hepatic vein is identified and marked on the liver surface to indicate the transaction line. Marking the hepatic and portal

M. D'Hondt, I. Sucandy (eds.), *Textbook of Robotic Liver Surgery*,
https://doi.org/10.1007/978-3-031-76536-0_18

veins on the surface of the liver will show the limit of segmental resection. These maneuvers are identical across all approaches to liver resection, whether open, laparoscopic, or robotic. In particular, during minimally invasive liver resection, important crossing hepatic and portal veins can be marked on the liver surface so they can be anticipated. US is a real-time examination performed during a real-time operation and therefore is the ideal modality to guide surgical procedures, in particular, liver resection. US is used actively during parenchymal transection and guides the resection by actively confirming major vascular structures and ensuring an appropriate margin of resection. Following completion of resection, US can then be repeated to confirm the completeness of resection. In these aspects, US is imperative to assist the surgeon to *guide* the strategy of the procedure.

Therefore, intraoperative liver US is indicated during any liver resection to determine the true extent of disease (*confirm* and *find*), to determine the strategy for resection (*decide*), and to safely conduct parenchymal transection (*guide*) (see Fig. 18.1).

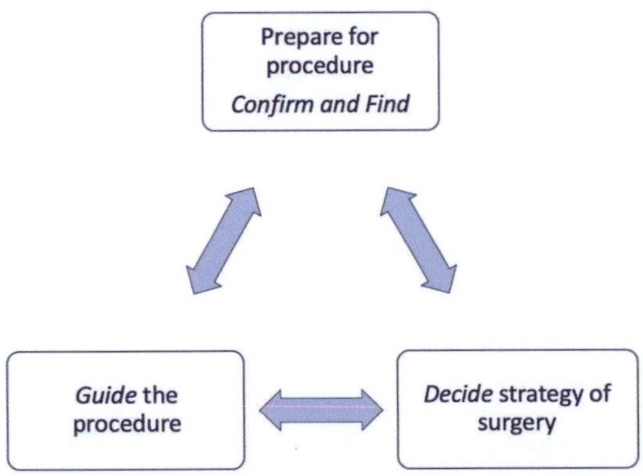

Fig. 18.1 Conceptual framework of the use of intraoperative liver ultrasound. The first step is for the surgeon to *prepare* for the procedure whereby the liver is fully examined with ultrasound (see Table 18.1). During this step, the surgeon confirms what pathology is known and potentially finds lesion(s) that were unknown on preoperative imaging. Using this information, the surgeon can then *decide* on the strategy of resection. In the final step, the surgeon will actively use ultrasound to *guide* the resection, including marking the limits of resection, important crossing vasculature, and confirming the completeness of resection. The interchanging arrows indicate the active use of ultrasound in examination across all steps of the procedure

18.3 Background

18.3.1 Early Evidence for US

There is a long history of intraoperative US use during liver surgery, which has had a substantial impact on the safety and quality of liver resection surgery [3]. Early reports by Makuuchi [4, 5] and Bismuth [6] of intraoperative US for liver resection reported its utility not only to define anatomy but also to guide resection, establishing its role in liver resection. Today, intraoperative US is considered a standard of care during liver resection across the international surgical community. In one of the original and largest reports of intraoperative US during abdominal, thoracic, cardiovascular, neurologic, and endocrine operations, Machi and Sigel found intraoperative US to be useful in 91.5% of cases [7]. Specifically, intraoperative liver US is highly accurate [8], and additional lesions are found in up to 27% cases [9–12]. Furthermore, the strategy (i.e., decision-making) of the operation is altered in up to approximately 40% of cases [9, 10, 13]. The utility of intraoperative liver US is well established [1].

18.3.2 Evolution of Robot-Assisted US

With the increasing use of minimally invasive approaches, both laparoscopic and robot-assisted, to liver resection, US continues to be an intraoperative imperative to evaluate the liver, its anatomy, and disease extent and to guide resection. Because of the lack of tactile feedback, US is a critical tool to identify pathology and critical structures during dissection and is extremely important to ensure a proper margin of resection.

With the development and evolution of robot-assisted liver resection, the use of intraoperative US faced multiple unique challenges, including the probe, assistant surgeon, and the display [14]. The first faced by surgeons was the *US probe and its operation.* In the initial experience, US performed during robot-assisted liver resection procedures was performed with rigid laparoscopic probes [14–16]. This posed several unique challenges. The need to operate the laparoscopic US probe at the bedside *requires a second assistant surgeon*, also with expertise in liver resection and US, during the procedure. The *inability of the operative surgeon to carry out the US procedure* creates several disadvantages. The direct performance of US allows the operative surgeon to build a mental image of the liver anatomy and pathology. This mental image of the liver is based on real-time US data while linking those data to other sensory feedback. The mental image is based on both visual and tactile feedback but also in the movement of the probe. The mental and spatial connection between the hand movement of the probe and the changing of the image on the ultrasound screen is extremely important to help the operative surgeon build the mental image. This is lost when the surgeon does not perform the ultrasound directly.

The US *image display* also posed a challenge in robot-assisted procedures [16]. Without an integrated display, US images are visualized only on the US system display. This takes the surgeon's eyes away from the 3D console to visualize the US image. Without the ability to simultaneously view the operative field and the US image, the surgeon's ability to build the metal image of liver anatomy and pathology and to perform simultaneous US with liver resection is severely limited.

The development of the robot-integrated US probe greatly enhances the surgeon's direct control of intraoperative US during robot-assisted procedures. Early work in the development of the robotic probe came in 2011 by Schneider and colleagues [16]. The early probe quickly evolved into the modern robot "drop-in" probe which has a flexible cable, offering probe flexibility and maneuverability, similar to open US procedures [14]. The robot US probe has small window ("foot"), which along with its flexible cable allows reaching difficult locations (see Fig. 18.2). Designed specifically to be operated with robotic instrumentation, the modern robot probe is operated directly by the surgeon seated at the console.

Furthermore, the integrated robot US probe allows for the US image display in the 3D viewer.

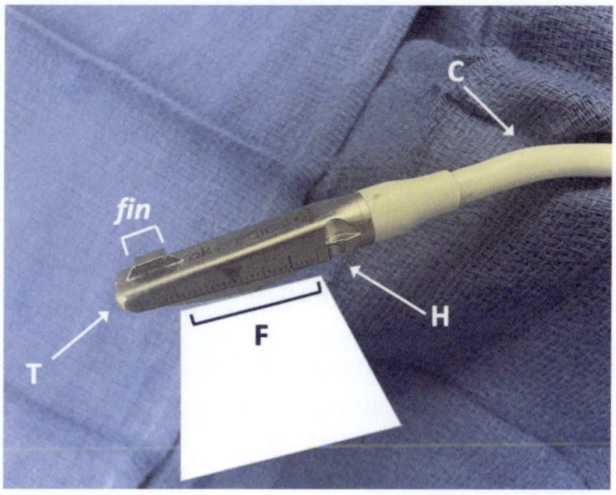

Fig. 18.2 Anatomy of the robot drop-in ultrasound probe. The robot drop-in probe has a flexible 12-mm-diameter cable (C), which can be inserted into any accommodating assistant port. The special fin on the opposite side of the window, or foot (F), is grasped by a robotic instrument which allows for multiple degrees of movement. The window is a curvilinear design which allows for a wide footprint (white box), extending from the probe's toe (T) to heel (H)

Using the TilePro™ function at the console, a simultaneous US image and operative view are shown, overcoming the challenges of the display [14, 17]. With a simultaneous display and the ability to control the US image in all degrees of freedom, the surgeon is able to build a mental image of the liver anatomy and pathology while also actively and simultaneously using US during parenchymal dissection. The modern integrated robot "drop-in" probe with simultaneous display using the TilePro™ function overcomes over the many early challenges, leading to the modern use of robot-assisted US during liver resection today.

18.4 Technical Aspects (i.e., *How I Do It*)

As detailed above, when performing US, the surgeon builds a mental image of the liver based on the real-time US findings. This is no different during robot-assisted procedures, where the mental image is based on visual and spatial feedback when the surgeon performs the US. These points emphasize the importance of the evolution of the modern robot-integrated US probe.

To be proficient in the US procedure and image interpretation, it is important to review some basics of the US console. For an in-depth review of US physics and other basic US content, the reader is directed to other sources [18, 19].

18.4.1 US Image Optimization

One of the major pitfalls of US is the risk of misinterpretation. Obtaining an optimal US image is essential for correct image interpretation. Misinterpretation of US images is directly related to patient outcome.

The US console may appear daunting and intimidating; however, there are a limited number of US controls that the surgeon needs to master to optimize the US image, including frequency, gain, depth, and focus.

Frequency can be thought of as the "resolution" of the ultrasound image. There is a direct relationship between the frequency and the resolution of the image such that as the frequency is increased, the resolution or clarity of the image is also increased. On the other hand, there is an inverse relationship between frequency and depth of penetration. A high frequency may provide better clarity, but it will provide less penetration into deeper tissue. Gain can be thought of as "volume." Turning up the gain is like turning up the volume on a radio. Increasing the gain will amplify the reflected sound, which in turn will increase the noise that the US probe hears or brightness of the image. For some US systems, the gain button is indicated as a "B," indicating brightness. Too much or two little gain can either produce too much or too little brightness (or whiteness of the image), interfering with the US interpretation. Depth can be thought of as "distance" and refers to the distance traveled by the US waves. The depth should be adjusted such that the center of the image is the area of interest. Focus should be thought of as the "point of interest," centering on the area of interest.

18.4.2 Robot-Integrated Probe

A robot-integrated US probe (i.e., "drop-in probe") became available in 2013 [14]. It is a convex multifrequency side-viewing transducer with a flexible long shaft, which allows it to be "dropped in" from any accommodating assistant laparoscopic trocar (see Fig. 18.2). Once inserted, the probe is brought into the operative view and is fully controlled by the console surgeon. As described above, the simultaneous real-time US image and the operative view through the stereo viewer facilitate the US procedure and image interpretation (see Fig. 18.3).

The probe is designed such that it can be controlled and maneuvered with a robotic instrument. It is held by a robotic grasper from a special "fin" located directly over the transducer array (see Fig. 18.2). The flexibility of the shaft of the robot probe allows movement in all degrees of freedom [14]. The freedom of movement is extremely advantageous and simulates the unrestricted movement of the US probe in the hand of the surgeon. Because of this freedom of movement, the robotic US procedure closely aligns with the open US method [1, 19].

18.4.3 Diagnostic Method

Intraoperative liver US should be the initial step to *prepare* for liver resection. Data obtained during the US procedure, including confirming what is known and finding what is unknown, provides critical information for the surgeon to *decide* the operative plan and to then *guide* the resection (see Fig. 18.1).

The robot-integrated probe should always be used, unless it is not available. In this circumstance, a laparoscopic probe can be used with the limitations described above. The robot-integrated probe is a multifrequency probe, typically ranging between 5 and 12 MHz. A starting frequency of 7.5 MHz is useful as this allows adequate depth penetration to view the entire organ while still detailing the intrahepatic vascular anatomy and parenchyma. A lower frequency (5 MHz) may be needed for larger, steatotic, or cirrhotic livers to allow deeper penetration. To visualize more superficial lesions and structures, a higher frequency, such as 10–12 MHz, is needed for better clarity.

Intraoperative US examination of the liver is typically performed using direct contact, whereby

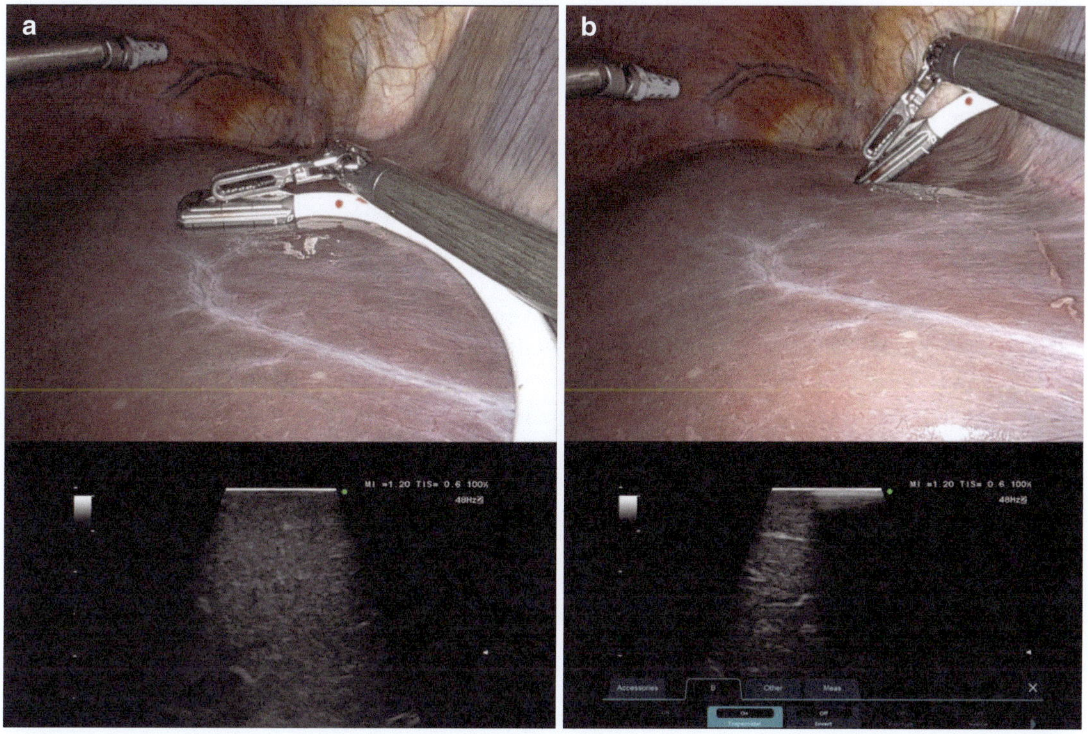

Fig. 18.3 Robot-integrated ultrasound (US) view using the TilePro™ function. (**a**) The modern integrated robot US probe allows for the simultaneous view of the US image and the operative view through the 3D stereo viewer, facilitating the US procedure and active use during liver resection. (**b**) Orientation of the probe is very important as a first step, even prior to optimizing the US image. Note that the image is oriented such that the "toe" of the probe (see Fig. 18.2) is oriented to the surgeon's left view (i.e., right side of screen) by adjusting the "invert" feature. This will facilitate liver US examination in the transverse axis

the probe is placed directly on the liver. Although liver contact scanning is standard for most situations, it may be necessary to use saline as an acoustic window to visualize very superficial liver pathology, as the probe has limited visualization 0.5–1.0 cm just beneath the probe [19].

As in any procedure, it is important to develop a systematic liver US scanning method. The author's preferred approach is divided into three tasks in the following order: (1) identification of the hepatic veins (outflow), (2) identification of the portal veins and their branches (inflow), and (3) systematic scanning of the whole liver (parenchyma) (see Table 18.1). During the US diagnostic procedure, the surgeon will *confirm* what is known and may *find* what may not be known.

The US procedure is begun by visualization of the hepatic veins. The flexibility of the robot-integrated US probe allows visualization in all planes, including transverse, oblique, and longitudinal. Grasping the US probe from a right-sided arm (arm 3 or 4) will afford the greatest flexibility when beginning the exam. It is the author's preference to begin the exam by placing the US

Table 18.1 Intraoperative liver ultrasound steps*

Hepatic veins (outflow)
Identify each hepatic vein's junction at the IVC
Define hepatic venous tributaries
Track the course of each hepatic vein from origin to termination
Portal veins and branches (inflow)
Identify the main portal vein and its bifurcation
Identify the right portal vein, its anterior and posterior branches, and follow to each one's segmental branches
Identify the left portal vein branch and follow to its segmental branches
Systematic liver scan (parenchyma)
Scan each liver section: Right posterior, right anterior, medial, left lateral
Identify and document any liver lesion's size, location, and echo characteristics
Define any tumor vascular or biliary involvement

*Adapted from Hagopian, 2020 [1]

probe in the transverse axis, positioning it in the center of liver, just to the right of the suspensory ligament (segment 4). To *identify the hepatic veins*, tilt the probe upward, and the junction of the hepatic veins at the suprahepatic vena cava is demonstrated. Visualize each hepatic vein along its longitudinal axis by rotating the US probe (from transverse to longitudinal) and following each hepatic vein from its origin to its termination. It is the author's preference to begin with the middle hepatic vein. The right hepatic vein is then similarly interrogated from origin to termination. The left hepatic vein can be visualized from the right side of the suspensory ligament, but to view and potentially mark the left hepatic vein, the surgeon should move the probe to the left of the suspensory ligament. Similarly, the left hepatic vein is followed peripherally to its terminus. Using the flexibility of the US probe cable and the EndoWrist capability of the robotic grasper, the hepatic veins are visualized in each axis and examined along each one's length (see Fig. 18.4).

Following this, the portal vein, portal vein branches, and segmental branches are examined. To *identify the portal vein* and its branches, place the probe again transversely in the center of segment 4 (the first position to identify the hepatic veins), and tilt inferiorly (caudally) to visualize the portal bifurcation just distal to the main portal vein. At times, this requires sliding the probe toward the lower aspect of segment 4 (see Fig. 18.5). Ensuring that the US probe toe is positioned laterally, follow the right portal vein out toward the periphery (laterally) and identify the right anterior and posterior sectoral branches (see Fig. 18.6). Each segmental branch is then visualized. Follow the anterior branch of the right por-

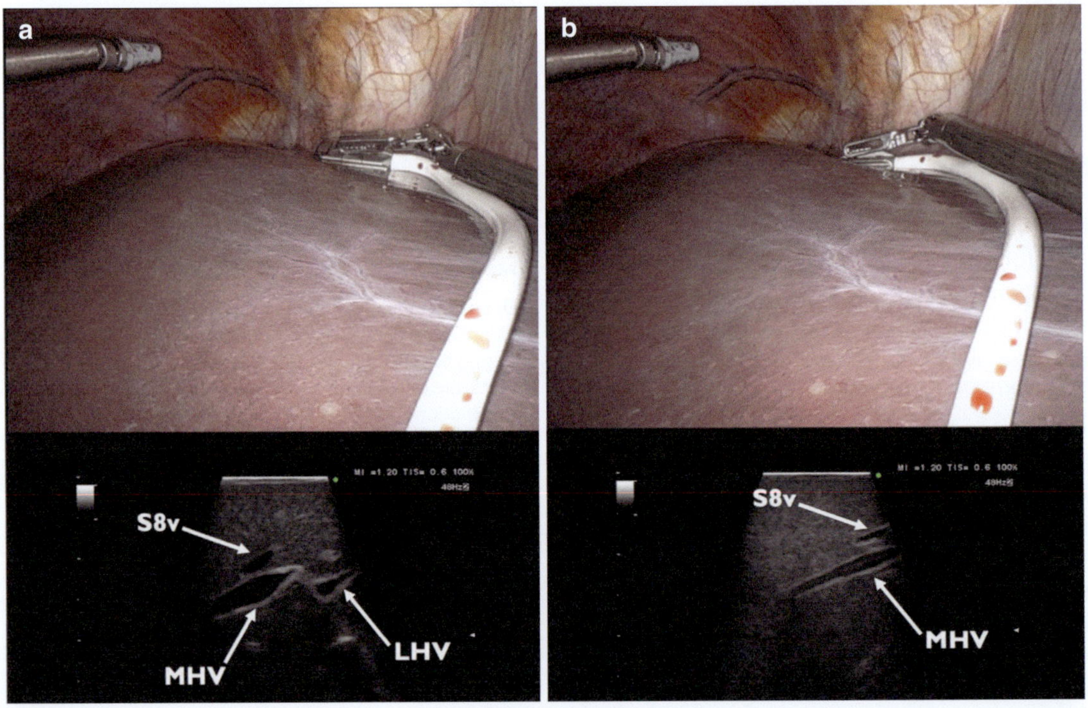

Fig. 18.4 Hepatic outflow. (**a**) The probe is placed on segment 4 in the transverse orientation and tilted up to visualize the hepatic veins and at their origin at the inferior vena cava. Depicted here are the middle hepatic (MHV) and left hepatic veins (LHV) in transverse and oblique axes, respectively, just distal to the common trunk (not shown). A small tributary from segment 8 is also seen (S8v). (**b**) Each hepatic vein is then examined along its length. In this view, the probe has been slightly rotated to view the MHV longitudinal axis. The segment 8 tributary is also seen in its longitudinal axis

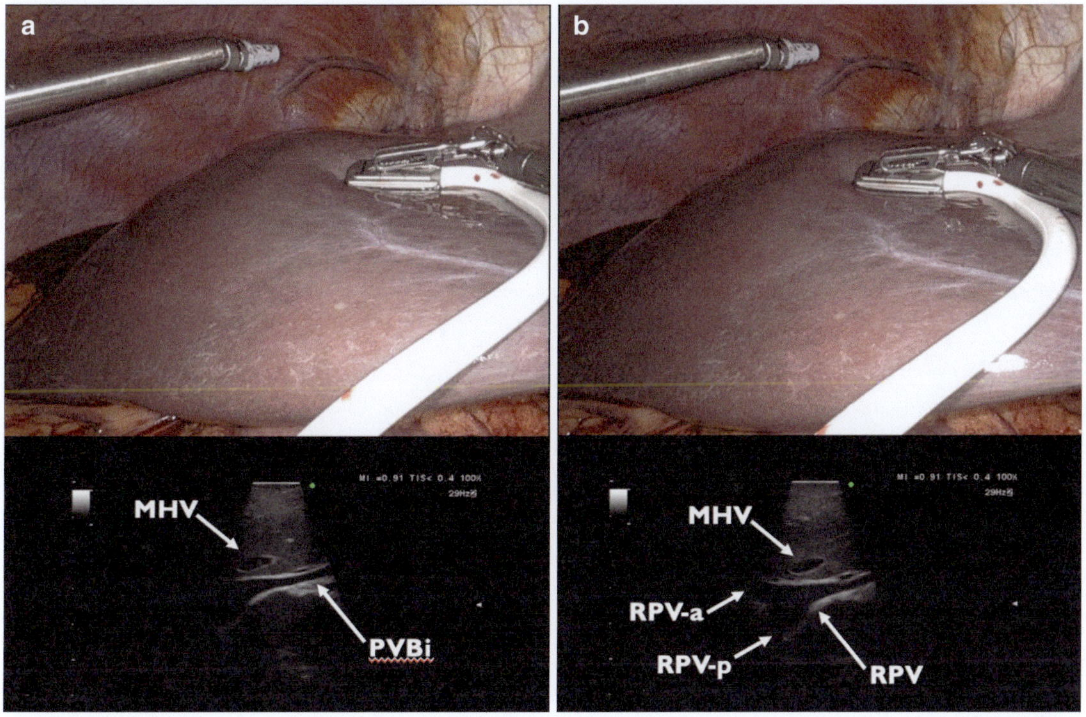

Fig. 18.5 Hepatic inflow [1]. (**a**) After each hepatic vein is examined, the probe is then moved back to segment 4 and is tilted down to visualize the portal vein bifurcation (PVBi). Note the MHV is seen in its transverse/oblique axis when viewing the portal bifurcation in the longitudinal axis. (**b**) Sliding the probe to the right will visualize the right portal vein (RPV). In this image, anterior (RPV-a) and posterior (RPV-p) sectoral branches of the right portal vein are just beginning to come into view

tal vein to its segmental branches, 5 (inferior) and 8 (superior). Slide the probe back to the right portal vein; next follow the posterior sectoral branch to its segmental branches, 6 (inferior) and 7 (superior). Note that the segment 6 branch may be variable and have multiple branches. Move the probe back toward the bifurcation and then track the horizontal left portal vein to the left. Rotate the probe to the longitudinal axis along the suspensory ligament and tilt the probe toward the left to visualize the ascending branch of the left portal vein. The left portal vein and its branches may be identified on either the right or left side of the suspensory ligament. Follow the left portal vein to identify its first branch, segment 2, just beyond the ligamentum venosum and terminal branches, segments 3 and 4 (see Fig. 18.6).

The systematic survey of the liver parenchyma completes the liver US procedure. Perform a systematic liver parenchymal scan using the sweep scan or lawnmower technique, as previously described by Hagopian and Machi [19]. Examine each individual section, including the right posterior, right anterior, left medial, and left lateral sections and identify any lesion or pathology in addition to its location, size, and sonographic features. Once the diagnostic US procedure is completed, the surgeon can then use this information to *decide* the operative plan (see Fig. 18.1).

18.4.4 Liver Parenchymal Transection

Once the decision for strategy of the liver resection is taken, US is used to *guide* the procedure. US aides to identify important vasculature, marks the limits of resection, and is actively used during parenchymal transection to ensure a proper mar-

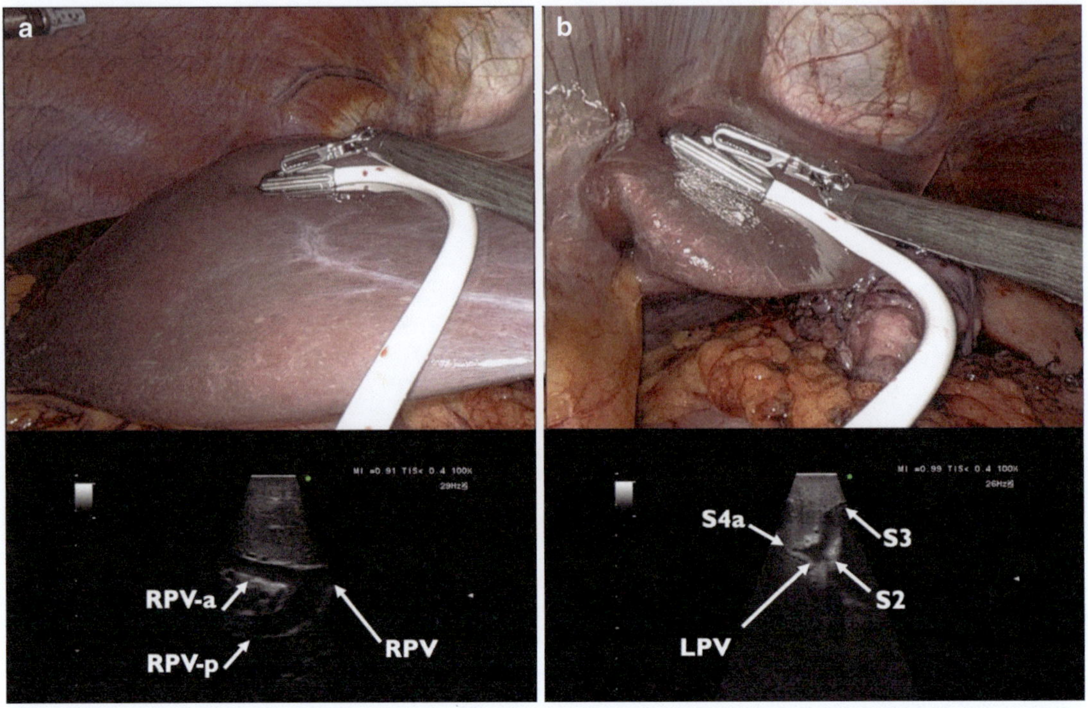

Fig. 18.6 Hepatic inflow [2]. (**a**) From the portal bifurcation (see Fig. 18.5), sliding the probe to the right while following the right portal vein will lead to the anterior and posterior sectoral branches of the right portal vein. (**b**) Moving the probe to the left of the falciform will visualize the left portal vein and its segmental branches. Depicted here is the ascending portion of the left portal vein (LPV) in the transverse orientation and the segmental 2 (S2), segment 4A (S4a), and segment 3 (S3) branches

gin and completeness of resection. For anatomic resection, it is the author's first preferred step to identify the hepatic vein(s) to mark the limits of resection. For example, for either right or left hepatectomy, the middle hepatic vein is identified in its longitudinal plane and is drawn on the liver surface along its course using cautery (see Fig. 18.7). The exact plane of dissection is chosen to preserve the vein whether performing either right or left resection and can be monitored under US during parenchymal transection. Similarly, for right posterior sectionectomy, the course of the right hepatic vein is marked, whereas for right anterior sectionectomy, the right and middle hepatic veins are drawn along their longitudinal axes to mark the lateral and medial lines (limits) of resection. For segmental resection, the portal vein bifurcation is identified and marked transversely on the liver, identifying the inferior and superior limits (i.e., segments) of each section (e.g., segments 5 and 8 of anterior section). It is furthermore the author's preference to mark vessels crossing the transection plane. For example, when performing a left lateral sectionectomy, the left hepatic vein and segment 2/3 branches are drawn on the liver surface to mark their locations so these structures can be anticipated during parenchymal transection.

During resection, US is actively used to monitor the transection line of the liver guiding the resection to ensure an adequate tumor margin. This active use of US during parenchymal transection ensures completeness of resection and a proper margin of resection. Given the loss of tactile feedback during minimally invasive procedures, this real-time continual feedback during resection is particularly important during robot-assisted liver resection.

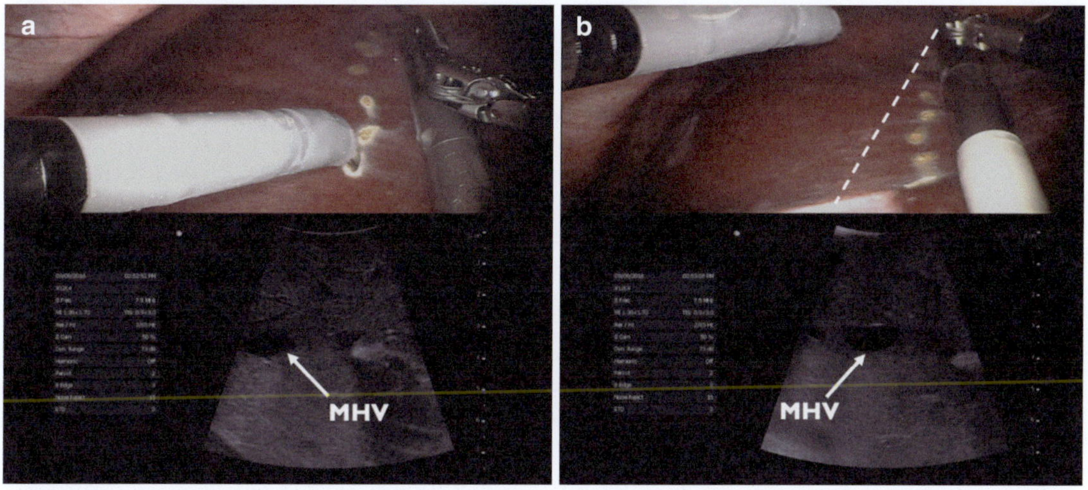

Fig. 18.7 Procedural guidance. Ultrasound is actively used to mark vasculature in preparation for parenchymal transection. Here, the middle hepatic vein (MHV) is marked with electrocautery (A) for left hepatectomy (**a**). Note that the line of resection is marked medial to the MHV (dashed line) to preserve the vein during parenchymal transection (**b**)

18.5 Robot-Assisted US Experience

Guerra and colleagues published the first series on the clinical application of the robotically integrated US probe during robot-assisted liver resection. In this report, they describe a series of ten patients and exemplify the ease of the robot "endowrist" in US liver assessment, concluding that the integrated probe offers significant advantages [14]. A few years later, Pesi and colleagues demonstrated the safety and feasibility US-guided robotic liver resection in 13 patients [20]. These reports and others describe the importance of the use of US during robot-assisted liver resection. Indeed, to the author's knowledge all published reports of robot-assisted liver resection have included the use of intraoperative US [14, 17, 20–24].

18.6 Conclusion

Liver US is a required competency of any modern liver surgeon, and it is an essential and necessary procedure during any robot-assisted liver procedure. The evolution of the robot-integrated US probe has been essential to the ease of incorporating the US procedure during robot-assisted liver procedures.

References

1. Hagopian EJ. Liver ultrasound: a key procedure in the surgeon's toolbox. J Surg Oncol. 2020;122:61–9. https://doi.org/10.1002/jso.25908.
2. Abu Hilal M, Aldrighetti L, Dagher I, Edwin B, Troisi RI, Alikhanov R, Aroori S, Belli G, Besselink M, Briceno J, Gayet B, D'Hondt M, Lesurtel M, Menon K, Lodge P, Rotellar F, Santoyo J, Scatton O, Soubrane O, Sutcliffe R, Van Dam R, White S, Halls MC, Cipriani F, Van der Poel M, Ciria R, Barkhatov L, Gomez-Luque Y, Ocana-Garcia S, Cook A, Buell J, Clavien PA, Dervenis C, Fusai G, Geller D, Lang H, Primrose J, Taylor M, Van Gulik T, Wakabayashi G, Asbun H, Cherqui D. The Southampton consensus guidelines for laparoscopic liver surgery: from indication to implementation. Ann Surg. 2018;268(1):11–8. https://doi.org/10.1097/sla.0000000000002524.
3. Kokudo N, Takemura N, Ito K, Mihara F. The history of liver surgery: achievements over the past 50 years. Ann Gastroenterol Surg. 2020;4(2):109–17. https://doi.org/10.1002/ags3.12322. PMID: 32258975; PMCID: PMC7105847.
4. Makuuchi M, Hasegawa H, Yamazaki S. Ultrasonically guided subsegmentectomy. Surg Gynecol Obstet. 1985;161(4):346–50. PMID: 2996162.

5. Makuuchi M, Hasegawa H, Yamazaki S, Takayasu K, Moriyama N. The use of operative ultrasound as an aid to liver resection in patients with hepatocellular carcinoma. World J Surg. 1987;11(5):615–21. PMID: 2823489. https://doi.org/10.1007/bf01655837.
6. Castaing D, Emond J, Kunstlinger F, Bismuth H. Utility of operative ultrasound in the surgical management of liver tumors. Ann Surg. 1986;204(5):600–5. PMID: 3021072. https://doi.org/10.1097/00000658-198611000-00015.
7. Machi J, Sigel B. Overview of benefits of operative ultrasonography during a ten year period. J Ultrasound Med. 1989;8(11):647–52. PMID: 2687497. https://doi.org/10.7863/jum.1989.8.11.647.
8. Arita J, Ono Y, Takahashi M, Inoue Y, Takahashi Y, Matsueda K, Saiura A. Routine preoperative liver-specific magnetic resonance imaging does not exclude the necessity of contrast-enhanced intraoperative ultrasound in hepatic resection for colorectal liver metastasis. Ann Surg. 2015;262(6):1086–91. PMID: 26010441. https://doi.org/10.1097/sla.0000000000001085.
9. Jrearz R, Hart R, Jayaraman S. Intraoperative ultrasonography and surgical strategy in hepatic resection: what difference does it make? Can J Surg. 2015;58(5):318–22. https://doi.org/10.1503/cjs.016914. PMID: 26384146; PMCID: PMC4599994.
10. Knowles SA, Bertens KA, Croome KP, Hernandez-Alejandro R. The current role of intraoperative ultrasound during the resection of colorectal liver metastases: A retrospective cohort study. Int J Surg. 2015;20:101–6. Epub 2015 Jun 10. PMID: 26070252. https://doi.org/10.1016/j.ijsu.2015.05.052.
11. Scaife CL, Ng CS, Ellis LM, Vauthey JN, Charnsangavej C, Curley SA. Accuracy of preoperative imaging of hepatic tumors with helical computed tomography. Ann Surg Oncol. 2006;13(4):542–6. Epub 2006 Feb 15. PMID: 16474907. https://doi.org/10.1245/aso.2006.07.009.
12. van Vledder MG, Pawlik TM, Munireddy S, Hamper U, de Jong MC, Choti MA. Factors determining the sensitivity of intraoperative ultrasonography in detecting colorectal liver metastases in the modern era. Ann Surg Oncol. 2010;17(10):2756–63. Epub 2010 Jun 2. PMID: 20517682. https://doi.org/10.1245/s10434-010-1108-y.
13. Sietses C, Meijerink MR, Meijer S, van den Tol MP. The impact of intraoperative ultrasonography on the surgical treatment of patients with colorectal liver metastases. Surg Endosc. 2010;24(8):1917–22. Epub 2010 Jan 29. PMID: 20112115; PMCID: PMC2895892. https://doi.org/10.1007/s00464-009-0874-8.
14. Guerra F, Amore Bonapasta S, Annecchiarico M, Bongiolatti S, Coratti A. Robot-integrated intraoperative ultrasound: initial experience with hepatic malignancies. Minim Invasive Ther Allied Technol. 2015;24(6):345–9. Epub 2015 Apr 2. PMID: 25835093. https://doi.org/10.3109/13645706.2015.1022558.
15. Casciola L, Patriti A, Ceccarelli G, Bartoli A, Ceribelli C, Spaziani A. Robot-assisted parenchymal-sparing liver surgery including lesions located in the posterosuperior segments. Surg Endosc. 2011;25(12):3815–24. Epub 2011 Jun 8. PMID: 21656067. https://doi.org/10.1007/s00464-011-1796-9.
16. Schneider CM, Peng PD, Taylor RH, Dachs GW 2nd, Hasser CJ, DiMaio SP, Choti MA. Robot-assisted laparoscopic ultrasonography for hepatic surgery. Surgery. 2012;151(5):756–62. Epub 2011 Oct 6. PMID: 21982071. https://doi.org/10.1016/j.surg.2011.07.040.
17. Guadagni S, Furbetta N, Franco GD, Palmeri M, Gianardi D, Bianchini M, Guadagnucci M, Pollina L, Masi G, Cremolini C, Falcone A, Mosca F, Di Candio G, Morelli L. Robotic-assisted surgery for colorectal liver metastasis: a single-centre experience. J Minim Access Surg. 2020;16(2):160–5. PMID: 30777992; PMCID: PMC7176011. https://doi.org/10.4103/jmas.jmas_265_18.
18. Hagopian EJ, Machi J, editors. Abdominal ultrasound for surgeons. New York: Springer Science+Business Media; 2014.
19. Hagopian EJ, Machi J. Chapter 17: Ultrasonography by surgeons. In: Ellison EC, Upchurch GR, editors. Fischer's mastery of surgery. 8th ed. Philadelphia: Wolters Kluwer; 2024. p. 155–73.
20. Pesi B, Moraldi L, Guerra F, Tofani F, Nerini A, Annecchiarico M, Coratti A. Surgical and oncological outcomes after ultrasound-guided robotic liver resections for malignant tumor. Analysis of a prospective database. Int J Med Robot. 2019;15(4):e2002. Epub 2019 May 27. PMID: 31022774. https://doi.org/10.1002/rcs.2002.
21. Guerra F, Guadagni S, Pesi B, Furbetta N, Di Franco G, Palmeri M, Annecchiarico M, Eugeni E, Coratti A, Patriti A, Morelli L. Outcomes of robotic liver resections for colorectal liver metastases. A multi-institutional analysis of minimally invasive ultrasound-guided robotic surgery. Surg Oncol. 2019;28:14–8. Epub 2018 Nov 1. PMID: 30851888. https://doi.org/10.1016/j.suronc.2018.10.011.
22. Sucandy I, Durrani H, Ross S, Rosemurgy A. Technical approach of robotic total right hepatic lobectomy: How we do it? J Robot Surg. 2019;13(2):193–9. Epub 2018 Oct 1. PMID: 30276634. https://doi.org/10.1007/s11701-018-0881-7.
23. Sucandy I, Gravetz A, Ross S, Rosemurgy A. Technique of robotic left hepatectomy : how we approach it. J Robot Surg. 2019;13(2):201–7. Epub 2018 Nov 8. PMID: 30406886. https://doi.org/10.1007/s11701-018-0890-6.
24. Zhu P, Liao W, Ding ZY, Luo HC, Zhang BH, Zhang WG, Zhang W, Zhang ZG, Zhang BX, Chen XP. Intraoperative ultrasonography of robot-assisted laparoscopic hepatectomy: initial experiences from 110 consecutive cases. Surg Endosc. 2018;32(10):4071–7. Epub 2018 Aug 27. PMID: 30151749. https://doi.org/10.1007/s00464-017-5854-9.

Robotic Resections in the Anterolateral Segments

19

Alessandro Parente and Robert P. Sutcliffe

Abbreviations

ICG	Indocyanine green
IOUS	Intraoperative ultrasound
LHV	Left hepatic vein
LDLT	Live donor liver transplantation
LLS	Left lateral sectionectomy
MHV	Middle hepatic vein
NAR	Non-anatomical resection
RHV	Right hepatic vein

19.1 Introduction

Minimally invasive liver surgery has gained popularity in recent decades and has been accepted as an alternative to open surgery for selected liver resections due to its short-term benefits, particularly for minor hepatectomy [1–3]. Widespread adoption of minimally invasive liver surgery has been hampered by high technical difficulty due to limited exposure, instrument motion and a steep learning curve. Although the learning curve for laparoscopic liver resection was reported to be in the range of 45–75 procedures by pioneering surgeons [4], more recent studies by second-generation laparoscopic liver surgeons have suggested that the learning curve has reduced to around 20–25 procedures [5, 6]. There is emerging data that the robotic platform may facilitate complex procedures, potentially shortening the learning curve for minimally invasive liver resection, particularly for accessing difficult segments or for major hepatectomy [7–11]. The anterolateral liver segments (segments 2, 3, 4b, 5 and 6) are generally considered readily amenable to a minimally invasive approach. In this chapter we provide a comprehensive overview of the technical aspects and outcomes of robotic resection of the anterolateral segments.

A. Parente
Institute of Liver Studies, Kings College Hospital, London, UK

R. P. Sutcliffe (✉)
Liver Unit, Queen Elizabeth Hospital, Birmingham, UK
e-mail: robert.sutcliffe@uhb.nhs.uk

19.2 Technical Aspects

A robotic approach may have several advantages compared to laparoscopy for application in liver surgery. First, the surgeon's precision is enhanced by the tenfold higher magnification and three-dimensionality which provides accurate high-quality and detailed images. Second, minimisation of a surgeon's natural tremor combined with improved degrees of instrument motion allows meticulous dissection and suturing that may be particularly relevant for hilar dissection and bile duct anastomosis. One of the main disadvantages

M. D'Hondt, I. Sucandy (eds.), *Textbook of Robotic Liver Surgery*,
https://doi.org/10.1007/978-3-031-76536-0_19

of robotic surgery is the lack of haptic feedback, but with experience the enhanced vision leads to the development of 'visual haptics' [11, 12].

As in laparoscopic liver surgery, correct patient positioning is paramount in robotic procedures [11]. The patient should be positioned supine on the operating table with legs up (French position) and a 15–20° reverse Trendelenburg position with the assistant positioned between the patient's legs. The right arm is normally adjoined to enable the robot to dock on both sides. Pneumoperitoneum (10–12 mmHg) is achieved either with Veress needle or open Hasson technique. Usually, the camera trocar should be placed 15–20 cm inferior to the xiphoid, and the first assistant port is positioned laterally to the camera to allow a wider motion range of the robotic arms [13]. Diagnostic laparoscopy is recommended before inserting additional ports in order to assess resectability and detect additional lesions or the presence of intra-abdominal adhesions that may preclude a minimally invasive approach. For the anterolateral segments, the placement of ports differs based on the localisation of the lesion and the planned extent of resection [14]. Indocyanine green (ICG) can be a useful adjunct to localise superficial tumours, identify transection margins, as well as delineate the biliary anatomy.

19.2.1 General Aspects of Robotic Liver Surgery

Intraoperative ultrasound (IOUS) is mandatory to confirm findings of preoperative imaging and ensure resectability. Next, the round ligament is taken down at its junction with the abdominal wall, and subsequently the falciform ligament is approached. This allows better exposure and can be used as a tool to mobilise the liver during resection. The liver is then mobilised based on the laterality of the resected segments. This chapter is focussed on anterolateral segments (segments 2, 3, 4b, 5 and 6); therefore, the following subchapters will be focused on left lateral sectionectomy (LLS), non-anatomical resection and segmentectomy/bisegmentectomy of segments 3/4b, 4b/5 and 5/6 (Fig. 19.1).

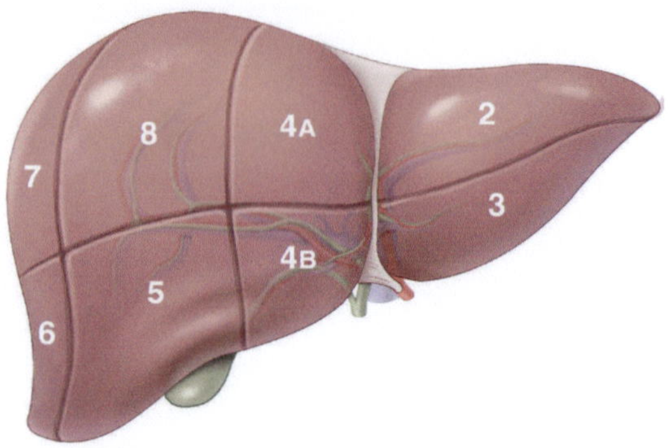

Fig. 19.1 The anterolateral segments of the liver segment 2, 3, 4b, 5 and 6

19.2.2 Left Lateral Sectionectomy

It is recommended to place the 12-mm camera port in the periumbilical region. The surgical target is positioned in the epigastrium, and three 8-mm ports for the robotic arms are positioned in the mid clavicular line both right and left side and left anterior axillary line. In addition, a 12-mm auxiliary port is placed between the camera and the left mid clavicular one (Fig. 19.2). After IOUS, the round as well as the falciform ligaments are divided using either ultrasonic or advanced bipolar energy. Subsequently, the left hepatic lobe is mobilised by dividing the left triangular ligament until the bare area is revealed, thus exposing the anterior portion of the left hepatic vein (LHV). Afterwards, the transection line is demarcated with diathermy on the Glissonian surface of the liver on the left side of the round and falciform ligaments. The robotic arms placed in the right mid clavicular line and left flank are used to retract the base of the round ligament to the right or left side, allowing a straight line for the transection margin accordingly. The parenchymal resection can be performed either with ultrasonic scalpel, bipolar and/or clamping-crushing technique. The inflow pedi-

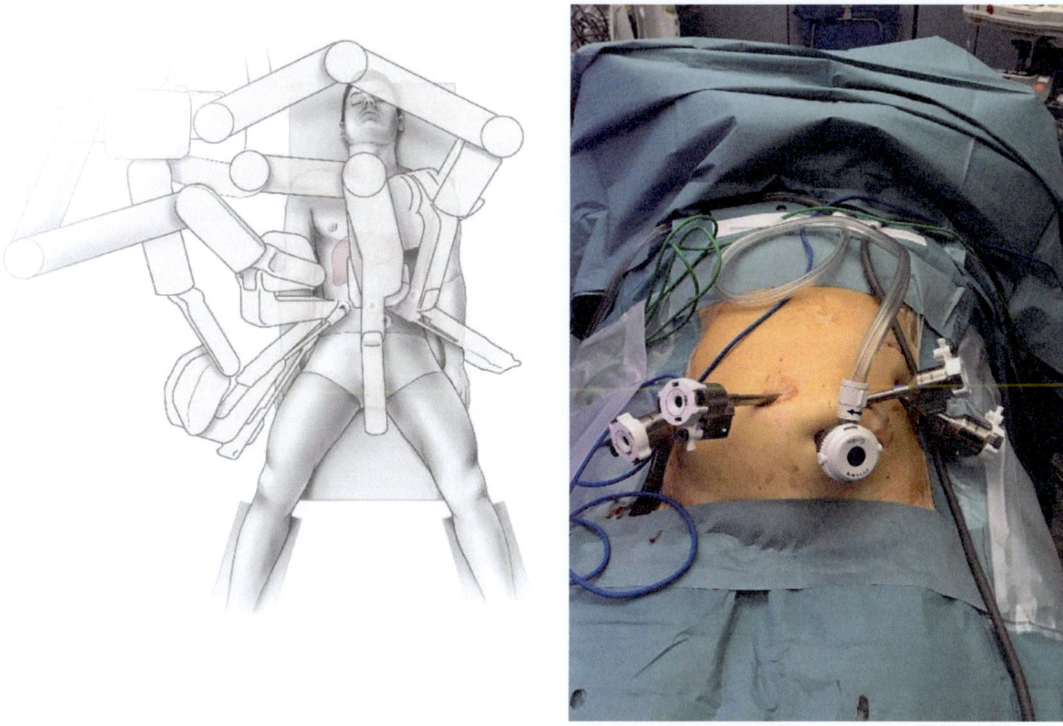

Fig. 19.2 Patient positioning and trocar placement for robotic left lateral sectionectomy

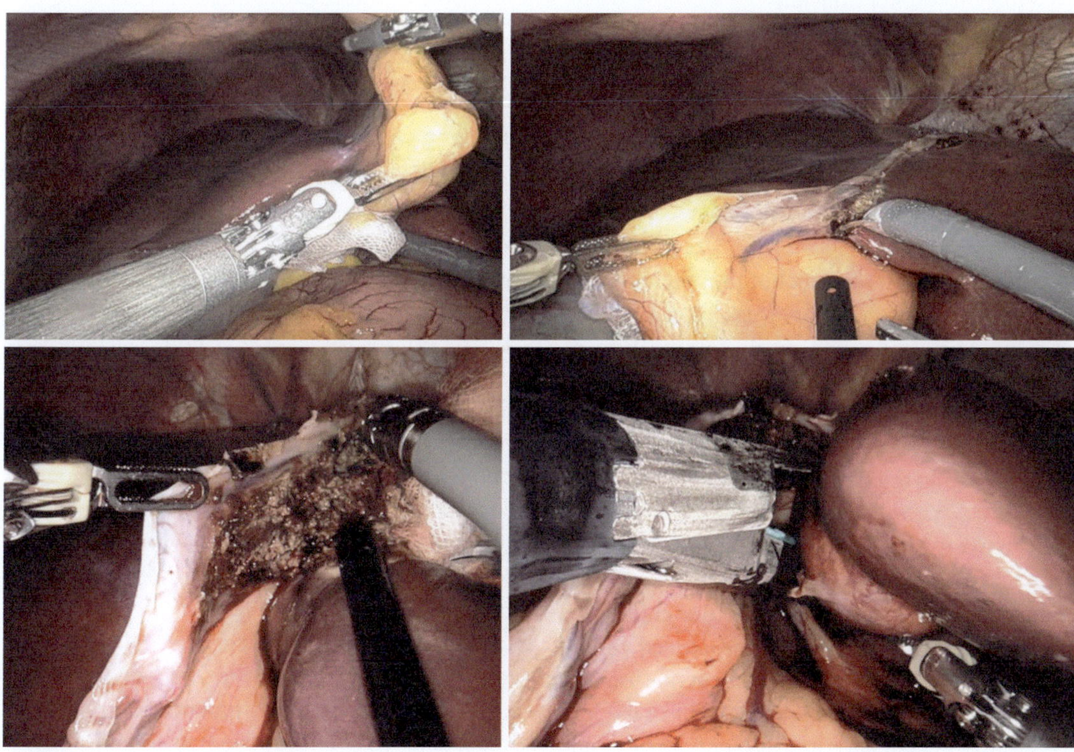

Fig. 19.3 Technique of robotic left lateral sectionectomy

cles of segments 2 and 3 are identified and divided with a stapler through the auxiliary port or switching the camera port (Fig. 19.3). Afterwards, the remnant parenchyma is transected until the LHV is identified and divided with a vascular stapler, thus completing the LLS procedure. Importantly, before dividing the LHV, the tip of the vascular stapler must be seen on top of the bare area of the liver left lobe previously identified during mobilisation. Haemostasis is achieved, and the specimen is delivered through a transverse suprapubic incision. Pringle manoeuvre is usually not necessary for LLS, although it can be performed based on surgeon's preferences.

Tips and Tricks for Robotic Left Lateral Sectionectomy

- Fully mobilise both falciform and left triangular ligaments to expose the left hepatic vein
- Parenchymal transection parallel and to left side of falciform ligament
- Fully expose segment 2 and 3 pedicles prior to stapler division
- Fully expose left hepatic vein prior to stapler division

19.2.3 Segmentectomy or Bisegmentectomy 3/4b

Port placements for approaching segments 3 and 4b broadly follow similar steps to LLS. The parenchyma is transected until identification of the segment 3 inflow pedicle which is divided with vascular stapler or clipped.

Subsequently, the remnant of the parenchyma is resected, sparing segment 2 pedicle as well as LHV (Fig. 19.4). When approaching segment 4b, the parenchyma is transected immediately to the right of the insertion of the round ligament. Towards the hilum at the base of segment 4b, the Glissonian pedicle to segment 4b is isolated and divided, resulting in ischemic delineation of segment 4b from 4a. In this regard, ICG can be used to confirm the line of ischaemic demarcation and also provides visualisation of the biliary anatomy. During parenchymal transection, it is important to perform IOUS to identify surgical margins and the hepatic vein anatomy, in particular, middle hepatic vein (MHV), in order to divide the segment 4b branches. This is achieved either with vascular stapler, clips or Hem-o-Lok® system. Usually, the specimen is delivered through a transverse suprapubic incision.

Non-anatomical resection of these segments can be performed when a parenchyma-sparing approach is indicated, for example, in cirrhotic patients. In this setting, the utilisation of IOUS and ICG is useful to avoid major vascular structures and minimise bleeding. A lifting suture inserted into the specimen may be used to elevate it, thereby facilitating adequate tumour clearance at the deep margin.

Tips and Tricks for Robotic Resection of Segments 3 and 4b

- Fully expose segmental pedicles prior to ligation/stapled division
- Determine optimal transection planes using intraoperative ultrasound
- Lifting suture to facilitate tumour clearance at deep margin of segment 4b

19.2.4 Segmentectomy or Bisegmentectomy 4b/5

Four trocars are placed with the camera in the periumbilical region, preferably below, and two robotic arms (right and left side) and a laparoscopic port for assistance. Cholecystectomy is performed to gain access to the cystic plate. The gallbladder fundus may be left attached to the liver and used as a traction tool for better exposure. Segment 4b and 5 inflow pedicles are identified and divided to demarcate the resection plane, which can be visualised with IOUS and ICG. The latter is also useful to identify biliary anatomy. The transection plane of segment 5 lies right to the gallbladder, whereas the margin of segment 4b is placed to the right of the round ligament. Subsequently, parenchyma is transected, and IOUS is utilised to identify MHV branches of both segments which are divided

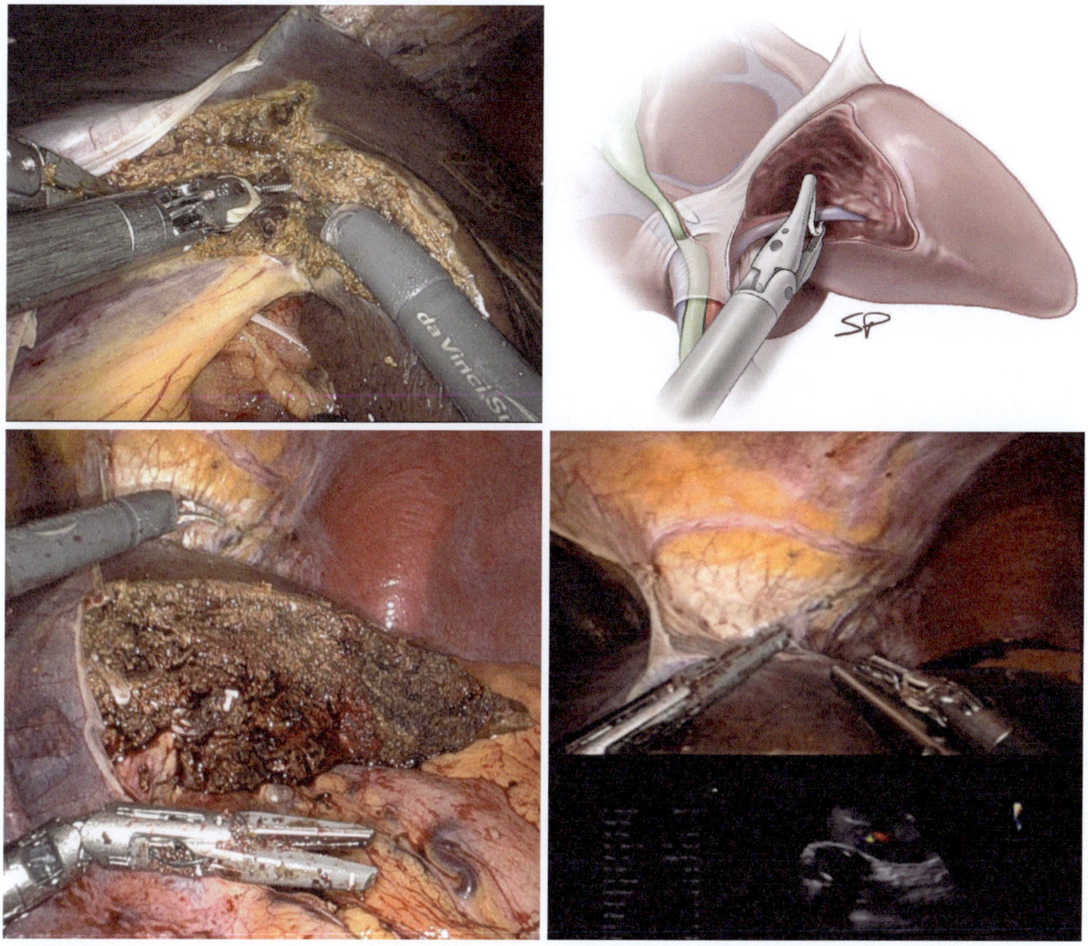

Fig. 19.4 Anatomical resection of segment 3. Isolation of pedicle of segment 3 and clipping. Final view after resection (segment 2) and IOUS and duplex of segment 2 pedicle

either with vascular stapler, clips or Hem-o-Lok® system. Also, right hepatic vein (RHV) branches to segment 5 can be encountered and divided. Usually, the specimen is delivered through a transverse suprapubic incision. Non-anatomical resection of these segments can be performed when deemed appropriate. A lifting suture can be used for traction to allow better exposure of the resection planes.

19.2.5 Segmentectomy or Bisegmentectomy 5/6

Port placement follows similar steps as for approaching segments 4b/5. An additional port can be placed in the right flank. Mobilisation of the right liver lobe by dividing the right triangular ligament is necessary to fully expose segment 6. As per segment 4b/5, cholecystectomy is performed, and the gallbladder is used for traction. The cystic plate is exposed, and the inflow pedicle of segment 5 is identified. At the level of Rouvière's sulcus, the inflow pedicle of the posterior liver segments (6 and 7) can be found. IOUS and ICG can help to guide the selective identification of inflow pedicles. Subsequently, parenchyma is transected, and IOUS is utilised to identify MHV branches of segment 5 and RHV branches of segment 6 which are divided either with vascular stapler, clips or Hem-o-Lok® system. With regard to the hepatic veins, a right inferior accessory vein of segment 6 is common and drains the posterior part of the right lobe [15]. Usually, the specimen is delivered through a transverse suprapubic incision. Non-anatomical resection of these segments can be performed when deemed appropriate. A lifting suture can be used for traction to allow better exposure of the margin planes.

Tips and Tricks for Robotic Resection of Segments 5 and 6

- The gallbladder is a key landmark and may be used as a handle to improve exposure and facilitate parenchymal transection.
- Identification of Rouvière's sulcus is critical to facilitate the approach to the segment 6 pedicle.
- Intraoperative ultrasound and ICG are useful tools to visualise tumours and vasculobiliary anatomy.

19.3 Literature Overview

19.3.1 Robotic Left Lateral Sectionectomy

Several studies failed to demonstrate any benefit of a robotic approach to left lateral sectionectomy compared to laparoscopy in terms of major complications or length of hospital stay, and the robotic approach was associated with longer operating times [16–20]. However, for complex cases such as intrahepatic duct stones, some authors have shown that robotic procedures may be indicated and associated with less blood loss [17, 19]. A recent meta-analysis illustrated that robotic LLS is associated with significantly longer operating time and higher costs and provides similar perioperative outcomes compared to the laparoscopic approach [21]. Some authors have reported a single-port robotic LLS with good outcomes; however, it is confined to case reports [22].

Robotic LLS has been utilised for procurement of grafts in the setting of live donor liver transplantation (LDLT). One recent study reported the short-term outcomes of 25 robotic LLS compared to 50 laparoscopic LLS for paediatric LDLT and found that robotic donor LLS was associated with reduced blood loss, lower rates of conversion to open surgery and reduced length of hospital stay [23]. However, warm ischemic time was longer in the robotic group. Overall, no differences in terms of donor morbidity and recipient complications were reported, thus concluding that the robotic approach appears to be safe for procurement of left lateral grafts in paediatric LDLT [23]. More recently, the same authors have reported good outcomes from a series of 177 left lateral grafts in LDLT [24]. Robotic procurement of left lateral grafts for LDLT is still confined to case series, and strong conclusions cannot be drawn.

19.3.2 Robotic Segmentectomy 2/3/4b/5/6

Several studies have focussed on robotic segmentectomy or bisegmentectomy of the anterior liver segments. Goh et al. published the results of their first series of robotic hepatectomy where they also compared anterolateral and posterosuperior segments [25]. Excellent outcomes were reported, and no significant differences were found between the locations of the resection.

In 2020, one study compared the long-term and oncological outcomes of robotic versus laparoscopic liver resection for colorectal cancer metastasis among six high-volume centres in the USA and Europe between 2002 and 2017 [26]. Robotic resection was performed in 115 patients as opposed to 514 patients who underwent laparoscopic resection. Among these, tumours were located in the anterolateral segments in 57 (49.6%) robotic and 245 (47.9%) laparoscopic cases. Although the analysis was not specifically performed for anterolateral segments, in the overall cohort the authors demonstrated that perioperative outcomes including mortality, morbidity, reoperation, readmission, intensive care requirement, length of hospital stay and oncological margin did not differ between robotic and laparoscopic. Therefore, they concluded that robotic resection for colorectal metastases is feasible and safe and does not compromise oncological outcomes and long-term survival [26].

Recently, another study reported the outcomes of robotic versus laparoscopic partial liver resections focussing solely on the anterolateral segments in a large dataset [27]. The authors analysed 2606 patients who underwent minimally invasive resection (laparoscopic versus robotic) of the anterolateral liver segments at 26 international centres between 2005 and 2020, comparing perioperative outcomes of the two techniques with two different statistical matching approaches. These authors demonstrated no significant differences in length of hospital stay, morbidity, mortality, post discharge readmission rates and oncological margins. However, although the robotic approach displayed a longer operating time, it was associated with significantly less blood loss and lower rates of conversion to open when compared to the laparoscopic approach, thus allowing more procedures to be completed in a minimally invasive fashion [27]. The results of this study appear in line with a recent research where other authors have analysed the outcomes of 288 robotic resections versus 864 laparoscopic resections (matched in a 1:3 ratio) with subgroup stratification according to location and difficulty scores. The robotic approach failed to show a benefit over laparoscopic procedure for low and intermediate difficulty, such as anterolateral segments. Nevertheless, in line with other reports, they demonstrated lower rates of conversion to open, less blood loss in resections using robotic technique with high difficulty scores [28].

Two other studies compared robotic and laparoscopic liver resections in the setting of major and minor hepatectomies [29, 30]. Despite not focusing solely on anterolateral segments, these studies were able to confirm the findings of previous reports as it was shown that the robotic approach is safe and feasible, allowing less blood loss, fewer rates of conversion to open and shorter length of hospital stay.

19.4 Conclusions

Robotic resection of the anterolateral segments has been demonstrated to be safe and feasible, even in the LDLT setting. The robot could be useful for more precise dissection of hepatic hilum, lymph node sampling and accurate anastomosis. The current positive findings are solely based on retrospective uncontrolled studies which displayed less blood loss, fewer rates of conversion to open and shorter length of hospital stay at the cost of longer operating times and higher procedures expenditures when compared to the laparoscopic approach. Furthermore, new implementations such as augmented reality, image-guided surgery and 3D ultrasound instruments are tools that could contribute to the continuous appeal for the future of robotic liver surgery. However, more research is still needed to broaden its widespread adoption. Currently, the lack of randomised controlled trials does not allow to draw robust recommendation, especially for low to intermediate difficulty procedures, such as left lateral sectionectomy of resection of anterolateral segments.

References

1. Papadopoulou K, Dorovinis P, Kykalos S, Schizas D, Stamopoulos P, Tsourouflis G, et al. Short-term outcomes after robotic versus open liver resection: a systematic review and meta-analysis. J Gastrointest Cancer. 2022;54:237.
2. Hildebrand N, Verkoulen K, Dewulf M, Heise D, Ulmer F, Coolsen M. Short-term outcomes of laparoscopic versus open hepatectomy in the elderly patient: systematic review and meta-analysis. HPB (Oxford). 2021;23(7):984–93.
3. Fretland ÅA, Dagenborg VJ, Bjørnelv GMW, Kazaryan AM, Kristiansen R, Fagerland MW, et al. Laparoscopic versus open resection for colorectal liver metastases: the OSLO-COMET randomized controlled trial. Ann Surg. 2018;267(2):199–207.
4. Nomi T, Fuks D, Kawaguchi Y, Mal F, Nakajima Y, Gayet B. Learning curve for laparoscopic major hepatectomy. Br J Surg. 2015 Jun;102(7):796–804.
5. Sultana A, Nightingale P, Marudanayagam R, Sutcliffe RP. Evaluating the learning curve for laparoscopic liver resection: a comparative study between standard and learning curve CUSUM. HPB (Oxford). 2019 Nov;21(11):1505–12.
6. Guilbaud T, Birnbaum DJ, Berdah S, Farges O, Beyer BL. Learning curve in laparoscopic liver resection, educational value of simulation and training programmes: a systematic review. World J Surg. 2019 Nov;43(11):2710–9.
7. Efanov M, Alikhanov R, Tsvirkun V, Kazakov I, Melekhina O, Kim P, et al. Comparative analysis of learning curve in complex robot-assisted and laparoscopic liver resection. HPB (Oxford). 2017 Sep;19(9):818–24.

8. D'Silva M, Han HS, Liu R, Kingham TP, Choi GH, Syn NLX, et al. International robotic and laparoscopic liver resection study group investigators. Limited liver resections in the posterosuperior segments: international multicentre propensity score-matched and coarsened exact-matched analysis comparing the laparoscopic and robotic approaches. Br J Surg. 2022;109(11):1140–9.
9. Sucandy I, Rayman S, Lai EC, Tang CN, Chong Y, Efanov M, et al. Robotic versus laparoscopic left and extended left hepatectomy: an international multicenter study propensity score-matched analysis. Ann Surg Oncol. 2022;29:8398.
10. Chong CC, Fuks D, Lee KF, Zhao JJ, Choi GH, Sucandy I, et al. International robotic and laparoscopic liver resection study group investigators. Propensity score-matched analysis comparing robotic and laparoscopic right and extended right hepatectomy. JAMA Surg. 2022;157(5):436–44.
11. Giulianotti PC, Bianco FM, Daskalaki D, Gonzalez-Ciccarelli LF, Kim J, Benedetti E. Robotic liver surgery: technical aspects and review of the literature. Hepatobiliary Surg Nutr. 2016;5(4):311–21.
12. Lafaro KJ, Stewart C, Fong A, Fong Y. Robotic liver resection. Surg Clin N Am. 2020;100:265–81.
13. Spiegelberg J, Iken T, Diener MK, Fichtner-Feigl S. Robotic-assisted surgery for primary hepatobiliary tumors—possibilities and limitations. Cancers. 2022;14:265.
14. Nota CL, Rinkes IHB, Hagendoorn J. Setting up a robotic hepatectomy program: a Western-European experience and perspective. Hepatobiliary Surg Nutr. 2017;6(4):239–45.
15. Buhe S, Miyaki T, Saito T, Sawuti A, Terayama H, Naito M, et al. A study of the accessory hepatic vein to segments VI and VII with a morphological reconsideration of the human liver. Surg Radiol Anat. 2008;30(3):201–7.
16. Salloum C, Lim C, Lahat E, Gavara CG, Levesque E, Compagnon P, et al. Robotic-assisted versus laparoscopic left lateral sectionectomy: analysis of surgical outcomes and costs by a propensity score matched cohort study. World J Surg. 2017;41(2):516–24.
17. Kim JK, Park JS, Han DH, Choi GH, Kim KS, Choi JS, et al. Robotic versus laparoscopic left lateral sectionectomy of liver. Surg Endosc. 2016;30(11):4756–64.
18. Packiam V, Bartlett DL, Tohme S, Reddy S, Marsh JW, Geller DA, et al. Minimally invasive liver resection: robotic versus laparoscopic left lateral sectionectomy. J Gastrointest Surg. 2012;16(12):2233–8.
19. Hu M, Liu Y, Li C, Wang G, Yin Z, Lau WY, et al. Robotic versus laparoscopic liver resection in complex cases of left lateral sectionectomy. Int J Surg. 2019;67:54–60.
20. Zhu L, Liu Y, Hu M, Zhao Z, Li C, Zhang X, et al. Comparison of robotic and laparoscopic liver resection in ordinary cases of left lateral sectionectomy. Surg Endosc. 2022;36(7):4923–31.
21. Hajibandeh S, Hajibandeh S, Dosis A, Qayum MK, Hassan K, Kausar A, Satyadas T. Level 2a evidence comparing robotic versus laparoscopic left lateral hepatic sectionectomy: a meta-analysis. Langenbeck's Arch Surg. 2022;407(2):479–89.
22. Kim WJ, Park PJ, Choi SB, Kim WB. Case report of pure single-port robotic left lateral sectionectomy using the da Vinci SP system. Medicine (Baltimore). 2021;100(51):e28248.
23. Troisi RI, Elsheikh Y, Alnemary Y, Zidan A, Sturdevant M, Alabbad S, et al. Safety and feasibility report of robotic-assisted left lateral sectionectomy for pediatric living donor liver transplantation: a comparative analysis of learning curves and mastery achieved with the laparoscopic approach. Transplantation. 2021;105(5):1044–51.
24. Schulze M, Elsheikh Y, Boehnert MU, Alnemary Y, Alabbad S, Broering DC. Robotic surgery and liver transplantation: a single-center experience of 501 robotic donor hepatectomies. Hepatobiliary Pancreat Dis Int. 2022;21(4):334–9.
25. Goh BKP, Lee LS, Lee SY, Chow PKH, Chan CY, Chiow AKH. Initial experience with robotic hepatectomy in Singapore: analysis of 48 resections in 43 consecutive patients. ANZ J Surg. 2019;89(3):201–5.
26. Beard RE, Khan S, Troisi RI, Montalti R, Vanlander A, Fong Y, et al. Long-term and oncologic outcomes of robotic versus laparoscopic liver resection for metastatic colorectal cancer: a multicenter, propensity score matching analysis. World J Surg. 2020;44(3):887–95.
27. Kadam P, Sutcliffe RP, Scatton O, Sucandy I, Kingham TP, Liu R, et al. An international multicenter propensity-score matched and coarsened-exact matched analysis comparing robotic versus laparoscopic partial liver resections of the anterolateral segments. J Hepatobiliary Pancreat Sci. 2022;29(8):843–54.
28. Cipriani F, Fiorentini G, Magistri P, Fontani A, Menonna F, Annecchiarico M, et al. Pure laparoscopic versus robotic liver resections: multicentric propensity score-based analysis with stratification according to difficulty scores. J Hepatobiliary Pancreat Sci. 2022;29(10):1108–23.
29. Schmelzle M, Feldbrügge L, Ortiz Galindo SA, Moosburner S, Kästner A, Krenzien F, et al. Robotic vs. laparoscopic liver surgery: a single-center analysis of 600 consecutive patients in 6 years. Surg Endosc. 2022;36(8):5854–62.
30. Yang HY, Choi GH, Chin KM, Choi SH, Syn NL, Cheung TT, et al. Robotic and laparoscopic right anterior sectionectomy and central hepatectomy: multicentre propensity score-matched analysis. Br J Surg. 2022;109(4):311–4.

Robotic Resections in the Posterosuperior Segments

20

Charles Chung-Wei Lin, Tzu-Jung Tsai, and Valerie Yu Chan

20.1 Introduction

Minimally invasive surgery in the liver has always been a challenge to surgeons. Although laparoscopic liver resection (LLR) has been proven feasible and safe, the steep learning curve indicates the difficulties in itself. To be competent in laparoscopic liver resection, surgeons must be familiar not only with the basic skills and knowledge of liver surgery but also in advanced laparoscopic techniques. Challenges include hand-eye coordination under a two-dimensional image in a restricted operation space, manipulation of straight and rigid instruments, precise liver parenchymal dissection with careful bleeding control, and intracorporeal suturing skills in difficult angles. The robotic system arises to overcome the limitations of laparoscopic surgery with its unique features. The high-definition, magnified three-dimensional imaging in the robotic system controlled by the surgeons themselves allows for more precise identification of bleeding vessels. The EndoWrist instruments provide the surgeon better dexterity with up to seven degrees of freedom, tremor filter, and motion scaling. This can also more accurately replicate surgeons' wrist and finger movements. These benefits of the surgical robot facilitate hepatic hilum dissection and intracorporeal suturing better, faster, and easier than the laparoscopic approach. Furthermore, surgeons will be seated comfortably in a console, diminishing the fatigue brought by prolonged operative time in hepatobiliary surgery.

For the varied types of hepatectomy, including the posterosuperior (PS) segments, the indication and degree of difficulty are not different between laparoscopic and robotic approach. Tumors located in peripheral segments (2, 3, 4b, 5, 6) are most amenable to minimally invasive approach. However, resection of the PS segments (4a, 7, 8) are considered particularly challenging due to the curved transection lines, poor laparoscopic view, multiplanar parenchymal dissection, narrow space at the subphrenic area, and difficulty in hemostasis [1–5] (Fig. 20.1).

The difficulty then leads to a more major hepatectomy, which sacrifices added substantial volume of normal liver for posterosuperior resections. Ideally, isolated resection of these segments serves patients the most benefit, but this may not be easily achieved. The features of the

C. C.-W. Lin (✉)
Ircad-Taiwan, Department of Surgery, Show Chwan Memorial Hospital, Changhua City, Taiwan

School of Medicine, National Yang-Ming Chiao-Tung University, Hsinchu, Taiwan
e-mail: charleslin@ircadtaiwan.com.tw

T.-J. Tsai
Department of Surgery, Koo Foundation Sun Yat-Sen Cancer Center, Taipei City, Taiwan

V. Y. Chan
Chinese General Hospital and Medical Center, Manila, Philippines

M. D'Hondt, I. Sucandy (eds.), *Textbook of Robotic Liver Surgery*,
https://doi.org/10.1007/978-3-031-76536-0_20

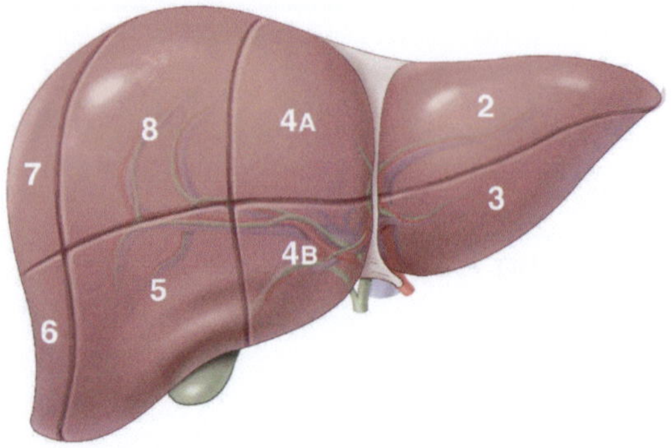

Fig. 20.1 The posterosuperior segments of the liver: segments 4A, 7, and 8

robotic system can aid to overcome the limitations aforementioned [6]. The wristed instruments can also facilitate the dissection along the convex diaphragm.

One critical step during resection of PS segments is the exposure of the major hepatic veins and the vena cava. The high-definition 3D videoscope and wristed instruments of the robotic system can help surgeons work more confidently and precisely. Even if bleeding occurs, surgeons can apply steady and delicate suturing to control the bleeding with little hesitation. In this chapter, the details of robotic liver resection of PS segments, including literature review and the surgical techniques, will be discussed.

20.2 Literature Overview

The second international consensus conference of minimally invasive liver resection was held in Morioka Japan in 2014. A study group focusing on the issue of robotic liver resection carefully reviewed the published literature and drew some conclusions [7]. Currently there is no randomized controlled trials available in terms of comparison between laparoscopic and robotic liver resection. Nevertheless, 17 major series were analyzed and yielded the feasibility and safety of robotic liver resection with zero mortality and comparable morbidity rates compared to laparoscopic liver resections. The robotic group had a significant higher proportion of major hepatectomies and resection of tumors located at posterosuperior segments compared with the laparoscopic group [8, 9]. That may also be attributed to longer operative time in robotic groups because surgeons prefer robotic resection in difficult cases. Theoretically, robotic technology has potential advantages to overcome the obstacles encountered in laparoscopic surgery. However, these are not translated into better perioperative outcomes since comparative studies currently do not show significant difference in factors such as blood loss, R0 resection rate, operation time, morbidity and mortality rate, and length of hospital stay between the two groups.

Nota et al. investigated robot-assisted liver resections for three subgroups of liver resection and concluded that the robotic platform is safe and suitable in all subgroups of liver resection and may be of particular advantage in resections of the PS segments [9]. Montalti et al. published a propensity score-matched comparison between robotic and laparoscopic resections of posterosuperior segments of the liver [10]. They found a higher rate of parenchymal-preserving resection in posterosuperior segments with the robotic techniques. Berardi et al. performed a multicentric analysis of the learning curve for laparoscopic liver resection of the PS segments [11]. The learning curve was estimated to be 40 procedures for wedge and 65 for anatomical resections. The laparoscopic approach to PS segments should be considered as a major procedure in terms of technical complexity; indeed, the proposed difficulty scoring systems for LLR grade these lesions with the highest score [12–15]. Nota et al. published an initial experience for the robotic liver resection including the PS segments [16] which showed that robotic-assisted liver resection is safe and feasible in selected patients. The procedure is associated with a long learning curve but is still significantly shorter than other complicated procedures such as pancreatic surgery. The implementation of minimally invasive liver resection will more likely be through a robotic approach [17].

There are a few disadvantages of robotic surgery that must be addressed, such as high costs and longer operation time [6]. With consideration of shorter hospital stay and lower conversion rate compared with laparoscopic approach, the higher costs of robotic surgery may not represent the total picture of minimally invasive liver surgery. In addition, the longer operation time may be due to the higher proportion of major liver resections and more difficult tumor location than in laparoscopic liver resection.

20.3 Technical Aspects

For robotic liver resection of PS segments, the patient is placed in reverse Trendelenburg position to decrease the central venous pressure and reduce the bowels to obtain better surgical space. The patient's legs could be split, and the bedside assistant can sit/stand between patient's legs during the operation (Fig. 20.2). It will be ergonomically easier to assist the operation, taking care of all four robotic arms and change the instruments. The surgical table is tilted approximately 30 degrees toward patient's left to obtain a better view of PS segments (Fig. 20.2). The curvilinear surface of the liver dome makes it difficult for straight laparoscopic view to approach the deeper hepatic parenchyma and the hepatic veins. The trocar for the videoscope is suggested to be positioned as high as possible at the right subcostal area, along with a 30-degree videoscope for a better view. A flexible laparoscope can also be beneficial with a vertical view of this area. However, it is still not available in the robotic system. The hepatic inflow control is performed by encircling the hepatoduodenal ligament with a 15-cm Fr.14 rubber Foley catheter, in preparation for total Pringle's maneuver [18].

20.3.1 Segment 4a

For segment 4 resection, the setting is similar to most of the upper abdominal surgeries (Fig. 20.3). The videoscope is inserted via an umbilical trocar and the working arms placed in bilateral upper abdomen. Patients with big bellies may have the umbilical trocar for the videoscope placed more superiorly along the abdominal midline for a better surgical view. The first hepatic parenchymal dissection plane is to open the liver between segment 4 and lateral segments along the falciform ligament. The Glissonean pedicle of segments 4b (G4b) and 4a (G4a) should be

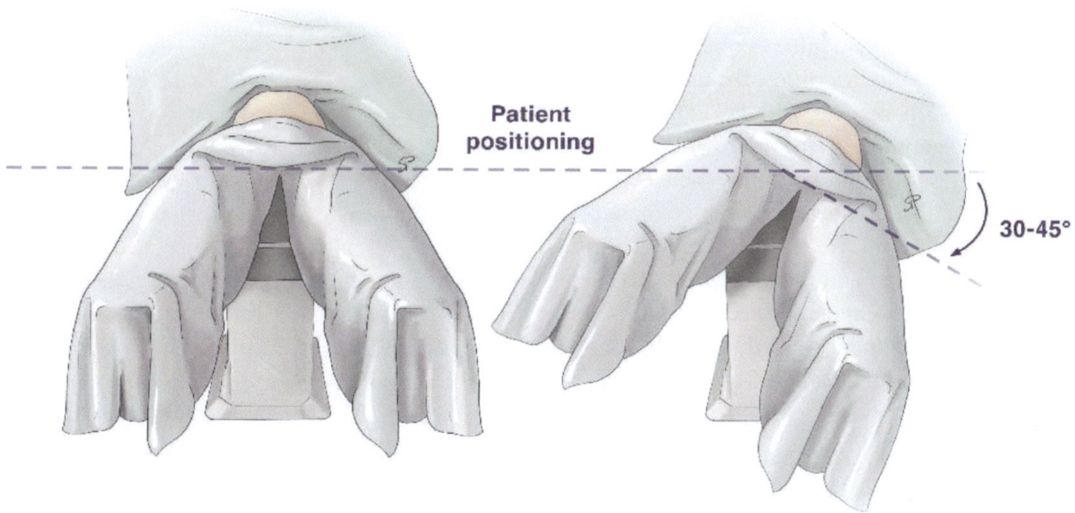

Fig. 20.2 Patient positioning. Patient is placed in a reverse Trendelenburg, legs split open. The assistant field surgeon can position in between the legs to assist in the procedure. Patient is tilted 30–45 degrees toward the left with a bolster for better visualization of PS segments

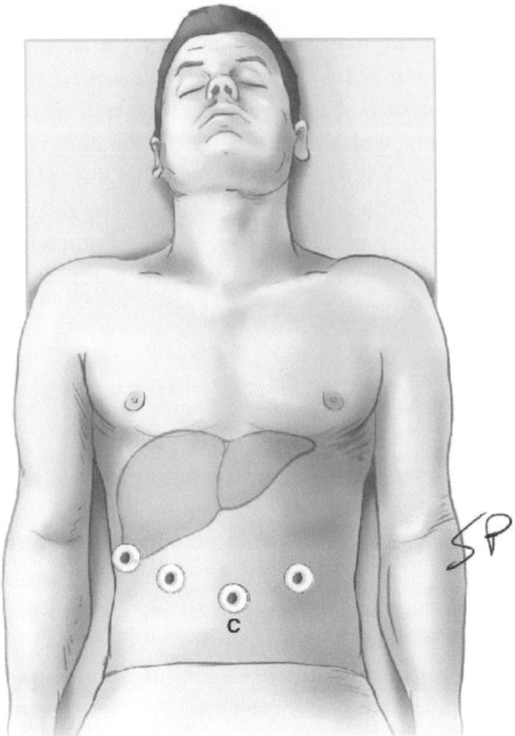

Fig. 20.3 Port site placement for hepatic resection, anterior approach. Robotic arm-1 is placed at the R subcostal anterior axillary line (AAL), robotic arm-2 at the R subcostal mid-clavicular line (MCL), robotic arm-3 videoscope at the umbilical port, and robotic arm-4 at the L subcostal MCL

identified here to obtain inflow control. Along the dissection from the insertion of round ligament, usually the G4b pedicle will be exposed first. After test clamping of the G4b, the ischemic demarcation shows the boundary between segments 4b and 4a. The dissection is then performed upward toward the confluence of the middle hepatic vein and the vena cava. The second plane of parenchymal dissection will be the left border of segment 4a, which is along the middle hepatic vein. Cranial-to-caudal dissection of the middle hepatic vein will decrease the risk of tearing the small venous branches from the main trunk, hence reducing the incidence of bleeding [19].

20.3.2 Segment 8

For tumors located in segment 8, the main parenchymal dissection plane is similar to hemi-hepatectomy between the right and left lobe of the liver, the so-called Cantlie line. The authors suggest to put the videoscope via a right subcostal trocar at the midclavicular line, instead of the midline abdomen, to obtain a better view parallel to middle hepatic vein and vena cava. The Glissonean pedicle of segment 8 (G8) is deeply located in central liver and therefore not easy to access. There are several approaches to G8 pedicle, roughly including ultrasound-guided direct approach after hepatic parenchymal transection, Glissonean pedicle approach from hepatic hilum, and a left and right hemi-hepatic splitting approach [20–22].

After ligating the G8 pedicle, the ischemic demarcation area should exactly be the anatomical area of segment 8, and this could be highlighted using fluorescence video system after indocyanine green intravenous injection (negative staining method). Then the parenchymal dissection is performed at the intersectional plane between segments 5 and 8. The later hepatic parenchymal dissection plane was performed toward the root of middle and right hepatic veins. Again, cranial-to-caudal dissection of hepatic veins offers the benefit of preventing tears in the small venous branches of right hepatic vein.

20.3.3 Segment 7

Segment 7, located in the most posterosuperior part of the liver, is difficult to approach with laparoscopy. Two times docking using the robotic setup is recommended for resection of the PS segments. Initial port site placement is done similar to the usual setup for upper abdominal surgeries (Fig. 20.3), for the complete anterior mobilization of the liver, and placement of catheter at hepatoduodenal ligament in preparation for Pringle maneuver. The videoscope is inserted from the umbilical trocar, and the working arms are positioned at bilateral upper abdomen. The surgical table is then tilted toward the patient's left side at 30–45 degrees (Fig. 20.2). With mobilization of liver by division of the falciform and round ligament, the liver will drop leftward with gravity, which will widen the working space around PS segment. For the re-docking of robotic arms, an additional 12-mm balloon trocar (Kii Balloon Blunt-Tip Access System, Applied Medical, CA, USA) is placed at the right ninth intercostal space along midaxillary line, in which an 8-mm robotic trocar is inserted into. This is the also known as the trocar-in-trocar method for robotic arm-1. The videoscope is then moved to the right subcostal anterior axillary line with arm-2, and trocars for arm-3 and arm-4 are placed at the bilateral subcostal midclavicular line. The traction by the robotic arm-4 controlled by the surgeon him/herself helps to maintain stability in this narrow working area around PS segment. With the steady traction of the liver by arm-4, posterior liver mobilization can be performed carefully along the vena cava. After dividing the coronary ligament attached to the right diaphragm, the liver can be rotated downward and leftward (Fig. 20.4). With this setup, the surgical view is from the patient's right caudal to cranial, and segment 7 will be just in the front of the videoscope (Fig. 20.5). Honda et al. published the

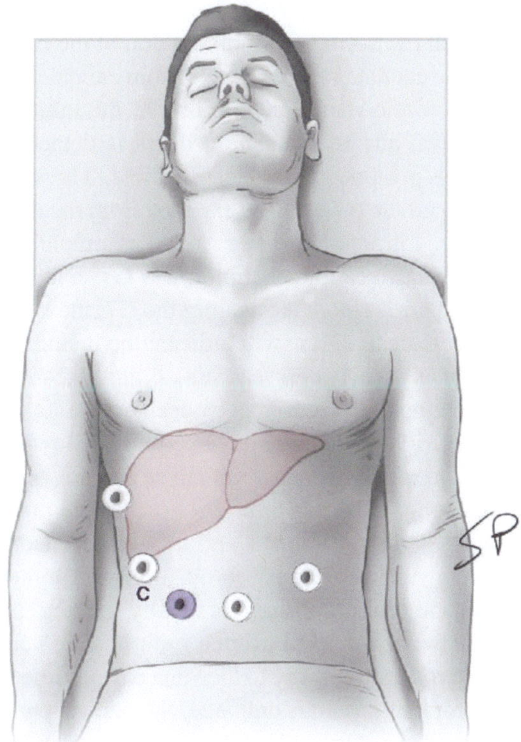

Fig. 20.4 Port site placement for PS segment hepatic transection. An additional port is placed at the ninth intercostal space midaxillary line (MAL) for robotic arm-1. The videoscope is moved to the R subcostal AAL with arm-2 and trocars for arm-3 placed at the umbilical port, with arm-4 still at L subcostal MCL. R subcostal MCL port will become assistant port site (purple)

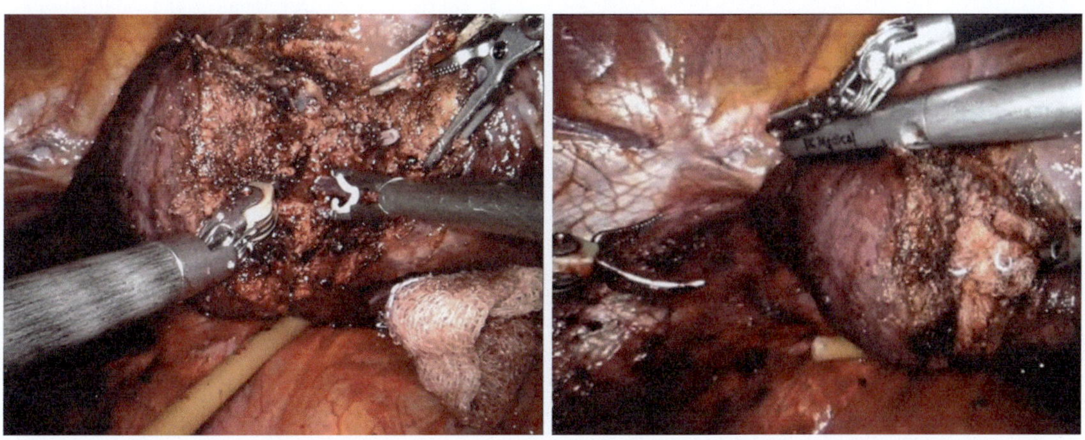

Fig. 20.5 Surgical view for PS segment resection with videoscope placed at the R subcostal AAL

resection of segment 7 by the caudate lobe first approach [23]. After full mobilization of the right liver, the caudate lobe is divided from caudal side parallel to the ventral central line of the inferior vena cava, and the caudate process is detached from the posterior Glissonean pedicle. The segment 7 pedicle (G7) can be exposed from the dorsal side. Intraoperative ultrasound is crucial to identify the location of tumor, the G7, and the right hepatic vein. After ligating the G7, the ischemic demarcation line will indicate the boundary of segment 7 between segments 6 and 8. When parenchymal dissection gets deeper, the main trunk of right hepatic vein is exposed. Following the plane of the hepatic veins, liver parenchymal dissection can be performed following the principle of ventral-to-dorsal and cranial-to caudal direction to avoid hepatic vein injury. Another advantage of this approach is that by moving the liver upward and leftward, the right hepatic vein will be rotated ventrally to the vena cava. Without compromising venous drainage, the venous pressure will be lowered, decreasing the risk of bleeding.

20.4 Conclusions

Minimally invasive liver resection of posterosuperior segments is technically feasible and safe. The features of robotic wristed instruments and steady three-dimensional videoscope may be beneficial to overcome the difficulties encountered in laparoscopy.

References

1. Cho JY, Han HS, Yoon YS, Shin SH. Feasibility of laparoscopic liver resection for tumors located in the posterosuperior segments of the liver, with a special reference to overcoming current limitations on tumor location. Surgery. 2008;144:32–8.
2. Cipriani F, Shelat VG, Rawashdeh M, Francone E, Aldrighetti L, Takhar A, et al. Laparoscopic parenchymal-sparing resections for nonperipheral liver lesions, the diamond technique: technical aspects, clinical outcomes, and oncologic efficiency. J Am Coll Surg. 2015;221:265–72.
3. Coles SR, Besselink MG, Serin KR, Alsaati H, Di Gioia P, Samim M, et al. Total laparoscopic management of lesions involving liver segment 7. Surg Endosc. 2015;29:3190–5.
4. Teo JY, Kam JH, Chan CY, Goh BK, Wong JS, Lee VT, et al. Laparoscopic liver resection for posterosuperior and anterolateral lesions—a comparison experience in an Asian centre. Hepatobiliary Surg Nutr. 2015;4:379–90.
5. Scuderi V, Barkhatov L, Montalti R, Ratti F, Cipriani F, Pardo F, et al. Outcome after laparoscopic and open resections of posterosuperior segments of the liver. Br J Surg. 2017;104:751–9.
6. Tsung A, Geller D, Sukato D, Sabbaghian S, Tohme S, Steel J, et al. Robotic versus laparoscopic hepatectomy: a matched comparison. Ann Surg. 2014;259(3):549–55.
7. Wakabayashi G, Cherqui D, Geller D, Buell J, Kaneko H, Han H, et al. Recommendations for laparoscopic liver resection: a report from the second international consensus conference held in Morioka. Ann Surg. 2015;261(4):619–29.
8. Lai E, Yang G, Tang C. Robotic-assisted laparoscopic liver resection for hepatocellular carcinoma: short-term outcome. Am J Surg. 2013;205(6):697–702.
9. Nota C, Rinkes I, Molenaar I, Santvoort H, Fong Y, Hagendoorn J, et al. Robotic-assisted laparoscopic liver resection: a systemic review and pooled analysis of minor and major hepatectomies. HBP (Oxford). 2016;18(2):113–20.
10. Montalti R, Scuderi V, Patriti A, Vivarelli TR. Robotic versus laparoscopic resections of posterosuperior segments of the liver: a propensity score-matched comparison. Surg Endosc. 2016;30(3):1004–13.
11. Berardi G, Aghayan D, Fretland A, Elberm H, Cipriani F, Spagnoli A, et al. Multicentre analysis of the learning curve for laparoscopic liver resection of the posterosuperior segments. BJS. 2019;106:1512–22.
12. Halls MC, Berardi G, Cipriani F, Barkhatov L, Lainas P, Harris S, et al. Development and validation of a difficulty score to predict intraoperative complications during laparoscopic liver resection. Br J Surg. 2018;105:1182–91.
13. Ban D, Tanabe M, Ito H, Otsuka Y, Nitta H, Abe Y, et al. A novel difficulty scoring system for laparoscopic liver resection. J Hepatobiliary Pancreat Sci. 2014;21:745–53.
14. Kawaguchi Y, Fuks D, Kokudo N, Gayet B. Difficulty of laparoscopic liver resection: proposal for a new classification. Ann Surg. 2018;267:13–7.
15. Hasegawa Y, Wakabayashi G, Nitta H, Takahara T, Katagiri H, Umemura A, et al. A novel model for prediction of pure laparoscopic liver resection surgical difficulty. Surg Endosc. 2017;31:5356–63.
16. Nota CLMA, Hillegersberg R, Rinkes I, Hagendoorn J. Robotic liver resection including the posterosuperior segments: initial experience. J Surg Res. 2016;206:133–8.

17. Zeh HJ, Bartlett D, Moser AJ. Robotic-assisted major pancreatic resection. Adv Surg. 2011;45:323–40.
18. Hunag J, Su W, Wang S. Alternative laparoscopic intracorporeal Pringle maneuver by Huang's loop. World J Surg. 2018;42:3312–5.
19. Honda G, Kurata M, Okuda Y, Kobayashi S, Sakamoto K, Takahashi K. Totally laparoscopic anatomical hepatectomy exposing the major hepatic veins from the root side: a case of the right anterior sectionectomy. J Gastrointest Surg. 2014;18:1379–80.
20. Berardi G, Wakabayashi G, Igarashi K, Ozaki T, Toyota N, Tsuchiya A, et al. Full laparoscopic anatomical segment 8 resection for hepatocellular carcinoma using the Glissonian approach with indocyanine green dye fluorescence. Ann Surg Oncol. 2019;26:2577–8.
21. Kishi Y, Hasegawa K, Kaneko J, Aoki T, Beck Y, Sugawara T, et al. Resection of segment VIII for hepatocellular carcinoma. Br J Surg. 2012;99:1105–12.
22. Kim J, Kim H. Pure laparoscopic anatomic resection of the segment 8 ventral area using the transfissural glissonean approach. Ann Surg Oncol. 2019;26:4608–9.
23. Li H, Honda G, Ome Y, Doi M, Yamamoto J, Muto J. Laparoscopic extended anatomical resection of segment 7 by the caudate lobe first approach: a video case report. J Gastrointest Surg. 2019;23:1084–5.

Robotic Right Hepatectomy

21

F. Krenzien, N. Raschzok, and J. Pratschke

21.1 Introduction

Right hepatectomy is a standardized procedure in which the liver parenchyma is resected on the Cantlie line or on the middle hepatic vein, respectively, removing segments 5–8 according to the Couinaud classification (Fig. 21.1).

Indication for right hepatectomy may be required for a number of conditions that cannot be removed by atypical resection or segmentectomy. Primary benign or malignant liver tumors such as hepatocellular carcinomas or adenomas, intrahepatic cholangiocellular carcinoma, and secondary malignant tumors like colorectal liver

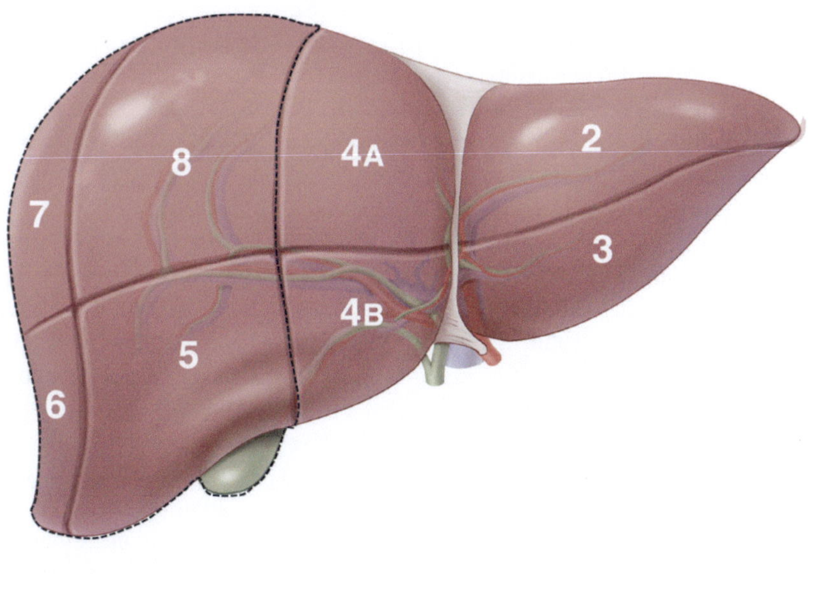

Fig. 21.1 Anatomical right hemihepatectomy (resection of segments 5–8)

F. Krenzien · N. Raschzok · J. Pratschke (✉)
Department of Surgery, Campus Charité Mitte/
Campus Virchow-Klinikum, Charité –
Universitätsmedizin Berlin, Berlin, Germany
e-mail: johann.pratschke@charite.de

M. D'Hondt, I. Sucandy (eds.), *Textbook of Robotic Liver Surgery*,
https://doi.org/10.1007/978-3-031-76536-0_21

metastases are the most common indications for a right hepatectomy.

In order to perform a right hepatectomy safely, the volume of the future liver remnant needs to be assessed preoperatively. A cutoff of 30% volume for the potential liver remnant is considered sufficient in case of normal liver parenchyma [1]. For cirrhotic livers and patients with steatosis or fibrosis, a higher cutoff point (35–40%) for the volume of the potential liver remnant might be necessary [2]. Preoperative liver function tests such as the LiMAx can be used to assess the enzymatic liver function and correlate with the future functional liver remnant [3]. Preoperative portal vein embolization may be indicated in selected cases to induce hypertrophy in the future liver remnant if the volume of the liver remnant is below the threshold.

Another important factor during a right hepatectomy is to minimize blood loss and the need for blood transfusions [4]. The Pringle maneuver, consisting of inflow occlusion of the portal vein and hepatic artery followed by sequential reperfusion, represents the most established method to minimize blood loss during liver resections. So far, intermittent occlusion intervals of 15 min followed by 5 min of reperfusion are considered the standard of care. Parenchymal transection can be performed using clamp-crush or with ultrasonic or radiofrequency energy-based devices. Small crossing vessels or bile ducts are either sealed with these devices or identified and closed with clips, while larger structures such as the pedicle or the right hepatic vein (RHV) need to be properly divided either by suturing or using a vascular stapler.

Robotic-assisted surgery is the logical evolution of laparoscopic liver resection. We here give an overview for robotic right hepatectomy, a step-by-step approach, and discuss relevant literature.

21.2 Technical Aspects

21.2.1 Patient Positioning

Analogous to laparoscopic liver surgery, different patient positions are used in practice. The simple supine position is universally applicable for all right- and left-sided liver resections, although an (over-rotated) left-sided position can be advantageous, e.g., for large hypertrophied livers [5]. In this case, the trocars are placed appropriately further to the right. A variation can be the positioning of the patient with spread legs, the so-called French positioning. Accordingly, the "bedside surgeon" sits between the patient's legs, which offers significant advantages for having good access during suction or exposure. A disadvantage of spread legs of the patient is a limited mobility of the robot arm in position 1 and a possible collision with the right patient leg. In this respect, this should be less spread compared to the left leg. Most centers position the robotic unit on the left side regardless of patient positioning. Accordingly, the left patient arm should be hinged and the right patient arm extended. Since the patient is later tilted into an reversed Trendelenburg position with a lateral left tilt, fixation of the patient is essential. Likewise, the application of lower leg compression devices is recommended for thrombosis prophylaxis.

21.2.2 Team Time-Out

The team time-out is performed immediately before the start of surgery to discuss the procedure. Communication with anesthesia is crucial here, as the patient management can be demanding for major resections such as hepatectomy. During team time-out, the following points should be discussed:

- Possibility of blood pressure shifts during hilar occlusion.
- Fluid management (250 mL/h) and low central venous pressure (CVP).
- Communication in case of bleeding or CO_2 embolism.
- Preparation of sutures (Prolene 4/0, 5/0; Vicryl 2/0 u. 3/0).
- Adjustment of pneumoperitoneum to 12–15 mmHg (depending on situs and patient).

21.2.3 Instruments, Materials, and Port Placement

The same instruments are required for right robotic hepatectomy as for other robotic liver resections. There are several possible settings, the most commonly used is as follows:

- Arm 1: Fenestrated bipolar forceps (optional arm 3)
- Arm 2: Camera
- Arm 3: Device for parenchymal dissection (optional arm 1)
  - Harmonic Ace Curved Shears
  - Vessel sealer Extend
  - Crush clamp with bipolar and scissors
- Arm 4: Tip-up Fenestrated Grasper

The use of an internal or external Pringle band is certainly useful, especially in patients in whom the CVP is not optimal or with signs of liver cirrhosis. The standard placement of the trocars is on an imaginary transverse line in the upper abdomen at a distance of approximately 7 cm from each other. The trocar placement is briefly explained below:

1. Gain access to the abdomen following surgeon's preference (Veress needle/optical trocar/open technique)
2. Establish pneumoperitoneum at 12–15 mmHg
3. Patient positioning
   (a) 10–20° reverse Trendelenburg
   (b) 5–10° tilt
4. On an oblique line of about 15–20° in the upper abdomen, place the robotic trocar
5. Port 1 on the right lateral side, minimally below segment 6, at the peritoneal fold
6. Port 2 at the level of the porta hepatis (approx. 6 cm right lateral), slightly below the imaginary line (see Fig. 21.2).
7. Port 3 (if a robotic stapler is used a 12-mm trocar) slightly lateral to the falciform ligament
8. Port 4 left on the imaginary line in the left upper abdomen
9. According to surgical strategy, a second assistant trocar may be beneficial.

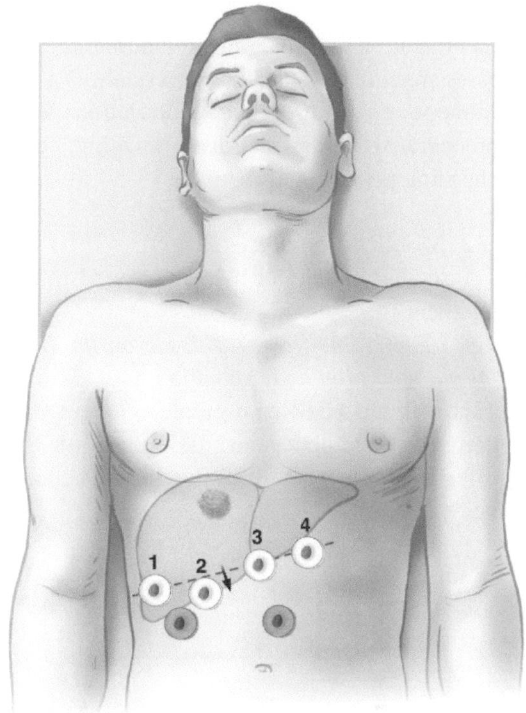

Fig. 21.2 Illustration of the possible trocar positions with Intuitive's da Vinci System. White trocars, robotic trocars; purple trocars, 12-mm assistant ports

10. Placement of a 5-mm trocar in the left upper abdomen used when applying an external Pringle band. Alternatively, the Pringle maneuver can also be performed internally.

21.2.4 Procedure Steps

In robotic surgery, the so-called caudal approach has become established for major anatomical resections. Here, the resection is performed from caudal to cranial, with the position of the camera from caudal. In the case of right hepatectomy, this means that the right pedicle is transected at an early stage of dissection, and transection of the cranially located RHV is the final step after parenchymal dissection. This principle, which was adopted from laparoscopic liver surgery, is of superior importance in robotic liver surgery because it is not possible to change the camera according to the fixed arrangement.

Step 1

- Identification of anatomical landmarks and assessment of liver parenchyma quality
- Intraoperative ultrasound, correlation with preoperative cross-sectional imaging, and checking perfusion of the liver

Step 2 (Mobilization of the Liver)

- Transection of falciform ligament and round ligament
- *Tip: Leave some ligament tissue on the liver for manipulation and exposition*
- Take off the right triangular ligament and coronary ligament and dissection of the inferior vena cava and the short hepatic veins.

Step 3

- Visualization of the RHV at its junction with the inferior vena cava (Fig. 21.3).
- Cholecystectomy

Step 4

- Using a fenestrated grasper with a gauze, elevate the liver cranially
- Set up the Pringle maneuver (absolutely necessary)
  - Extracorporeally: this involves placing a band (analogous to open surgery) around the hilus and tightening it later with a tube (sectioned drainage) Fig. 21.4a.
  - Intracorporeal: apply bulldog clamps or Foley catheter (Fig. 21.4b).

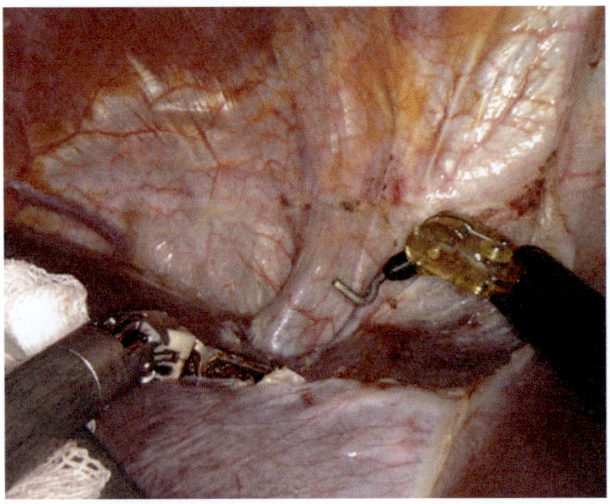

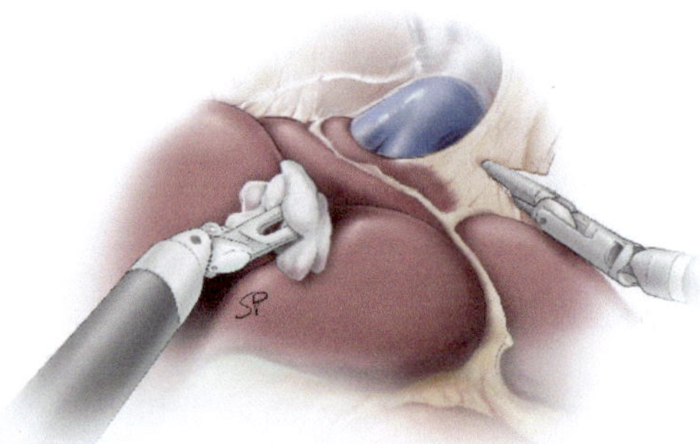

Fig. 21.3 Dissection at the hepatocaval confluence

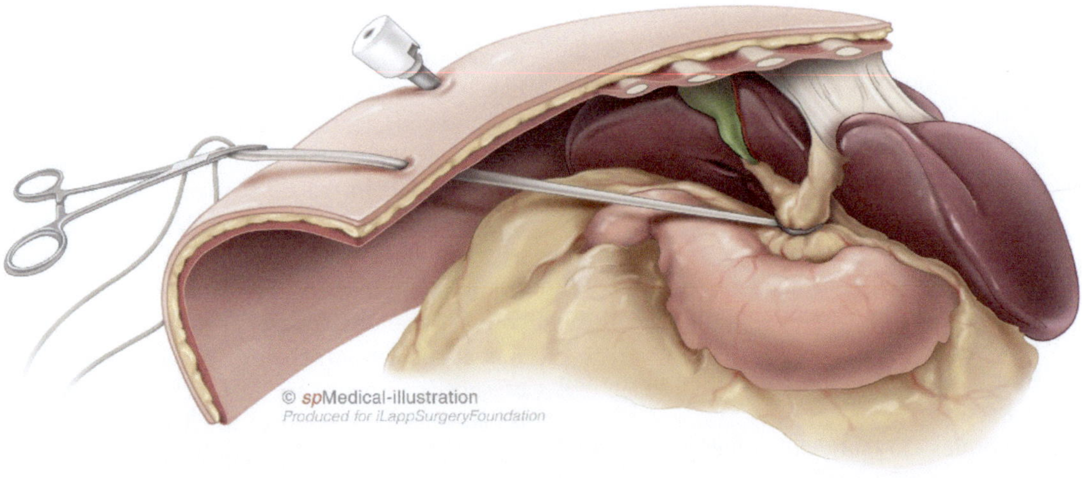

Fig. 21.4a Extracorporeal Pringle maneuver

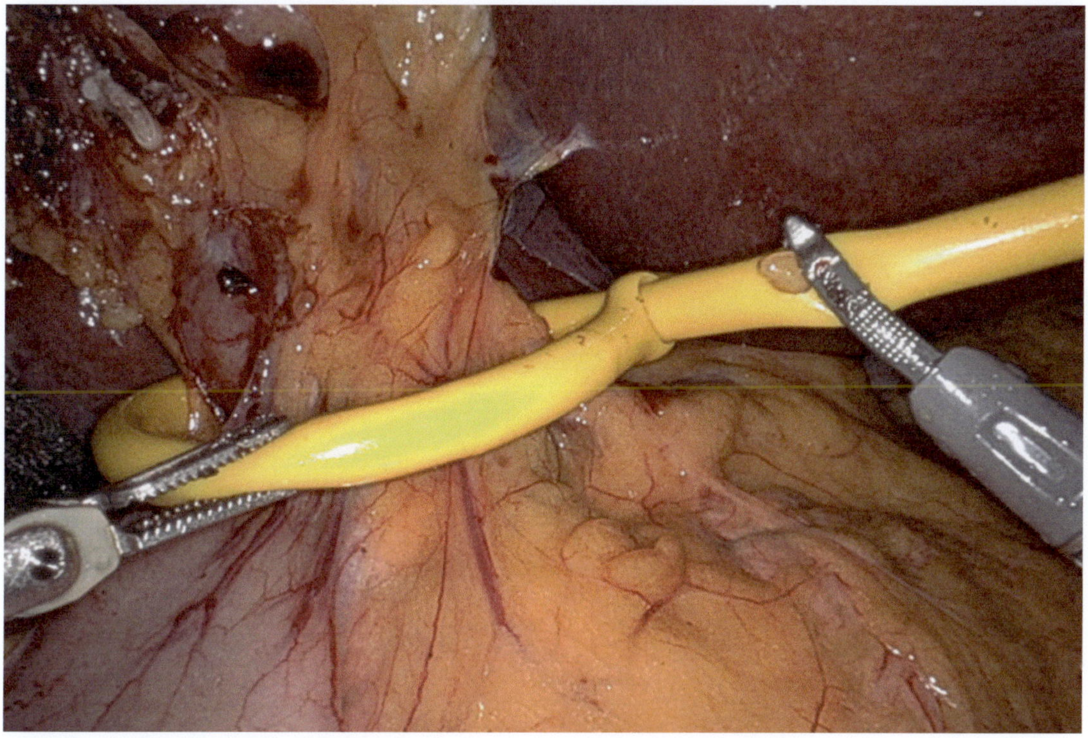

Fig. 21.4b Placement of Foley catheter around the hepatoduodenal ligament to perform intracorporeal Pringle maneuver (Huang loop technique)

- Dissect and loop the right hepatic artery (Fig. 21.5a).
- Dissect and loop the right portal vein (Fig. 21.5b).
- Lymphadenectomy if necessary
- Test clamp of right hepatic artery and right portal vein, ultrasound, and check of perfusion of the left liver.
- Ligate and divide the right hepatic artery and right portal vein

Tips for Hilar Preparation

For hilar preparation, two surgical approaches can be distinguished: the intrafascial and the extrafascial approach. Defined by Couinaud, the classic intrafascial approach with hilar preparation, isolation of hepatic artery and the portal vein, and eventually the bile duct can be distinguished from the extrafascial intraparenchymal approach with parenchymal dissection and secondary intraparenchymal dissection of the hepatic pedicle. The conventional intrafascial extrahepatic approach allows early vascular control and parenchymal dissection following a demarcated line but bears the risk of injuring the vessels or the bile duct of the wrong side with fatal outcome. The intraparenchymal approach bypasses the difficult hilar dissection and may be even more appropriate after portal vein embolization.

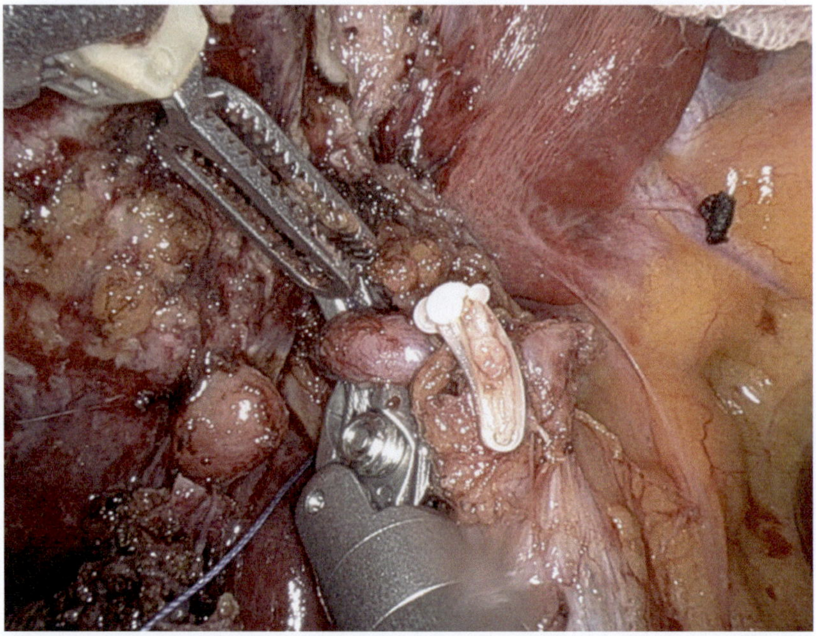

Fig. 21.5a Right hepatic artery encircled

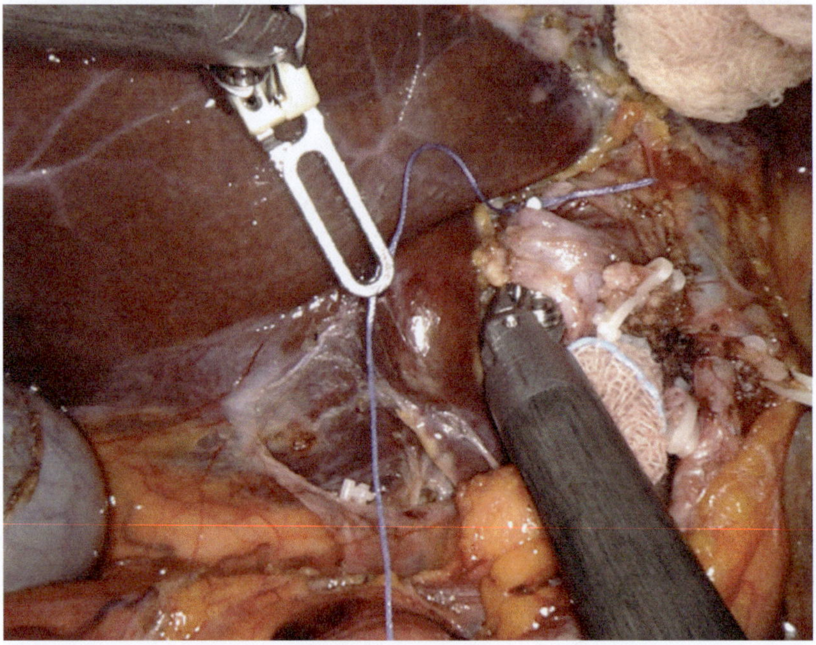

Fig. 21.5b Isolation of the right portal vein

Step 5 (Definition of Transection Line)

- Definition of the parenchymal transection line (Cantlie line) by inferior perfusion of the right lobe of the liver (after ligation of the hilar vessels) or by visualization of the middle hepatic vein with ultrasound (which runs directly along the Cantlie line).
- Marking of the transection line on the liver surface with monopolar energy.

Step 6 (Parenchymal Transection)

- Place energy device in position 3
  - Harmonic ACE
  - Vessel sealer Extend of SynchroSeal

- Crush clamp with bipolar and scissors

  Some centers use a laparoscopic cavitron ultrasonic surgical aspirator (CUSA) for parenchymal dissection, but this does not correspond to classic robotic surgery, as it is used by the bedside surgeon. Note that to date, there is no robotic CUSA.

- Use bipolar energy in position 1 to assist in hemostasis
- Perform the transection from caudal to cranial and from superficial to deep (Fig. 21.6).

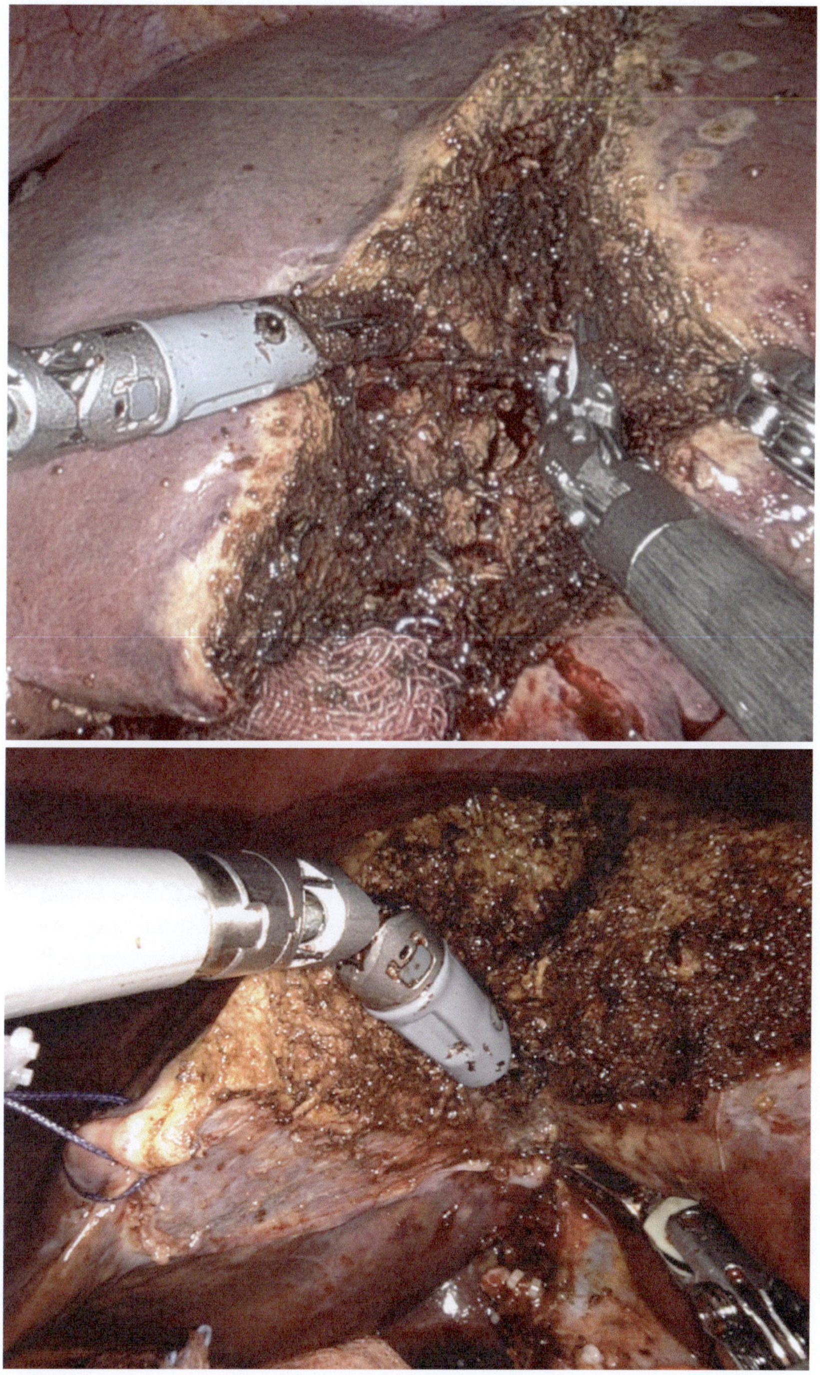

Fig. 21.6 Parenchymal transection (SynchroSeal)

- Apply the Pringle maneuver (intermittently for a maximum of 15 min followed by 5-min reperfusion).
- Staple or ligate the right pedicle
- Clamp/seal crossing vessels
  Split the liver down to the IVC
- Staple the right hepatic vein with a white reload

Tips for Parenchymal Dissection

Dissection of the superficial 2 cm of parenchyma is usually uneventful in terms of intraoperative bleeding, CO_2 emboli, or postoperative bile leaks, and the opposite is the case for deeper parenchyma. Superficial dissection is performed analogously to laparoscopic liver surgery using an ultrasound device, the Harmonic Ace Curved Shears or Vessel Sealer (both from Intuitive). While the techniques offered for parenchymal dissection in laparoscopic liver surgery (CUSA, Hydrojet) reflect the entire spectrum of open liver surgery, the technical options in robotic liver surgery are (still) limited. In contrast to superficial, less selective parenchymal dissection, a modified clamp-crush technique is necessary to safely visualize crossing vessels and structures. The subtlety of the dissection is greatly enhanced by the intermittent hilar occlusion described above, and this technique is now routinely used in most centers. In this context, it also appears interesting to note a "visual haptic" confirmed by several robotic liver surgeons that develops with experience and further improves the quality of dissection. While smaller crossing vessels and bile ducts can usually be dissected easily between clips, dissection of large pedicles or the hepatic veins is performed using staple suture devices or classic sutures. Placement of clips or use of linear staple suture devices can be done entirely through the robot. Alternatively, the bedside surgeon can place the clips and staplers laparoscopically, thereby increasing the speed of surgery due to faster or unnecessary instrument changes.

Step 7

- Lower pressure of pneumoperitoneum to 7 mmHg and perform a Valsalva maneuver for 15 s to check for bleeding
- Place the specimen in an endoscopic retrieval bag
- Remove the specimen by extending the assistant port according to the specimen size
- Close the fascia with running suture
- Reestablish pneumoperitoneum, final check for hemostasis and potential bile leakage, and remove all remaining trocars under visual control
- Close the fascia of 12-mm trocar insertion sites
- Close skin with subcutaneous sutures and tissue glue or intracutaneous suture

21.3 Literature Overview

Robotic right hepatectomy has been established and performed for several years now, although literature regarding the open vs. laparoscopic approach has not even been definitively clarified. Thus, the superiority of the robotic technique for right hepatectomy compared to the open or laparoscopic approach has never been studied in a randomized controlled setting. The multicenter international ORANGE II PLUS trial has been constructed to provide evidence of the laparoscopic approach versus open hemihepatectomy (NCT01441856). Although the results are not yet fully published, the data indicate that the laparoscopic approach is superior to open surgery in terms of time to functional recovery and lengths of hospital stay [6, 7]. The advantage of the robotic system for major liver resections such as right hepatectomy has so far only been investigated by retrospective series, registry analysis, and meta-analyses.

The most reliable data regarding robotic vs. laparoscopic right hepatectomy is available from a multicenter case-control study involving 989 patients from 29 international centers [8]. Of the 989 patients who met study criteria, 220 underwent surgery using the robotic approach, and 769 underwent laparoscopic right hepatectomy. Propensity score matching analysis was performed to minimize selection bias. The authors demonstrated that robotic right hepatectomy/extended right hepatectomy was associated with a lower conversion rate (19 of 220 [8.6%] vs 39 of 220 [17.1%]; P = 0.01) and shorter hospital stay compared with laparoscopic right hepatectomy/extended right hepatectomy (median [IQR], 7.0 [5.0–10.0] days; vs median [IQR], 7.0 [5.75–10.0] days; $P = 0.048$). After reduction of the learning curve effect, the robotic approach was associated with a shorter hospital stay and similar perioperative outcomes. These findings suggest that the robotic platform may offer some benefits especially at the initial period of the learning curve.

A systematic review and meta-analysis of 525 patients found no significant difference between robotic and laparoscopic liver resection for resection of three or more liver segments in terms of overall morbidity and mortality rates, conversion to open hepatectomy, blood loss, operative time, and length of hospital stay [9]. The authors concluded that both techniques are equivalent when performed in selected patients and in high-volume centers.

In a meta-analysis of 6 studies with 1093 patients investigating the effectivity and safety of robotic-assisted versus laparoscopic hepatectomy for liver tumors, Hu et al. found that the robotic approach was more commonly used for right hepatectomy and is associated with longer operative time, larger tumor size, and lower open conversion rate [10]. The authors concluded that for malignancies requiring hepatectomy, robotic-assisted liver resections demonstrated similar safety and feasibility to laparoscopy, with a lower rate of conversion.

From today's perspective, robotic liver surgery leads to higher perioperative costs compared to open surgery due to the high equipment expenditure and the prolonged operation time. A similar circumstance already existed in the 1990s when laparoscopy entered the field of visceral surgery. Thus, this aspect probably also has a short half-life, and new products and competition in the market will lower costs in the future. Currently, the increased costs are partially compensated by faster convalescence and shortened length of stay [11–13].

21.4 Conclusions

Robotic right hepatectomy is challenging but a highly standardized procedure. The key to establishing such an operation is the precise definition of the procedure steps, intraoperative management, and standardization. However, robotic systems are subject to rapid change due to new emerging models and innovations, so the present description of robotic right hepatectomy is a snapshot in time.

References

1. Khan AS, et al. Assessment and optimization of liver volume before major hepatic resection: Current guidelines and a narrative review. Int J Surg. 2018;52:74–81.
2. Clavien P-A, Petrowsky H, DeOliveira ML, Graf R. Strategies for safer liver surgery and partial liver transplantation. N Engl J Med. 2007;356:1545–59.
3. Jara M, et al. Reductions in post-hepatectomy liver failure and related mortality after implementation of the LiMAx algorithm in preoperative work-up: a single-centre analysis of 1170 hepatectomies of one or more segments. HPB. 2015;17:651–8.
4. Ortiz Galindo SA, et al. Safety of intermittent Pringle maneuver during minimally invasive liver resection in patients with hepatocellular carcinoma with and without cirrhosis. Langenbeck's Arch Surg. 2022;407:235–44.
5. Schmelzle M, Schöning W, Pratschke J. [Liver surgery – setup, port placement, structured surgical steps – standard operating procedures in robot-assisted liver surgery]. Zentralbl Chir. 2020;145:246–251.
6. 384O Laparoscopic versus open hemihepatectomy: The ORANGE II PLUS multicenter randomized controlled trial. Ann Oncol. 2021;32:S531.
7. Laparoscopic versus open hemihepatectomy: The ORANGE II PLUS multicenter randomized controlled trial. Eur J Surg Oncol. 2022;48:e40.
8. Chong CC, et al. Propensity score-matched analysis comparing robotic and laparoscopic right and extended right hepatectomy. JAMA Surg. 2022;157:436–44.
9. Ziogas IA, et al. Laparoscopic versus robotic major hepatectomy: a systematic review and meta-analysis. Surg Endosc. 2021;35:524–35.
10. Hu Y, et al. Robotic versus laparoscopic hepatectomy for malignancy: a systematic review and meta-analysis. Asian J Surg. 2021;44:615–28.

11. Hawksworth J, et al. Robotic Hepatectomy Is a Safe and Cost-Effective Alternative to Conventional Open Hepatectomy: a Single-Center Preliminary Experience. J Gastrointest Surg. 2021;25:825–8.
12. Daskalaki D, et al. Financial impact of the robotic approach in liver surgery: a comparative study of clinical outcomes and costs between the robotic and open technique in a single institution. J Laparoendosc Adv Surg Tech A. 2017;27:375–82.
13. D'Hondt M, et al. Transition from laparoscopic to robotic liver surgery: clinical outcomes, learning curve effect, and cost-effectiveness. J Robot Surg. 2023;17(1):79–88. https://doi.org/10.1007/s11701-022-01405-w.

22 Robotic Left Hemihepatectomy

Mathieu D'Hondt and Dennis A. Wicherts

22.1 Introduction

Minimally invasive liver surgery has evolved significantly in recent years. The laparoscopic resection of liver segments II and III (left lateral sectionectomy) was one of the first procedures to become standard of care in preference to open surgery [1, 2]. With improved surgical techniques and experience, laparoscopic major hepatectomies are currently also considered feasible and safe in selected patients [3–6].

Left hemihepatectomy is considered a procedure of medium complexity according to a recent classification system for hepatectomy [7, 8]. In comparison with other medium-complexity procedures, laparoscopic left hemihepatectomy has shown to have the shortest operative time, least amount of blood loss, and a rapid learning curve, suggesting that it is a feasible first step in laparoscopic major hepatectomy [9]. Some technical and anatomical aspects of a left hemihepatectomy contribute to this finding. First, dissection of the left hepatic artery and portal vein is relatively easy due to their anatomical course and position in the hepatoduodenal ligament. Furthermore, the parenchymal transection plane is smaller compared to a right hemihepatectomy. Also, the hepatic venous pressure is lower on the left side, reducing the risk of bleeding.

Compared to conventional laparoscopic liver surgery, robotic liver surgery has the advantages of a three-dimensional view, tremor filtration, and improved dexterity. In this chapter, we aim to provide a detailed technical description of robotic left hemihepatectomy.

22.2 Technical Aspects

22.2.1 Patient Positioning and Trocar Placement

Patients are generally placed in supine French position on a vacuum mattress with the arms alongside the body. The table is tilted in slight reverse Trendelenburg (approximately 10–20 degrees). Pneumoperitoneum (12 mmHg) is established using a Verres needle in the left hypochondrium on the midclavicular line. Three robotic trocars are placed in a straight line two hands below the xiphoid. The 8 mm trocar for the first robotic arm is placed in the right midclavicular line. An 8 mm robotic trocar is placed at the level of the umbilicus for the second

M. D'Hondt (✉)
Department of Digestive and Hepatobiliary/Pancreatic Surgery, Groeninge Hospital, Kortrijk, Belgium

D. A. Wicherts
Department of Abdominal and Hepatobiliary/Pancreatic Surgery, Ziekenhuis Oost-Limburg, Genk, Belgium
e-mail: DennisAndre.Wicherts@zol.be

M. D'Hondt, I. Sucandy (eds.), *Textbook of Robotic Liver Surgery*,
https://doi.org/10.1007/978-3-031-76536-0_22

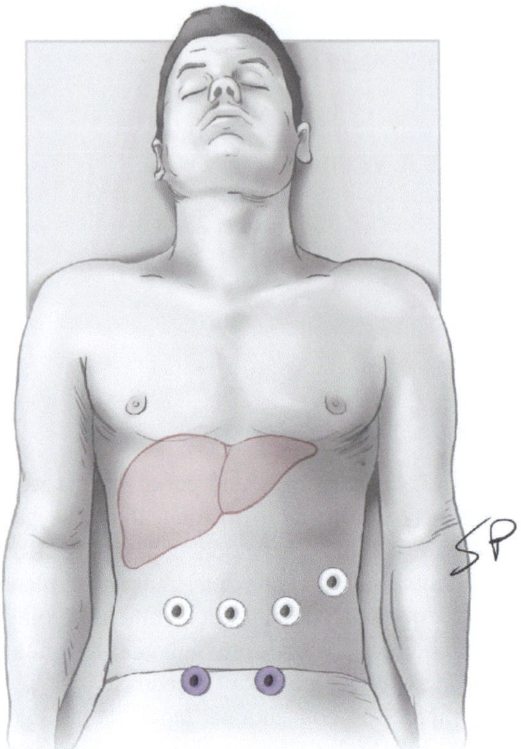

Fig. 22.1 Patient positioning and trocar placement

robotic arm (camera port) and the trocar for the third robotic arm is placed in the left midclavicullar line. A fourth 8-mm trocar is placed on the left side of the patient slighthly more cranally at the left anterior axillary line. A 12-mm laparoscopic assistant trocar is inserted in the lower abdomen in between two robotic trocars, usually on the right side of the patient, ensuring that the assistant can reach the transection area. The robot is docked coming from the right side of the patient. Patient installation is illustrated in Fig. 22.1.

22.2.2 Mobilization

Mobilization starts with dividing the round and falciform ligaments using the robotic vessel sealer and is continued by dividing the left coronary and triangular ligaments. Dissection from medial to lateral, after identification of the suprahepatic inferior vena cava, is recommended to avoid injury to the left hepatic vein or branches of the phrenic vein [10].

22.2.3 Cholecystectomy

Lift the gallbladder cranially by using the fourth robotic arm. Identify the cystic duct and artery using the robotic hook. After achieving a critical view of safety, clip both structures using Hem-o-lok clips and divide the cystic duct and artery using the robotic scissors. Alternatively, the cystic artery can be divided by using the robotic vessel sealer. Then dissect the gallbladder from the liver and place it outside the working field.

22.2.4 Hilar Dissection

Two methods of hilar dissections are generally used for a left hemihepatectomy: extrahepatic Glissonian approach with separate dissection and ligation of the left hepatic artery and portal vein or an intrahepatic Glissonian approach with encircling and stapling of the complete left pedicle.

For the first approach, the left hepatic artery is identified and isolated in the hepatoduodenal ligament by using the robotic hook (Fig. 22.2).

Always identify the origin of the right hepatic artery to make sure it is preserved. Clip the left hepatic artery with Hem-o-lok clips and divide it using the robotic scissors. Following arterial division, the left portal vein can be isolated (Fig. 22.3).

Small portal vein branches to segment 1 can be divided using the robotic vessel sealer to create more space to isolate the left portal vein. Clip and divide the left portal vein in a similar way as the left hepatic artery. The left hepatic duct is usually not divided in this phase of the procedure.

For an intrahepatic Glissonean approach, the left pedicle is encircled with robotic forceps. Temporary clamping is performed to ensure ischemic demarcation of the left hemiliver. The left hepatic pedicle is then divided using a robotic or laparoscopic stapler device (Fig. 22.4).

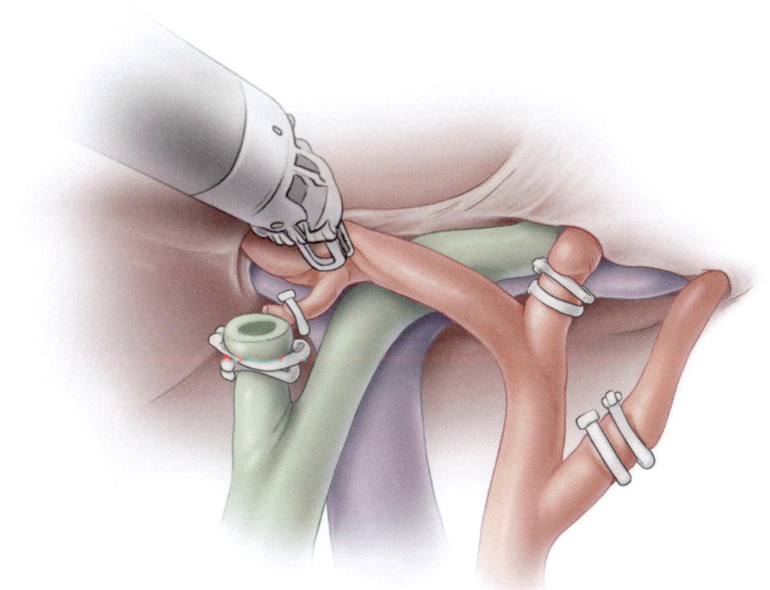

Fig. 22.2 Hilar dissection: isolation of the left hepatic artery (branch to left lateral sector and branch to segment 4 clipped). Note: the right hepatic artery runs anteriorly to the bile duct in this patient

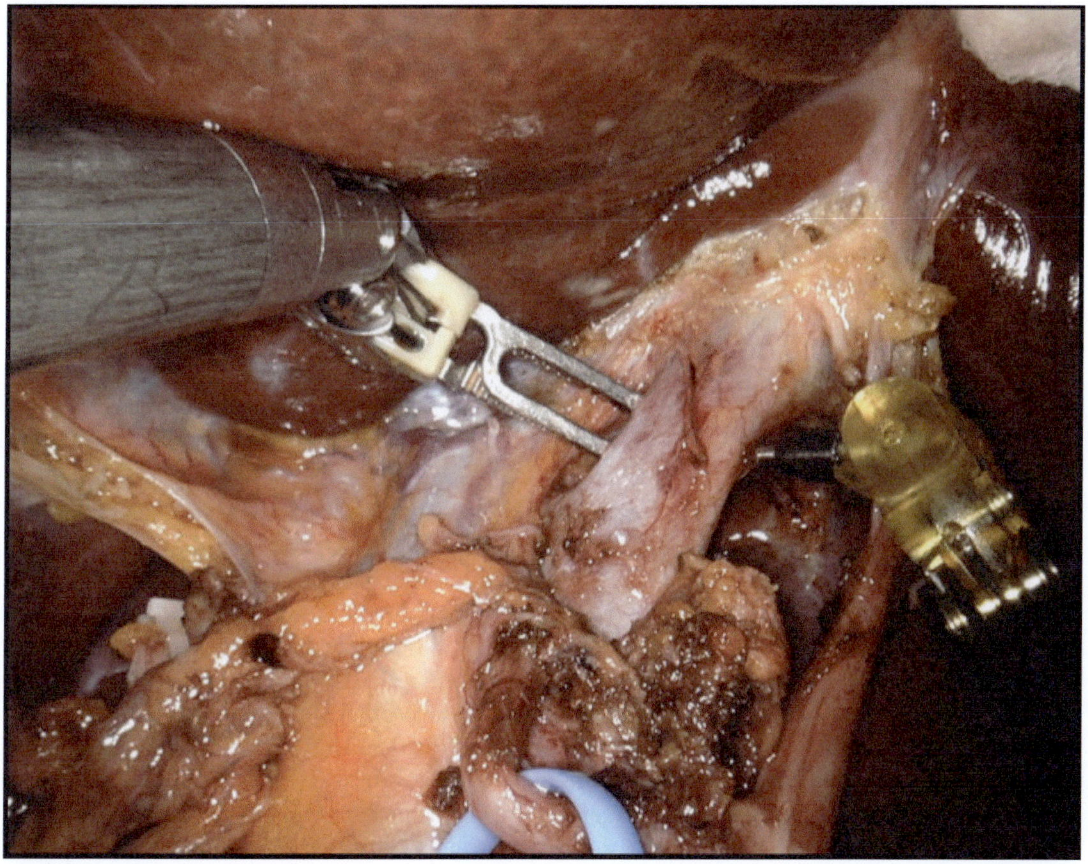

Fig. 22.3 Hilar dissection: isolation of the left portal vein

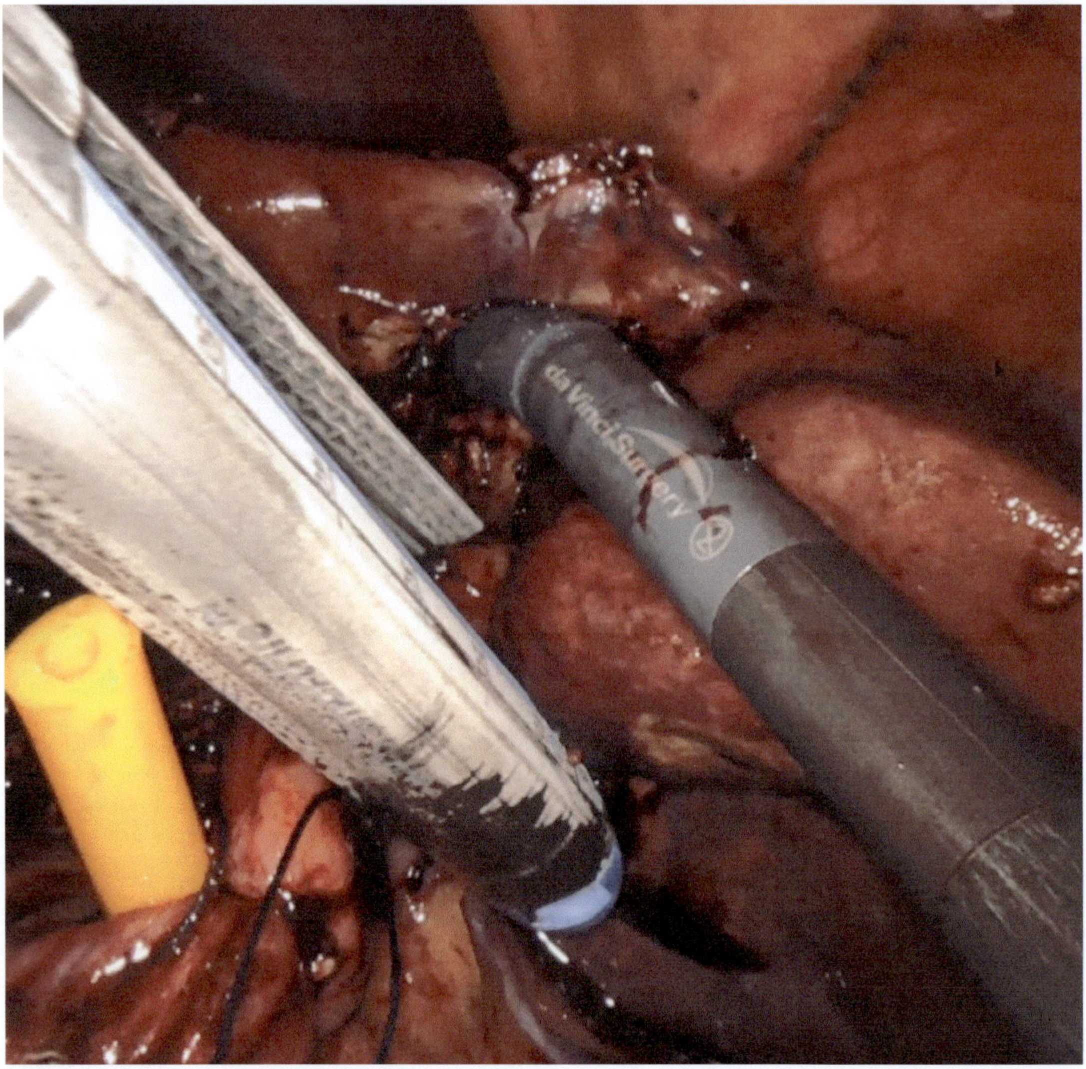

Fig. 22.4 Intrahepatic Glissonean approach: en bloc stapling of pedicle of the left hemiliver

Preoperative imaging is essential to be aware of anatomical variations risking injury to sectoral branches of the right hemiliver.

22.2.5 Pringle Maneuver

In general, it is recommended to prepare a Pringle loop for vascular inflow occlusion during parenchymal transection. Different techniques have been described for this purpose. In robotic liver surgery, we prefer an intracorporeal tourniquet method using a shortened Foley catheter (approximately 12-cm length) [11]. After opening the lesser omentum, the porta hepatis is encircled by the catheter, and the tail is pulled through the side hole to make a loop. During transection, the loop can be tightened and secured with a Hem-o-lok clip or laparoscopic bulldog clamp. However, following selective hilar dissection and ligation of the left hepatic artery and portal vein prior to liver transection, a Pringle maneuver is often not necessary.

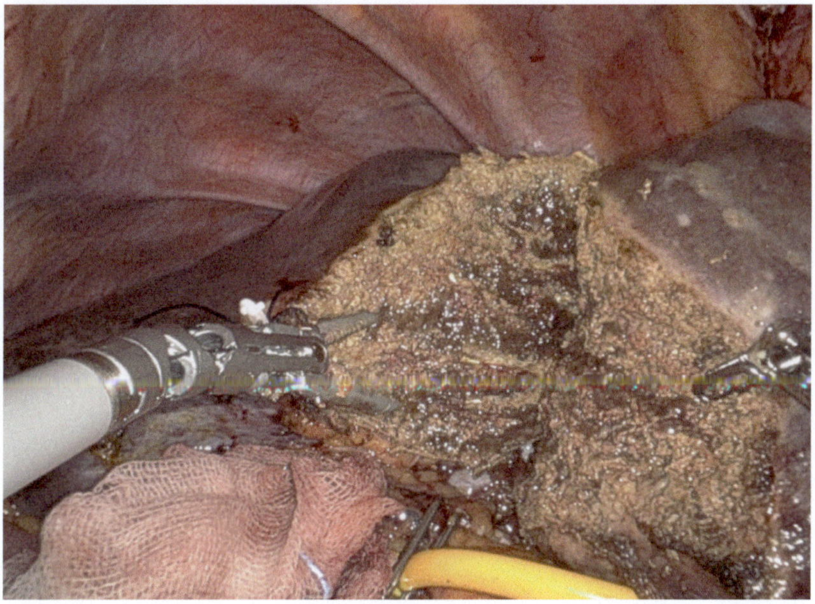

Fig. 22.5 Parenchymal transection using the robotic vessel sealer. Note: the Pringle maneuver is performed using a Huang loop (Foley catheter)

22.2.6 Parenchymal Transection

Following hilar transection of the left hepatic artery and portal vein, an ischemic demarcation line is visible over the liver surface. The transection line is marked slightly on the left side of the demarcation line using the robotic hook. The round ligament can be used to pull the liver to the left side of the patient using the fourth robotic arm, improving visualization of the transection plane. It is recommended to perform an intraoperative ultrasound of the liver prior to parenchymal transection to confirm the localization of the lesion to be resected.

The vessel sealer is generally used for parenchymal transection (Fig. 22.5).

A bipolar device is used in addition to secure hemostasis. First, superficial transection of the parenchyma is performed along the total length of the transection line. Deeper intrahepatic vascular and biliary structures can be controlled by the vessel sealer. Parenchymal transection is continued until the hilum is reached. At that point, the left hepatic duct is transected using a robotic or laparoscopic stapler device. In selective cases where the bifurcation of the bile duct is low the left hepatic duct can be ligated or clipped and cut during the hilar dissection (Fig. 22.6).

Finally, the left hepatic vein is also ligated or clipped and transected. The left hepatic vein can also be stapled (Fig. 22.7).

After completion of the transection, the liver surface is checked for hemostasis and biliostasis. Different hemostatic agents can be used to secure hemostasis. The resected specimen and gallbladder are placed in an extraction bag and removed through a Pfannenstiel incision. In general, no intra-abdominal drains are placed.

22.2.7 Tips and Tricks

- *Use intraoperative Doppler ultrasound to confirm preserved inflow to the right hemiliver following selective clamping of the left hepatic artery and vein.*
- *A Vicryl suture can be used to ligate the left hepatic artery and portal vein distally to avoid interference with the stapler device when dividing the left hepatic duct.*
- *Set intra-abdominal pressure to 10 mmHg during parenchymal transection to lower the risk of gas embolism.*
- *Use saline irrigation (applied by the assistant) to avoid carbonization of the vessel sealer and bipolar forceps and enhance parenchymal transection.*

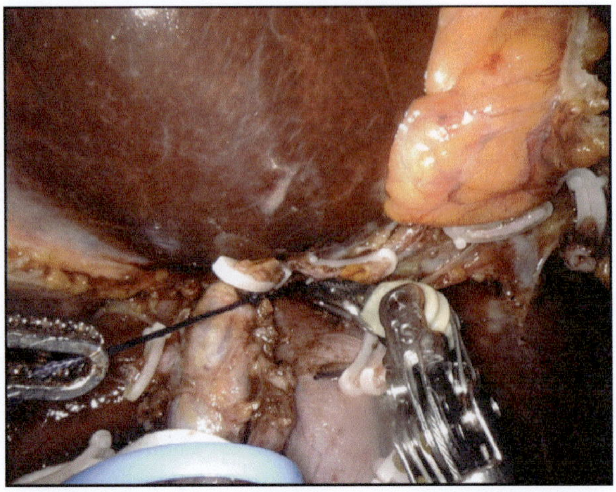

Fig. 22.6 Extrahepatic ligation and clipping of the left hepatic duct

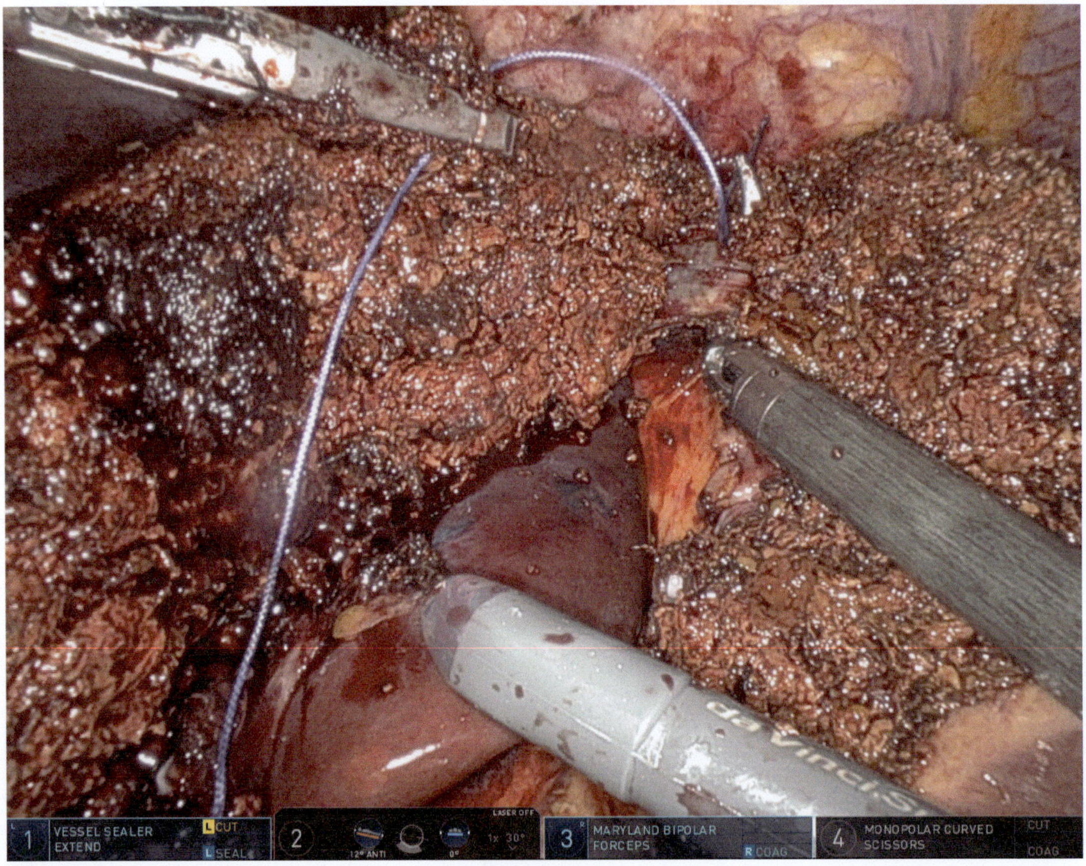

Fig. 22.7 Isolating the left hepatic vein

22.3 Literature Overview

The available literature concerning robotic left hemihepatectomy is still limited. Published reports can be divided into technical descriptions of the procedure and studies comparing the robotic approach with conventional laparoscopic left hemihepatectomy.

Recently, the perioperative outcomes of robotic left and extended left hepatectomy were compared with a laparoscopic approach in a large international multicenter study [12]. After propensity score matching, 164 robotic patients were included for analysis. The results showed less blood loss, a lower rate of conversion to an open approach, and shorter hospital stay in favor of the robotic approach. These results most likely reflect the technical advantages of the robotic platform for complex major liver resections. For example, the ease of suturing during robotic surgery may facilitate a rapid control of major bleeding, limiting the need for urgent conversion. Similarly, Cai et al. reported less blood loss in the robotic group but similar rates of postoperative morbidity as compared to a laparoscopic left hemihepatectomy group [13].

Some other papers, including a variety of liver resections, have reported no significant differences in short-term outcome between a robotic and laparoscopic approach. A recent systematic review and meta-analysis showed similar rates of blood transfusion and length of stay with but a lower conversion rate for the robotic group [14]. Cipriani et al. stratified resected patients according to different levels of difficulty for liver resection [15]. They concluded that for highly difficult hepatectomies, the robotic approach results in less blood loss and a lower conversion rate. This result suggests that the robotic platform may contribute to a higher rate of successful minimally invasive surgery especially in complex liver resections. This was confirmed by Tsung et al., reporting a greater proportion of major hepatectomies that was completed minimal invasively by a robotic approach [16].

A single-center case series from Magistri et al. recently reported a median estimated blood loss of 300 mL for robotic left hemihepatectomy with a median hospital stay of 3 days [17]. In addition, a few other technical case reports on robotic left hemihepatectomy have been published so far [18–20]. All these papers show limited amounts of blood loss (less than 50 mL) and patient discharge on day 3 or 4 postoperatively. No conversion was needed in these cases.

22.4 Conclusions

Robotic left hemihepatectomy is a technical feasible and safe procedure in experienced hands. It is a suitable procedure in the intermediate phase of the learning curve for robotic liver surgery. The benefits of a robotic approach for major liver resections are illustrated by a lower conversion rate as compared to conventional laparoscopic surgery.

References

1. Chang S, Laurent A, Tayar C, et al. Laparoscopy as a routine approach for left lateral sectionectomy. Br J Surg. 2007;94:58–63.
2. Dokmak S, Raut V, Aussilhou B, et al. Laparoscopic left lateral resection is the gold standard for benign liver lesions: a case-control study. HPB. 2014;16:183–7.
3. Pearce NW, Di Fabio F, Teng MJ, et al. Laparoscopic right hepatectomy: a challenging, but feasible, safe and efficient procedure. Am J Surg. 2011;202:e52–8.
4. Belli G, Gayet B, Han H-S, et al. Laparoscopic left hemihepatectomy a consideration for acceptance as standard of care. Surg Endosc. 2013;27:2721–6.
5. Valente R, Sutcliffe R, Levesque E, et al. Fully laparoscopic left hepatectomy – a technical reference proposed for standard practice compared to the open approach: a retrospective propensity score model. HPB. 2018;20:347–55.

6. Dagher I, Gayet B, Tzanis D, et al. International experience for laparoscopic major liver resection. J Hepatobiliary Pancreat Sci. 2014;21:732–6.
7. Lee MK 4th, Gao F, Strasberg SM. Perceived complexity of various liver resections: results of a survey of experts with development of a complexity score and classification. J Am Coll Surg. 2015;220:64–9.
8. Lee MK 4th, Gao F, Strasberg SM. Completion of a liver surgery complexity score and classification based on an international survey of experts. J Am Coll Surg. 2016;223:332–42.
9. Hasegawa Y, Nitta H, Takahara T, et al. Laparoscopic left hemihepatectomy is suitable as a first step in pure laparoscopic major hepatectomy. Ann Gastroenterol Surg. 2018;2:376–82.
10. Ikoma N, Itano O, Oshima G, Kitagawa Y. Laparoscopic liver mobilization: tricks of the trade to avoid complications. Surg Laparosc Endosc Percutan Tech. 2015;25:e21–3.
11. Huang JW, Su WL, Wang SN. Alternative laparoscopic intracorporeal Pringle maneuver by Huang's loop. World J Surg. 2018;42:3312–5.
12. Sucandy I, Rayman S, Lai EC, et al. Robotic versus laparoscopic left and extended left hepatectomy: an international multicenter study propensity score-matched analysis. Ann Surg Oncol. 2022;29:8398–406.
13. Cai J-P, Chen W, Chen L-H, et al. Comparison between robotic-assisted and laparoscopic left hemihepatectomy. Asian J Surg. 2022;45:265–8.
14. Hu Y, Guo K, Xu J, et al. Robotic versus laparoscopic hepatectomy for malignancy: a systematic review and meta-analysis. Asian J Surg. 2022;44:615–28.
15. Cipriani F, Fiorentini G, Magistri P, et al. Pure laparoscopic versus robotic liver resections: multicentric propensity score-based analysis with stratification according to difficulty scores. J Hepatobiliary Pancreat Sci. 2022;29:1108–23.
16. Tsung A, Geller DA, Sukato DC, et al. Robotic versus laparoscopic hepatectomy: a matched comparison. Ann Surg. 2014;259:549–55.
17. Magistri P, Assirati G, Ballarin R, et al. Major robotic hepatectomies: technical considerations. Updat Surg. 2021;73:989–97.
18. Sucandy I, Gravetz A, Ross S, Rosemurgy A. Technique of robotic left hepatectomy: how we approach it. J Robot Surg. 2019;13:201–7.
19. Machado M, Mattos BH, Lobo Filho M, Makdissi F. Intrahepatic Glissonian approach for robotic left hepatectomy. Surg Oncol. 2021;38:101579.
20. Görgec B, Zonderhuis BM, Besselink MG, et al. Robotic left hepatectomy using indocyanine green fluorescence imaging for an intrahepatic complex biliary cyst. J Vis Exp. 2022;(184):e63265.

Robotic Extended Hepatectomy

23

Kristin E. Goodsell, Lindsay K. Dickerson, and James O. Park

23.1 Introduction

The first surgical robotic system was FDA approved in 2000, and the first robotic partial hepatectomy-a left lateral sectorectomy-was performed in 2003, more than 20 years ago [1]. Over the last two decades, as robotic hepatectomy has gained wider adoption, there has been significant progress toward utilizing robotic systems for major liver resections. Evidence from oncologic hepatectomies demonstrates that major resections can be performed robotically with outcomes comparable or superior to open resections [2]. The robotic system enhances surgical dexterity by providing greater range and degrees of freedom of motion, magnified three-dimensional (3D) visualization with excellent depth perception of the operative field, and precise control of delicate surgical maneuvers [3]. Compared to laparoscopic major resections, robotic major resections are associated with reduced blood loss [4]. Utilizing validated liver resection difficulty scores, there is evidence that the robotic approach may enable minimally invasive execution of more challenging resections [5, 6]. Based on the recently published International Experts Consensus Guidelines, robotic hepatectomy been deemed feasible, safe, and effective for all indications, including minor and major hepatectomy and live donor hepatectomy [7]. Furthermore, it has been demonstrated to be non-inferior in cost effectiveness, advantageous in parenchymal sparing hepatectomy around major structures, and confers an improved learning curve.

According to the Brisbane 2000 nomenclature, extended left hepatectomy/hemihepatectomy encompasses resection of Couinaud segments 2, 3, 4a, 4b, 5, and 8 +/− 1; extended right hepatectomy/hemihepatectomy includes resection of segments 4a, 4b, 5, 6, 7, and 8 +/− 1. The inclusion of the caudate lobe (segment 1), depends on the disease distribution. The term "extended hemihepatectomy" may also describe any resection extending beyond the midplane which may incorporate a complete sector/section with division of the middle hepatic vein or include a only a portion of a sector/section or segment [8].

Indications for extended hepatectomy include both benign and malignant diseases, with malignant conditions being more common [9]. Malignant indications include hepatocellular carcinoma (HCC), cholangiocarcinoma (CCA), colorectal liver metastases (CRLM), or metastases from other sites [10]. Extended resections,

K. E. Goodsell · L. K. Dickerson
Department of Surgery, University of Washington, Fred Hutch Cancer Center, Seattle, WA, USA

J. O. Park (✉)
Department of Surgery, University of Washington, Fred Hutch Cancer Center, Seattle, WA, USA

Department of Surgery, Mount Sinai Health System, Tisch Cancer Institute, New York, NY, USA,
e-mail: jopark@uw.edu; james.oh.park@mountsinai.org

M. D'Hondt, I. Sucandy (eds.), *Textbook of Robotic Liver Surgery*,
https://doi.org/10.1007/978-3-031-76536-0_23

including those requiring biliary reconstruction (e.g., hilar CCA), are technically feasible with the robotic approach [11, 12]. These resections may be undertaken by surgeons with appropriate expertise in robotic liver surgery, with outcomes comparable to or better than open or laparoscopic techniques [7, 11]. However, given the complexity of these extended resections, careful assessment of the liver remnant is essential, along with thorough preparations for safe and efficient conversion to an open approach if required. As such, the surgical teams must possess a high level of expertise in open resection and reconstruction techniques before undertaking extensive robotic resections.

23.2 Technical Approach

23.2.1 Preoperative Considerations

Thorough preoperative planning is essential prior to undertaking an extended hepatectomy, regardless of the surgical approach. High-quality cross-sectional imaging performed with contrast-enhanced, multiphase computed tomography (CT) or magnetic resonance imaging (MRI) is critical in the evaluation of surgical liver disease. Preoperative imaging should characterize vascular and biliary anatomy, including any anatomic variations, to effectively guide operative planning [13]. Given the extent of resection with resultant limited functional liver remnant and the potential need for vascular and biliary resection and reconstruction, it is imperative that all blood vessels and bile ducts are accounted for and patent. Nuclear imaging with fluorodeoxyglucose (FDG) positron emission tomography (PET) or DOTATATE PET may also be completed in cases of metastatic colorectal or neuroendocrine tumors [14]. Additionally, MRI with gadoxetate disodium (EOVIST) gadolinium-based contrast can be particularly useful in identifying small-volume disease in metastatic colorectal and neuroendocrine tumors, and cholangiocarcinoma. Surgical resectability is determined based on the comprehensive assessment of physiologic factors (patient comorbidities and preoperative performance status), pathologic factors (tumor biology or disease characteristics), and technical considerations (feasibility of obtaining a negative margin while preserving adequate liver remnant) [15].

Especially important in the preparation for extended hepatectomy is the careful assessment of future liver remnant (FLR) expected after the planned resection. FLR can be estimated using various techniques, including volumetry, scintigraphy, and 3D volumetric analysis software. Current practice at our institution favors high fidelity CT or MR volumetric analysis. Risk of post-hepatectomy liver failure—the result of inadequate FLR and a source of major morbidity and mortality—can largely be mitigated through meticulous preoperative planning. Portal vein embolization (PVE), associating liver partition and portal vein ligation for staged hepatectomy (ALPPS), or two-stage hepatectomy (TSH) procedures may be considered in scenarios where FLR is predicted to be inadequate without additional hypertrophy [16, 17]. When preoperative chemotherapy has been administered, FLR calculations must consider potential hepatotoxic effects, such as chemotherapy-associated steatohepatitis or sinusoidal dilatation [18].

A clear understanding of the patient's vascular and biliary anatomy is imperative in planning extended resections. Specific considerations for vascular variants in extended hepatectomy include presence of replaced and/or accessory right or left hepatic arteries, portal vein trifurcation or early right posterior portal vein branching, anomalous takeoff of the right anterior sectoral portal vein from the left portal vein, presence of an accessory inferior right hepatic vein, and the relationship of lesions to the middle hepatic vein and its branches [13]. Biliary anatomical variations include aberrant drainage of the right posterior duct into the left hepatic duct and triple confluence of the right anterior, right posterior, and left hepatic ducts.

23.2.2 Operative Considerations

23.2.2.1 Patient Preparation

At our institution, all patients undergoing hepatectomy have their cycle of care outlined in an Enhanced Recovery After Surgery protocol.

Preoperatively, patients are provided protein shakes supplemented with glutamine and omega-3 fatty acids for five days leading up to the operation and a carbohydrate load, consisting of 8 ounces of apple juice, is given the night before and two hours prior to induction of anesthesia. Standard venous thromboembolism and antibiotic prophylaxis is administered 60 min prior to incision. After standardized surgical time-out confirms patient identity, procedure, and key operative considerations, general anesthesia with endotracheal intubation is induced. Vascular access with large bore peripheral venous and hemodynamic monitoring with arterial catheters are secured. A Foley catheter is placed to monitor urine output and end organ perfusion. Recent imaging studies are displayed in the operating room and reviewed to confirm pertinent vascular and biliary anatomy.

Patients are typically placed with both arms abducted, in modified French, split-leg position to allow for an assistant to stand at the inferior edge of the surgical field. In this position, the assistant has the best access to exchange robotic instruments, deliver sutures and additional instruments, e.g., ultrasound probe, and assist, e.g. provide suction or operate the cavitron ultrasonic surgical aspirator (CUSA) through the assistant port (Fig. 23.1). Sterile prepping and draping are performed, exposing pertinent anatomic landmarks (including xiphoid process, costal margins, anterior superior iliac spine) within the operative field.

23.2.2.2 Port and Instrument Placement

Access to the peritoneal cavity is typically obtained through a 12-mm periumbilical port, with abdominal insufflation to 15 mmHg initially, which is decreased to 6–9 mmHg during the hepatic venous dissection. In cases of malignancy, diagnostic laparoscopy is performed to exclude peritoneal or other distant metastases that would preclude resection. Under direct visualization, four 8-mm robotic ports are placed in the left and right subcostal positions, modified based on laterality of resection and patient's body habitus. The patient is placed in reverse Trendelenburg (20°–30°) with slight right or left lateral decubitus position prior to docking of the robotic arms, with the patient cart to either the patient's right or left side. For select right

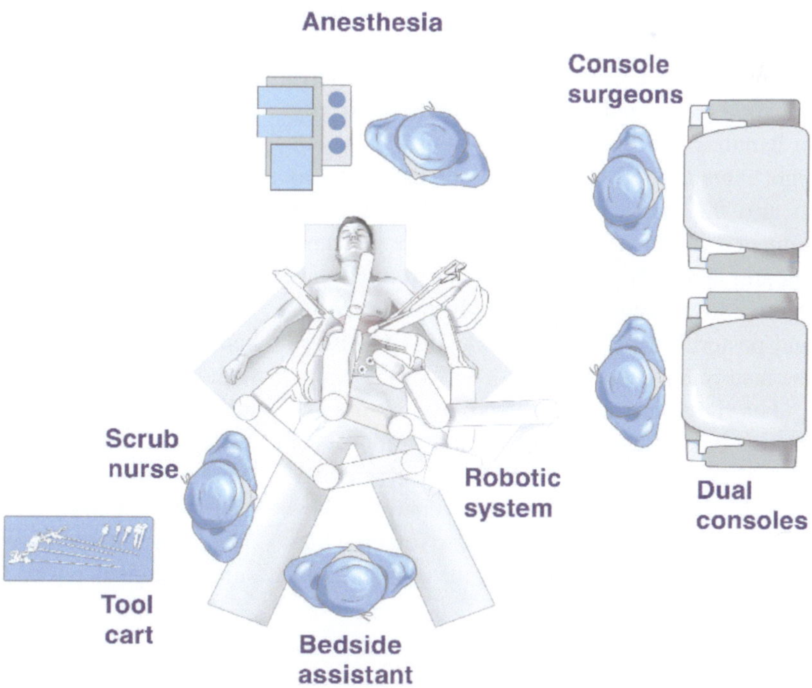

Fig. 23.1 OR setup with robot docked from patient's left side

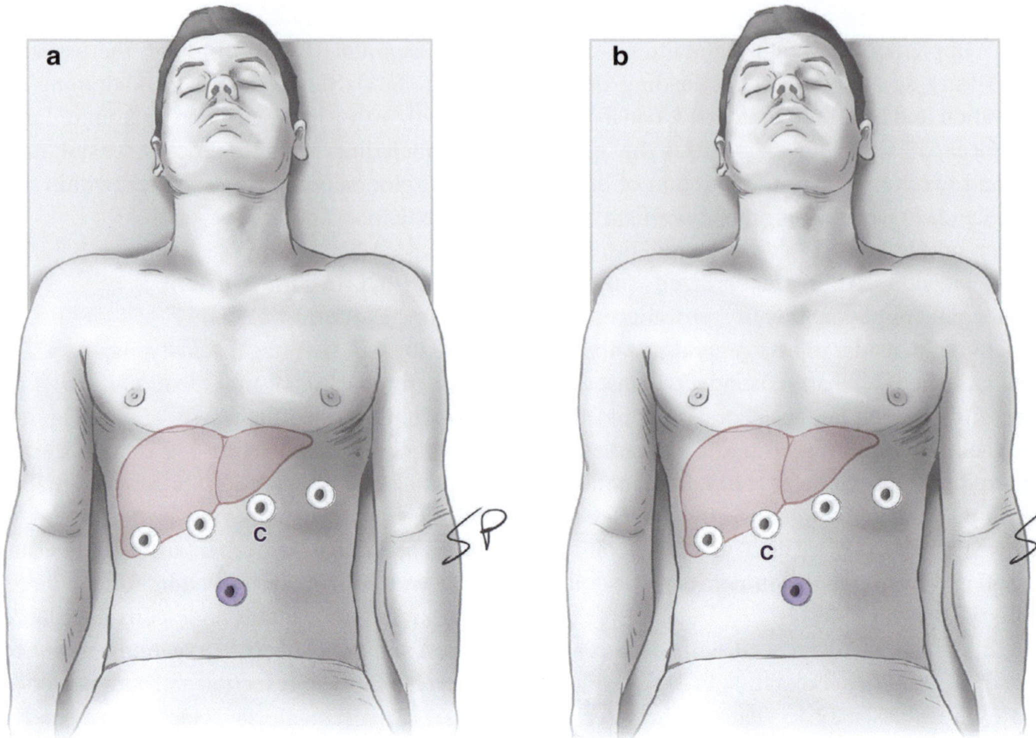

Fig. 23.2 Port placement for extended left (**a**) and extended right (**b**) hepatectomy 1, R anterior axillary; 2, R parasternal; 3, L parasternal; 4, L anterior axillary; C, camera; purple port, assistant

extended hepatectomy cases, a more pronounced left lateral decubitus position may, facilitate exposure and the retro-hepatic dissection.

For extended hepatectomy, the robotic ports are arranged for maximal exposure and dexterity in handling the target anatomy (Fig. 23.2). For right extended hepatectomy, the camera is placed in arm #2; for left extended hepatectomy, the camera is placed in arm #3. The port for the camera arm is ideally placed 10–15 cm from the target anatomy taking into account the most superior and posterior aspect of the dissection to ensure the instrument's reach.

For right extended hepatectomy, we use the Tip-up fenestrated grasper through arm #1 for retraction using vaginal packs for additional padding against the liver capsule, an energy device (monopolar curved shears, bipolar forceps, harmonic scalpel, vessel sealer, or SynchroSeal device) or the suture-cut needle driver in arm #3, and the Cadiere forceps in arm #4. For left extended hepatectomy, the Cadiere forceps are placed in arm #1, an energy device or the needle driver in arm #2, and the Tip-up fenestrated grasper in arm #4. The instruments in arms #1 and #4 are used in concert to provide an optimized plane for parenchymal transection, especially for the harmonic scalpel as this is a non-articulating instrument.

The periumbilical access port is utilized by the bedside assistant to help the operating surgeon with retraction, including the liver and surrounding organs such as intestine and omentum; manipulate additional instruments, including the suction irrigator, cavitron ultrasonic surgical aspirator (CUSA), and vascular stapler; and to introduce intraoperative equipment, including suture, vaginal packs for retraction or bleeding control, and the robotic drop-in ultrasound probe. When the Pringle maneuver is applied through an external approach, this is performed through or in parallel to this port.

Key considerations for port and instrument placement in robotic extended hepatectomy include ensuring optimal visualization of and instrument reach to the most technically complex and critical component of the case but also considering the most superior and posterior aspects of the parenchymal transection. Furthermore, if significant adhesions are encountered on initial laparoscopic entry, an additional port can be placed for adhesiolysis with laparoscopic shears with electrocautery prior to robotic port placement and docking. Finally, the patient's body habitus and adequacy of insufflation should be evaluated laparoscopically prior to robotic port placement to assess if optimal exposure of target anatomy can be achieved robotically [12]. For example, an extended left hepatectomy in a patient with low costal margins may be challenging.

23.2.2.3 Liver Mobilization

For extended left hepatectomy, the monopolar curved shears are used to divide the ligamentum teres hepatis, the falciform ligament, the left coronary and triangular ligaments, and the ligamentum venosum (Arantius ligament) to enable maximal mobilization of the left liver. For extended right hepatectomy, the falciform ligament, and the right coronary and triangular ligaments are divided to mobilize the right liver.

Retro-hepatic dissection with division of the retro-hepatic IVC branches (short hepatic veins) between ties and clips and division of the IVC ligament (Makuuchi ligament) is performed for additional mobilization. Circumferential dissection encircling the right hepatic vein to gain entry between the middle and right hepatic veins can be achieved to perform the hanging maneuver which can facilitate parenchymal transection of the superior segments. Mobilization can be facilitated with optimal use of the fourth arm for retraction and assistance from the bedside assistance using the suction device.

23.2.2.4 Ultrasound

Intraoperative ultrasound examination is performed following liver mobilization to confirm the number and distribution of lesions and define the surgical margins before further dissection is performed. Ultrasound is also utilized for intraoperative identification of vascular anatomy, in particular, the tumor location relative to vascular structures including the middle hepatic vein and its branches. Furthermore, intraoperative ultrasound may rarely offer additional diagnostic capabilities, such as the identification of small lesions not seen on preoperative imaging [3, 12, 19].

23.2.2.5 Inflow Control for Extended Left Hepatectomy

For extended left hepatectomy, the liver is mobilized as described above. Cholecystectomy is performed if the gallbladder is present. Monopolar curved shears are used to enter and divide the gastrohepatic ligament to expose the left hepatic artery. If present, an accessory left hepatic artery is divided between Hem-o-lok clips. The left hepatic artery and portal vein are identified. After ultrasound confirms inflow preservation to the contralateral liver with these structures occluded sequentially using the Cadiere forceps, the left hepatic artery and portal vein are divided between sutures and Hem-o-lok clips or using the Endo GIA or the robotic SureForm stapler with a vascular staple load, based on the caliber of the vessels. The left bile duct is similarly identified, encircled, and divided between ties and Hem-o-lok clips. Intraoperative cholangiogram can be performed if there is a need to clarify the biliary anatomy. If additional inflow control is necessary, the Pringle maneuver can be applied intermittently (15-min intervals). A Glissonian approach may be performed using a stapler if the portal pedicle structures are fused as in repeat hepatectomy cases. In this case, hepatotomies can be made at the base of segment 4B and lateral to the ligamentum venosum to encircle the left portal pedicle.

23.2.2.6 Inflow Control for Extended Right Hepatectomy

For extended right hepatectomy, the liver is first mobilized as described above. Cholecystectomy is performed if the gallbladder is present. The gastrohepatic ligament is entered and divided. A replaced or accessory left hepatic artery within the gastrohepatic ligament is identified and preserved if present. The right hepatic artery and right portal vein are identified and occluded sequentially using the Cadiere forceps. After ultrasound confirms inflow preservation to the left liver, the right hepatic artery and portal vein are divided between ties and Hem-o-lok clips. A large portal vein may be divided using an Endo GIA or robotic stapler. The right hepatic duct is identified and encircled. Before division, an additional ductal structure to the left must be confirmed to avoid inadvertently transecting or compromising the common hepatic duct. If additional inflow control is necessary, the Pringle maneuver can be intermittently applied (15-min intervals). Of note, while a Glissonian approach can be performed using a stapler to the anterior and posterior sectoral pedicles separately, it is generally avoided when applied to the main right portal pedicle given the higher risk of biliary injury due to its proximity.

23.2.2.7 Parenchymal Transection

Following division of hepatic inflow, intravenous indocyanine green (ICG) is administered. This enables a more accurate vascular demarcation and enhanced evaluation of ischemic versus preserved hepatic parenchyma prior to transection. Under ultrasound guidance, the surgical margin over the liver capsule is scored to mark the transection plane, with special attention to the tumor relationship to the portal bifurcation and middle hepatic vein.

Parenchymal transection is accomplished with a combination of energy devices such as the monopolar scissors, bipolar forceps, harmonic scalpel, vessel sealer, and SynchroSeal device with application of the clamp-crush technique using the latter four instruments (Table 23.1). CUSA can also be used and is performed by the bedside assistant. Smaller (1–2-mm diameter) vessels may be sealed and transected with the harmonic scalpel, vessel sealer, or SynchroSeal device. The majority of vessels encountered during transection are controlled and divided between clips (titanium or Hem-o-lok) or suture ligated. Suture ligation of vessels is facilitated by the dexterity of robotic intracorporeal suturing. Larger main hepatic veins may be divided using the Endo GIA or robotic SureForm stapler with a vascular staple load. As bleeding is encountered during parenchymal transection, vaginal packs may be used to effect venous tamponade.

As the parenchymal transection progresses superiorly and posteriorly in the region of the major branches of the hepatic veins, the insufflation pressure is lowered (6–9 mmHg) to minimize clinically significant CO_2 embolism, which is detected using end-tidal CO_2 measurement (abrupt decrease in $EtCO_2$). Additionally, the hanging maneuver can tamponade hepatic venous bleeding by applying pressure on the veins generated from the weight of the parenchyma on either side of the umbilical tape lifted anteriorly between the middle and right hepatic veins.

Intraoperative ultrasound is again utilized to evaluate the relationship between the transection plane and any identified lesions as well as the course of the middle hepatic vein. Hepatic veins are divided between Hem-o-lok clips and suture or alternatively with the Endo GIA or robotic SureForm stapler.

23.2.2.8 Specimen Retrieval

Surgical specimens are retrieved in an EndoCatch bag and delivered through the assistant port incision or through a separate Pfannenstiel incision. The assistant port incision is extended based on the size of the retrieved specimen, and retrieval can be hand assisted. Following this, the robot may be undocked. Following a period of relative desufflation to ensure complete hemostasis at lower intra-abdominal pressures, the ports are removed and the incisions are closed.

23.2.2.9 Conversion to Open

Indications for conversion to an open approach include both patient and operative factors. Patient factors include anatomic challenges, e.g., signifi-

Table 23.1 Tips and tricks for extended hepatectomy

Surgical Step	Challenges, Considerations, and Recommendations
Port/Instrument placement	Ports: Optimal placement must consider patient habitus, laterality of resection, and ability for robotic instruments to adequately expose and access anatomy during dissection. Attention should be paid not only to the most technically complex/critical component of the case but also to the most superior and posterior aspects of the parenchymal transection, when planning port placement. When indicated, an additional port can be placed for adhesiolysis with laparoscopic shears with electrocautery prior to robotic port placement and docking. In cases of challenging body habitus, adequacy of insufflation should be evaluated laparoscopically prior to robotic port placement to assess if a minimally invasive approach is feasible (e.g., extended left hepatectomy in a patient with low costal margins)
Liver mobilization	Liver retraction: Retraction should performed carefully during mobilization to avoid parenchymal fracturing especially for patients with steatosis. Vaginal pack in the tip-up grasper may be useful
Parenchymal transection	Instrument selection: Parenchymal transection may be performed with a combination of CUSA or energy devices including bipolar forceps, monopolar scissors, harmonic scalpel, or vessel sealer. Limitations of the harmonic scalpel and CUSA include lack of intracorporeal wrist articulation Ultrasound: Drop-in ultrasound probe introduced into the 12-mm assistant port can be manipulated using the Cadiere forceps, with real-time images projected on TilePro to the surgeon console and laparoscopic monitors ICG: Should be administered 48 h preoperatively for identification of tumors; may be administered intraoperatively for identification of surgical margins and remnant perfusion Division of portal pedicles/vessels: Smaller structures may be directly divided with the harmonic scalpel, vessel sealer, or between titanium clips; larger structures may be divided between Hem-o-lok clips, reinforced with suture or tie; main portal and hepatic veins may be divided with Endo GIA or robotic staplers if too large for clips Carbon dioxide embolism: Insufflation pressure should be lowered to 7–9 mmHg to decrease risk of CO_2 embolism, and venotomies should be promptly repaired to decrease risk
Extended left hepatectomy	Challenges: Hepatotomies to access the left portal pedicle when dissection is challenging
Extended right hepatectomy	Inflow access: When hilar dissection for exposure of right portal pedicles is difficult, parenchymal dissection can be incorporated for additional exposure. It is important to control and divide the anterior and posterior sectoral portal pedicles separately whenever possible to avoid injury to the left hepatic duct
Conversion to open	Hemorrhage: If intraoperative bleeding cannot be controlled robotically, the assistant should maintain direct pressure (with a vaginal pack) to the area of hemorrhage while conversion laparotomy is performed by the primary surgeon
Complications and management	Early detection and low threshold for suspecting carbon dioxide embolism via continuous monitoring of $EtCO_2$ and rapid communication of significant drop in EtCO2 are key to prevent hemodynamic collapse. When $EtCO_2$ drop is detected, insufflation should be stopped, and venotomies should be quickly identified and closed

cant obesity or narrow/low costal margin, prohibitive adhesions, or cardiopulmonary intolerance of insufflation. Operative factors include failure to progress, hemorrhage control, carbon dioxide embolism, or anatomy that cannot be safely visualized or delineated.

In non-emergent conversion, for example, in the case of failure to progress, existing port sites may be incorporated into the laparotomy incision, e.g., J-type, subcostal, or Chevron incision. In cases of emergent conversion due to hemorrhage, priority should be given to maintaining direct pressure to the area of uncontrolled bleeding. When possible, this control is transferred to the bedside assistant maintaining pressure with a vaginal pack through the assistant port. With the pressure maintained, laparotomy is performed concurrent to the undocking of the robot. It is imperative that operating room teams maintain excellent communication to optimally support the patient's hemodynamic and transfusion needs during this conversion.

23.2.2.10 Reconstruction

With the enhanced visualization and dexterity, perhaps the most significant advantage of robotic compared with laparoscopic minimally invasive extended hepatectomy is the ability to perform complex vascular and biliary reconstructions when needed. For instance, this may involve segmental portal venous resection and reconstruction when the tumor is focally abutting or encasing the vein or complex biliary reconstruction with Roux-en-Y intrahepatic biliary reconstruction for cholangiocarcinoma. It may also involve patch reconstruction of the IVC or hepatic vein to prevent narrowing with bovine pericardium.

23.3 Literature Overview

The use of robotic platforms for minimally invasive hepatectomy has expanded significantly, with an increasing number of major hepatectomies performed using this approach. Based on the 2018 International Consensus Statement and the 2023 International Experts Consensus Guidelines on robotic hepatectomy, robotic technology is most effective in enhancing our ability to perform complex hepatectomies, e.g., parenchymal sparing near major vasculature and major hepatectomy [7, 20].

The robotic platform offers several advantages over conventional laparoscopic techniques, including enhanced visualization and greater degrees of freedom in operator range of motion. These features enable more precise dissection and improved management of intraoperative hemorrhage through superior suturing capabilities. Furthermore, this enhanced dexterity facilitates minimally invasive vascular or biliary reconstructions, further broadening its application [11].

Despite these advantages, some limitations remain. For example, a skilled bedside assistant is required to manipulate and exchange instruments not yet adapted for the robotic platform, such as the laparoscopic cavitron ultrasonic surgical aspirator (CUSA) for suction, irrigation, and manipulation of hepatic parenchyma. However, the growing array of robot compatible instruments being developed is expected to further facilitate complex resections and in particular, hemostatic parenchymal transections [21].

While data on outcomes of robotic extended hepatectomies is relatively limited, international multicenter analyses comparing robotic and laparoscopic hepatectomies, including extended resections, have shown promising results. In left and extended left hepatectomies, the robotic approach was associated with reduced blood loss (100 ml vs 200 ml), shorter hospital stays (6 days vs 7 days), and lower conversion rates to open surgery (2.4% vs 7.9%) [22]. Similarly, in right and extended right hepatectomies, robotic surgery demonstrated lower conversion rates (8.6% vs 17.1%) as well as reduced overall (30.9% vs 38.4%) and major (10.9% vs 16.8%) morbidity [23]. A 2022 meta-analysis by Aboudou et al. found no significant differences in blood loss, transfusion requirements, complications, conversion rates, length of hospital stay, or reoperation rates between robotic and laparoscopic hepatectomy, including major resections [24].

In terms of oncologic outcomes, robotic hepatectomy shows no significant differences have been found in margin status, recurrence rates, or overall survival when compared to laparoscopic or open hepatectomy [20]. While robotic hepatectomies are associated with longer operative times compared to laparoscopic hepatectomies, conversion rates to open surgery are consistently lower for the robotic compared to laparoscopic approach (left: 2.4% vs 7.9%; right: 8.6% vs 17.1%) [22, 23]. Notably, conversion rates for robotic right and extended right hepatectomies remain higher than those for left-sided resections, with rates approaching 10% [23]. This underscores the importance of the surgeon being proficient in both robotic and open techniques to manage these complex cases.

Finally, there is evidence that the advanced capabilities of the robotic platform enable more difficult resections to be completed minimally invasively, offering potential benefits in patient recovery and outcomes [5, 25].

23.4 Conclusions

In conclusion, extended hepatectomy can be performed safely and effectively using a robotic approach. Key considerations for these complex extended resections include strategic port placement tailored to the target anatomy, careful selection of robotic or laparoscopic instruments for parenchymal transection, strategic use of an assistant port for additional instrument manipulation, and liberal use of intraoperative ultrasound to identify lesions and assess their proximity to critical biliary and vascular structures during transection.

The robotic approach offers additional advantages, such as enhanced suturing capabilities for managing hemorrhage, performing complex reconstructions and tissue repairs. It also appears that the robotic platform enables more difficult resections to be performed in a minimally invasive manner, which is associated with earlier initiation of adjuvant systemic therapies following resection of malignancy. Finally, as with any minimally invasive surgical approach, the operating team must be comfortable with either planned or emergent conversion to an open approach if necessary. Therefore, it is essential for the operating team to have substantial expertise in open resection and reconstruction techniques before undertaking extensive robotic resections.

References

1. Giulianotti PC, Coratti A, Angelini M, et al. Robotics in general surgery: personal experience in a large community hospital. Arch Surg. 2003;138(7):777–84. https://doi.org/10.1001/archsurg.138.7.777
2. Sucandy I, Shapera E, Syblis CC, Crespo K, Przetocki VA, Ross SB, et al. Propensity score matched comparison of robotic and open major hepatectomy for malignant liver tumors. Surg Endosc. 2022;36:6724.
3. Guerra F, Amore Bonapasta S, Annecchiarico M, Bongiolatti S, Coratti A. Robot-integrated intraoperative ultrasound: initial experience with hepatic malignancies. Minim Invasive Ther Allied Technol. 2015;24(6):345–9.
4. Zhang L, Yuan Q, Xu Y, Wang W. Comparative clinical outcomes of robot-assisted liver resection versus laparoscopic liver resection: a meta-analysis. PLoS One. 2020;15(10):e0240593.
5. Chong CCN, Lok HT, Fung AKY, Fong AKW, Cheung YS, Wong J, et al. Robotic versus laparoscopic hepatectomy: application of the difficulty scoring system. Surg Endosc. 2020;34(5):2000–6.
6. Labadie KP, Droullard DJ, Lois AW, Daniel SK, McNevin KE, Gonzalez JV, et al. IWATE criteria are associated with perioperative outcomes in robotic hepatectomy: a retrospective review of 225 resections. Surg Endosc. 2022;36(2):889–95.
7. Liu R, Abu Hilal M, Wakabayashi G, Han HS, Palanivelu C, Boggi U, et al. International experts consensus guidelines on robotic liver resection in 2023. World J Gastroenterol. 2023;29(32):4815–30.
8. Strasberg SM, Belghiti J, Clavien PA, Gadzijev E, Garden JO, Lau WY, et al. The Brisbane 2000 terminology of liver anatomy and resections. HPB. 2000;2(3):333–9.
9. Reddy SK, Barbas AS, Turley RS, Steel JL, Tsung A, Marsh JW, et al. A standard definition of major hepatectomy: resection of four or more liver segments. HPB (Oxford). 2011;13(7):494–502.
10. Dimitroulis D, Tsaparas P, Valsami S, Mantas D, Spartalis E, Markakis C, et al. Indications, limitations and maneuvers to enable extended hepatectomy: current trends. World J Gastroenterol. 2014;20(24):7887–93.
11. Giulianotti PC, Sbrana F, Bianco FM, Addeo P. Robot-assisted laparoscopic extended right hepatectomy with biliary reconstruction. J Laparoendosc Adv Surg Tech A. 2010;20(2):159–63.
12. Newton AD, Newhook TE, Ikoma N, Tzeng CD, Chun YS, Vauthey JN, et al. Robotic extended right hepatectomy for colorectal liver metastasis. Ann Surg Oncol. 2022;29(13):8455.
13. Faria LL, Darce GF, Bordini AL, Herman P, Jeismann VB, de Oliveira IS, et al. Liver surgery: important considerations for pre- and postoperative imaging. Radiographics. 2022;42(3):722–40.
14. Winslow E. Invited commentary: issues at the Interface of hepatic imaging and hepatic surgery. Radiographics. 2022;42(3):E92–e3.
15. Labadie KP, Dickerson LK, Park JO. Robotic hepatectomy. In: Patti MG, Zureikat AH, Fichera A, Schlottmann F, editors. Techniques in minimally invasive surgery. Cham: Springer International Publishing; 2021. p. 149–63.

16. Schnitzbauer AA, Lang SA, Goessmann H, Nadalin S, Baumgart J, Farkas SA, et al. Right portal vein ligation combined with in situ splitting induces rapid left lateral liver lobe hypertrophy enabling 2-staged extended right hepatic resection in small-for-size settings. Ann Surg. 2012;255(3):405–14.
17. Sandström P, Røsok BI, Sparrelid E, Larsen PN, Larsson AL, Lindell G, et al. ALPPS improves Resectability compared with conventional two-stage hepatectomy in patients with advanced colorectal liver metastasis: results from a Scandinavian multicenter randomized controlled trial (LIGRO trial). Ann Surg. 2018;267(5):833–40.
18. Abdalla EK, Adam R, Bilchik AJ, Jaeck D, Vauthey JN, Mahvi D. Improving resectability of hepatic colorectal metastases: expert consensus statement. Ann Surg Oncol. 2006;13(10):1271–80.
19. Guerra F, Guadagni S, Pesi B, Furbetta N, Di Franco G, Palmeri M, et al. Outcomes of robotic liver resections for colorectal liver metastases. A multi-institutional analysis of minimally invasive ultrasound-guided robotic surgery. Surg Oncol. 2019;28:14–8.
20. Liu R, Wakabayashi G, Kim HJ, Choi GH, Yiengpruksawan A, Fong Y, et al. International consensus statement on robotic hepatectomy surgery in 2018. World J Gastroenterol. 2019;25(12):1432–44.
21. Hawksworth J, Radkani P, Nguyen B, Belyayev L, Llore N, Holzner M, et al. Improving safety of robotic major hepatectomy with extrahepatic inflow control and laparoscopic CUSA parenchymal transection: technical description and initial experience. Surg Endosc. 2022;36(5):3270–6.
22. Sucandy I, Rayman S, Lai EC, Tang CN, Chong Y, Efanov M, et al. Robotic versus laparoscopic left and extended left hepatectomy: an international multicenter study propensity score-matched analysis. Ann Surg Oncol. 2022;29(13):8398–406.
23. Chong CC, Fuks D, Lee KF, Zhao JJ, Choi GH, Sucandy I, et al. Propensity score-matched analysis comparing robotic and laparoscopic right and extended right hepatectomy. JAMA Surg. 2022;157(5):436–44.
24. Aboudou T, Li M, Zhang Z, Wang Z, Li Y, Feng L, et al. Laparoscopic versus robotic hepatectomy: a systematic review and meta-analysis. J Clin Med. 2022;11(19):5831.
25. Tsung A, Geller DA, Sukato DC, Sabbaghian S, Tohme S, Steel J, et al. Robotic versus laparoscopic hepatectomy: a matched comparison. Ann Surg. 2014;259(3):549–55.

24 Robotic-Assisted Laparoscopic Resection of Klatskin Tumors

Trenton Lippert and Iswanto Sucandy

24.1 Introduction

The utilization of the robotic platform in hepatobiliary surgery has marked a significant advancement in the field, maintaining the core principles, aims, and indications of the operations [1]. Minimally invasive approaches necessitated adaptation of the technical approach while achieving the same goals. This shift toward minimally invasive approaches to complex hepatobiliary operations offers substantial benefit by reducing the physiologic stress patients endure during curative-intent operations [2–4]. Notably, in liver surgery, the use of the minimally invasive approaches eliminates the need for rib retraction and minimizes large surgical incisions, leading to significant decreases in wound-related complications, postoperative respiratory issues, and enhanced perioperative pain management [5]. These improvements underscore the value of minimally invasive techniques in enhancing patient recovery while maintaining the curative ability of the operations.

This chapter addresses the intricate surgical management of perihilar cholangiocarcinoma, a formidable malignancy requiring a comprehensive treatment plan and often technically heroic operative steps [6]. Over the past three decades, advancements in diagnostic modalities and surgical techniques have led to a notable improvement in the 5-year survival rate, which has increased two to threefold [7–10]. Despite these advances, the 5-year survival rate post-resection for extrahepatic cholangiocarcinoma remains below 50%, underscoring the tumor's propensity for aggressive invasion of critical structures in the hepatic hilum [10]. A key challenge in resecting these tumors is their frequent involvement of the portal vein and hepatic artery, necessitating intricate dissection and often vascular resection with reconstruction. This chapter provides a comprehensive overview of each step in the surgical process. Furthermore, the implementation of an Enhanced Recovery After Surgery (ERAS) protocol is discussed as a strategy to decrease postoperative hospital stays while maintaining low unplanned readmission rates.

In the context of perihilar cholangiocarcinoma management, liver transplantation, though an established treatment for early-stage and limited cases, is not universally recognized as a definitive curative strategy, unlike for hepatocellular carcinoma [11]. The cornerstone of curative treatment remains achieving R-0 resection margins through meticulous surgical resection, followed by complex biliary reconstruction [12]. The decision-making process for adjuvant and neoadjuvant

T. Lippert
University of South Florida Morsani College of Medicine, Tampa, FL, USA

I. Sucandy (✉)
Digestive Health Institute, AdventHealth Tampa, Tampa, FL, USA
e-mail: Iswanto.Sucandy.MD@AdventHealth.com

M. D'Hondt, I. Sucandy (eds.), *Textbook of Robotic Liver Surgery*,
https://doi.org/10.1007/978-3-031-76536-0_24

therapies is multifactorial, influenced by patient-specific factors, preoperative staging, response to preoperative treatments, intraoperative findings, and final postoperative pathological staging. While the selection of the optimal preoperative biliary drainage technique remains a subject of debate, the choice is highly institution dependent, hinging on the availability and expertise of endoscopists and interventional radiologists. Each center must weigh its resources and capabilities in managing the potential complications associated with biliary decompression in this patient population. This chapter delves into these nuanced considerations, offering insights into the complex decision-making process and technical intricacies involved in the management of perihilar cholangiocarcinoma.

In the surgical management of perihilar cholangiocarcinoma, the traditional approach has been an open resection. However, the advent of robotic-assisted surgery, when executed by a skilled robotic hepatobiliary surgeon, has emerged as a safe and viable alternative for select patients [7]. The robotic platform offers enhanced dexterity compared to conventional laparoscopy, and studies have shown a superior yield in portal lymphadenectomy when compared to both open and laparoscopic methods [13]. Our experience suggests that perioperative outcomes in patients undergoing robotic-assisted resection for perihilar cholangiocarcinoma are at least similar and may be superior to those achieved through traditional open or laparoscopic approaches while maintaining the oncologic integrity of the operation [7].

24.2 Indications

The initial step in the management of perihilar cholangiocarcinoma involves its accurate diagnosis, which is challenging as a definitive (histological/cytological) preoperative cancer diagnosis is achieved in only approximately two-thirds of patients. The suspicion of extrahepatic and perihilar cholangiocarcinoma is primarily based on clinical symptoms and high-resolution imaging findings. For the robotic approach, our protocol involves stringent patient selection, recommending it primarily for Klatskin tumors classified as type 1–3 based on the Bismuth-Corlette classification. Currently, there is insufficient clinical evidence to support the minimally invasive approach for Klatskin type 4 tumors.

Patients typically present with clinical manifestations of jaundice and dark urine. Efforts to obtain preoperative tissue diagnosis usually involve biliary endoscopy with endoluminal biopsies or brushings, although the latter has a limited diagnostic yield of only around 30% due to the paucity of cells characteristic of perihilar cholangiocarcinoma [14–16]. Patients presenting with locally advanced tumors or suspected metastatic disease are subjected to neoadjuvant systemic chemotherapy or chemoradiation, followed by a comprehensive restaging. The criteria for resectability are variable but generally exclude patients with retropancreatic, para-celiac, and aortocaval lymph node involvement and extensive local organ invasion beyond the liver. While contralateral portal vein and hepatic artery involvement traditionally preclude surgical resection, specialized centers with experienced hepatobiliary surgeons can consider an en bloc resection with vascular reconstruction of the involved segments [7, 17].

Delays in surgery are sometimes necessary to address other uncontrolled medical comorbid conditions such as diabetes and cardiopulmonary issues or to achieve weight loss. Additional preoperative interventions, including biliary decompression, cardiac clearance, pulmonary function testing, and coronary angiography, may be required to optimize the patient for the perioperative period and enhance postoperative recovery.

24.3 Preoperative Management

In the preoperative evaluation for perihilar cholangiocarcinoma resection, obtaining high-definition imaging is paramount. This involves 1-mm cut computed tomography (CT) scans or magnetic resonance imaging (MRI), with CT scans often preferred for their cost-effectiveness, detailed vascular imaging, and widespread familiarity among surgeons. MRI, however, holds distinct value in identifying small intrahepatic lesions indicative of potential occult hepatic metastases. Magnetic resonance cholangiopancreatography (MRCP) plays a critical role in delineating the perihilar and intrahepatic biliary architecture, crucial for identifying any anatomical variations that could complicate later biliary reconstruction. For instance, preoperative identification of atypical biliary anatomy, such as an aberrant insertion of the right posterior hepatic duct, is essential in planning the resections in a Klatskin type 3B tumor, where a complete anatomical left hepatectomy with en bloc caudatectomy and extrahepatic biliary resection are required. Failure to recognize these variations can lead to significant postoperative complications, including persistent bile leaks, sepsis, or the need for additional surgical interventions. Additionally, a CT scan of the chest is integral to rule out the presence of extrahepatic metastases, which would be a contraindication for surgical resection. While positron emission tomography (PET) scans can be beneficial in further excluding metastatic disease, their use is generally reserved for cases where a tissue diagnosis of malignancy has been confirmed due to the high cost and strict scrutiny by insurance carriers in the United States regarding preoperative PET scan utilization [18].

Three-dimensional liver volumetric reconstructions, performed using sophisticated imaging software, are essential when major hepatectomy is anticipated, as in type 3 and 4 Klatskin tumor resections. This is particularly important in the context of future liver remnant (FLR) volume assessment. In patients with liver remnant volumes less than 30%, especially those with underlying hepatic parenchymal dysfunction, there is an increased risk of postoperative hepatic insufficiency and death [19]. This risk is even more pronounced in patients with a history of liver cirrhosis, hepatic steatosis, steatohepatitis, chemotherapy-induced liver damage, or cholestatic liver disease, where the FLR volume threshold should ideally exceed 45% [19, 20]. To mitigate the risk of mortality from post-hepatectomy liver failure, it is critical to identify patients who may benefit from preoperative interventions to induce hypertrophy of the FLR. These interventions include portal vein embolization, associating liver partition and portal vein ligation for staged hepatectomy (ALPPS), or hepatic venous deprivation (double embolization) techniques [21].

24.4 Patient Optimization

Special attention should be given to patients presenting with hyperbilirubinemia exceeding 3 mg/dL, subclinical symptoms of cholangitis, or those requiring concomitant major hepatectomy where the expected future liver remnant (FLR) volume is 35% or less. These patients are typically referred for preoperative biliary drainage, as hyperbilirubinemia and cholestatic liver disease significantly elevate the risk of post-hepatectomy liver failure and death [22]. At our institution, we employ endoscopic retrograde cholangiopancreatography (ERCP) with endobiliary plastic stent placement as the primary modality for biliary decompression, an approach geared toward the availability of advanced endoscopists in our center. This approach also facilitates the collection of brushings or endoluminal biopsies for tissue diagnosis when necessary. In cases where ERCP is unsuccessful or if the patient has persistent hyperbilirubinemia post-ERCP, percutaneous transhepatic cholangiography (PTC) can then be considered as an alternative.

Following successful biliary drainage, we prefer to delay undertaking the operation until the total serum bilirubin level decreases to 3 mg/dL or lower. However, it is noteworthy that a bilirubin range of 3–10 mg/dL, while not ideal, does not constitute an absolute contraindication to proceed with surgery.

24.5 Patient Preparation

Obtaining thorough informed consent is a critical step in preparing patients for resection of perihilar cholangiocarcinoma. This process involves not only the patients but also, ideally, their family members, ensuring they fully understand the nature and implications of the surgery. It is essential to communicate the details of the operation and its inherent risks in an accessible manner, including the use of diagrams to illustrate the surgical steps and a comprehensive discussion of potential complications. A key aspect of this conversation is the acknowledgment of the risk of mortality and the need for postoperative rehabilitation services, presented in a clear manner yet sensitive to the patient's emotional well-being.

Patients must be made aware that while the minimally invasive robotic approach offers benefits in terms of postoperative recovery, it does not diminish the complexity or seriousness of this operation. Robotic resection of perihilar cholangiocarcinoma remains a major procedure, carrying similar risks of postoperative complications as the open approach. The expected duration of the operation ranges from 6 to 9 h, with an average hospital stay of about 5–6 days [7, 17]. Postsurgical home recovery typically spans around 4 weeks, and some patients, especially older or less physically robust individuals, may require home healthcare or a stay in an acute rehabilitation center for 2–3 weeks. In our experience, the recovery from Klatskin tumor resections is generally more extensive and much longer than that of Whipple procedures.

Before departing the office after their surgical consultation, patients are briefed to our comprehensive Enhanced Recovery After Surgery (ERAS) protocol. This protocol is designed to mitigate complications associated with preoperative nutritional deficits and physical deconditioning. Key elements include the intake of immune-enhancing shakes for 5 days before surgery, a prescribed home exercise regimen, and the cessation of smoking and alcohol consumption. Postoperatively, the focus shifts to early ambulation, effective pain management through multimodal analgesia with minimal opioid use, prompt nutritional intake, and intensive physical therapy.

24.6 Operating Room Setup and Patient Positioning

The patient is rolled to the operating room, positioned supine on the operating table, with arms abducted at 90° on arm boards. This is followed by the induction of general anesthesia. Our team does not employ the French position as we have not identified a clear advantage to its use. Crucial for intraoperative management, a central arterial line and central venous catheter are placed by our anesthesia team for optimal hemodynamic monitoring. A nasogastric tube (NGT) is inserted for gastric decompression, and a Foley catheter is placed for urinary bladder decompression. The abdomen is then prepared with an alcohol-based solution, and a betadine-based plastic drape is applied after a 3-min drying period.

The operation commences with the patient in a 15° reverse Trendelenburg and 5° right side up position. The da Vinci Xi® robotic surgical system (Intuitive Surgical, Sunnyvale, CA, USA), equipped with a dual console, is positioned approximately 10 ft. from the operating table. This arrangement ensures ample working space for the bedside surgeon, scrub technician, and the anesthesia team, avoiding any encumbrance by the consoles. The robotic system is docked over the patient's right shoulder. During the operation, the bedside surgeon stands to the right of the patient, while the scrub technician is positioned on the left, facilitating efficient and unobstructed surgical workflow (Fig. 24.1).

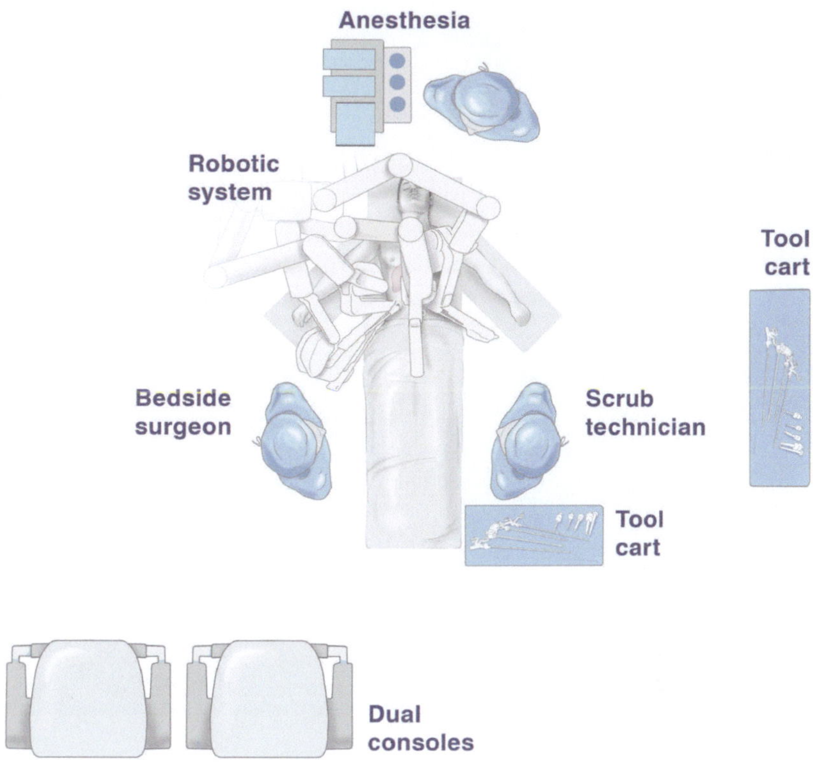

Fig. 24.1 Operating room diagram

24.7 Key Operative Steps

24.7.1 Port Placement and Robotic Arm Placement

Local anesthesia is administered at each incision site. The initial incision is made at the periumbilical area for the insertion of an 8-mm blunt robotic port, ensuring the incision is just wide enough to accommodate the port and prevent air leakage during pneumoperitoneum, which is established and maintained at 15 mmHg. The robotic 30-degree camera is inserted through the umbilical port for an initial diagnostic laparoscopy to rule out obvious peritoneal metastases. Any suspicious nodules encountered are biopsied and sent for immediate frozen section examinations by our pathology colleagues.

After ruling out peritoneal metastases, we proceed to place additional ports: two 8-mm ports along the right anterior axillary and left anterior axillary lines and a 12-mm port on the left midclavicular line at the umbilical level for the robotic stapler. A 3–5-cm incision is made for the insertion of an Access Gelport® (Applied Medical, Rancho Santa Margarita, California, USA) slightly below and between the right midclavicular line and the umbilicus (Fig. 24.2). This Gelport® allows for manual assistance during the surgery, including suction, exposure, and eventual specimen extraction. An extra 5-mm incision is made in the right upper quadrant for a 5-mm AirSeal® (ConMed, Largo, Florida, USA) port, facilitating smoke evacuation, rapid insufflation, and, if needed, the insertion of a laparoscopic liver retractor (Fig. 24.2). The da Vinci Xi® robotic system is synchronized with the operating table and docked (Fig. 24.1).

Instrument organization for the robotic system is crucial to minimize instrument clashing that

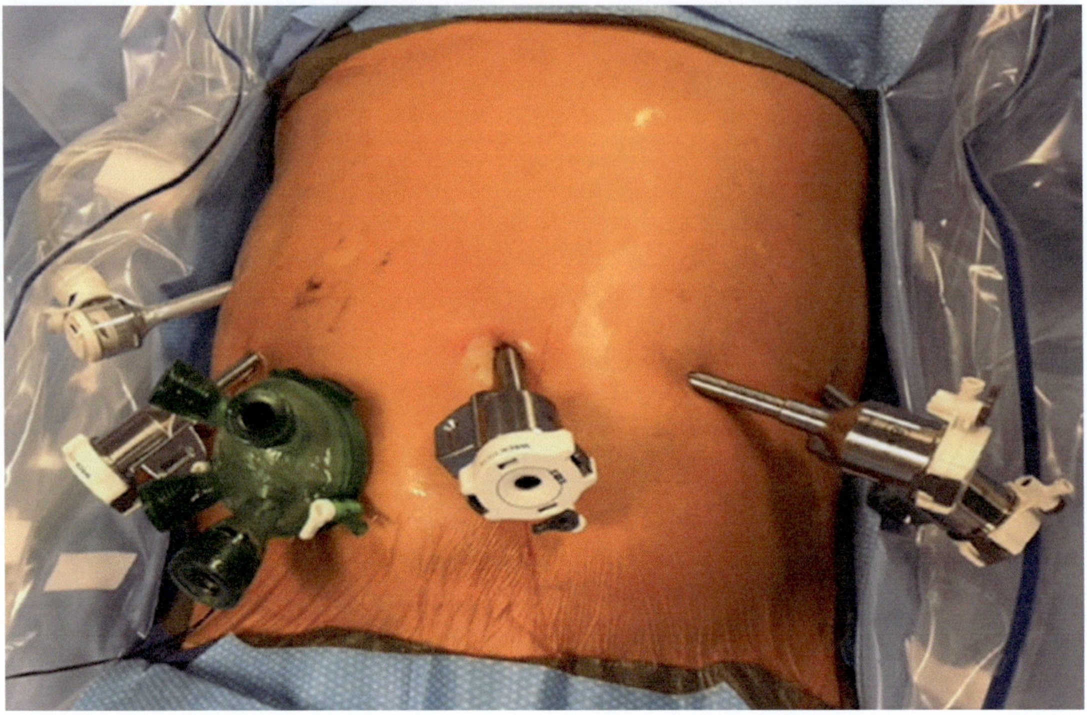

Fig. 24.2 Robotic port placement

can increase operative time and frustration. Arm one holds a robotic fenestrated bipolar forceps, arm two controls the camera, arm three is equipped with a robotic hook cautery or monopolar curved scissors, and arm four uses an atraumatic bowel grasper. At the Gelport®, a laparoscopic suction device and an atraumatic bowel grasping tool are introduced by the bedside surgeon.

The operation begins with the division of the falciform ligament to expose the liver's anterior surface and check for subcapsular occult hepatic metastases. Dissection continues cephalad to reveal the hepatocaval confluence, and the base of the hepatic veins is exposed for planned hemihepatectomies in Klatskin types 3A and 3B. The liver retractor, inserted via the AirSeal® port, lifts the liver cephalad to allow adequate exposure of the porta hepatis. In patients with smaller livers, this can be achieved by using the atraumatic bowel grasper in robotic arm four, gently retracting segment 4B cephalad. This setup offers the flexibility to adjust retraction as needed throughout the operation for optimal exposure of the hilar plate.

24.7.2 Systematic Radical Portal Lymphadenectomy

We perform a thorough dissection within the porta hepatis and the adjacent areas for meticulous removal of periportal lymph nodes while carefully skeletonizing the hepatic artery, portal vein, and common bile duct (CBD). The primary tools for these steps of the operation are the robotic fenestrated bipolar forceps and hook cautery or scissor cautery. The bedside surgeon assists in this step by providing traction to the transverse colon, duodenum, or major omentum for better exposure using a laparoscopic suction device.

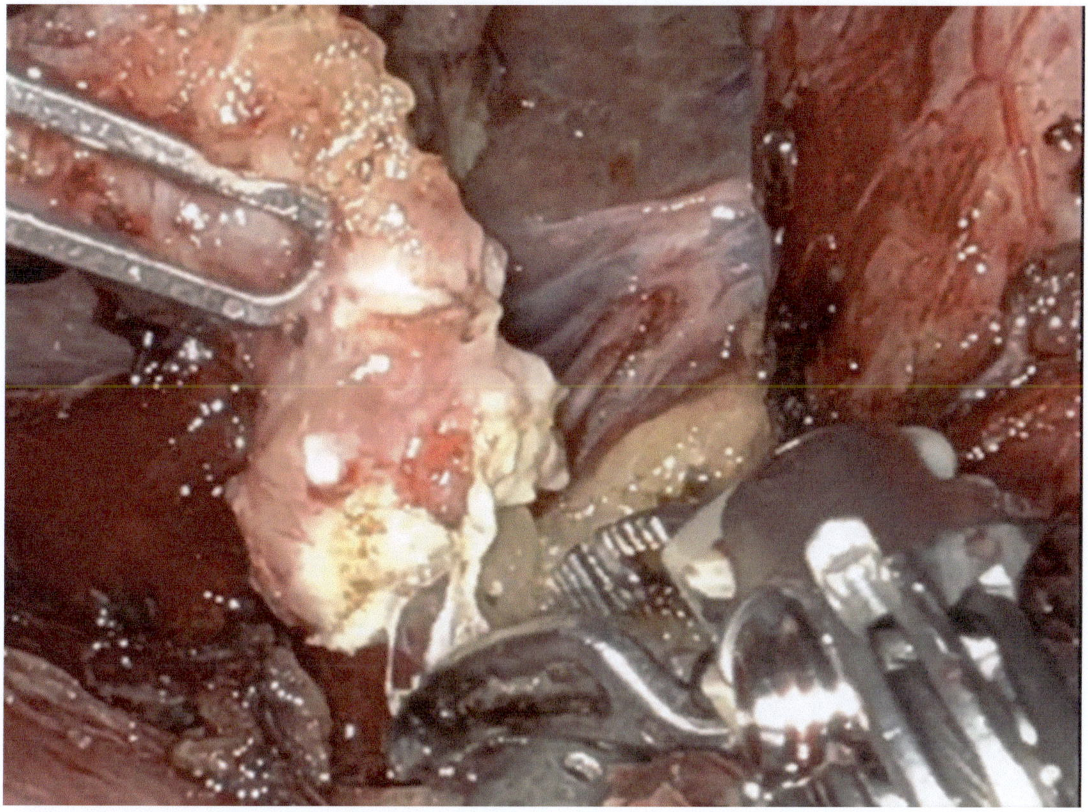

Fig. 24.3 Excision of common hepatic artery lymph node

Our lymphadenectomy steps begin with the division of the gastrohepatic ligament to gain access to the caudate lobe of the liver. We then expose the origin of the common hepatic artery from the celiac trunk and carefully remove all lymph nodes along its course, ensuring preservation of the portal vein (Fig. 24.3). The superior border of the head, neck, and proximal body of the pancreas is subsequently exposed.

As the dissection progresses cephalad toward the hilar plate, we remove all lymphatic-bearing tissues anterior to the hepatoduodenal ligaments, resulting in clear exposure of the bile duct and the right and left hepatic arteries. To access and harvest the retroduodenal lymph nodes, mobilization of the duodenum is helpful. This is achieved through Kocher maneuver, exposing the inferior vena cava and the foramen of Winslow. The duodenum is mobilized to facilitate exposure of the distal CBD and to enable the removal of retroduodenal, retropancreatic, and retroportal lymph nodes while preserving the ligament of Treitz.

24.7.3 Transection of Distal Common Bile Duct

Following the retroduodenal lymphadenectomy, we proceed to the crucial step of transecting the distal common bile duct (CBD). This transection is performed as low as feasibly possible at/into the level of the pancreatic head, ensuring careful avoidance of the pancreatic duct junction to prevent inadvertent injury (Fig. 24.4). Upon transecting the distal CBD, a distal CBD specimen is immediately sent for frozen section analysis to assess the distal margins for potential tumor involvement.

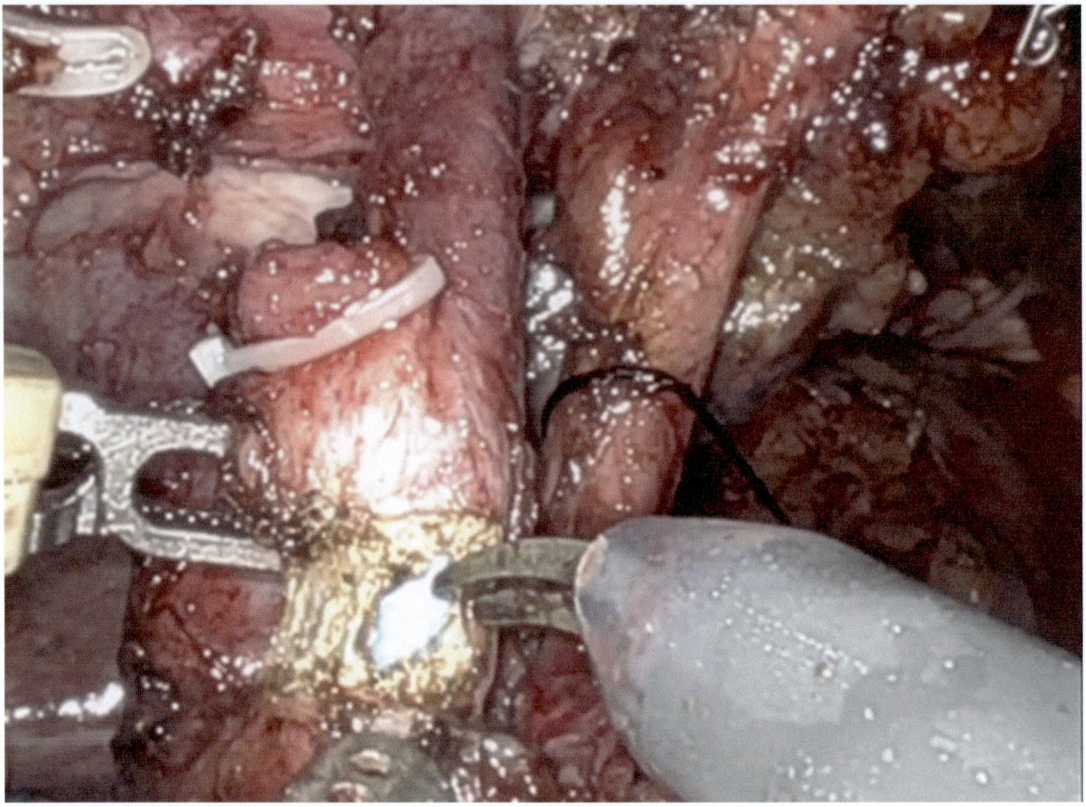

Fig. 24.4 Distal common bile duct transection

While waiting for the pathology results, we utilize intraoperative cholangioscopy to endoluminally visualize the proximal extent of the tumor in the common hepatic duct and the biliary confluence. This assessment aids in determining the precise location for the proximal bile duct transection and provides intraoperative hints to the specific type of Klatskin tumor we are addressing. If a cholecystectomy has not yet been performed, we remove the gallbladder en bloc with the extrahepatic bile duct. In cases where the tumor is identified as a Klatskin type 3, we now plan the ipsilateral hemi-hepatectomy including caudate lobectomy based on the tumor's extension into the secondary and tertiary biliary radicles.

Once the pathology report confirms that the distal CBD margins are clear of tumor, we proceed to close the lumen of the distal CBD at the pancreatic head entrance. This closure is performed using a running 4-0 polypropylene suture to prevent any retrograde pancreatic leakage through the distal CBD lumen.

24.7.4 Transection of Proximal Hepatic Duct

After completing the transection and closure of the distal CBD, we place a temporary suture on the proximal segment of the common hepatic duct (CHD) to both halt the ongoing bile leakage and to provide a handle for cephalad traction. This maneuver facilitates better exposure of the main portal vein, allowing for the completion of retroportal lymphadenectomy and ensuring the foramen of Winslow is cleared of any remaining lymphatic tissues.

The dissection then focuses on the dorsal aspect of the CHD, particularly around the right hepatic artery, which typically crosses posteriorly to the CHD as it enters the Glissonean pedicle of the right hepatic lobe. Given the complexity of this dissection, the surgeons must be prepared to perform arteriorrhaphy in cases of right hepatic artery involvement. It is helpful to place a curved vascular bulldog clamp in the proximal right hepatic artery for inflow control prior to an aggressive arterial dissection of the common hepatic duct. We prefer to use 5-0 or 6-0 polypropylene sutures when this is required.

Once the CHD is fully dissected and freed from surrounding structures, we proceed to prepare the bile duct bifurcation for proximal hepatic duct transection. At the hilar plate, we expose Laennec's capsule at the inferior aspect of segment 4B by carefully lowering the hilar plate. This step allows for a more radical resection proximally toward both the intrahepatic left and right hepatic ducts, aiming to achieve negative proximal margins.

The level of proximal bile duct transection is determined based on the Bismuth-Corlette classification and findings from intraoperative cholangioscopy. Our preferred approach is to dissect up to the secondary biliary radicles when possible, striving to avoid accepting R-1 resection margins. Specimens from both the right and left hepatic duct margins are then sent for separate frozen section analysis to confirm the absence of tumor cells.

24.7.5 Ipsilateral Hepatectomy

This section applies to Bismuth classifications three and above. We perform a concomitant hepatectomy with en bloc caudate lobectomy. Our approach to robotic major hepatectomy has been outlined in previous publications, so we will focus on the aspects unique to the robotic resection of Klatskin type 3 and 4 tumors [1, 3, 23]. Given the prior dissection and clearance of the porta hepatis from lymphatic nodal tissue, we can easily isolate and ligate the inflow vessels to either the right or left hemiliver (Figs. 24.5 and 24.6). In a case of right or left portal vein tumor invasion, a vascular bulldog clamp can be placed proximally (at the main portal vein) and distally

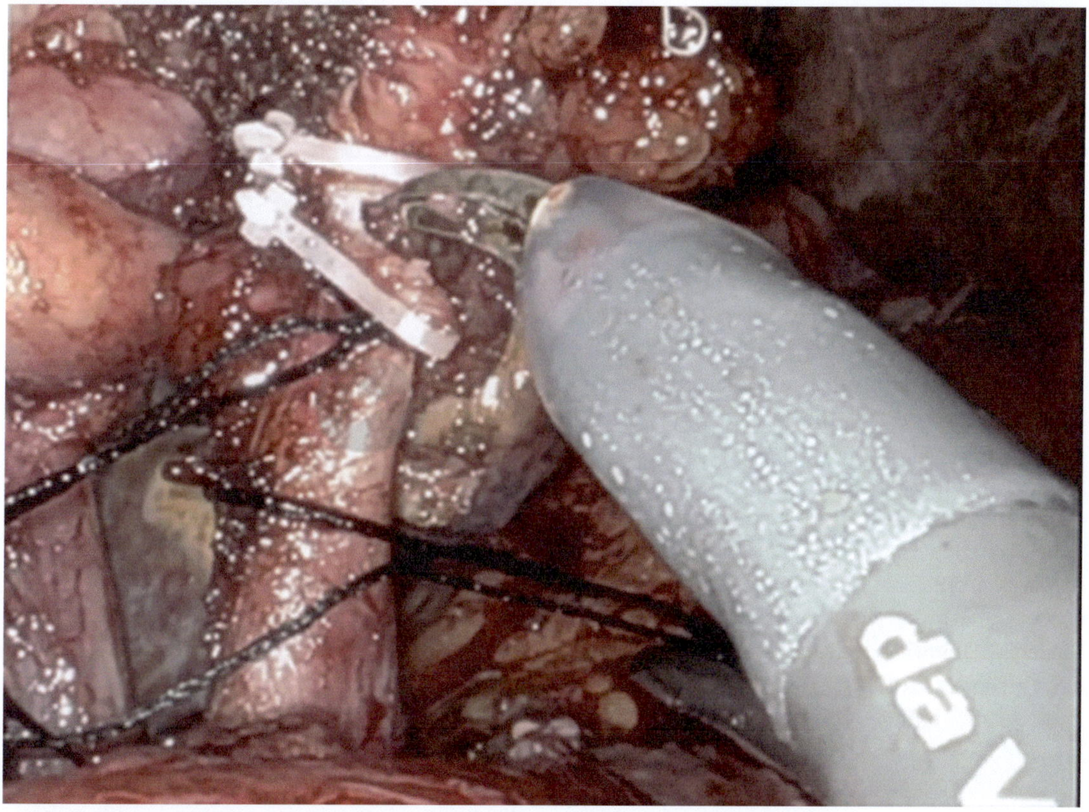

Fig. 24.5 Left hepatic artery transection

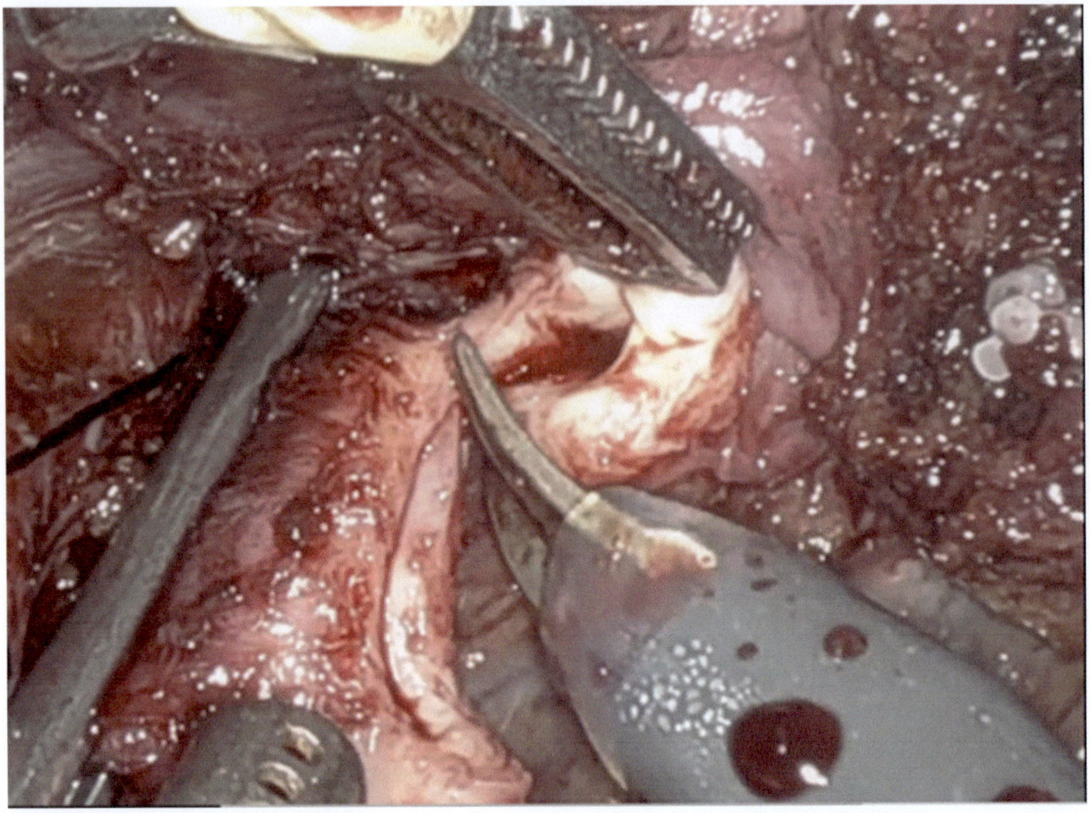

Fig. 24.6 Left portal vein transection

(at the right or left portal vein) to isolate the portal vein bifurcation. This way, a more aggressive portal vein resection can be done to include resection part of the portal vein confluence. The portal vein bifurcation was then closed/repaired with a running 5-0 or 6-0 polypropylene sutures with or without a vein patch (Fig. 24.6).

Once inflow control is established, we proceed with the parenchymal transection. The initial superficial liver parenchyma (approximately 1–2-cm depth) is divided using monopolar scissor cautery (Fig. 24.7). For the deeper parts of the liver parenchyma, our preferred technique involves a crush and clamp method, employing robotic fenestrated Maryland forceps and a Vessel Sealer® (Intuitive Surgical, Sunnyvale, CA, USA). Larger crossing vessels exceeding 7-mm and bile duct branches are divided between locking clips. The final step involves transection of the hepatic vein at its origin, flush to the inferior vena cava (IVC), using a robotic vascular stapler (Fig. 24.8). We do not adopt the practice of routine hanging maneuver since we do not feel this technique is needed and helpful for every right hepatectomy.

In cases of robotic resection for Bismuth type 3A and 3B perihilar cholangiocarcinoma, routine removal of the caudate lobe is crucial, given the tendency of local tumor recurrence in the remaining caudate lobe is the highest among other places [24]. The caudate lobe's bile duct branches, which often drain directly into the biliary bifurcation, present a common pathway for tumor spread and invasion. The dissection and division of the short hepatic veins, separating the caudate lobe from the IVC, are facilitated by the robotic platform from the left and right side of the IVC (Fig. 24.9). A large right inferior hepatic vein and hepatocaval ligament (Makuuchi ligament) must be handled with care to avoid unnecessary caval bleeding. Any minor bleeding from the IVC is otherwise often easily managed with robotic

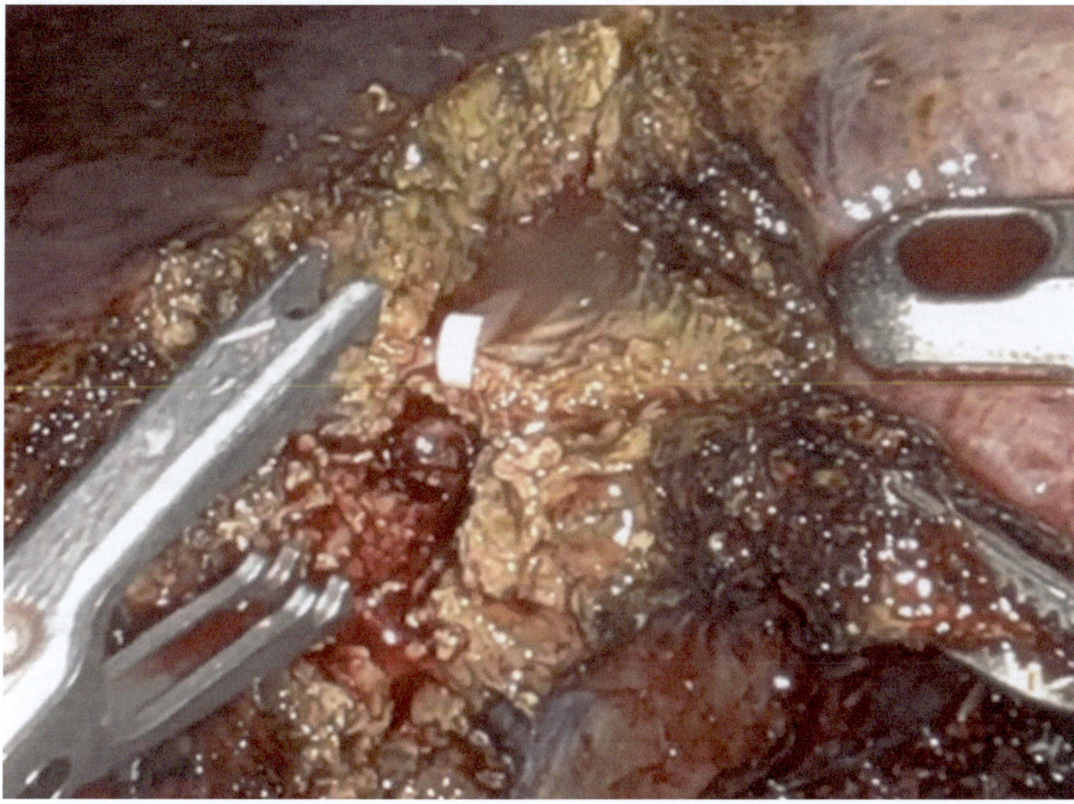

Fig. 24.7 Liver transection, initial

Fig. 24.8 Liver transection, complete

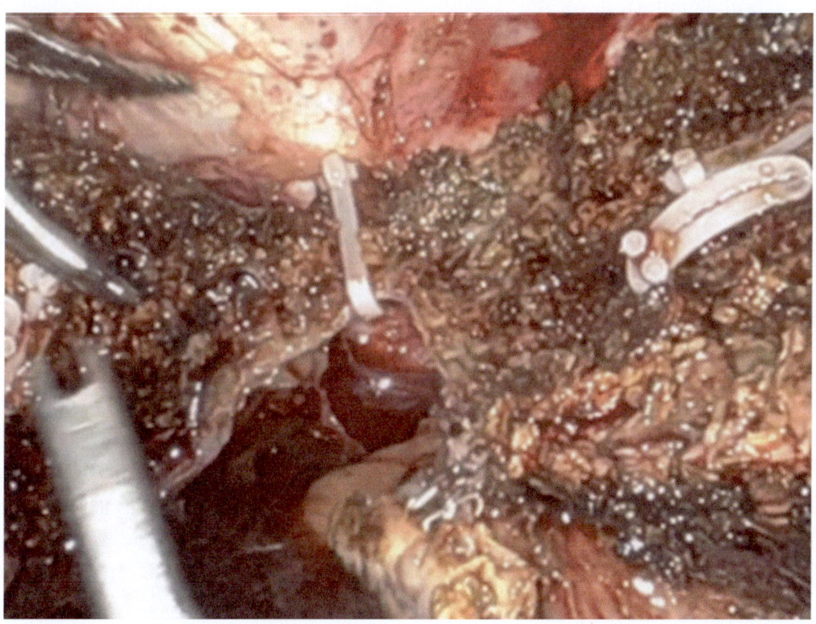

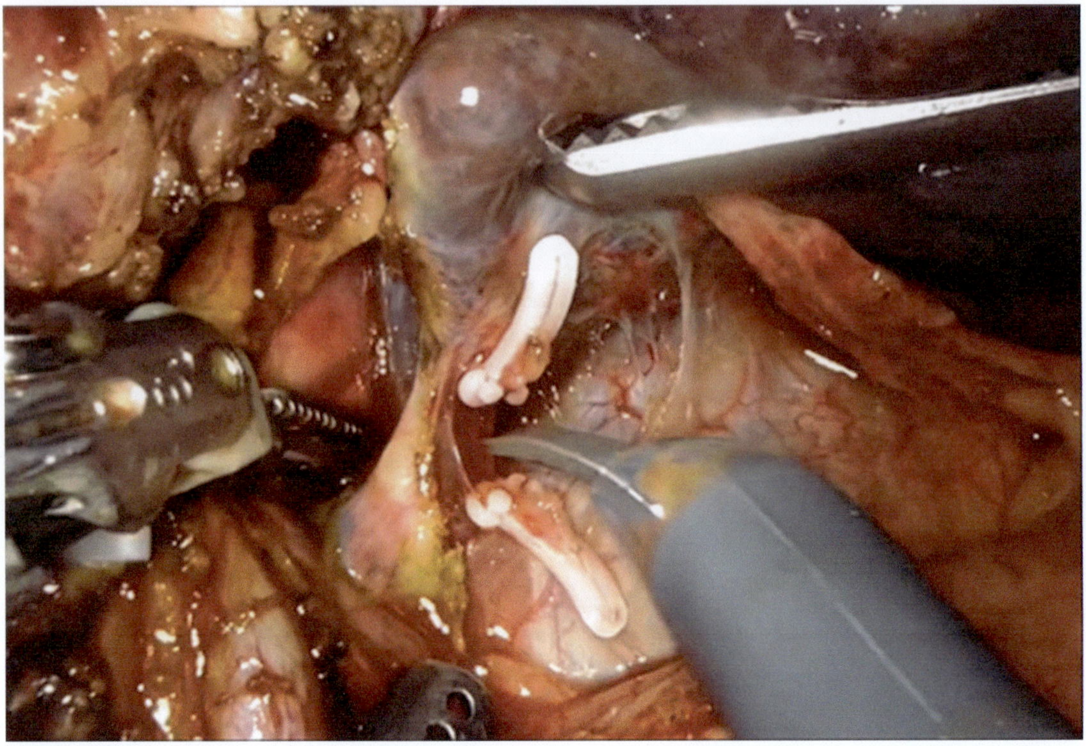

Fig. 24.9 Dissection of caudate lobe from inferior vena cava

suturing under low central venous pressure environment.

Upon separation of the specimen from the FLR, it is enclosed in a laparoscopic extraction bag and removed through the right lower quadrant incision, which previously housed the Gelport®. In the specific context of type 3B Klatskin tumor resections, we perform a unification ductoplasty, joining the right anterior and posterior sectoral bile ducts. This technique mirrors those used in liver transplantation, allowing for a single classical end-to-side hepaticojejunostomy anastomosis. The ductoplasty is executed using 5-0 or 6-0 absorbable monofilament sutures, ensuring a tension-free approximation (Figs. 24.10 and 24.11).

24.7.6 Roux-En-Y Hepaticojejunostomy Biliary Reconstruction

After completion of the resection phase of the operation, with verification of negative resection margins, we proceed to the critical phase of biliary reconstruction. This begins with the creation of a jejunal mesenteric window approximately 35 cm distal to the ligament of Treitz, followed by the division of the proximal jejunum using a 45-mm robotic blue load stapler (Fig. 24.12). We then measure a 60-cm length of a Roux limb and construct a standard stapled side-to-side jejunojejunostomy anastomosis. The common enterotomy created during this process is then closed with running 3-0 barbed sutures to ensure a watertight anastomosis.

The next step involves transposing the Roux limb toward the hilar plate to create the hepaticojejunostomy. Our preferred method for this anastomosis is a single-layer, running hepaticojejunostomy using two 9-inch 4-0 absorbable barbed sutures. The initial suture is placed at the nine o'clock position of the anastomosis, and the posterior layer is performed first (Fig. 24.13). This way, the Roux limb is now fixed in place without the manual support from the bedside assistant, which technically facilitates the placement of the anterior layer of the

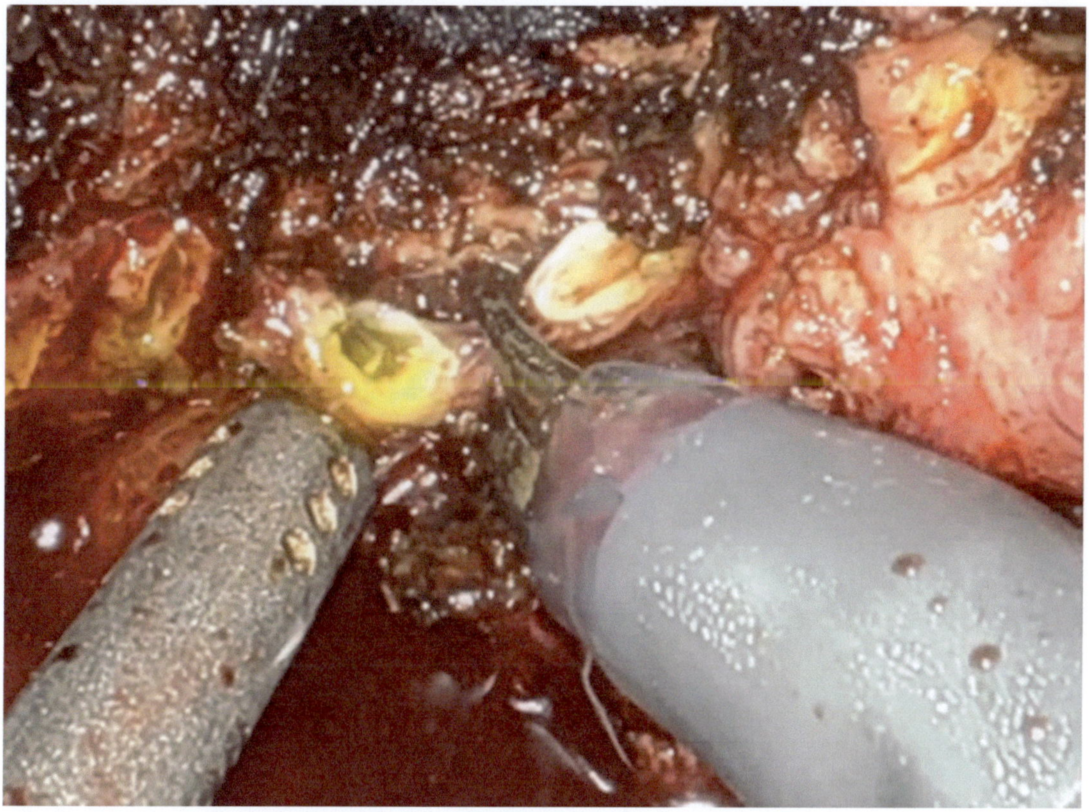

Fig. 24.10 Proximal bile duct transection

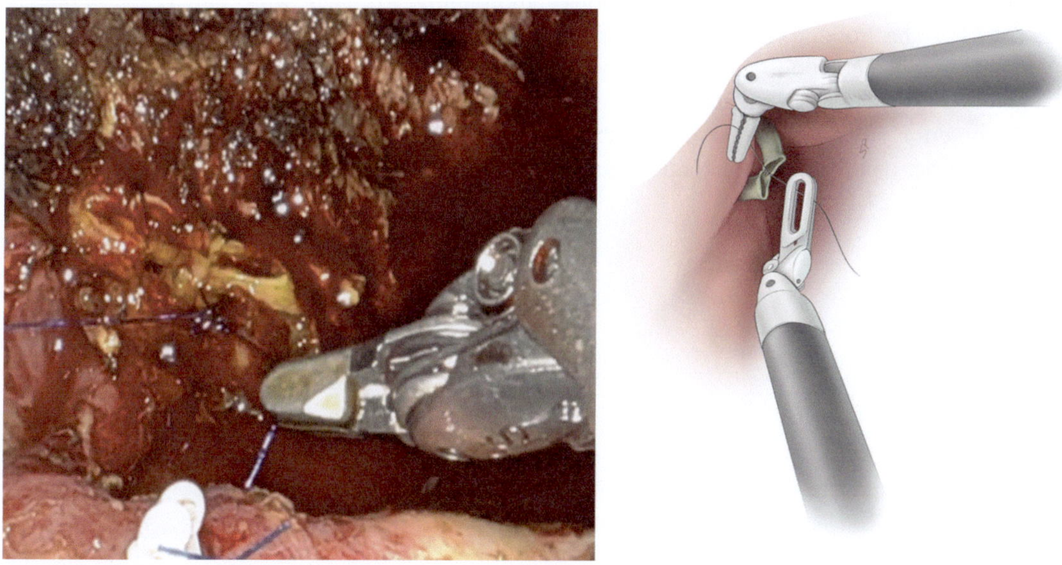

Fig. 24.11 Ductoplasty, right anterior and posterior sectoral ducts

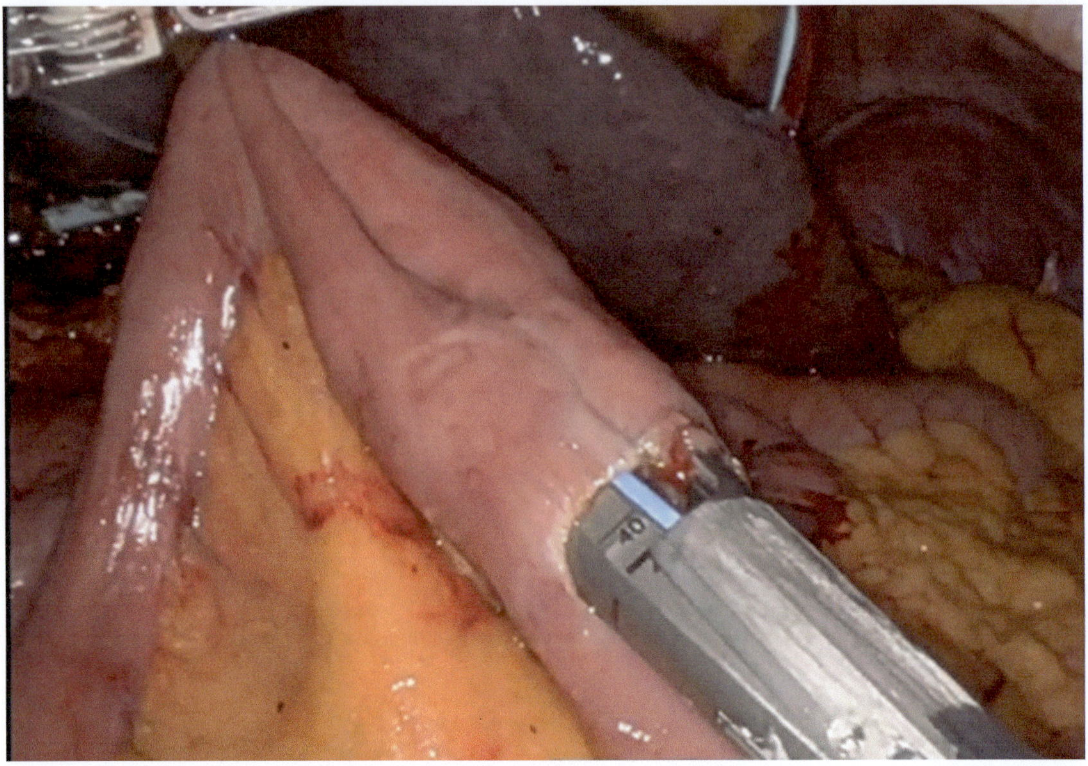

Fig. 24.12 Side-to-side stapled jejunojejunostomy anastomosis

anastomosis. The anterior layer is next sutured from the nine to three o'clock positions (Fig. 24.14). The robotic system's precision greatly aids in this meticulous suturing, allowing for ambidextrous hand motions. Both the posterior and anterior sutures are tied at the three o'clock position, completing the hepaticojejunal anastomosis (Fig. 24.15).

Following the anastomosis, the abdomen is thoroughly irrigated with 1–2 L of saline. A closed suction drain is placed near the hepaticojejunostomy site and exited through the skin at the site of the robotic arm one incision. The NGT is secured in place prior to the patient's departure from the operating room to the surgical floor.

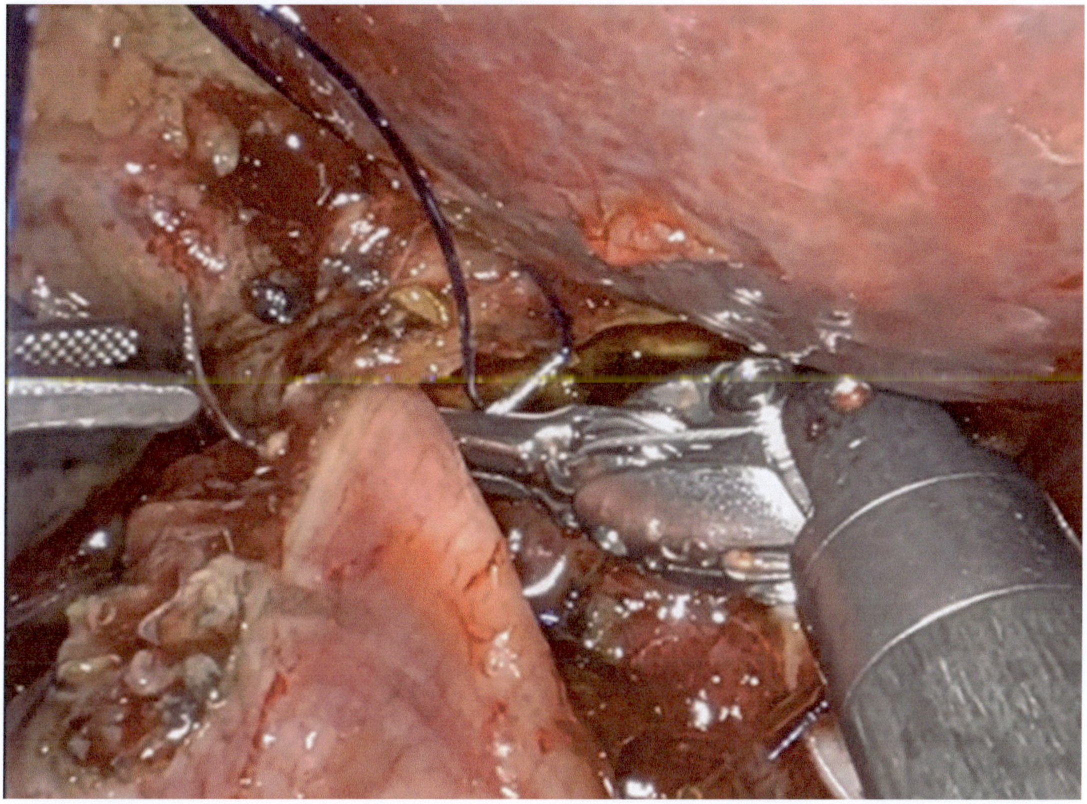

Fig. 24.13 Hepaticojejunostomy, suture posterior wall of the anastomosis

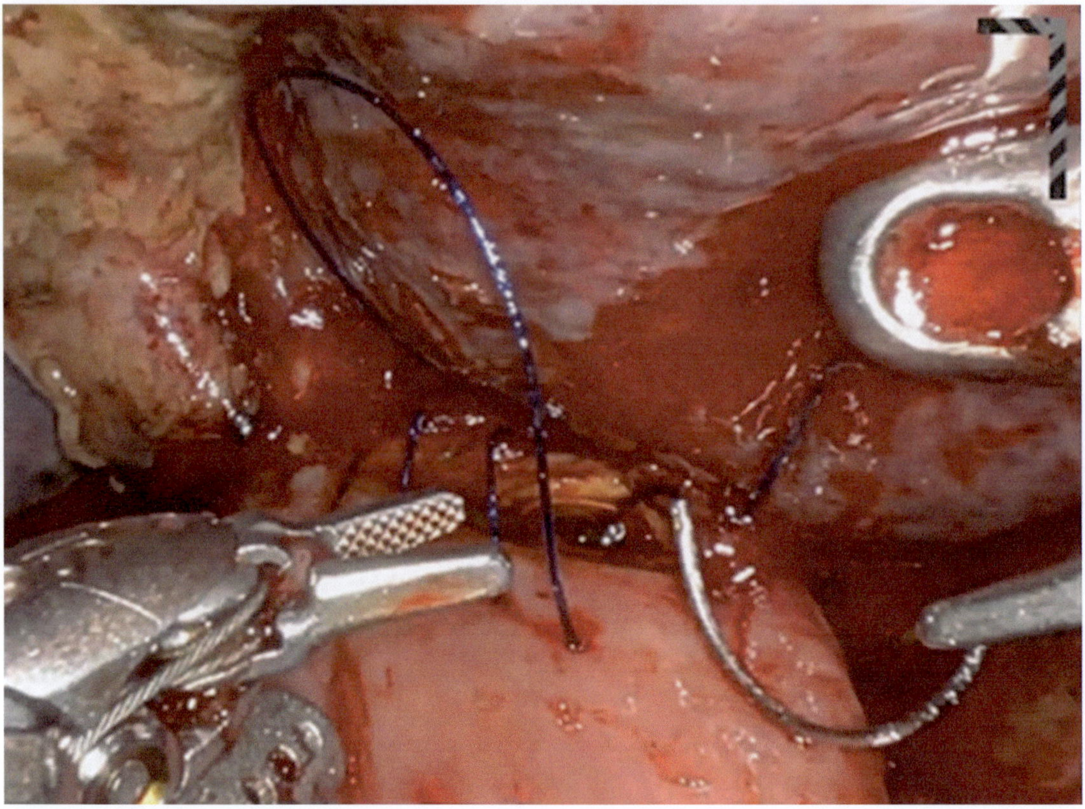

Fig. 24.14 Hepaticojejunostomy, suture anterior wall of the anastomosis

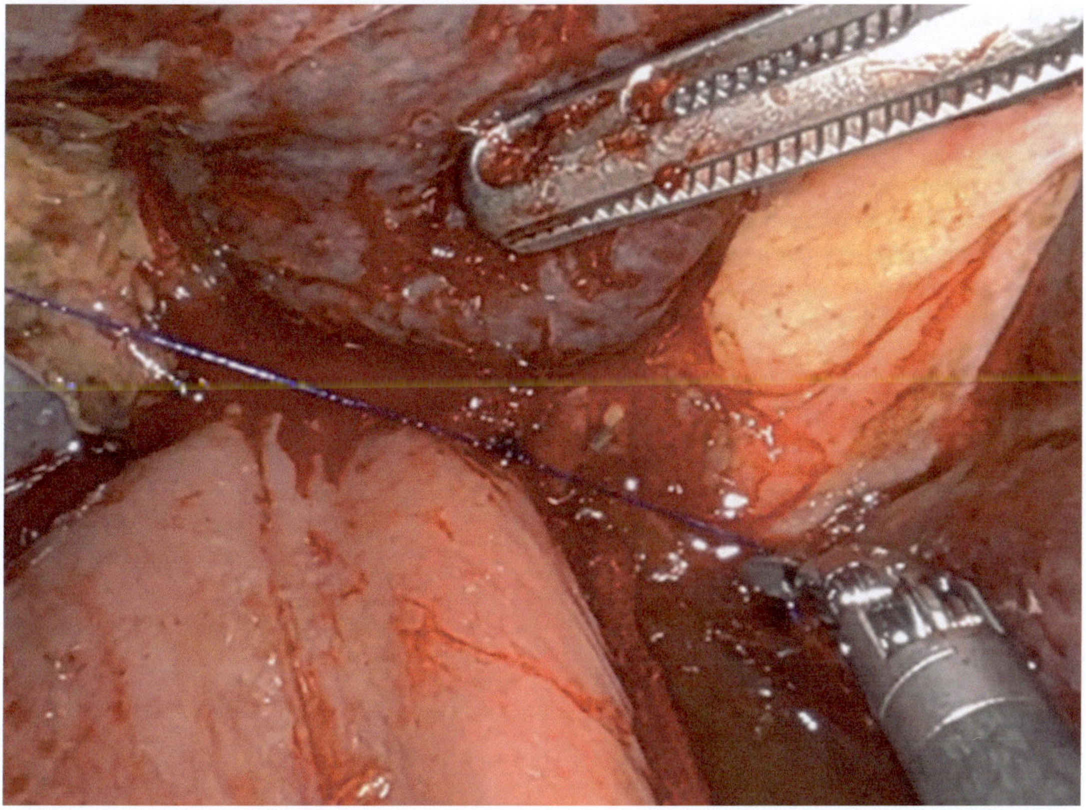

Fig. 24.15 Hepaticojejunostomy, complete

24.8 Postoperative Instructions and Clinical Outcomes

Postoperative patient care adheres to our ERAS protocol. Initially, patients are kept nil per os (NPO, nothing by mouth) for the first 24 h post-surgery while beginning early ambulation protocol with the aid of nursing and physical therapy team. If the NGT output is unremarkable during the first postoperative day (POD), it is removed along with the central line and Foley catheter, provided the patient passes a voiding trial.

On POD two, patients are transitioned to a clear liquid diet and engaged in inpatient physical therapy sessions at least twice daily during their inpatient stay. If they tolerate the clear liquid diet without issues, we advance their diet to full liquids by POD three. The goal is to discharge patients' home between POD four and six, ensuring they are achieving appropriate ambulation and pain control. We prefer to remove the closed suction drain before hospital discharge in the absence of bile leak.

Post-discharge care includes close monitoring, with a scheduled follow-up appointment in the clinic within 1–2 weeks of discharge. Additionally, our nurse practitioners conduct a follow-up phone call within 2–3 days of the patient's discharge to check on their recovery and address any clinical concerns.

24.9 Looking Forward

Since 2019, our team has successfully performed 30 robotic resections of perihilar cholangiocarcinoma, building on our extensive experience of over 200 robotic major hepatectomies for liver tumors, as well as our proficiency in robotic pancreaticoduodenectomies dating back to 2016. This progression was strategically planned to ensure that we had thoroughly navigated the initial learning curve associated with robotic major hepatectomies before tackling the intricate challenges posed by perihilar cholangiocarcinoma resections. To date, we have completed more than 500 robotic hepatectomies for treatment of liver tumors.

Our findings thus far affirm that the robotic approach for perihilar cholangiocarcinoma resection is both safe and effective, particularly when applied to carefully selected patients and performed by skilled robotic hepatobiliary surgeons and experienced operating room team [7]. As our experience grows, we are increasingly convinced that the robotic approach may become the preferred method for undertaking these resections in patients with Klatskin types 1–3 due to the numerous advantages associated with minimally invasive surgery.

However, alongside these advancements, we are committed to ongoing research into long-term oncologic outcomes. It is imperative to us that the short-term benefits provided by the minimally invasive nature of the robotic approach do not compromise the essential long-term oncological efficacy. This balance is a cornerstone of our practice and research in the field of robotic surgery for any hepatopancreatobiliary cancers.

References

1. Sucandy I, Shapera E, Jacob K, et al. Robotic resection of extrahepatic cholangiocarcinoma: institutional outcomes of bile duct cancer surgery using a minimally invasive technique. J Surg Oncol. 2022;125(2):161–7. https://doi.org/10.1002/jso.26674.
2. McCarron F, Cochran A, Ricker A, et al. 10 years, 100 robotic major hepatectomies: a single-center experience. Surg Endosc. 2023;38(2):902–7. https://doi.org/10.1007/s00464-023-10459-2.
3. Sucandy I, Luberice K, Lippert T, et al. Robotic major hepatectomy: an institutional experience and clinical outcomes. Ann Surg Oncol. 2020;27(13):4970–9. https://doi.org/10.1245/s10434-020-08845-4.
4. Zhang XP, Jiang N, Zhu L, et al. Short- and long-term outcomes after robotic versus open hepatectomy in patients with large hepatocellular carcinoma: a multicenter study. Int J Surg. 2023; https://doi.org/10.1097/JS9.0000000000000873.
5. Bagante F, Spolverato G, Strasberg SM, et al. Minimally invasive vs. open hepatectomy: a comparative analysis of the National Surgical Quality Improvement Program Database. J Gastrointest Surg. 2016;20(9):1608–17. https://doi.org/10.1007/s11605-016-3202-3.
6. Sharma P, Yadav S. Demographics, tumor characteristics, treatment, and survival of patients with Klatskin tumors. Ann Gastroenterol. 2018;31(2):231–6. https://doi.org/10.20524/aog.2018.0233.
7. Sucandy I, Marques HP, Lippert T, et al. Clinical outcomes of robotic resection for perihilar cholangiocarcinoma: a first, multicenter, trans-Atlantic, expert-center, collaborative study. Ann Surg Oncol. 2024;31(1):81–9. https://doi.org/10.1245/s10434-023-14307-4.
8. Jang JY, Kim SW, Park DJ, et al. Actual long-term outcome of extrahepatic bile duct cancer after surgical resection. Ann Surg. 2005;241(1):77–84. https://doi.org/10.1097/01.sla.0000150166.94732.88.
9. Boerma EJ. Research into the results of resection of hilar bile duct cancer. Surgery. 1990;108(3):572–80.
10. Marsh Rde W, Alonzo M, Bajaj S, et al. Comprehensive review of the diagnosis and treatment of biliary tract cancer 2012. Part II: multidisciplinary management. J Surg Oncol. 2012;106(3):339–45. https://doi.org/10.1002/jso.23027.
11. Tan EK, Taner T, Heimbach JK, Gores GJ, Rosen CB. Liver transplantation for perihilar cholangiocarcinoma. J Gastrointest Surg. 2020;24(11):2679–85. https://doi.org/10.1007/s11605-020-04721-4.
12. Hewitt DB, Brown ZJ, Pawlik TM. Surgical management of cholangiocarcinoma. Hepatoma Research. 2021;7:75. https://doi.org/10.20517/2394-5079.2021.83.
13. Ratti F, Cipriani F, Ingallinella S, Tudisco A, Catena M, Aldrighetti L. Robotic approach for lymphadenectomy in biliary tumours: the missing ring between the benefits of laparoscopic and reproducibility of open approach? Ann Surg. 2022; https://doi.org/10.1097/SLA.0000000000005748.
14. Mansour JC, Aloia TA, Crane CH, Heimbach JK, Nagino M, Vauthey JN. Hilar cholangiocarcinoma: expert consensus statement. HPB (Oxford). 2015;17(8):691–9. https://doi.org/10.1111/hpb.12450.
15. Jarnagin W, Winston C. Hilar cholangiocarcinoma: diagnosis and staging. HPB (Oxford). 2005;7(4):244–51. https://doi.org/10.1080/13651820500372533.
16. Dondossola D, Ghidini M, Grossi F, Rossi G, Foschi D. Practical review for diagnosis and clinical man-

agement of perihilar cholangiocarcinoma. World J Gastroenterol. 2020;26(25):3542–61. https://doi.org/10.3748/wjg.v26.i25.3542.
17. Magistri P, Pang NQ, Guidetti C, et al. Robotic approach for perihilar cholangiocarcinoma: from Bismuth 1 to vascular resection. Eur J Surg Oncol. 2023;49(11):107002. https://doi.org/10.1016/j.ejso.2023.107002.
18. Horny M, Burgess JF Jr, Cohen AB. Advanced imaging utilization trends in privately insured patients from 2007 to 2013. J Am Coll Radiol. 2015;12(12 Pt B):1380–1387 e4. https://doi.org/10.1016/j.jacr.2015.09.013.
19. Piccus R, Joshi K, Hodson J, et al. Significance of predicted future liver remnant volume on liver failure risk after major hepatectomy: a case matched comparative study. Front Surg. 2023;10:1174024. https://doi.org/10.3389/fsurg.2023.1174024.
20. Milana F, Famularo S, Diana M, et al. How much is enough? A surgical perspective on imaging modalities to estimate function and volume of the future liver remnant before hepatic resection. Diagnostics (Basel). 2023;13(17):2726. https://doi.org/10.3390/diagnostics13172726.
21. Mehrabi A, Golriz M, Ramouz A, et al. Promising outcomes of modified ALPPS for staged hepatectomy in cholangiocarcinoma. Cancers (Basel). 2023;15(23) https://doi.org/10.3390/cancers15235613.
22. Kauffmann R, Fong Y. Post-hepatectomy liver failure. Hepatobiliary Surg Nutr. 2014;3(5):238–46. https://doi.org/10.3978/j.issn.2304-3881.2014.09.01.
23. Sucandy I, Gravetz A, Ross S, Rosemurgy A. Technique of robotic left hepatectomy : how we approach it. J Robot Surg. 2019;13(2):201–7. https://doi.org/10.1007/s11701-018-0890-6.
24. Capobianco I, Rolinger J, Nadalin S. Resection for Klatskin tumors: technical complexities and results. Transl Gastroenterol Hepatol. 2018;3:69. https://doi.org/10.21037/tgh.2018.09.01.

Robotic Liver Resection for Gallbladder Cancer, Hilar Lymphadenectomy, and Biliary Reconstruction

25

Frances N. McCarron and John B. Martinie

25.1 Introduction

Gallbladder carcinoma although rare (1.2% of cancer diagnoses worldwide) is an especially aggressive malignancy with 84,695 deaths in 2020 according to the World Health Organization [1]. Four out of five cases have spread beyond the gallbladder at the time of diagnosis, and the overall estimated 5-year survival for patients with gallbladder cancer is 19%. Survival rates drastically decline when tumor has metastasized to regional lymph nodes from 65% 5-year survival for localized disease to 28% with regional nodal spread. Those with distant metastasis at the time of diagnosis have about a 2% 5-year survival rate [2]. Treatment options are patient dependent and include surgical resection, chemotherapy, radiation, or a combination. Currently, except for few smaller studies, there is little research to support neoadjuvant vs adjuvant systemic therapy, and surgical resection is the mainstay for resectable disease [3].

Since survival depends on the stage and biology of the cancer, a detailed workup is imperative [4, 5]. Unfortunately, gallbladder cancer can present to the hepatobiliary surgeon in a few different ways, making its workup and treatment complicated. One common scenario is discovering gallbladder cancer on the pathologic examination after a cholecystectomy. In these cases, the most important factor is a detailed pathology report. If not already documented, the surgeon should have pathology slides sent to their institutional pathologist who routinely analyzes hepatobiliary specimen to report on the cystic duct margin, T stage of the tumor, and any lymph node spread of disease if sampled, as well as the liver/cystic plate margin. If a T1a lesion was resected with negative margins and no spread to Calot's node, no further surgical resection is required, and the patient may go into a period of observation. For T1b and higher disease, positive cystic duct margin, or lymph node spread, a complete staging evaluation should ensue prior to any other treatments. Cross-sectional imaging with multiphase CT or MRI and tumor markers should be obtained. Operative details are also important to identify spillage of the bile, disruption of the gallbladder wall, subtotal cholecystectomy, or conversion to an open procedure because of severe "inflammation." In these cases, a PET/CT or diagnostic laparoscopy may be helpful in evaluating for residual disease or peritoneal spread. If resectable disease is diagnosed, then surgical re-resection of the gallbladder fossa to include segments 4b and 5 and hilar lymphadenectomy are indicated.

Gallbladder lesions can also present on imaging or at the time of surgery for another reason. If it is found during surgery and the surgeon is not trained in gallbladder cancer operations, the case

F. N. McCarron · J. B. Martinie (✉)
Atrium Health Carolinas Medical Center, Charlotte, NC, USA
e-mail: John.Martinie@atriumhealth.org

M. D'Hondt, I. Sucandy (eds.), *Textbook of Robotic Liver Surgery*,
https://doi.org/10.1007/978-3-031-76536-0_25

should be aborted, and the patient should be referred to a hepatobiliary expert. A suspicious gallbladder mass on imaging should prompt a more thoughtful evaluation. This again includes multiphase CT scans or MRI of the liver and its vasculature noting associations with the lesion in the gallbladder. Chest imaging and tumor markers are also part of the algorithm for confirmed malignancy. Endoscopic evaluation with both endoscopic ultrasound (EUS) and endoscopic retrograde cholangio-pancreatography (ERCP) can be helpful to provide information about portal lymphadenopathy and extrahepatic/intrahepatic ductal involvement, and FNA biopsies can be obtained in these cases. Although EUS-FNA has a sensitivity of 80–100%, its role is controversial and not currently listed in NCCN guidelines [6]. PET/CT can also detect occult peritoneal metastases with a sensitivity of about 56% [7]. Laparoscopy is a useful tool and is the most sensitive modality to diagnose peritoneal spread and resectability [8, 9]. For resectable lesions, the indicated operation is a cholecystectomy, lymphadenectomy, segment 4b and 5 liver resection, and possible extended hepatectomy or biliary excision with reconstruction with the goal of achieving R0 status.

The authors of this chapter prefer to use the algorithm below for dealing with the complexities of an

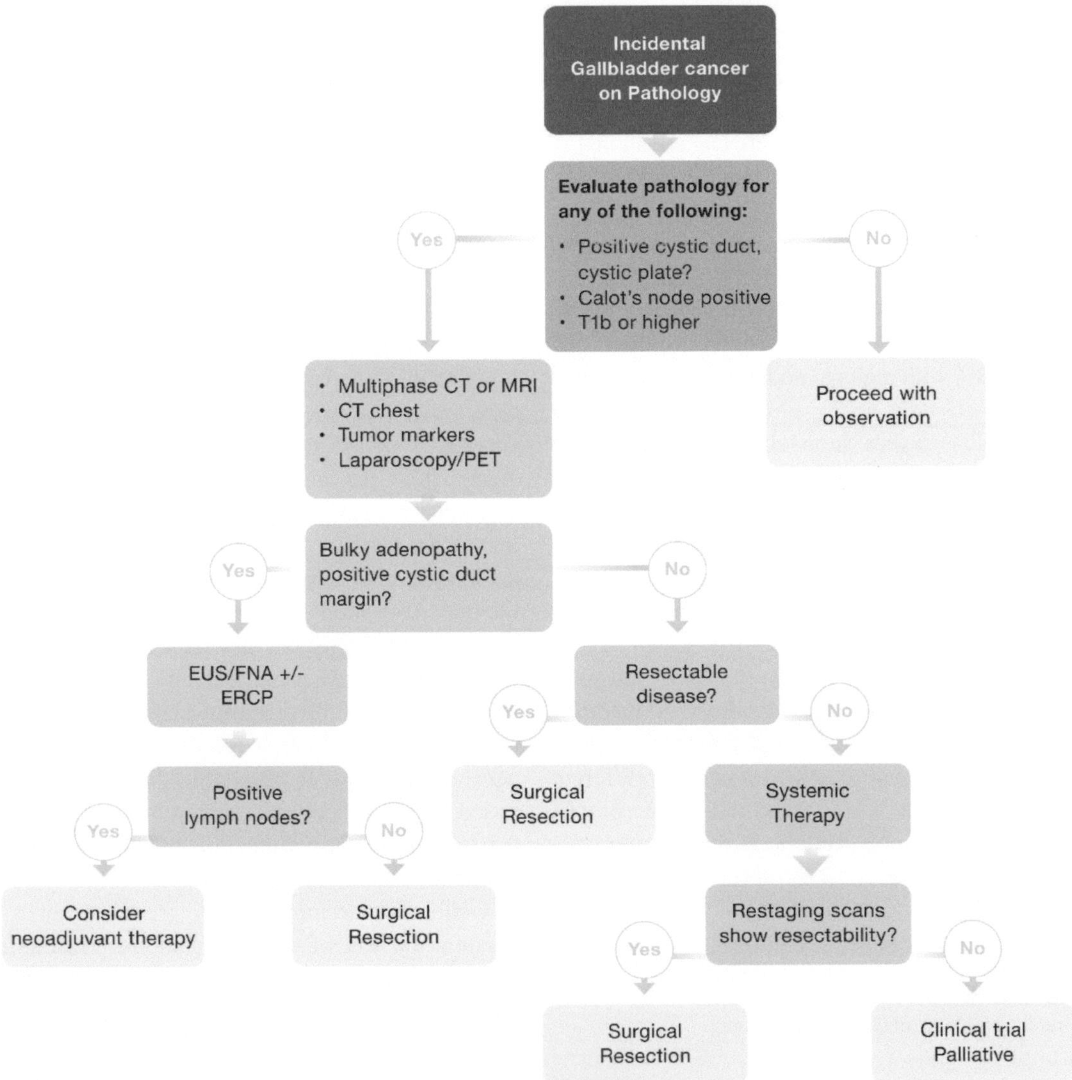

Fig. 25.1 Algorithm for management of incidental gallbladder cancer found post cholecystectomy

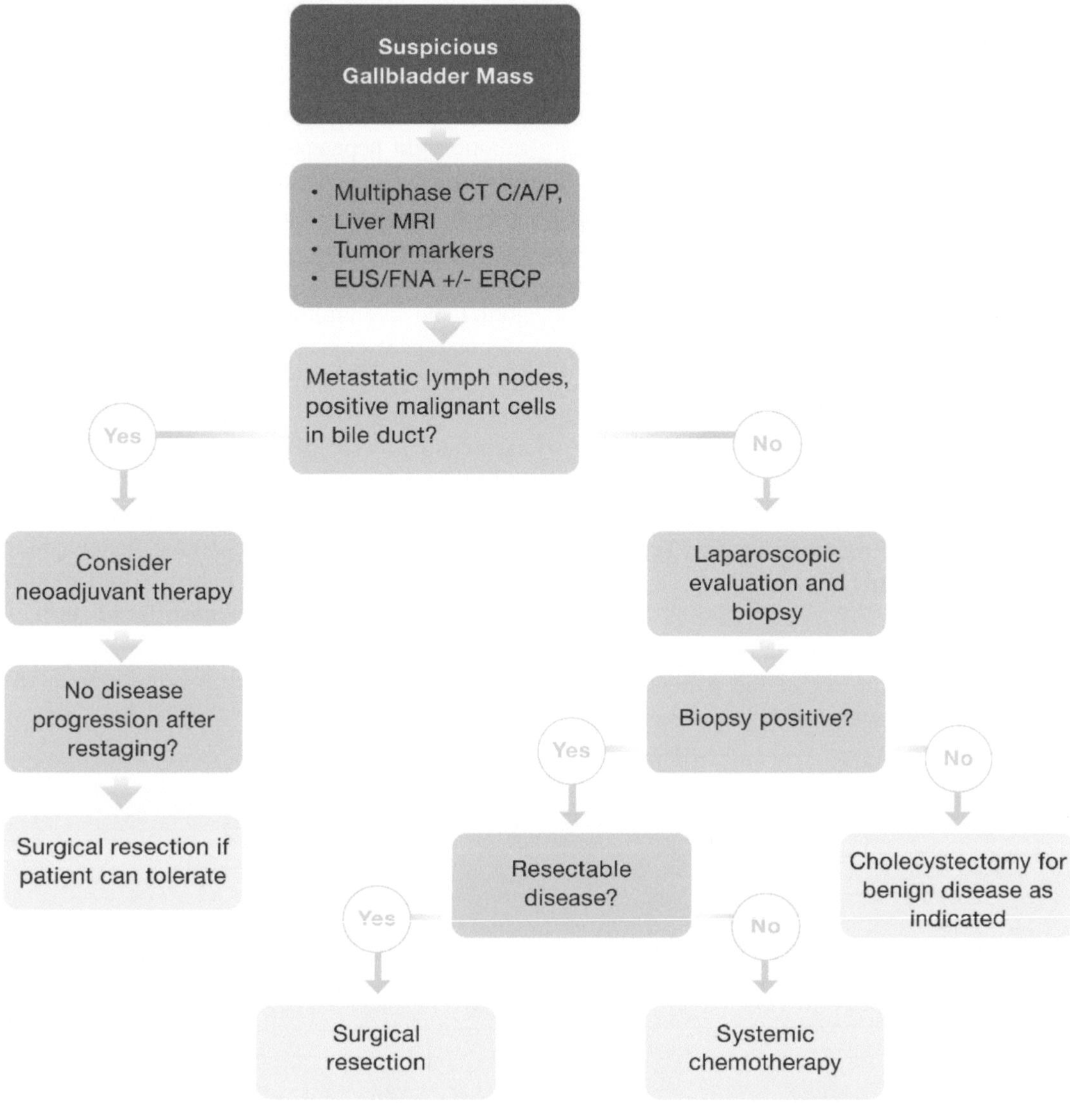

Fig. 25.2 Algorithm for management of a suspicious gallbladder mass seen on preoperative imaging

incidental pathologic diagnosis of gallbladder cancer (Fig. 25.1) and suspicious gallbladder mass on imaging (Fig. 25.2). In the remainder of this chapter, we will discuss the technical aspects of performing robotic liver resection for gallbladder cancer, hilar lymphadenectomy, and biliary reconstruction.

25.2 Technical Aspects

The robotic platform provides many benefits for gallbladder cancer surgery. The wristed instruments and high-definition magnification of the robotic camera are especially useful when performing hilar dissections and biliary anastomoses which require precision. With advancements in the production of a robotic bipolar, vessel sealing device, liver parenchymal transection has also become much simpler. The authors of this chapter solely use the DaVinci Xi robotic platform (Intuitive Surgical Inc.), and instruments and trocar positioning will reflect this modality.

25.3 Robotic Liver Resection for Gallbladder Cancer

When performing an operation for gallbladder cancer, one must remember the importance of oncologic outcome. For this reason, the authors have divided this operation into six steps: [1] patient positioning and trocar placement, [2] diagnostic laparoscopy and intraoperative ultrasound, [3] hepatic artery and celiac lymph node evaluation, [4] hepatoduodenal ligament dissection and lymphadenectomy, [5] cystic duct stump evaluation and cholecystectomy, and [6] hepatectomy.

25.3.1 Patient Positioning and Trocar Placement

The patient is positioned supine with arms out. Prior to docking the robot, the patient is then placed in approximately 10–15 degrees of reverse Trendelenburg and 8–10 degrees of rotation with the left side down (Fig. 25.3). The robotic trocar insertion pictured below is similar to our trocar placement for a robotic cholecystectomy or robotic right hepatectomy (Fig. 25.4). We place two 8-mm robotic trocars to the right of the umbilicus at the anterior-axillary and midclavicular lines and two 8-mm robotic trocars to the left of the umbilicus slightly higher on the abdomen targeting the liver hilum. An assistant 12-mm or 15-mm laparoscopic port is placed infraumbilical for assistant suctioning, ultrasound, or stapler use if needed (Fig. 25.5). In an effort to streamline cases and reduce inefficiencies, we have minimized the number and types of instruments for most cases. These include a fenestrated bipolar grasper in arm 1, robotic camera is positioned in arm 2, monopolar scissors in arm 3, and another grasping instrument (ProGrasp™) in arm 4 (Fig. 25.6). Almost all instrument exchanges occur through arm 3, which is referred to as the primary operative arm.

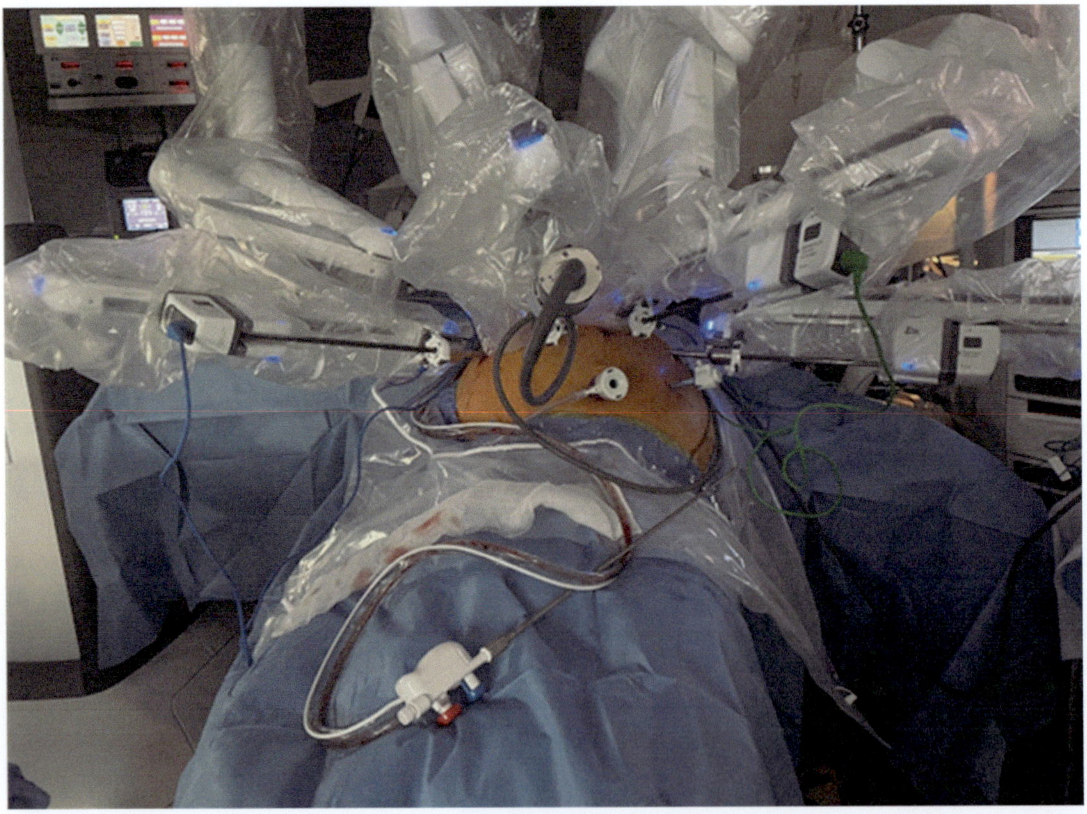

Fig. 25.3 Patient positioning. 10–15 degrees reverse Trendelenburg, 8–10 degrees left lateral decubitus

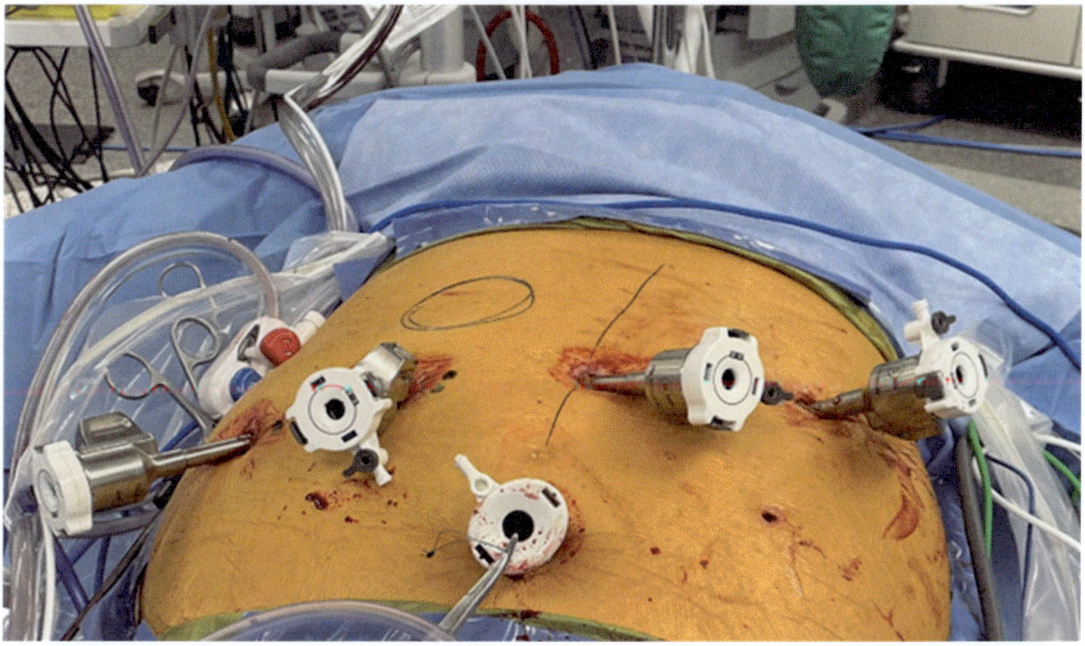

Fig. 25.4 Trocar placement. 15-mm assistant port placed midline infraumbilical robotic trocars triangulating toward the right upper quadrant, and the liver mass which is palpable on exam

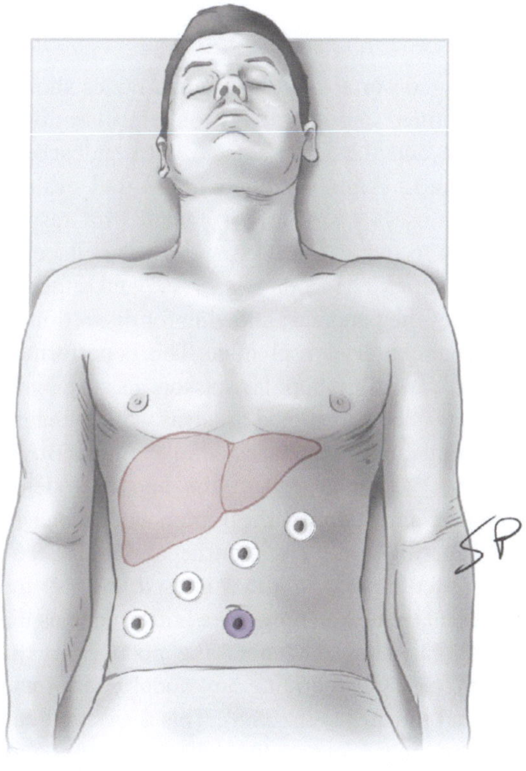

Fig. 25.5 Trocar placement for robotic gallbladder cancer resection

25.3.2 Diagnostic Laparoscopy and Intraoperative Ultrasound

The procedure begins with a thorough laparoscopic evaluation of the abdomen to rule out peritoneal or unresectable liver metastases. Generally, we use the robotic camera to examine the entire abdominal cavity, prior to placing additional trocars, docking the robot, or placement of any instruments. This allows for evaluation of all four quadrants and the pelvis as peritoneal carcinomatosis is common in gallbladder cancer. Any suspicious nodules are biopsied and sent for frozen pathologic evaluation prior to proceeding. If none are found, we proceed with docking of the robot and instruments in the configuration mentioned above. Often, a cholecystectomy has already been performed and possibly converted to an open cholecystectomy by a prior surgeon. In these cases, adhesiolysis is performed, to fully expose the anatomy of the right upper quadrant. An ultrasound probe is inserted through the 12- or 15-mm assistant port. We use an ultrasound probe (BK Medical™) which is compatible with

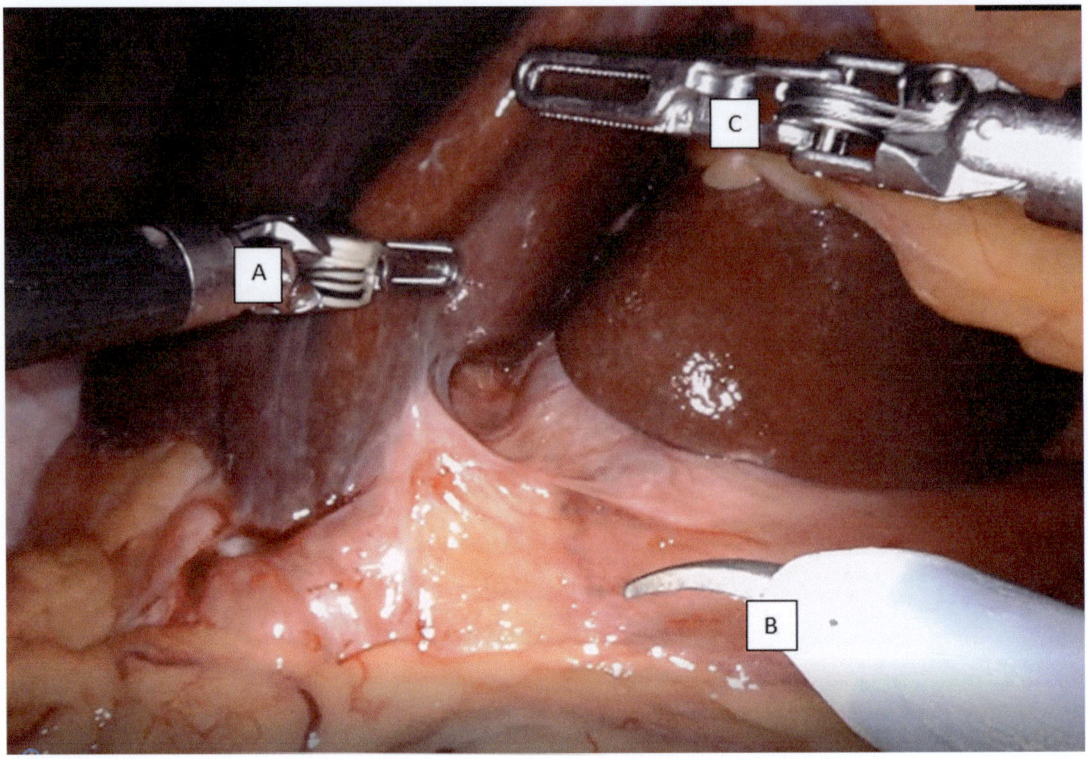

Fig. 25.6 Robotic instruments used for the majority of the case include fenestrated bipolar forceps in arm 1 (**a**), monopolar scissors in arm 3 (**b**), and ProGrasp™ in arm 4 (**c**). Camera is used in arm 2

the DaVinci Xi™ system, allowing for a "picture-in-picture" view. A thorough ultrasound evaluation is performed of the entire liver identifying the tumor and any satellite lesions in the liver as well as involvement of any vascular pedicles (Fig. 25.7). We also perform an ultrasound evaluation of the hepatoduodenal ligament to identify lymph nodes, vascular anatomy, and the extrahepatic bile ducts for tumor involvement.

25.3.3 Hepatic Artery and Celiac Lymph Node Evaluation

Involvement of the hepatic artery (station 8) and/or celiac lymph nodes (station 9) is a predictor of extremely poor prognosis in gallbladder cancer [10]. For this reason, we aim to evaluate the station 8 node before any further step in the sequence is performed, such as portal lymphadenectomy or resection of the gallbladder fossa. (Furthermore, it should go without saying that grossly abnormal or enlarged portal or celiac lymph nodes should be identified preoperatively on cross-sectional imaging and further evaluated with endoscopic ultrasound.) After thorough ultrasound evaluation, we use a grasping instrument (ProGrasp™) in arm 4 to retract the liver, and the pars flaccida is opened, exposing the lesser sac and superior border of the pancreas, allowing for dissection of the hepatic artery lymph node. This is performed primarily with monopolar scissors in arm 3 utilizing its ability to spread tissue as well as coagulate to maintain hemostasis similar to the Bovie cautery. We move toward the hepatoduodenal ligament and identify the lymph node anterior and superior to the common hepatic artery. The fenestrated bipolar in arm 1 is used to grasp the perinodal tissue and retract the lymph node as the monopolar scissors dissect the node from the hepatic artery cauterizing any nodal arterial and venous branches (Fig. 25.8). This dissection can continue down toward the celiac axis removing any nodes along this path. The nodes should be

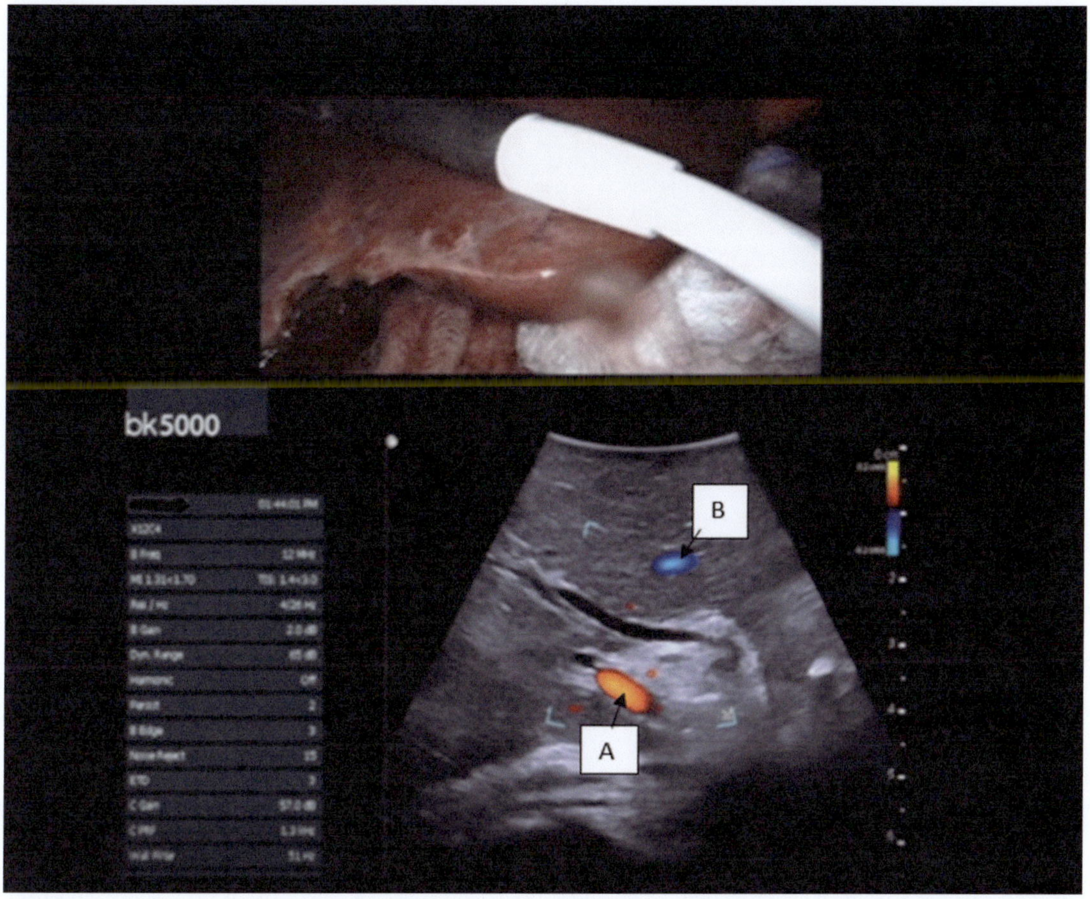

Fig. 25.7 Intraoperative ultrasound evaluation. Right portal pedicle (**a**) and middle hepatic vein (**b**)

sent for frozen pathologic evaluation prior to completing the gallbladder and liver resections.

25.3.4 Hepatoduodenal Ligament Dissection and Lymphadenectomy

Now that the hepatic artery is exposed, the dissection of the hepatoduodenal ligament can take place. For reasons listed above, the authors highly recommend using the monopolar scissors for complete hilar dissection. Using ultrasound aids in identification of major vascular structures (Fig. 25.9). The common bile duct, common hepatic duct, common hepatic artery, hepatic artery proper, right and left hepatic arteries, and the main portal vein should be skeletonized of all fatty tissue. This tissue contains portal lymph nodes which should be sent for pathologic evaluation but do not necessarily require frozen sectioning. A more complete description of the technique for robotic hilar lymphadenectomy will be discussed later in this chapter (Fig. 25.10).

25.3.5 Cystic Duct Stump Evaluation and Cholecystectomy

Step 5 of the operation may be the most variable due to the different presentations of gallbladder cancer. In many cases the cholecystectomy may have already been performed either laparoscopically or open or attempted and aborted. In cases where the gallbladder remains in situ, the next step is dissection of the hepato-cystic triangle and

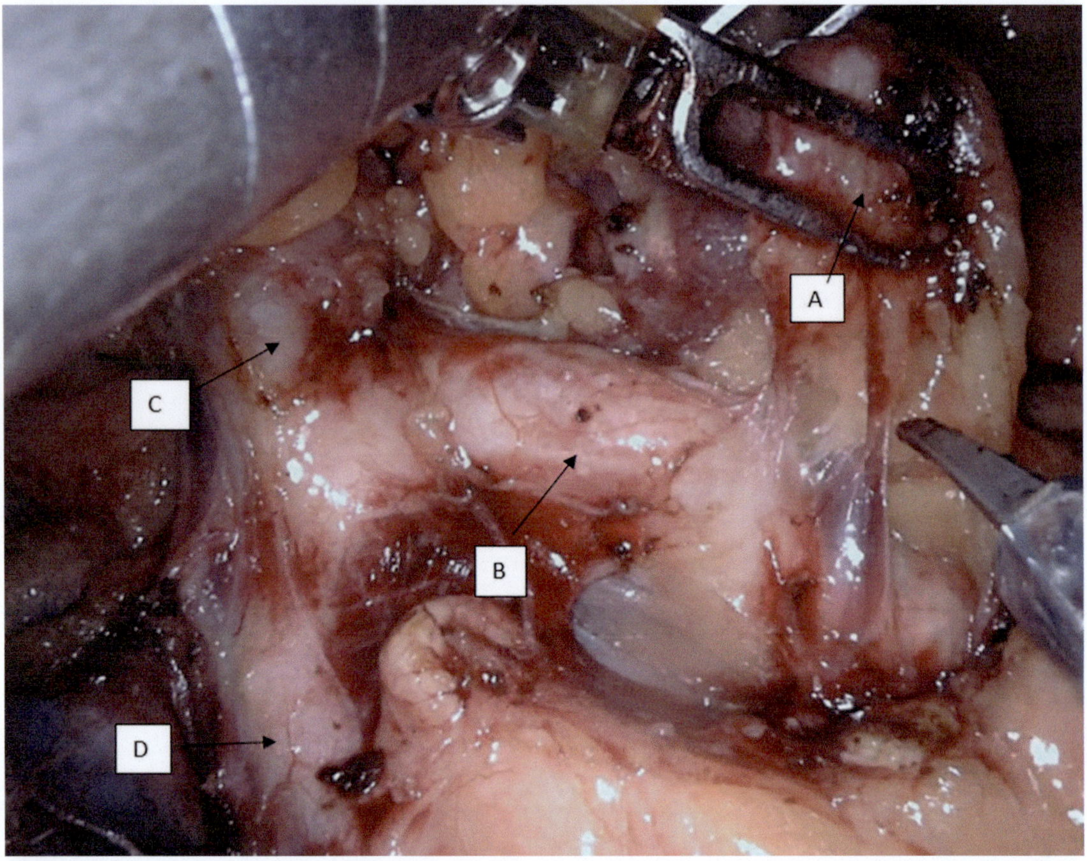

Fig. 25.8 Station 8 lymph node dissection. Lymph node (**a**), common hepatic artery (**b**), hepatic artery proper (**c**), gastroduodenal artery (**d**)

obtaining the "critical view of safety." The cystic duct is dissected all the way to its emergence from the common bile duct. Again, ICG fluorescence should be utilized to assist with this dissection (Fig. 25.11). The authors of this chapter use the monopolar scissors for this portion as well again for the same reasons listed above. Once the cystic duct and artery are skeletonized, the cystic artery is clipped and divided. The duct is clipped and divided as close to the common bile duct as safely feasible (Fig. 25.12). The margin of the cystic duct is then sent for frozen pathologic evaluation. Care must be taken to avoid any disruption of the gallbladder wall or spillage of bile or stones during this portion of the operation. If the gallbladder can then be removed completely from the cystic plate of the liver, then it is prudent to do so to visualize the cystic plate for hepatic transection. If the gallbladder mass is at the fundus and invading into segment 4B or 5 as often it does, it should be left attached and resected en bloc with the liver resection.

This step becomes more complicated in those who have had a prior cholecystectomy. In this case, utilization of ICG becomes even more important. Often the cystic duct is scarred to the cystic plate or even the hilar plate where the right main or right anterior and posterior hepatic ducts coalesce. Precise dissection is key in this circumstance to avoid unwanted biliary or vascular injury. The cystic duct stump must be mobilized to its emergence from the common bile duct and a margin sent for frozen pathology. A positive cystic duct margin may mean an extrahepatic bile

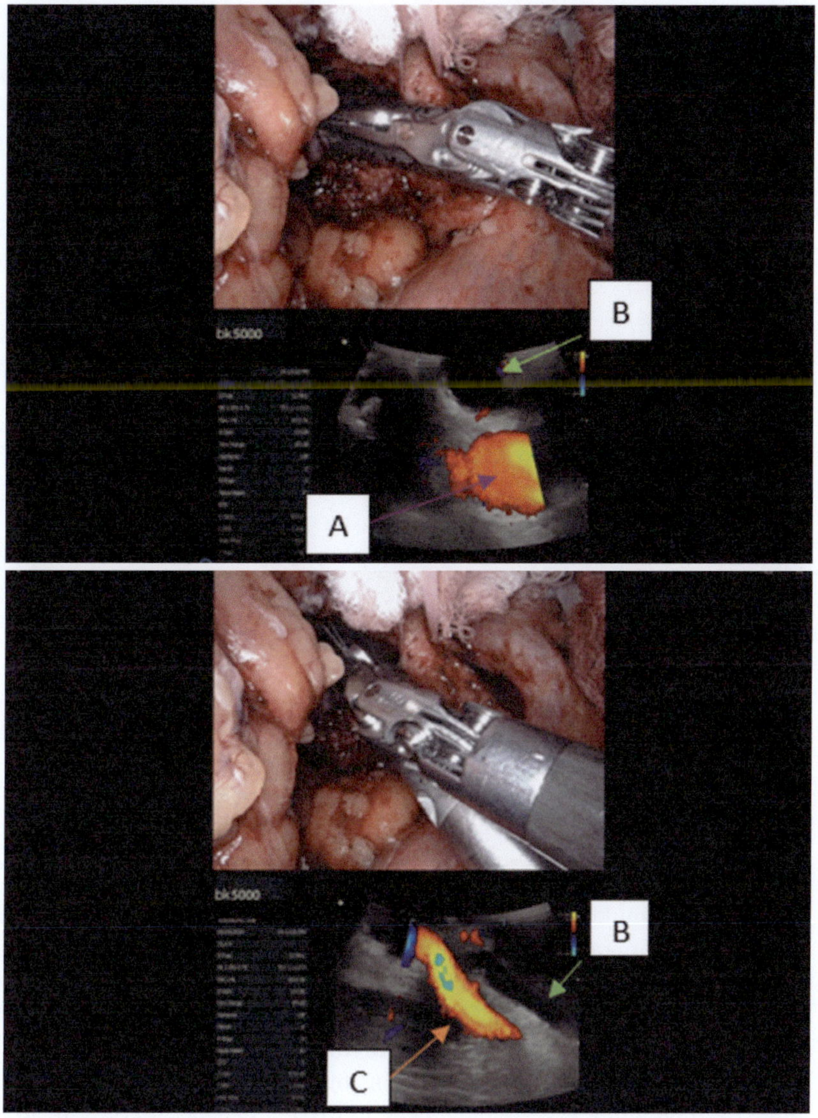

Fig. 25.9 Intraoperative ultrasound of the hepatoduodenal ligament. The portal vein shows doppler flow (**a**), as opposed to the common bile duct (**b**), and the location of aberrant anatomy, in this case a replaced right hepatic artery (**c**) posterior and lateral to the common bile duct (**b**)

duct resection is required. Ultrasound should then be used to survey the bile ducts for evidence of tumor invasion. If an extrahepatic bile duct resection is required, the dissection has already been performed as part of the hilar lymphadenectomy. Patients who require an extrahepatic bile duct resection and reconstruction have not only a much higher morbidity and mortality from the operation but also have a much poorer oncologic prognosis especially in those who are jaundiced [11, 12, 13]. For this reason, we recommend proceeding with caution if a bile duct excision is required for negative margins and strongly consider the patient's ability to tolerate such a procedure. Biliary reconstruction will be addressed later in this chapter.

25.3.6 Hepatectomy

The final step of the operation should be the hepatectomy. Ultrasound should be used to identify

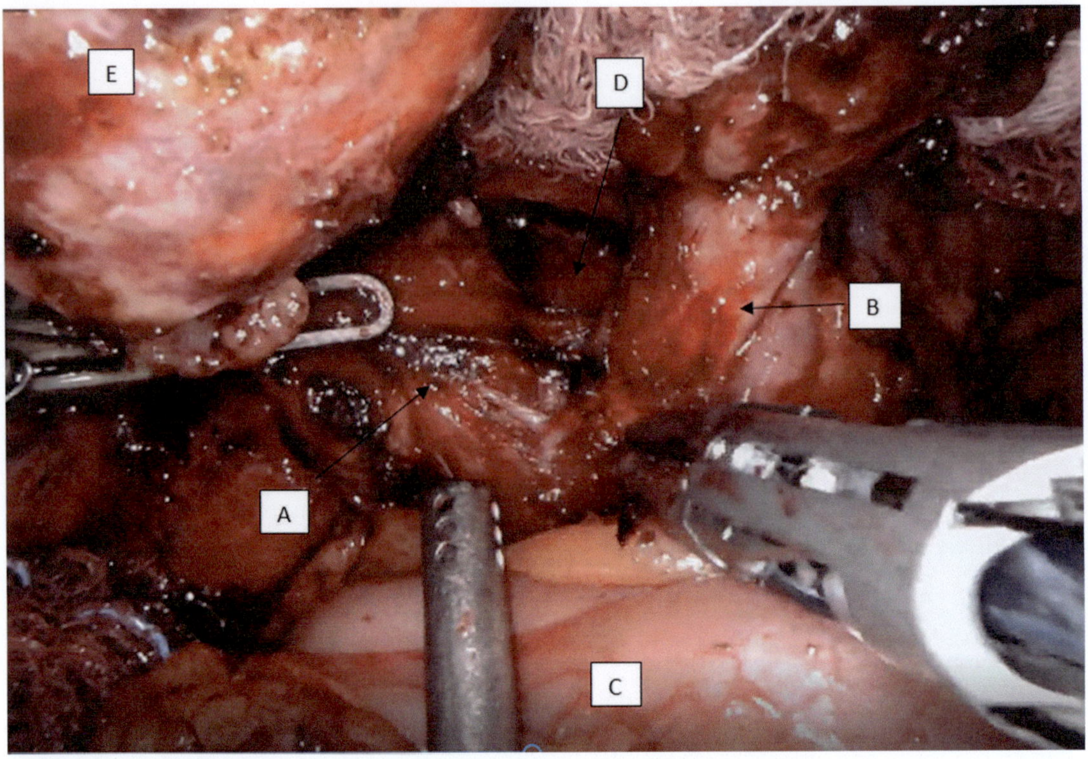

Fig. 25.10 Lateral portal lymph node dissection (station 12b). Station 12b nodes (**a**), common bile duct (**b**), duodenum (**c**), portal vein (**d**), gallbladder mass (**e**)

margins of the tumor within the liver and their relation to the middle hepatic vein, right portal pedicle, and their branches (Fig. 25.13). The middle hepatic vein will be coursing through the gallbladder fossa, and the surgeon should know its exact location prior to transecting parenchyma. It is then recommended to mark the margins of the tumor or, if no tumor is seen in the liver, to mark the boarders of segments 4b and 5 noting the areas where one will come in contact with these vascular structures. Prior to transection, we recommend preparation for inflow control. This can be done in several ways, and we prefer two of these. The first is a pringle maneuver using either a Foley catheter or umbilical tape around the porta hepatis (Fig. 25.14). The second is to perform isolated inflow control by dissecting out the right hepatic artery and right portal vein at the hilum and placing a vascular clamp on both vessels. This is the authors' preferred mechanism since it is relatively simple to place clamps on and take off when needed. In some cases, there can be extensive involvement of the liver outside of segments 4b and 5 which may require a more formal right hepatectomy. In this case, if one has already prepared for isolated inflow control, they may simply ligate the right portal vein and right hepatic artery prior to transection. If this step is necessary, we recommend ligation with silk ties followed by locking clips over stapling.

Parenchymal transection can now be initiated by incising the liver capsule with monopolar scissors. At this portion of the operation, we then switch to using the robotic vessel sealer to transect the liver parenchyma along the pre-marked margins (Fig. 25.15). The vessel sealer can apply bipolar (cauterizing) energy while performing a crush-clamp maneuver and sealing small vessels at the same time. We generally will place our clamps for isolated inflow control prior to tran-

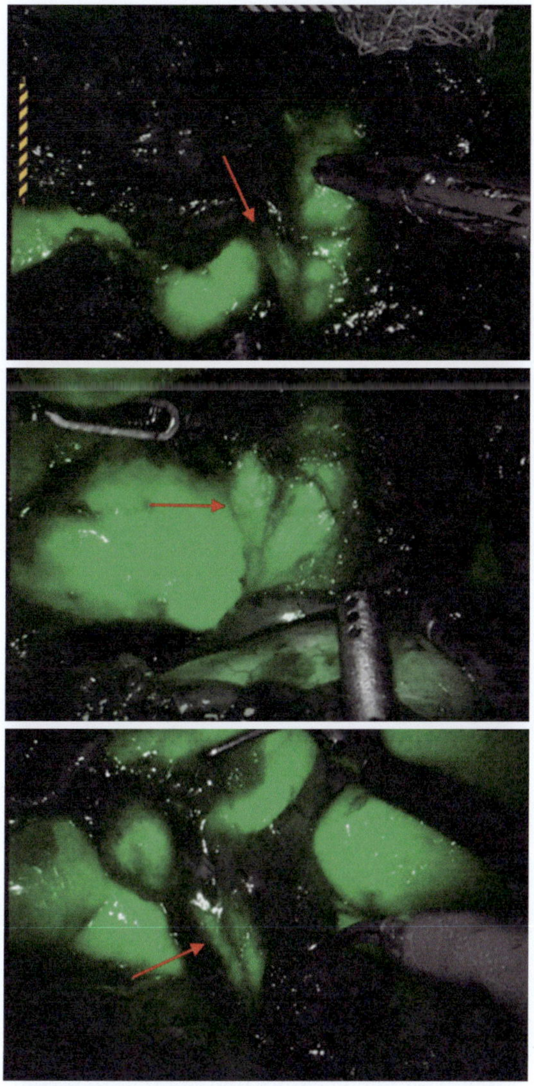

Fig. 25.11 ICG-assisted evaluation of the cystic duct stump (red arrows)

section to minimize blood loss. As the liver parenchyma is transected, the surgeon should be carefully aware of where the middle hepatic vein and right portal pedicle branches are located and use caution while dissecting near them. Once they are identified, they may be ligated with clips or suture and divided. The hepatic parenchyma resected should allow for a clear margin around the disease. We place the specimen in a large lap sack and obtain hemostasis prior to removing it from the assistant midline port which is extended slightly (Fig. 25.16). A drain is usually placed in the liver bed if there is concern for possible bile leak post operatively.

25.3.7 Tips and Tricks

1. The authors of this chapter do not routinely use staplers for liver parenchymal transection; however, if needed, a 15-mm assistant port is used, allowing for a 12-mm robotic stapler cannula to be inserted through the laparoscopic trocar. The robotic arm is then moved to the 12-mm cannula for stapling. This avoids the need for a port larger than 8 mm positioned off midline which would necessitate fascial closure at the end of the case.
2. Ultrasound is extremely useful and integral to any robotic liver procedure. We recommend routine use to gain expertise as well as precise anatomical evaluation and guided liver resection.
3. The authors prefer to use 4 × 8 sponges, which are placed on the undersurface of segment 4b of the liver and retracted with arm 4. Another 4 × 8 sponge is placed in Morrison's pouch to collect any spillage of bile or stones during the cholecystectomy if the gallbladder has not already been removed. Simply placing a sponge under the right or left lobes creates enough retraction to visualize the hepatoduodenal ligament.
4. Monopolar scissors are much more versatile than other dissecting instruments but takes time and practice to master. They allow for use of cutting current for finer dissection when bleeding is not a concern, coagulating current, and cold cutting in the same instrument. Using one tine to "hook" behind tissue and coagulate simulates a hook cautery if desired. The fine point allows for precise dissection similar to a fine Metzenbaum or tenotomy scissors.
5. We recommend administering ICG in the preoperative area or 30 min before incision to have uptake into the biliary tree by the time one begins the hilar dissection.

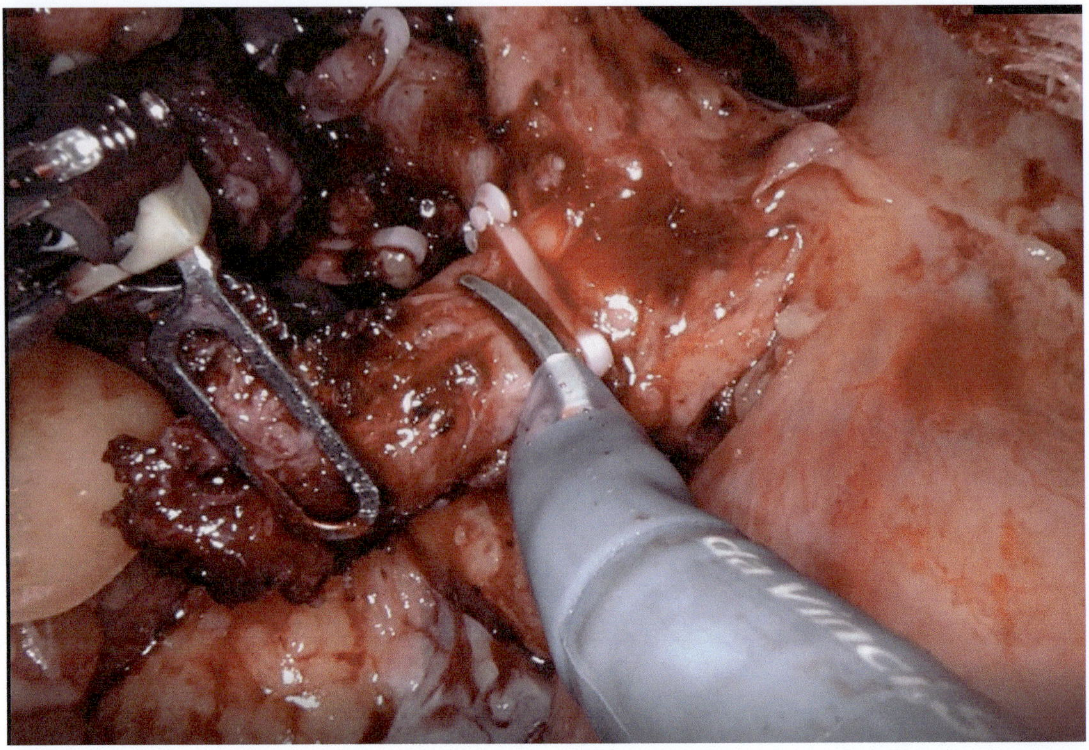

Fig. 25.12 Cystic duct stump evaluation. Margin is sent for frozen pathology

6. Umbilical tapes or vessel loops can be placed around the structures of the hepatoduodenal ligament for easy identification and gentle retraction with the grasper in arm 4.
7. To create a Pringle with a Foley, cut a Foley catheter to about 12–15 cm, removing the side with the ports. Wrap the catheter around the porta hepatis and pull the cut end of the catheter through the window at the tip of the Foley. Then tighten the loop down on the porta hepatis. A clip can be placed to hold the loop in place. To create a Pringle with umbilical tape, wrap the umbilical tape around the porta hepatis and secure with a Hem-o-lok™ clip to desired tightness or secure the umbilical tape extracorporeally with a hemostat on tension to occlude the vessels.
8. For liver parenchymal transection, we recommend "precoagulation" by activating the vessel sealer device before clamping fully on the tissue.

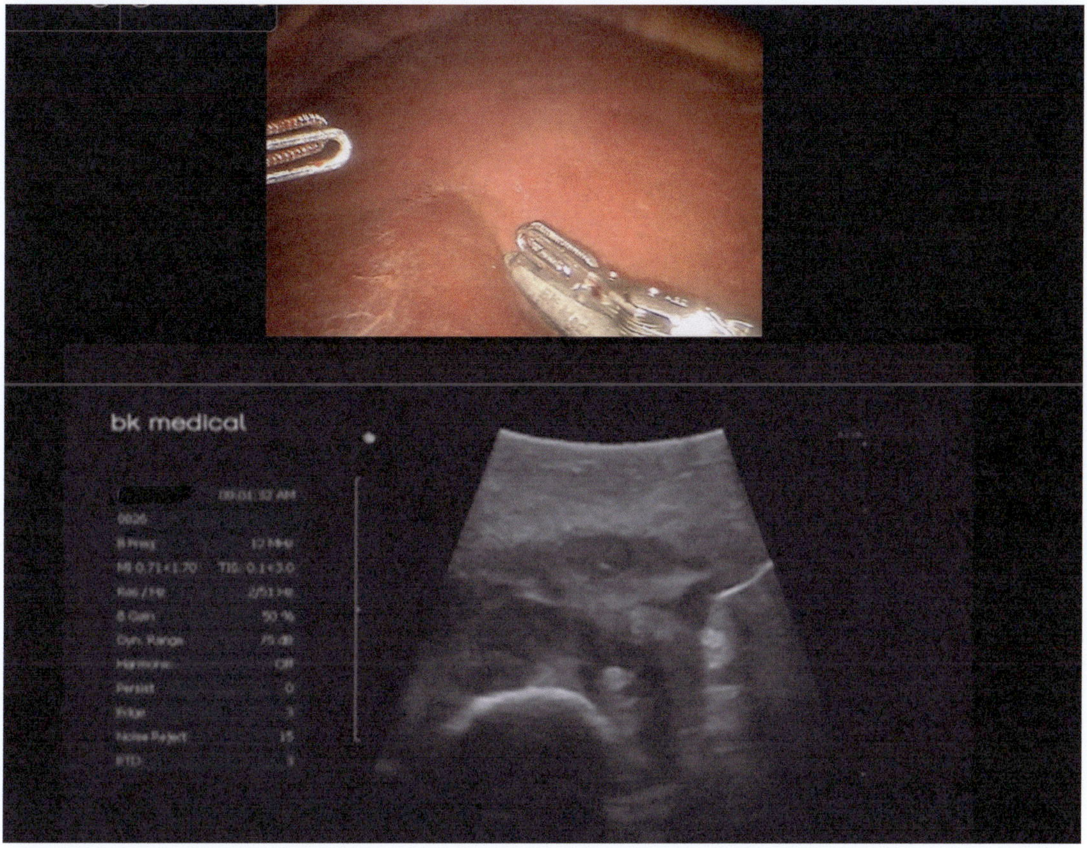

Fig. 25.13 Ultrasound evaluation of tumor margins shows extension into the liver parenchyma

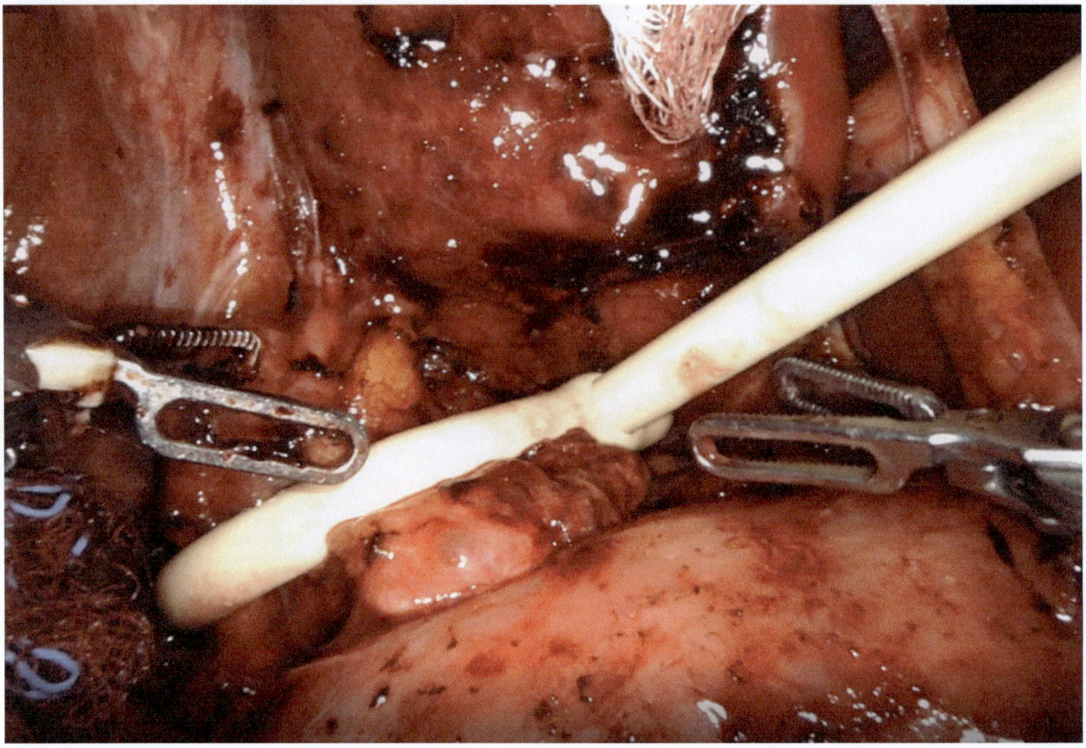

Fig. 25.14 Pringle maneuver with a Foley catheter wrapped around the porta hepatis

Fig. 25.15 Parenchymal transection with the robotic vessel sealer. A prior cholecystectomy had been performed. The margins of segments 4b and 5 were marked with cautery to guide transection

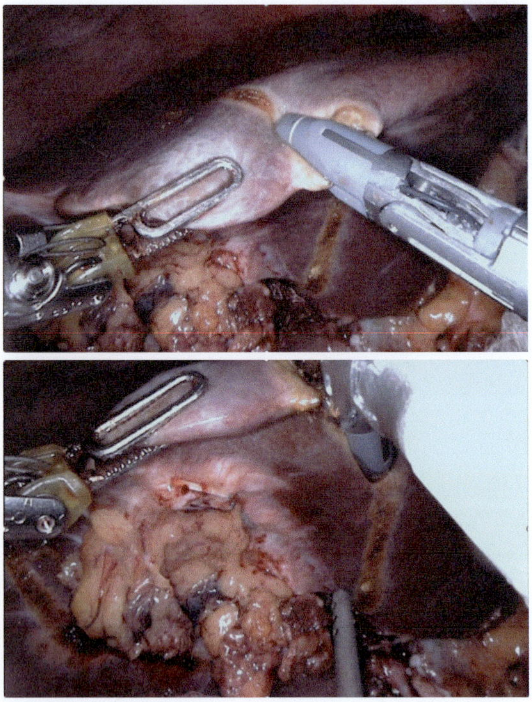

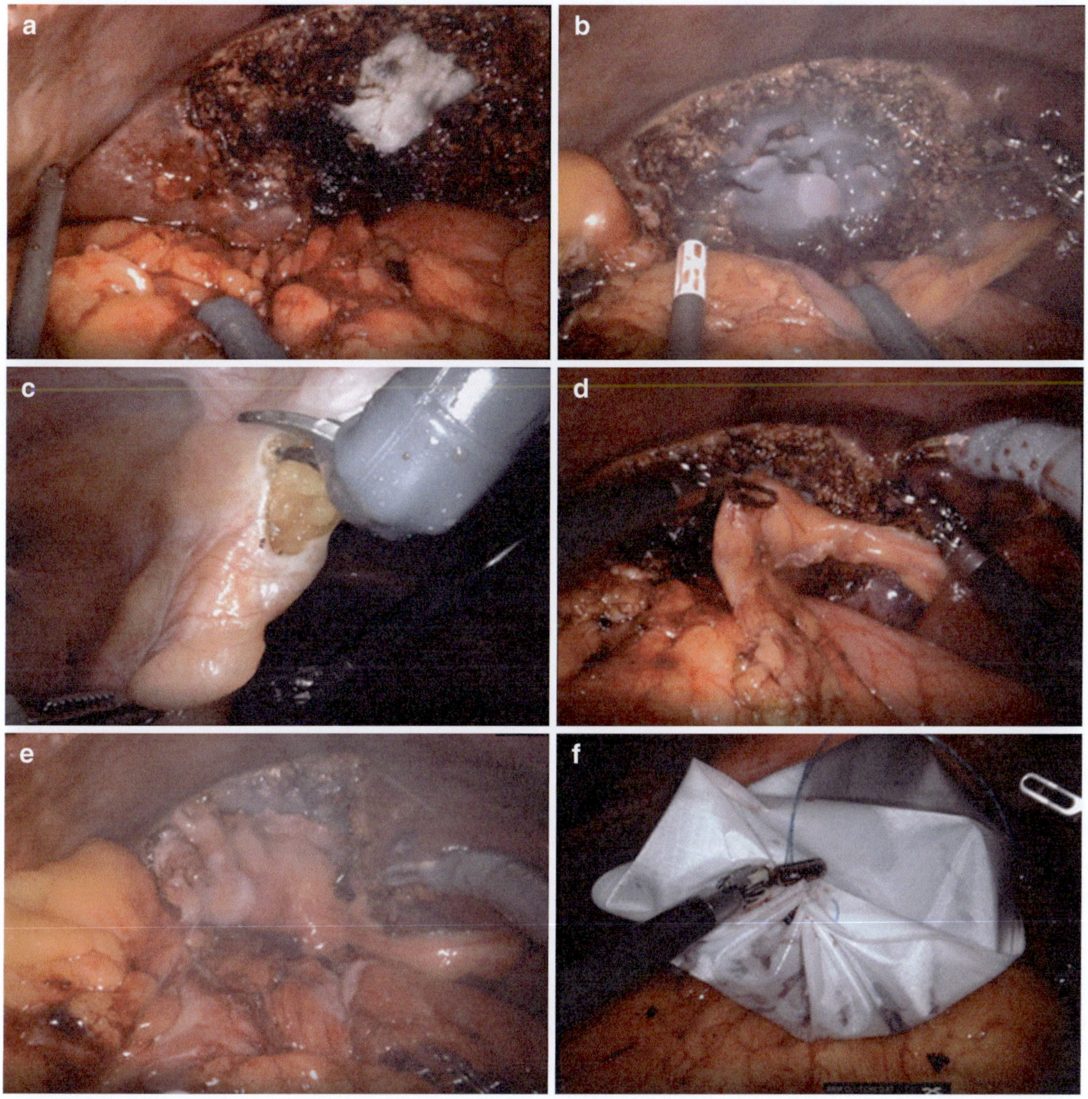

Fig. 25.16 Completed segment 4b/5 partial hepatectomy (**a**), hemostasis (**B**), round ligament flap harvest (**c**, **d**, and **e**), and placement of the specimen into a lap sack (**f**)

25.4 Hilar Lymphadenectomy

Evaluation of the hepatoduodenal ligament is especially important in gallbladder and other biliary cancers because survival is significantly decreased with positive nodal disease. Despite this, hilar lymphadenectomy is underperformed in many centers [13]. In previous studies on robotic pancreaticoduodenectomy for pancreas cancer, the authors of this chapter have found a much higher lymph node yield compared to open pancreaticoduodenectomy which we attributed to a much more precise portal lymph node dissection [14]. For this reason, we believe that the robotic platform provides a superior lymphadenectomy whether performed for pancreas, hepatobiliary, or gallbladder cancer since this dissection is similar.

The hilar dissection should include all nodes of the porta hepatis including stations 8, 9, 12(a, p, and b), and 13 [5] (Fig. 25.17). As described in the previous section of this chapter, we prefer to

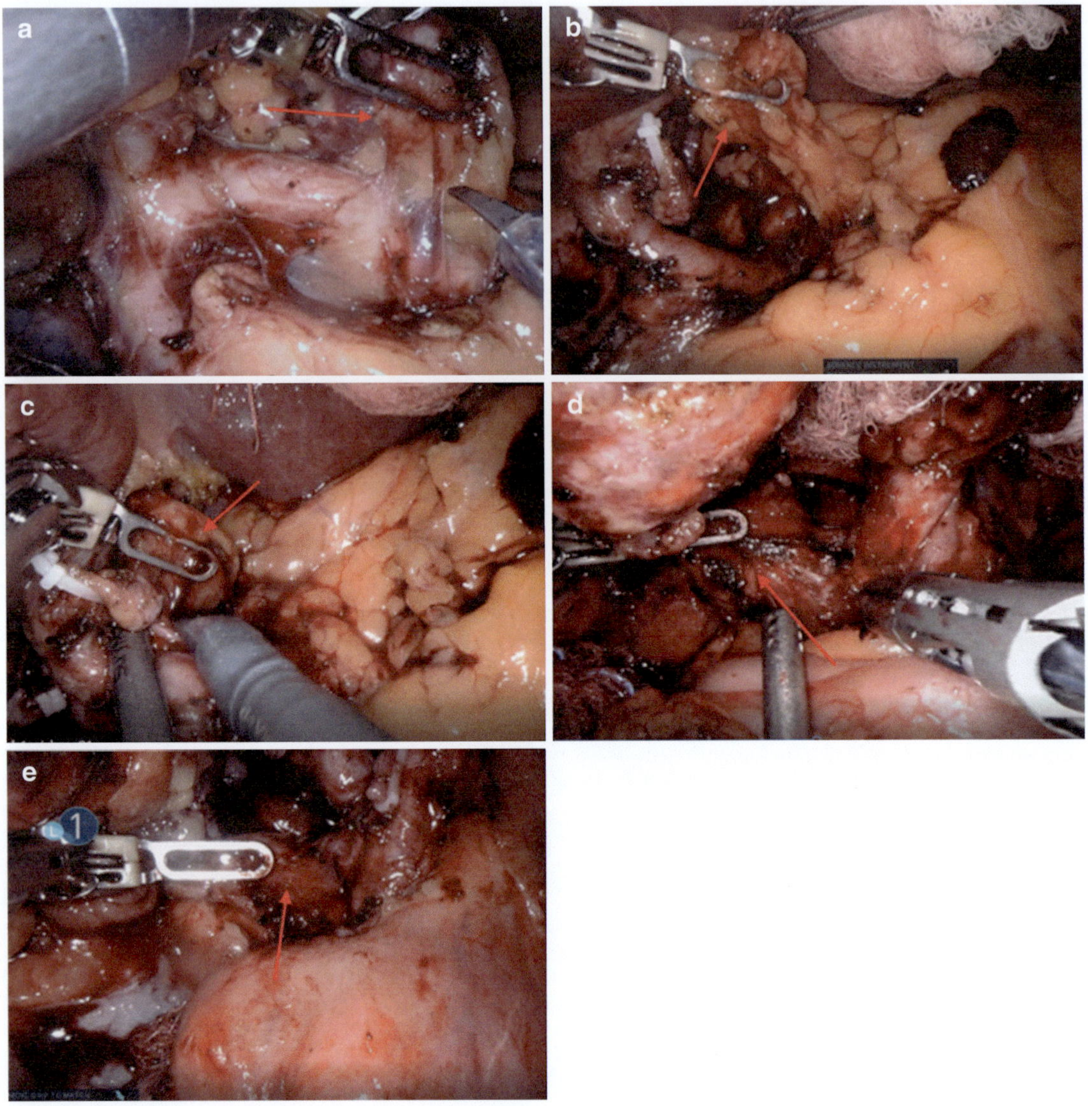

Fig. 25.17 Hilar lymphadenectomy. Red arrows mark the nodal tissue. (**a**) Hepatic artery lymph node (station 8), (**b**) celiac Lymph nodes (station 9), (**c**) medial portal lymph nodes (station 12a), (**d**) lateral/posterior portal lymph nodes (station 12b), (**e**) posterior duodenal lymph nodes (station 13)

begin with evaluation of the station 8 lymph node since metastasis outside the porta hepatis is a significant prognostic factor. This can be found by incising the pars flaccida along the lesser curve of the stomach and dissecting toward the hepatoduodenal ligament. The lymph node will be found just medial to the porta hepatis and anterior to the common hepatic artery. We prefer to use the monopolar scissors for fine dissection of this node to avoid inadvertent injury to the hepatic artery and its branches. There are usually small nodal branches of the hepatic artery and coronary vein which can be taken with a vessel sealer. Occasionally, there will be more lymph nodes heading down toward the celiac axis that can be removed with the same specimen. One must take care not to injure the coronary vein, left gastric vein, or left gastric artery when excising the nodes in this area. Once station 8 lymph node has been excised, the dissection continues toward the celiac axis to remove station 9 lymph nodes in a similar fashion. Next incising the peritoneum over the porta hepatis allows for identification of the right gastric and gastroduodenal arteries as they originate from the common hepatic artery. We prefer to preserve the right gastric artery and GDA in this case; however, if they are impeding visualization, they may be sacrificed without much consequence. If the GDA is ligated, the portal vein will then become visible posterior to it. The hepatic artery and its branches should be traced up to the liver hilum and removing all the lymph nodal tissue of stations 12a and 12p which are between these structures and the bile duct. Indocyanine green fluorescence when administered preoperatively helps to identify the bile duct. Again, care should be taken not to injure the anterior wall of the portal vein at the posterior aspect of the porta hepatis. Lateral portal lymph nodes will be found on the lateral edge of the bile duct (station 12b). Use caution to avoid injury to a replaced right hepatic artery at this juncture (Fig. 25.9). The vessel sealer is useful at this stage to dissect these nodes away from the bile duct and portal vein. Medial retraction on the bile duct by either placement of an umbilical tape or retracting from the cystic duct stump helps to gain access to the portal vein posteriorly and the posterior portal lymph nodes. The duodenum should be partially kocharized at the superior border to aid with visualization of these nodes and the posterior duodenal nodes (station 13) which should then be carefully dissected. With the angle of the robotic camera looking up toward the liver hilum, this nodal dissection should take place from inferior to superior starting at the edge of the duodenum and finishing near the hilar plate "de-fatting" the porta and skeletonizing the structures (Fig. 25.18).

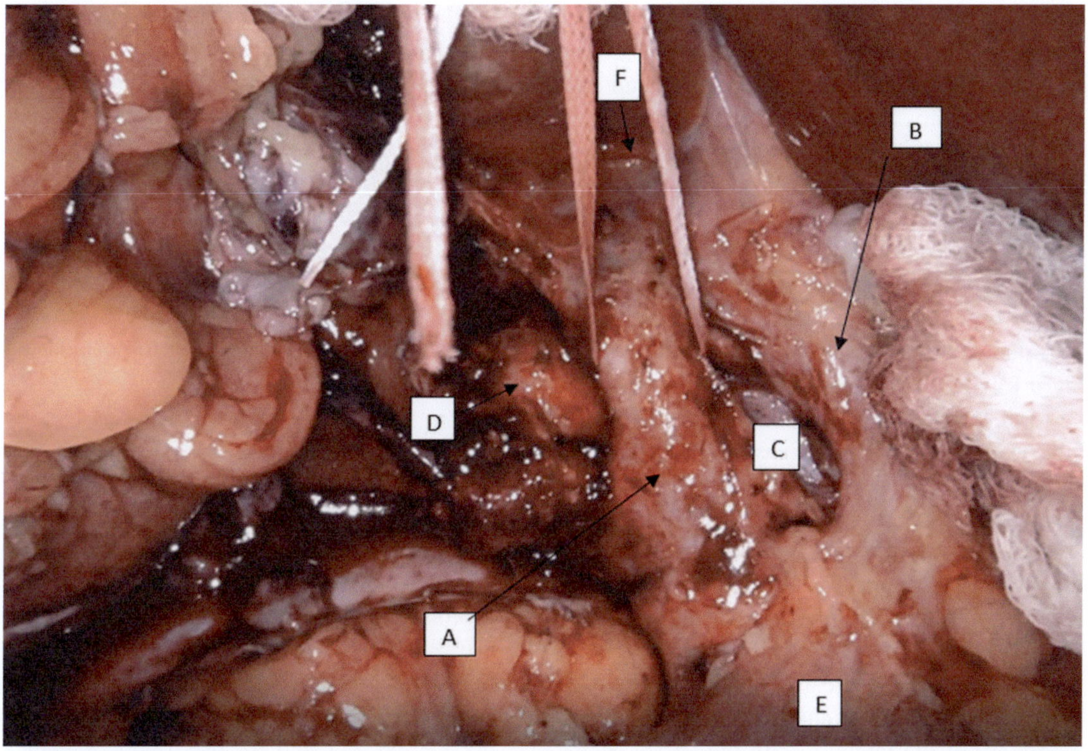

Fig. 25.18 Hilar lymphadenectomy. Hepatoduodenal structures are skeletonized with removal of all lymphatic tissue. Common bile duct (**a**), hepatic artery (**b**), portal vein (**c**), replaced right hepatic artery (**d**), duodenum (**e**), hilar plate (**f**)

25.4.1 Tips and Tricks

1. Whether performing hilar lymphadenectomy for gallbladder cancer, cholangiocarcinoma, pancreas cancer, or any other pathology, begin with the station 8 lymph node. This helps to identify the hepatic artery which can then be traced to its branches and safely expose other structures within the hepatoduodenal ligament.
2. One can use vessel loops or umbilical tape to retract different structures in the hepatoduodenal ligament for maximum exposure. We recommend using arm 4 as a "double retractor" by lifting the left lobe of the liver with the arm and grasping a vessel loop or umbilical tape with the instrument for retraction.
3. Partially kocharize the superior portion of the duodenum to gain access to station 13 nodes.

25.5 Biliary Reconstruction

It should be noted that gallbladder cancers directly extending into the porta hepatis and bile duct are usually also involving the vasculature and therefore typically unresectable, from both technical and oncological standpoints. If deemed necessary and feasible to perform a bile duct resection, one must be prepared to reconstruct the biliary system. The robotic approach provides many advantages in this situation. The wristed instruments are especially valuable when suturing a fragile anastomosis. We always choose to perform a Roux en Y hepaticojejunostomy for biliary reconstruction for reasons of feasibility and lower infectious complications associated with choledochal or hepaticoduodenostomy.

Reconstruction begins with preparing a roux limb of small bowel. We divide the jejunum approximately 20–40 cm from the ligament of Treitz where the mesentery allows for a tension free anastomosis. To streamline this process with minimal instrument exchanges, first incise the mesentery of the jejunum with the vessel sealer prior to transecting the bowel. We find that the robotic stapler is a great tool for bowel transection. A 12-mm robotic canula can be placed through the 15-mm assistant port, described in the patient positioning and trocar placement segment of the section on "Robotic Liver Resection for Gallbladder Cancer." This obviates the need for upsizing one of the 8-mm robotic trocars off midline. Arm 3 is then re-docked to this cannula, and the stapler can be used. The position of this trocar is also more amenable to stapling the jejunum in this area than any of the other robotic trocar sites. After stapling, arm 3 can then be re-docked to its original trocar position. A retrocolic Roux limb is brought through the bare area of the transverse colon mesentery to the right of the middle colic vessels and positioned near the bile duct for an end-to-side reconstruction. We recommend using monopolar scissors to create an enterotomy and switching to a needle driver in arm 3 to suture the back wall of the bile duct to the jejunum. Hepaticojejunostomy can be performed with either running or interrupted sutures (Fig. 25.19). We prefer a running technique for larger ducts over 8 mm in diameter. Interrupted anastomoses are performed with an absorbable monofilament suture such as Monocryl™ or PDS™. Sutures are placed in the back wall first and continue around the duct circumferentially with the last stitch at the 12 o'clock position on the duct. For a running anastomosis, we recommend using an absorbable barbed suture for a thicker duct or an absorbable monofilament such as Monocryl™ or PDS™ for thinner ducts. We prefer to place a drain near the anastomosis at the end of the case.

25.5.1 Tips and Tricks

1. Place 4 × 8 sponges assist with holding the transverse colon up to identify the ligament of Treitz, thereby minimizing tissue trauma.
2. We recommend transecting the jejunum and performing the jejunojejunostomy prior to the biliary anastomosis so this process does not need to be repeated for the robotic stapler a second time and streamlines the process (Fig. 25.20).

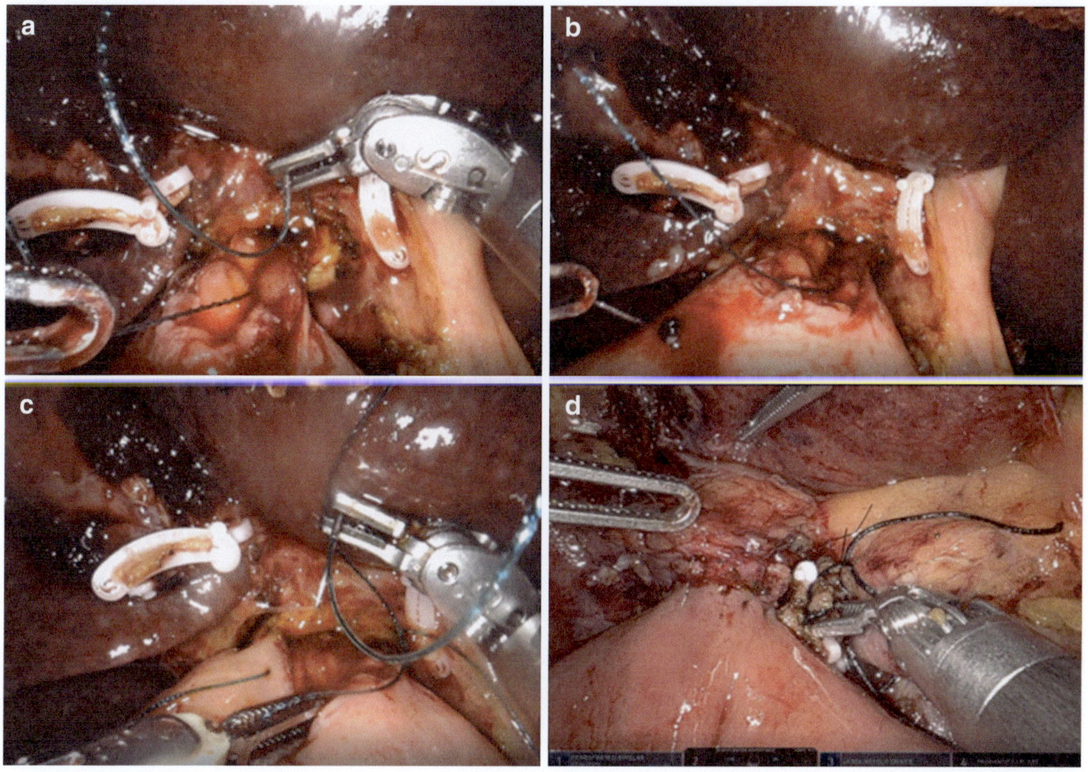

Fig. 25.19 Hepaticojejunostomy is performed either running (**a**, **b**, and **c**) for larger ducts or interrupted (**d**) for smaller ducts

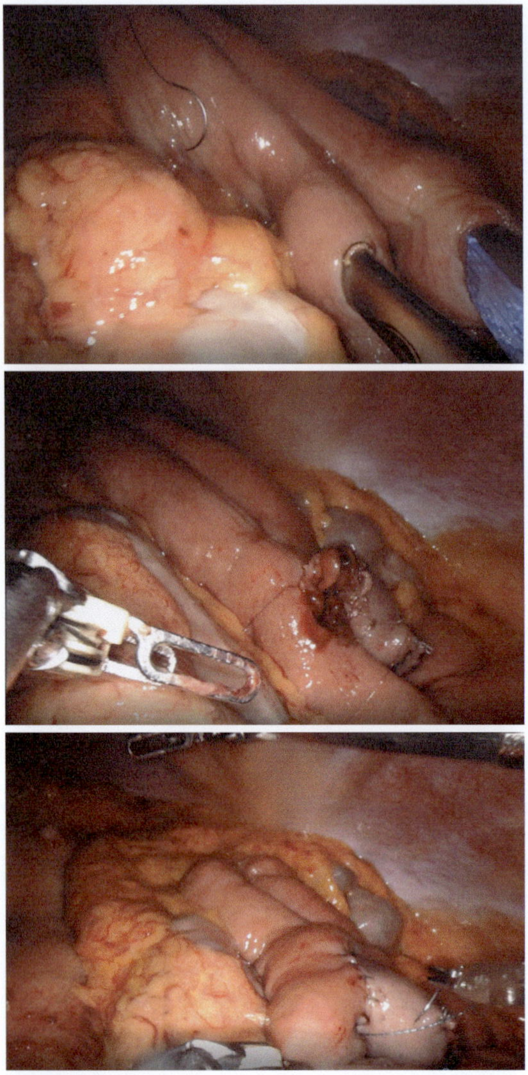

Fig. 25.20 Creation of a stapled side to side jejunojejunostomy after mobilization of a Roux limb for biliary reconstruction. The common channel is closed with running barbed suture

25.6 Literature Review

25.6.1 The Robotic Approach to Gallbladder Cancer and Biliary Malignancies

There has been a global evolution of the surgical management of hepatobiliary cancers over the last few decades. Prior to the 1990s, liver resections were largely performed open. The first reports of laparoscopic hepatobiliary surgery began to emerge at this time, and minimally invasive liver resections for biliary cancers were reported in the early 2000s [15–19]. While laparoscopic hepatectomy began to gain popularity, robotic surgery was emerging for other abdominal operations [20]. Robotic approaches to hepatobiliary malignancies steadily progressed over the following 20 years and continue to grow alongside laparoscopy [21, 22, 23].

Surgery for gallbladder cancer is considered much more complex than simple cholecystectomy or even minor liver resections. As described in the technical aspects above, it is a multistep procedure which requires attention to oncologic detail. One of the more difficult aspects is the lymphadenectomy and hilar dissection. Consequently, minimally invasive approaches to gallbladder cancer operations and radical cholecystectomy only began to surface in the last decade [24]. Classical teaching involved an open resection for gallbladder cancer due to the theoretical risk of port site tumor seeding with a minimally invasive approach. However, this notion has been largely dismissed with many studies showing similar rates of port site recurrence in both MIS and open cohorts and lack of improved outcomes with port site resection [25–28].

With minimally invasive approaches to gallbladder cancer surgery becoming more accepted, surgeons began to realize the benefit of the robot in the liver hilum. Multiple reports of improved fine dissection within the porta hepatis during robotic resections for gallbladder and biliary tract cancers have been published [29]. Robotic resections have been documented as safe and effective means for such complex operations. More recently comparative studies between robotic, laparoscopic, and open resections for gallbladder cancer began to emerge showing improvements in operative blood loss, length of stay, and postoperative morbidity for robotic resections [30, 31, 32]. Despite these reports, there remains an ongoing need for prospective data regarding comparative outcomes.

25.7 Conclusions

Gallbladder cancer is an extremely complex malignancy with a prognosis that is dictated by stage and lymph nodal involvement. As stated in the current National Comprehensive Cancer Network guidelines, resectable disease should be addressed by a surgeon with expertise in gallbladder cancer resections and involves a cholecystectomy with resection of the cystic plate, hilar lymphadenectomy, and liver resection of segments 4b and 5 with possible extended hepatectomy or bile duct excision with the goal of achieving R0 status [33]. Initial workup for gallbladder cancer depends on the presentation. A suspicious gallbladder mass should prompt imaging with multiphase CT or MRI as well as tumor markers. We also recommend EUS if any suspicious lymphadenopathy is suggested on imaging or it appears that the lesion may involve a significant portion of the liver, bile ducts, or adjacent organs. In these cases, our institution utilizes chemotherapy on a neoadjuvant basis prior to resection. Laparoscopy is an extremely useful tool to diagnose unresectable disease and peritoneal spread and is recommended as part of the algorithm for diagnosis based on multiple studies documenting its usefulness [7, 8, 33]. In patients who have an incidental pathologic diagnosis of gallbladder cancer after cholecystectomy, pathologic evaluation is most important in this case to document the cystic duct and cystic plate margins, location, and T stage of the tumor and any lymph node involvement if sampled. In cases of T1a disease with R0 resection and no tumor in lymph nodes, no further surgical resection is required. In T1b and above disease, further evaluation with cross-sectional imaging to include chest imaging is warranted for staging. If resectable disease is diagnosed, then one can proceed with a staging operation to include liver resection, portal lymphadenectomy, and bile duct excision if warranted after inspection of the cystic duct stump. Medical oncology referral for adjuvant chemotherapy is then pursued. If unresectable disease is identified on imaging or laparoscopy, then systemic therapy is preferred [33].

The robotic approach to gallbladder cancer resections has shown to be effective and safe and may provide improvements in certain outcomes such as hospital length of stay, estimated blood loss, and postoperative morbidity. As outlined above we recommend approaching this operation in six steps: initial patient positioning and trocar placement, laparoscopy and intraoperative ultrasound, hepatic artery and celiac lymph node evaluation, hepatoduodenal ligament dissection and lymphadenectomy, cystic duct stump evaluation and cholecystectomy, and hepatectomy. If necessary, a bile duct excision can take place at the time of cystic duct stump evaluation to achieve R0 status if the patient is deemed fit enough to withstand this operation and any potential complications. We also describe the preferred technique for robotic biliary reconstruction in this chapter. It should be emphasized that robotic radical resection for gallbladder cancer is a complex operation and prior robotic expertise is absolutely necessary.

References

1. Rawla P, Sunkara T, Thandra KC, Barsouk A. Epidemiology of gallbladder cancer. Clin Exp Hepatol. 2019;5(2):93–102. https://doi.org/10.5114/ceh.2019.85166. Epub 2019 May 23. PMID: 31501784; PMCID: PMC6728871.
2. Patel JD, LLson D, Kamath S, et al., editors. Cancer. Net: Gallbladder Cancer. Alexandria, VA: American Society of Clinical Oncology (ASCO); 2005–2022. Available from: https://www.cancer.net/cancer-types/gallbladder-cancer . Accessed 15 Dec 2022.
3. Williams TM, Majithia L, Wang SJ, Thomas CR Jr. Defining the role of adjuvant therapy: cholangiocarcinoma and gall bladder cancer. Semin Radiat Oncol. 2014;24(2):94–104. https://doi.org/10.1016/j.semradonc.2014.01.001. PMID: 24635866.
4. Butte JM, Kingham TP, Gönen M, D'Angelica MI, Allen PJ, Fong Y, DeMatteo RP, Jarnagin WR. Residual disease predicts outcomes after definitive resection for incidental gallbladder cancer. J Am Coll Surg. 2014;219(3):416–29. https://doi.org/10.1016/j.jamcollsurg.2014.01.069. Epub 2014 May 16. PMID: 25087941; PMCID: PMC4143454.
5. Bartlett DL, Fong Y, Fortner JG, Brennan MF, Blumgart LH. Long-term results after resection for gallbladder cancer. Implications for staging and management. Ann Surg. 1996;224(5):639–46. https://doi.org/10.1097/00000658-199611000-00008. PMID: 8916879; PMCID: PMC1235441

6. Cassese G, Han HS, Yoon YS, Lee JS, Cho JY, Lee HW, Lee B, Troisi RI. Preoperative assessment and perioperative management of resectable gallbladder cancer in the era of precision medicine and novel technologies: state of the art and future perspectives. Diagnostics (Basel). 2022;12(7):1630. https://doi.org/10.3390/diagnostics12071630. PMID: 35885535; PMCID: PMC9320561.
7. Leung U, Pandit-Taskar N, Corvera CU, D'Angelica MI, Allen PJ, Kingham TP, DeMatteo RP, Jarnagin WR, Fong Y. Impact of pre-operative positron emission tomography in gallbladder cancer. HPB (Oxford). 2014;16(11):1023–30. https://doi.org/10.1111/hpb.12282. Epub 2014 Jun 4. PMID: 24894161; PMCID: PMC4487754
8. Agrawal S, Sonawane RN, Behari A, Kumar A, Sikora SS, Saxena R, Kapoor VK. Laparoscopic staging in gallbladder cancer. Dig Surg. 2005;22(6):440–5. https://doi.org/10.1159/000091447. Epub 2006 Feb 10. PMID: 16479113.
9. Agarwal AK, Kalayarasan R, Javed A, Gupta N, Nag HH. The role of staging laparoscopy in primary gall bladder cancer—an analysis of 409 patients: a prospective study to evaluate the role of staging laparoscopy in the management of gallbladder cancer. Ann Surg. 2013;258(2):318–23. https://doi.org/10.1097/SLA.0b013e318271497e. PMID: 23059504.
10. Tran Cao HS, Zhang Q, Sada YH, Chai C, Curley SA, Massarweh NN. The role of surgery and adjuvant therapy in lymph node-positive cancers of the gallbladder and intrahepatic bile ducts. Cancer. 2018;124(1):74–83. https://doi.org/10.1002/cncr.30968. Epub 2017 Aug 25. PMID: 28841223.
11. D'Angelica M, Dalal KM, DeMatteo RP, Fong Y, Blumgart LH, Jarnagin WR. Analysis of the extent of resection for adenocarcinoma of the gallbladder. Ann Surg Oncol. 2009;16(4):806–16. https://doi.org/10.1245/s10434-008-0189-3. Epub 2008 Nov 5. PMID: 18985272.
12. Dasari BVM, Ionescu MI, Pawlik TM, Hodson J, Sutcliffe RP, Roberts KJ, Muiesan P, Isaac J, Marudanayagam R, Mirza DF. Outcomes of surgical resection of gallbladder cancer in patients presenting with jaundice: a systematic review and meta-analysis. J Surg Oncol. 2018;118(3):477–85. https://doi.org/10.1002/jso.25186. PMID: 30259519.
13. Kim BJ, Newhook TE, Tzeng CD, Ikoma N, Chiang YJ, Chun YS, Vauthey JN, Tran Cao HS. Lymphadenectomy and margin-negative resection for biliary tract cancer surgery in the United States-differential technical performance by approach. J Surg Oncol. 2022;126(4):658–66. https://doi.org/10.1002/jso.26924. Epub 2022 May 16. PMID: 35578764.
14. Baimas-George M, Watson M, Murphy KJ, Iannitti D, Baker E, Ocuin L, Vrochides D, Martinie JB. Robotic pancreaticoduodenectomy may offer improved oncologic outcomes over open surgery: a propensity-matched single-institution study. Surg Endosc. 2020;34(8):3644–9. https://doi.org/10.1007/s00464-020-07564-x. Epub 2020 Apr 23. PMID: 32328825.
15. Stoker ME, Leveillee RJ, McCann JC Jr, Maini BS. Laparoscopic common bile duct exploration. J Laparoendosc Surg. 1991;1(5):287–93. https://doi.org/10.1089/lps.1991.1.287. PMID: 1834281.
16. Ferzli G, David A, Kiel T. Laparoscopic resection of a large hepatic tumor. Surg Endosc. 1995;9(6):733–5. https://doi.org/10.1007/BF00187953. PMID: 7482177.
17. Harimoto N, Shimada M, Tsujita E, Maehara S, Rikimaru T, Yamashita Y, Maeda T, Tanaka S, Shirabe K, Sugimachi K. Laparoscopic hepatectomy and dissection of lymph nodes for intrahepatic cholangiocarcinoma. Case report. Surg Endosc. 2002;16(12):1806. https://doi.org/10.1007/s00464-002-4511-z. Epub 2002 Sep 23. PMID: 12239645.
18. Harimoto N, Shimada M, Tsujita E, Maehara S, Rikimaru T, Yamashita Y, Maeda T, Tanaka S, Shirabe K, Sugimachi K. Laparoscopic hepatectomy and dissection of lymph nodes for intrahepatic cholangiocarcinoma. Case report Surg Endosc. 2002;16(12):1806. https://doi.org/10.1007/s00464-002-4511-z. Epub 2002 Sep 23. PMID: 12239645.
19. Nguyen KT, Gamblin TC, Geller DA. World review of laparoscopic liver resection-2,804 patients. Ann Surg. 2009;250(5):831–41. https://doi.org/10.1097/SLA.0b013e3181b0c4df. PMID: 19801936.
20. Sheetz KH, Claflin J, Dimick JB. Trends in the adoption of robotic surgery for common surgical procedures. JAMA Netw Open. 2020;3(1):e1918911. https://doi.org/10.1001/jamanetworkopen.2019.18911. PMID: 31922557; PMCID: PMC6991252.
21. Berber E, Akyildiz HY, Aucejo F, Gunasekaran G, Chalikonda S, Fung J. Robotic versus laparoscopic resection of liver tumours. HPB (Oxford). 2010;12(8):583–6. https://doi.org/10.1111/j.1477--2574.2010.00234.x. PMID: 20887327; PMCID: PMC2997665.
22. Nota CL, Rinkes IHB, Molenaar IQ, van Santvoort HC, Fong Y, Hagendoorn J. Robot-assisted laparoscopic liver resection: a systematic review and pooled analysis of minor and major hepatectomies. HPB (Oxford). 2016;18(2):113–20. https://doi.org/10.1016/j.hpb.2015.09.003. Epub 2015 Dec 11. PMID: 26902129; PMCID: PMC4814602.
23. Fruscione M, Pickens R, Baker EH, Cochran A, Khan A, Ocuin L, Iannitti DA, Vrochides D, Martinie JB. Robotic-assisted versus laparoscopic major liver resection: analysis of outcomes from a single center. HPB (Oxford). 2019;21(7):906–11. https://doi.org/10.1016/j.hpb.2018.11.011. Epub 2019 Jan 5. PMID: 30617001.
24. Gumbs AA, Hoffman JP. Laparoscopic radical cholecystectomy and roux-en-Y choledochojejunostomy for gallbladder cancer. Surg Endosc. 2010;24(7):1766–8. https://doi.org/10.1007/s00464-009-0840-5. Epub 2010 Jan 7. PMID: 20054570.
25. Paolucci V. Port site recurrences after laparoscopic cholecystectomy. J Hepato-Biliary-Pancreat Surg. 2001;8(6):535–43. https://doi.org/10.1007/s005340100022. PMID: 11956905.

26. Suzuki K, Kimura T, Ogawa H. Long-term prognosis of gallbladder cancer diagnosed after laparoscopic cholecystectomy. Surg Endosc. 2000;14(8):712–6. https://doi.org/10.1007/s004640000145. PMID: 10954815.
27. Ricardo AE, Feig BW, Ellis LM, Hunt KK, Curley SA, MacFadyen BV Jr, Mansfield PF. Gallbladder cancer and trocar site recurrences. Am J Surg. 1997;174(6):619–22; discussion 622–3. https://doi.org/10.1016/s0002-9610(97)00178-5. PMID: 9409585.
28. Maker AV, Butte JM, Oxenberg J, Kuk D, Gonen M, Fong Y, Dematteo RP, D'Angelica MI, Allen PJ, Jarnagin WR. Is port site resection necessary in the surgical management of gallbladder cancer? Ann Surg Oncol. 2012;19(2):409–17. https://doi.org/10.1245/s10434-011-1850-9. Epub 2011 Jun 23. PMID: 21698501.
29. Shen BY, Zhan Q, Deng XX, Bo H, Liu Q, Peng CH, Li HW. Radical resection of gallbladder cancer: could it be robotic? Surg Endosc. 2012;26(11):3245–50. https://doi.org/10.1007/s00464-012-2330-4. Epub 2012 May 31
30. Tschuor C, Pickens RC, Isenberg EE, Motz BM, Salibi PN, Robinson JN, Murphy KJ, Iannitti DA, Baker EH, Vrochides D, Martinie JB. Robotic resection of gallbladder cancer: a single-center retrospective comparative study to open resection. Am Surg. 2021;15:31348211047491. https://doi.org/10.1177/00031348211047491. Epub ahead of print. PMID: 34652250.
31. Byun Y, Choi YJ, Kang JS, Han Y, Kim H, Kwon W, Jang JY. Robotic extended cholecystectomy in gallbladder cancer. Surg Endosc. 2020;34(7):3256–61. https://doi.org/10.1007/s00464-020-07554-z. Epub 2020 Apr 18. PMID: 32306112.
32. Goel M, Khobragade K, Patkar S, Kanetkar A, Kurunkar S. Robotic surgery for gallbladder cancer: operative technique and early outcomes. J Surg Oncol. 2019;119(7):958–63. https://doi.org/10.1002/jso.25422. Epub 2019 Feb 25. PMID: 30802316.
33. Benson AB, D'Angelica MI, Abbott DE, et al. Hepatobiliary cancers. National Comprehensive Cancer Network. NCCN Guidelines Version 4.2022. Available from: https://www.nccn.org/professionals/physician_gls/pdf/hepatobiliary.pdf. Accessed on 28 Dec 2022.

Robotic ALPPS Procedure

26

Marcel Autran Machado

26.1 Introduction

The ALPPS procedure, acronym for associating liver partition and portal vein ligation for staged hepatectomy, is a useful strategy to treat patients with advanced primary or metastatic liver tumors and small future liver remnants (FLR) [1, 2]. This technique induces greater liver hypertrophy, although initial reports indicated that it results in significant morbidity and mortality [1, 3] when compared to portal vein embolization (PVE). Since then, better patient selection and precise indications have led to better results [4–7]. Simultaneously, important technical modifications [8–13] have reduced complications, and today, the ALPPS procedure is established and accepted worldwide [7]. The use of minimally invasive surgery (in our view) is an excellent way to reduce surgical severity, and since 2012, we have been applying minimally invasive techniques for ALPPS procedure [8, 13]. This chapter shows the technique and current status of robotic ALLPS procedure.

M. A. Machado (✉)
Sirio Libanes Hospital, São Paulo, Brazil
e-mail: dr@drmarcel.com.br

26.2 Surgical Technique

26.2.1 Positioning of the Patient and Placement of the Ports

The patient is positioned supine and 30° inverted Trendelenburg position. Robotic surgery is performed using the da Vinci Xi robotic platform (Intuitive Surgical Inc., Sunnyvale, CA). In this technique, five trocars are used (Fig. 26.1). A pneumoperitoneum is created through the umbilical port using an open technique. The pneumoperitoneum is created at 14 mmHg. The remaining trocars are inserted under direct vision. The surgeon sits at the robotic console, and the assistant

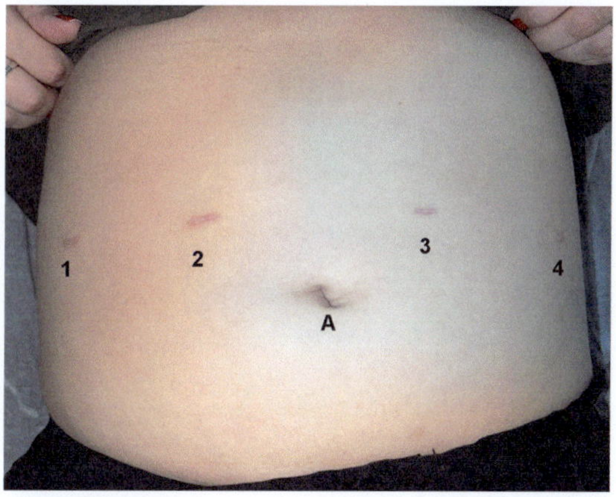

Fig. 26.1 Robotic ALPPS. Placement of the trocar. Postoperative image of the abdominal wall with incisions for the robotic arms (1–4) and the auxiliary port (A)

M. D'Hondt, I. Sucandy (eds.), *Textbook of Robotic Liver Surgery*,
https://doi.org/10.1007/978-3-031-76536-0_26

surgeon stands on the left side of the patient. The assistant surgeon performs retraction, suction, clipping, and stapling and changes robotic instruments.

26.2.2 First Stage

The first stage consists of preparation for future liver remnant (FLR) growth. In cases where tumors are located in the FLR, intraoperative ultrasonography and resection of the lesions from the FLR are performed (Fig. 26.2). After this task, or if there are no lesions in the FLR, the next step is dissection and ligation of the right portal vein, followed by the bipartition of the liver, usually at the line of the falciform ligament, once the FLR will be liver segments 2 and 3 (± S1) as originally described [1]. There are several other types of ALPPS that are derived from the original and may be important to mention. In reversed ALPPS [14], the FLR is the right posterior sector (S6–S7), and monosegmental ALPPS is named after ALPPS in which only one liver segment (+/− s1) remains as the FLR [15].

The hepatoduodenal ligament is dissected, and the right portal vein is dissected and encircled (Fig. 26.3). If the gallbladder is present, cholecystectomy may be useful to expose the portal vein and spare the hepatic artery.

The next step is to divide the liver in two, which can be done with a combination of robotic

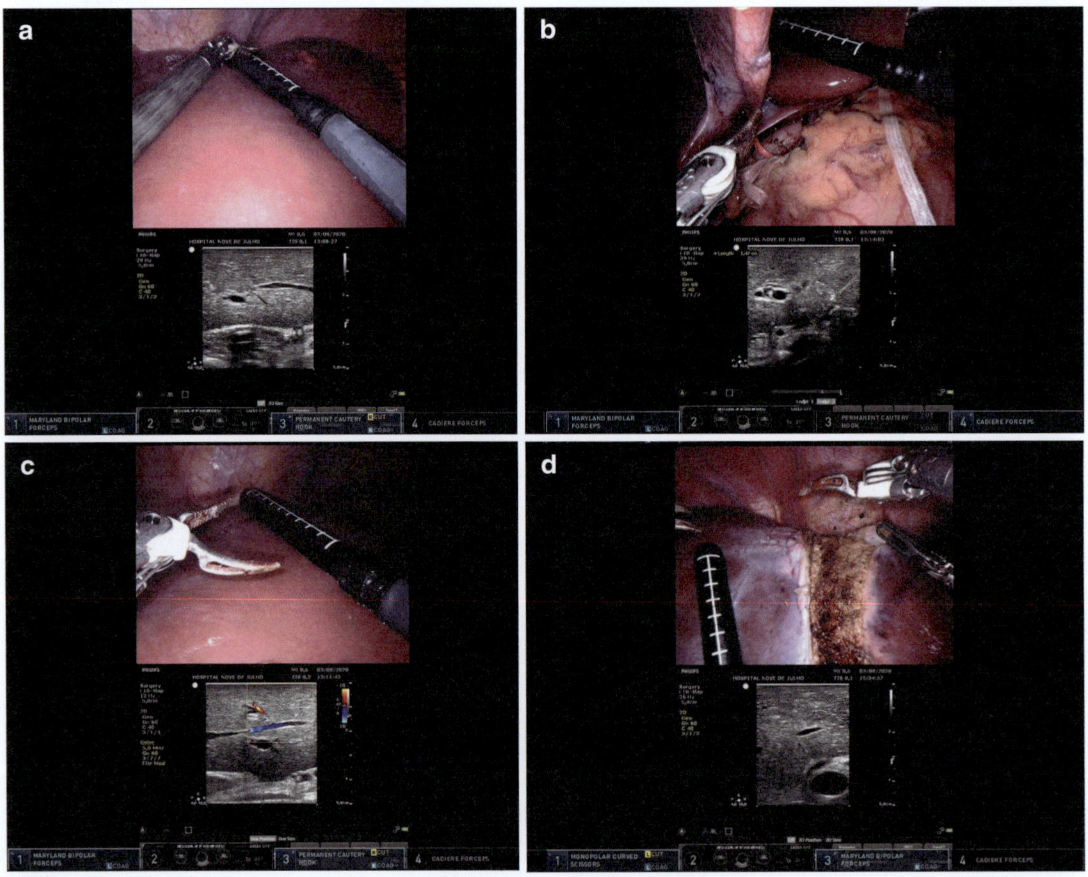

Fig. 26.2 Robotic ALPPS (first stage). Intraoperative ultrasound (IOUS). (**a**) IOUS is used to screen the FLR for occult metastases. (**b**) IOUS is used to localize and define the surgical margin during resection of a known metastasis. (**c**). IOUS is used to determine the intrahepatic course of the left hepatic vein to avoid damage during clearing of the FLR. (**d**) IOUS is used to assess the anatomy during liver division

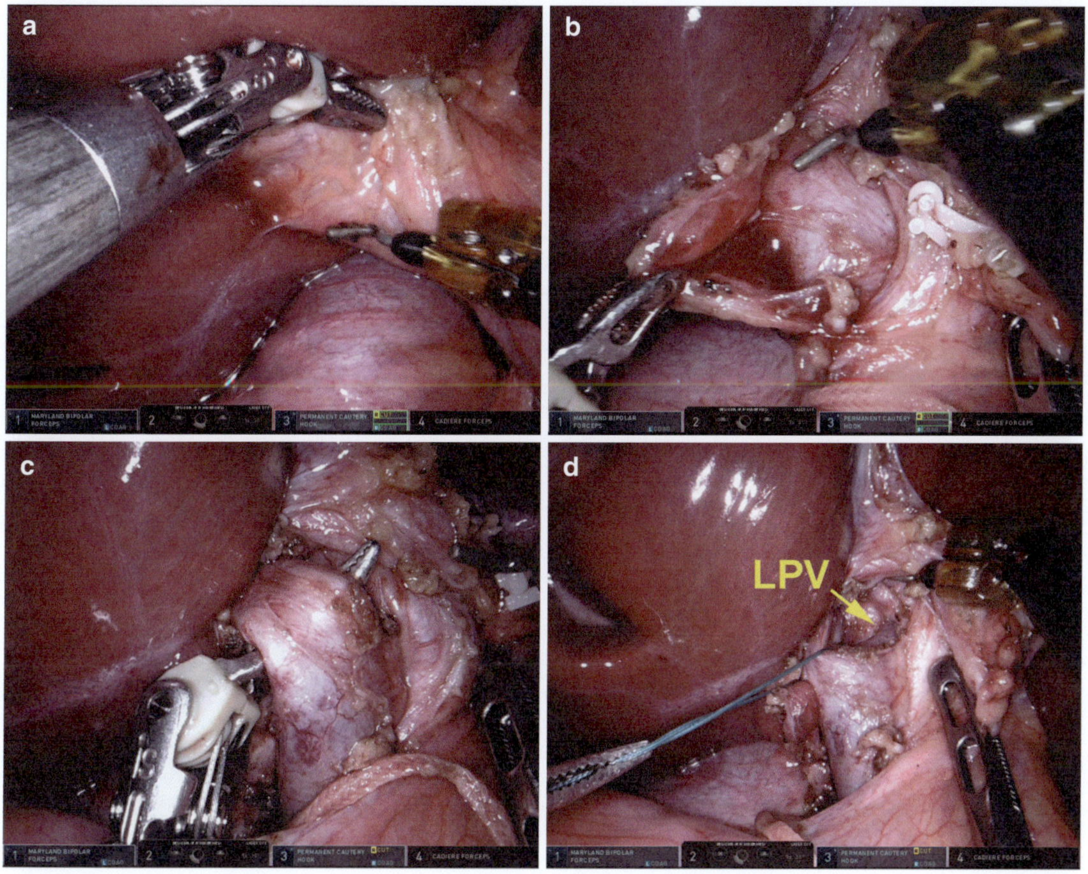

Fig. 26.3 Robotic ALPPS (first stage). Dissection of the right portal vein. (**a**) The common bile duct and hepatic artery are pulled up to expose the right portal vein after cholecystectomy. (**b**) The right portal vein is dissected with Maryland bipolar forceps (**c**) The right portal vein is encircled for further ligature. (**d**) The right portal vein is encircled for further ligation. Note that the left portal vein (LPV) is visible and preserved

bipolar forceps and scissors under saline irrigation (Fig. 26.4). There are some centers where liver division is performed with CUSA, which is guided by the assistant surgeon. The bipartition can be performed up to the retrohepatic vena cava, or we can stop about 2 cm above the vena cava, a strategy known as partial ALPPS [11]. During liver division at the level of the falciform ligament, the Glisson pedicles of segment 4 are transected (Fig. 26.4b). The liver is transected in the direction of the middle hepatic vein, which should be preserved, whenever possible, to avoid congestion of the FLR. The right portal vein (already encircled) is then ligated (there are some surgeons who prefer transection with a stapler). The procedure is completed with removal of the specimen and drainage of the abdominal cavity. Usually, some hemostatic agent is left in the raw area of the liver to assist in hemostasis and to keep the two sides of the liver separated, which will be helpful in the second stage.

To facilitate identification of the structures of the liver hilum in the second stage, some authors use to dissect the liver triad in the first stage and leave a band around the main structures [16]. We used the same principle, but we encircle the entire Glissonian pedicle instead of dissecting individual structures.

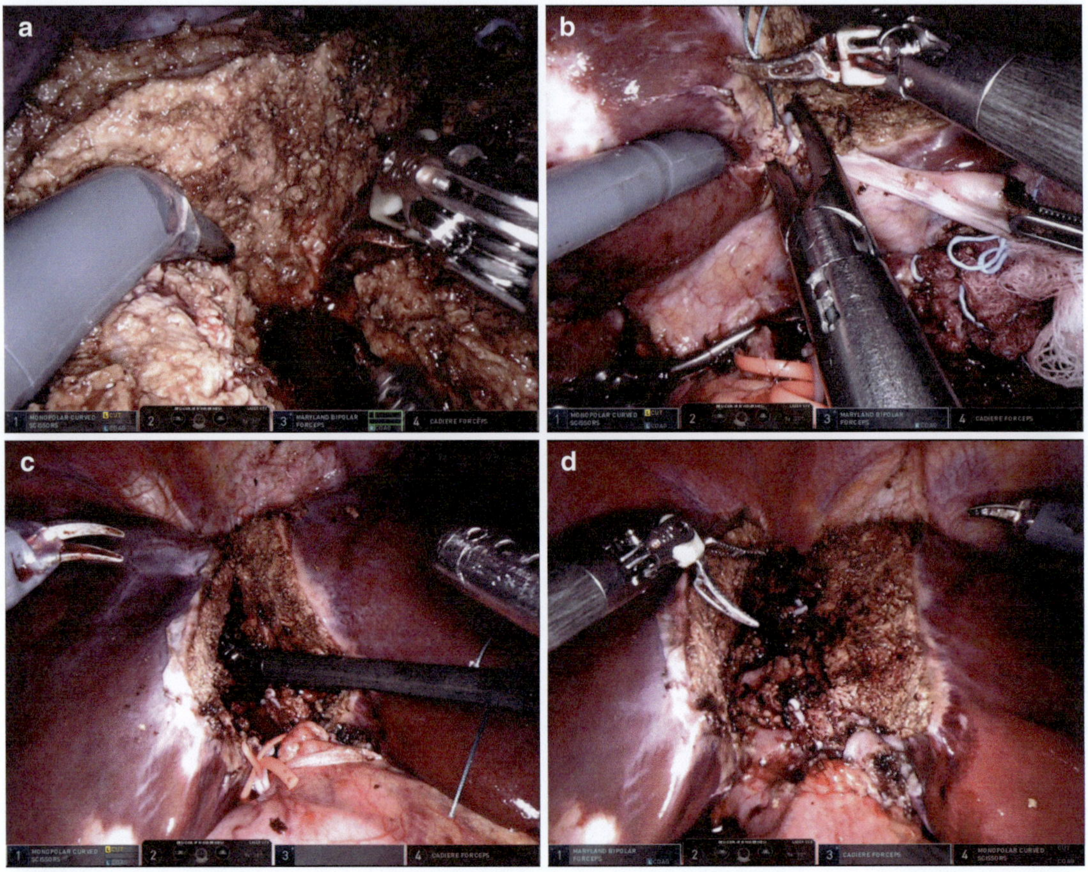

Fig. 26.4 Robotic ALPPS (first stage). Liver bipartition. (**a**) The liver parenchyma is divided using a combination of bipolar forceps and scissors under saline irrigation. (**b**) When dividing the liver, the Glisson pedicle for segment 4 (4a and 4b) is encountered and divided between Hem-o-loks. (**c**) The liver is divided along the falciform ligament. (**d**) Final aspect after liver division

26.2.3 Second Stage

After 7 days, a CT scan is performed to determine sufficient hypertrophy of the FLR. The second stage is performed with the da Vinci Xi robotic platform (Intuitive Surgical Inc., Sunnyvale, CA), using the same number and position of trocars as in the first stage (Fig. 26.5a–b). We usually wait 14–21 days to perform the second stage. Since the entire procedure is performed with minimally invasive technique and pneumoperitoneum, no major or hard adhesions are expected, and we can perform the second stage later if the FLR hypertrophy or the patient's condition is not ideal without jeopardizing the entire ALPPS procedure. In fact, in cases where open ALPPS is indicated, we perform pneumoperitoneum pretreatment to reduce adhesions [17].

During the second stage operation, there are usually no adhesions other than those that occur in the partition area (Fig. 26.5). The adhesions are divided, and the partition area is exposed with blunt dissection. The right liver (+S4) is then

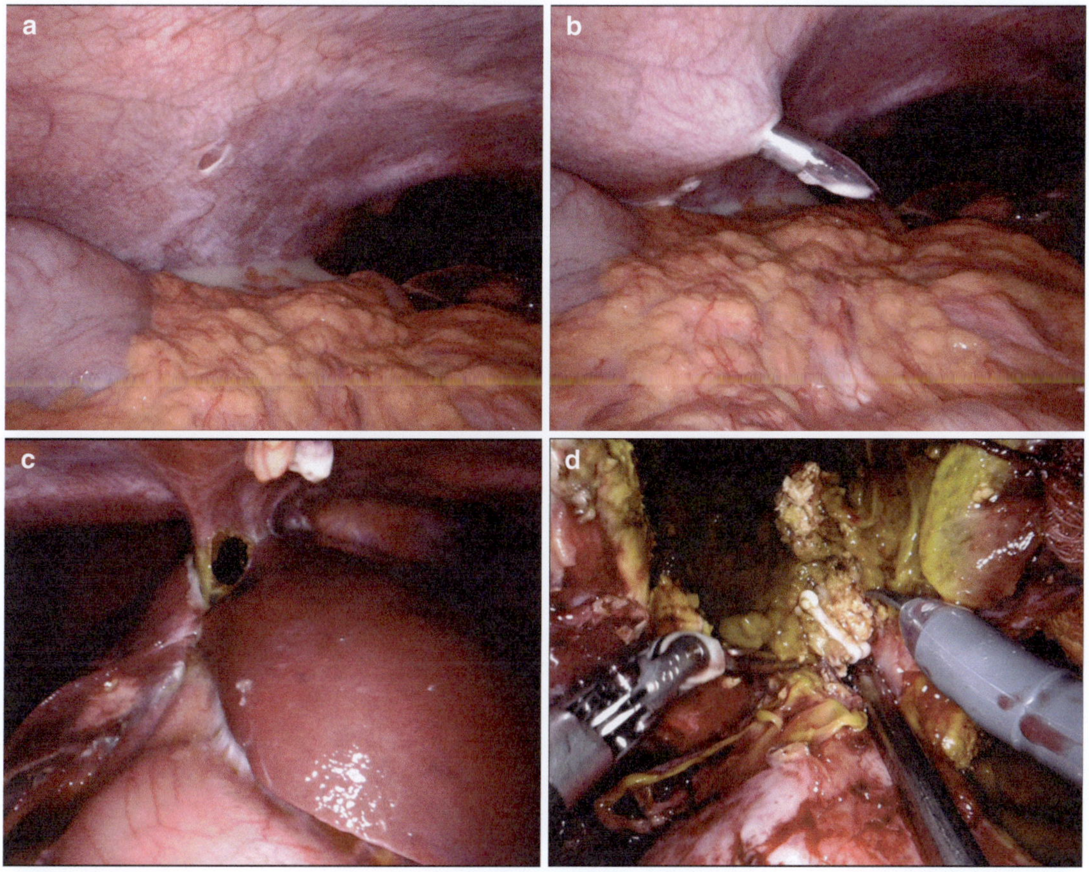

Fig. 26.5 Robotic ALPPS. Second stage. (**a**) Incisions from first stage are seen. (**b**) The same trocar incisions are used for the second stage. (**c**) Few adhesions are seen on laparoscopy, mainly in the bipartition and hepatic hilum (above the dissected area—right portal vein dissection). (**d**) The bipartition area is opened with a blunt maneuver. We can see the Hem-o-loks securing the pedicle of segment 4

fully mobilized. The next step is to access the right Glissonian pedicle, which can be seen after removal of the liver tissue around the pedicle (Fig. 26.6). The right pedicle is transected with a stapler, resulting in ischemic discoloration of the right liver (Fig. 26.6). Fluorescence imaging after infusion of indocyanine green should disclose adequate perfusion of the future liver remnant. Right trisectionectomy begins with division of the right portion of the caudate lobe and placement of an umbilical tape around the right liver on the right side of the inferior vena cava, a modified hanging liver maneuver [18]. The tape is pulled upward to expose the liver partitioned area (Fig. 26.7). The liver is carefully divided using a combination of bipolar robotic forceps and scissors under saline toward the right hepatic vein. Finally, the hepatic vein is divided with a stapler, and the second stage is completed (Fig. 26.7c). The procedure is completed with removal of the specimen through a suprapubic incision and drainage of the abdominal cavity.

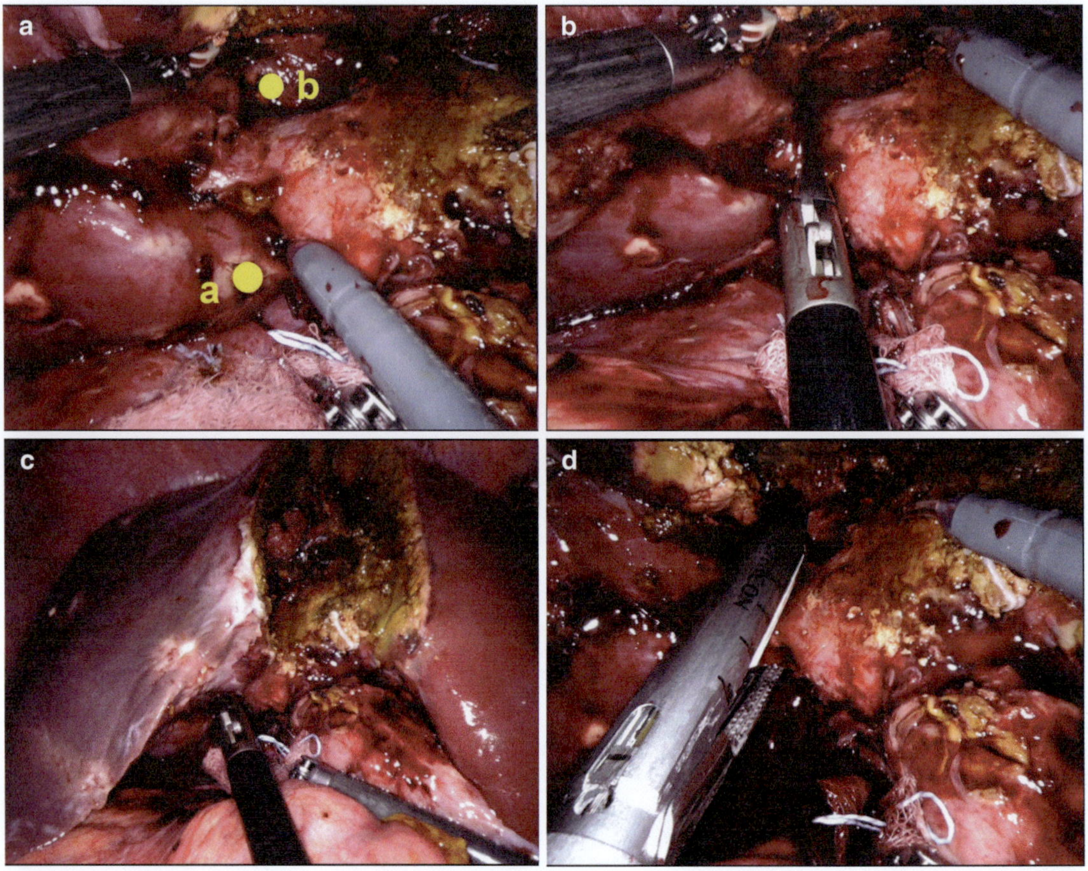

Fig. 26.6 Robotic ALPPS (second stage). Glissonian approach to the right pedicle. (**a**) Two small incisions (a and b, yellow circles) are used to encircle the right Glissonian pedicle. (**b**) A vascular clamp is used to temporarily occlude the right pedicle. (**c**) The ischemic demarcation of the right liver (+S4) is seen. (**d**) The right Glissonian pedicle is transected with a stapler

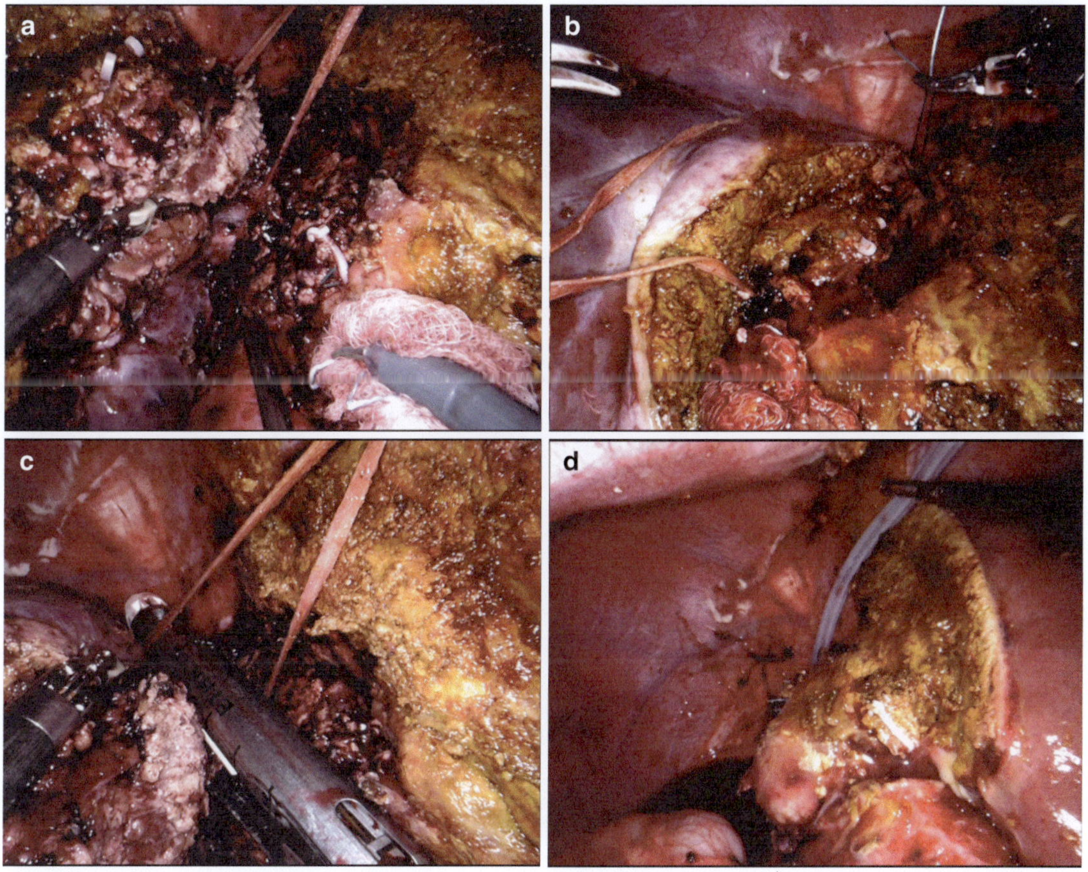

Fig. 26.7 Robotic ALPPS (second stage). Right trisectionectomy. (**a**) An umbilical tape is placed around the right liver (+S4) after mobilization. (**b**) The right hepatic vein is dissected and encircled. (**c**) The liver is divided using the hanging maneuver and the right hepatic vein is transected with a stapler. (**d**) Final aspect after right trisectionectomy

26.3 Worldwide Experience with Robotic ALPPS

The first minimally invasive ALPPS procedure was published in 2012 [8]. A letter to the editor described a fully laparoscopic first- and second-stage ALPPS [8]. A recent literature review on this topic analyzed 119 cases from 6 series [19]. According to this review, minimally invasive ALPPS appears to be a feasible and safe procedure, with potentially better short-term outcomes in terms of blood loss, morbidity, and mortality compared with open series. One possible explanation is better patient selection and underlying pathology with a preponderance of colorectal liver metastases. However, the preponderance of colorectal metastases is explained by the original analysis of the ALPPS registry [3]. Here, better outcomes were found in this subset of patients and a relative contraindication in some types of pathology, such as biliary tumors [3].

In a study by our team comparing 10 laparoscopic ALPPS procedures with 20 open cases, we found better outcomes in the laparoscopic series [13]. There are several reasons for this better outcome. One of them is patient selection, because not every patient is suitable for a minimally invasive approach. Another reason, which has been neglected, is the longer interval between the two steps that minimally invasive procedures allow by reducing adhesions, which can be extremely dangerous after open surgery. In fact, a longer interval has been associated with better outcomes after ALPPS [20], suggesting that the large hypertrophy of the future liver remnant may take some time to reflect the actual liver function; this has already been discussed by other authors [21]. Another reason is that liver division performed by laparoscopy is usually partial, because it is difficult to perform the classic hanging maneuver by laparoscopy. Partial ALPPS [11] and other techniques such as mini-ALPPS [12, 22] have been described to reduce the surgical trauma of the first stage, increase the safety of ALPPS, and reduce the number of patients who could not undergo the second stage because of complications. We likely performed partial ALPPS without realizing it when a minimally invasive approach was used. In one systematic review, all minimally invasive cases of ALPPS resulted in adequate hypertrophy of the future liver remnant, and the second stage was possible in every case [23].

Robotic-assisted surgery has gained increasing acceptance and expanded to liver resection in recent years [24, 25]. ALPPS is one of the most complex procedures in surgery and may be associated with increased morbidity [1, 3, 4]. A variety of techniques and a high level of skill are required for safe robotic-assisted ALPPS [18]. There are few reports of fully robotic ALPPS in the literature [18, 26–32]. The latest robotic platform, the da Vinci Xi, has the advantage of interchangeable cameras between the four robotic arms. Therefore, it is possible to perform the entire treatment of synchronous colorectal liver metastases, including primary ones, with the same trocars without increasing the surgical difficulty [27]. Several techniques and approaches used in liver surgery are useful to safely perform robotic ALPPS, such as liver first [33], selective hepatic artery clamping [34], partial ALPPS [11], Glissonian approach [35], and modified liver hanging maneuver [36]. After sufficient experience with open and laparoscopic ALPPS as well as robotic liver resection, the move toward robotic ALPPS seemed obvious. As predicted by some authors [37], we, too, are "convinced that robotic approach is a valuable option to expand the application of minimally invasive surgery, even for very challenging procedures such as the ALPPS procedure.

To date, only 11 complete robotic ALPPS have been described in the literature from 6 different centers worldwide [18, 26–32]. However, several centers use minimally invasive techniques, including the robotic approach, for the first stage, whereas the second stage is performed with an open approach [38, 39]. Although this may seem reasonable, we believe that the first stage is much more difficult to perform than the second stage, especially in patients with colorectal metastases and multiple metastases in the FLR. The main obstacle to the use of the robotic approach in these patients is the ability to find and remove all liver metastases in the FLR after neoadjuvant chemotherapy, because there is a risk that not all metastases will be removed. After the first stage is adequately performed, the second stage would be easier because only a small portion of the liver needs to be divided and most of the liver partition, which is the crucial part of the major liver resection, has already been done.

Our experience with robotic ALPPS with those in the current literature concludes that it is feasible and safe and may have some advantages over laparoscopic and open ALPPS.

References

1. Schnitzbauer AA, Lang SA, Goessmann H, et al. Right portal vein ligation combined with in situ splitting induces rapid left lateral liver lobe hypertrophy enabling 2-staged extended right hepatic resection in small-for-size settings. Ann Surg. 2012;255:405–14.
2. de Santibañes E, Clavien PA. Playing Play-Doh to prevent postoperative liver failure: the "ALPPS" approach. Ann Surg. 2012;255:415–7.
3. Schadde E, Ardiles V, Robles-Campos R, et al. Early survival and safety of ALPPS: first report of the international ALPPS registry. Ann Surg. 2014;260:829–38.
4. Schadde E, Raptis DA, Schnitzbauer AA, et al. Prediction of mortality after ALPPS Stage-1: an analysis of 320 patients from the international ALPPS registry. Ann Surg. 2015;262:780–5.
5. Linecker M, Stavrou GA, Oldhafer KJ, et al. The ALPPS risk score: avoiding futile use of ALPPS. Ann Surg. 2016;264:763–71.

6. Alvarez FA, Ardiles V, de Santibañes M, Pekolj J, de Santibañes E. Associating liver partition and portal vein ligation for staged hepatectomy offers high oncological feasibility with adequate patient safety: a prospective study at a single center. Ann Surg. 2015;261:723–32.
7. Linecker M, Björnsson B, Stavrou GA, et al. Risk adjustment in ALPPS is associated with a dramatic decrease in early mortality and morbidity. Ann Surg. 2017;266:779–86.
8. Machado MA, Makdissi FF, Surjan RC. Totally laparoscopic ALPPS is feasible and may be worthwhile. Ann Surg. 2012;256:e13.
9. Robles R, Parrilla P, Lopez-Conesa A, et al. Tourniquet modification of the associating liver partition and portal ligation for staged hepatectomy procedure. Br J Surg. 2014;101:1129–34.
10. Gall TM, Sodergren MH, Frampton AE, et al. Radio-frequency-assisted Liver Partition with Portal vein ligation (RALPP) for liver regeneration. Ann Surg. 2015;261:e45–6.
11. Petrowsky H, Gyori G, de Oliveira M, et al. Is partial-ALPPS safer than ALPPS? A single-center experience. Ann Surg. 2015;261:e90–2.
12. de Santibanes E, Alvarez FA, Ardiles V, et al. Inverting the ALPPS paradigm by minimizing first stage impact: the mini-ALPPS technique. Langenbeck's Arch Surg. 2016;401:557–63.
13. Machado MA, Makdissi FF, Surjan RC, et al. Transition from open to laparoscopic ALPPS for patients with very small FLR: the initial experience. HPB (Oxford). 2017;19:59–66.
14. Machado MA, Surjan R, Basseres T, Makdissi F. Total Laparoscopic Reversal ALPPS. Ann Surg Oncol. 2017;24(4):1048–9.
15. Schadde E, Malagó M, Hernandez-Alejandro R, Li J, Abdalla E, Ardiles V, Lurje G, Vyas S, Machado MA, de Santibañes E. Monosegment ALPPS hepatectomy: extending resectability by rapid hypertrophy. Surgery. 2015;157(4):676–89.
16. Brustia R, Scatton O, Perdigao F, El-Mouhadi S, Cauchy F, Soubrane O. Vessel identifications tags for open or laparoscopic associating liver partition and portal vein ligation for staged hepatectomy. J Am Coll Surg. 2013;217(6):e51–5.
17. Machado MA, Makdissi FF, Surjan RC. ALPPS procedure with the use of pneumoperitoneum. Ann Surg Oncol. 2013;20(5):1491–3.
18. Machado MA, Surjan RC, Makdissi F. Robotic ALPPS. Ann Surg Oncol. 2020;27(4):1174–9.
19. Cioffi L, Belli G, Izzo F, Fantini C, D'Agostino A, Russo G, Patrone R, Granata V, Belli A. Minimally invasive ALPPS procedure: a review of feasibility and short-term outcomes. Cancers (Basel). 2023;15(6):1700.
20. Kumar N, Duncan T, O'Reilly D, et al. Partial ALPPS with a longer wait between procedures is safe and yields adequate future liver remnant hypertrophy. Ann Hepatobiliary Pancreat Surg. 2019;23:13–9.
21. Dokmak S, Belghiti J. Which limits to the "ALPPS" approach? Ann Surg. 2012;256:e6.
22. Pekolj J, Alvarez FA, Biagiola D, et al. Totally laparoscopic mini-ALPPS using a novel approach of laparoscopic-assisted transmesenteric portal vein embolization. J Laparoendosc Adv Surg Tech A. 2018;28:1229–33.
23. Melandro F, Giovanardi F, Hassan R, Larghi Laureiro Z, Ferri F, Rossi M, Mennini G, Pawlik TM, Lai Q. Minimally invasive approach in the setting of ALPPS procedure: a systematic review of the literature. J Gastrointest Surg. 2019;23(9):1917–24.
24. Tsung A, Geller DA, Sukato DC, et al. Robotic versus laparoscopic hepatectomy: a matched comparison. Ann Surg. 2014;259:549–55.
25. Machado MA, Lobo-Filho MM, Mattos BH, Ardengh AO, Makdissi FF. Robotic liver resection. Report of the first 50 cases. Arq Gastroenterol. 2021;58(4):514–9.
26. Vicente E, Quijano Y, Ielpo B, Fabra I. First ALPPS procedure using a total robotic approach. Surg Oncol. 2016;25(4):457.
27. Masetti M, Lombardi R, Romboli A, Jovine E. Fully robotic ALPPS and simultaneous left colectomy for synchronous colorectal liver metastases. J Laparoendosc Adv Surg Tech A. 2020;30(10):1106–9.
28. Solomonov E, Tzadok I, Stemmer S, Biswas S. Case report: robotic ALPPS procedure for hepatocellular carcinoma in the right lobe of the liver. Front Surg. 2021;13(8):655683.
29. Hu MG, Wang J, Yin ZZ, Liu R. First two-stage robotic ALPPS in HCC patients with hepatic vein invasion: a step-by-step procedure from a clinical case. World J Surg Oncol. 2021;19(1):58.
30. Fernandes ES, de Barros F, Magistri P, Di Sandro S, Rezende de Carvalho P, Roza da Silva F, Andrade RO, Pimentel LS, Girão CL, Pedreira de Mello F, Torres OJM, Di Benedetto F. Total robotic ALPPS approach for hepatocellular carcinoma in cirrhotic liver. Int J Med Robot. 2021;17(3):e2238.
31. Di Benedetto F, Magistri P. First case of full robotic ALPPS for intrahepatic cholangiocarcinoma. Ann Surg Oncol. 2021;28(2):865.
32. Di Benedetto F, Magistri P, Guerrini GP, Di Sandro S. Robotic liver partition and portal vein embolization for staged hepatectomy for perihilar cholangiocarcinoma. Updat Surg. 2022;74(2):773–7.
33. Verhoef C, van der Pool AE, Nuyttens JJ, et al. The "liver-first approach" for patients with locally advanced rectal cancer and synchronous liver metastases. Dis Colon Rectum. 2009;52:23–30.
34. Surjan RC, Makdissi FF, Basseres T, et al. First totally laparoscopic ALPPS procedure with selective hepatic artery clamping: case report of a new technique. Medicine (Baltimore). 2016;95:e4236.
35. Machado MA, Makdissi F, Surjan R. Laparoscopic glissonean approach: making complex something easy or making suitable the unsuitable? Surg Oncol. 2020;33:196–200.

36. Chu H, Cao G, Tang Y, et al. Laparoscopic liver hanging maneuver through the retrohepatic tunnel on the right side of the inferior vena cava combined with a simple vascular occlusion technique for laparoscopic right hemihepatectomy. Surg Endosc. 2018;32:2932–8.
37. Levi Sandri GB, Guerra F. Are we ready to perform fully minimally invasive ALPPS? Ann Surg. 2017;266:e56–7.
38. Krishnamurthy J, Naragund AV, Mahadevappa B. First ever robotic stage one ALPPS procedure in India: for colorectal liver metastases. Indian J Surg. 2018;80:269–71.
39. Li J, Yang GS, Sun KJ, Ma Y, Bi XW, Han X. Clinical evaluation of modified ALPPS procedures based on risk-reduced strategy for staged hepatectomy. Ann Hepatol. 2021;20:100245.

Robotic Living Donor Liver Procurement for Transplantation

27

Phillipe Abreu, Dimitri A. Raptis, and Dieter C. Broering

27.1 Introduction

Robotic-assisted liver procurement from living donors represents a significant advancement in the field of transplant surgery, offering precision, minimal invasiveness, and better outcomes for donors (1). The three main operations performed for this purpose are right lobe donor hepatectomy, left lobe donor hepatectomy, and left lateral donor hepatectomy. This chapter describes the step-by-step technical aspects of robotic living donor liver procurement for different liver segments, as well as summarizes the current literature on outcomes of robotic donor hepatectomy.

27.2 Technical Aspects

27.2.1 Patient Positioning and Ports

The patient is positioned supine in the French position, and a pneumoperitoneum of 10–12 mmHg is created using an open technique via the umbilical port. Four 8-mm ports are then inserted, each 8 cm apart, with two on each side along the midclavicular and anterior axillary lines, slightly angled toward the right upper quadrant, forming a "smiling" line arrangement. For left lobe or left lateral hepatectomies (when stapling the left hepatic vein will be necessary), the right flank port is upgraded to a 12-mm size for the robotic stapler device (Fig. 27.1). The robot (da Vinci Xi, Intuitive Surgical, Sunnyvale, USA) is brought in from the patient's left side.

27.2.2 Right Lobe Donor Hepatectomy

27.2.2.1 Liver Mobilization

The initial step involves mobilizing the right lobe of the liver. This is achieved by transecting the teres ligament and the falciform ligament up to the hepatic veins and identifying the groove between the right hepatic vein (RHV) and middle hepatic vein (MHV). Subsequently, the uppermost 5 cm of this groove is marked with monopolar cautery, using the Cantlie line as the first fixed landmark for the subsequent parenchymal transection.

Next, step-by-step mobilization of the right liver lobe is achieved by using the robotic arm number 4 with a ProGrasp™ Forceps (Intuitive Surgical, Sunnyvale, USA) for retraction toward the left. During the initial phase, a laparoscopic

P. Abreu
University of Colorado Anschutz Medical Campus, Colorado Center for Transplantation Care, Research and Education, Denver, CO, USA

D. A. Raptis · D. C. Broering (✉)
Organ Transplant Center of Excellence, King Faisal Specialist Hospital & Research Center, Riyadh, Saudi Arabia
e-mail: dbroering@kfshrc.edu.sa

M. D'Hondt, I. Sucandy (eds.), *Textbook of Robotic Liver Surgery*,
https://doi.org/10.1007/978-3-031-76536-0_27

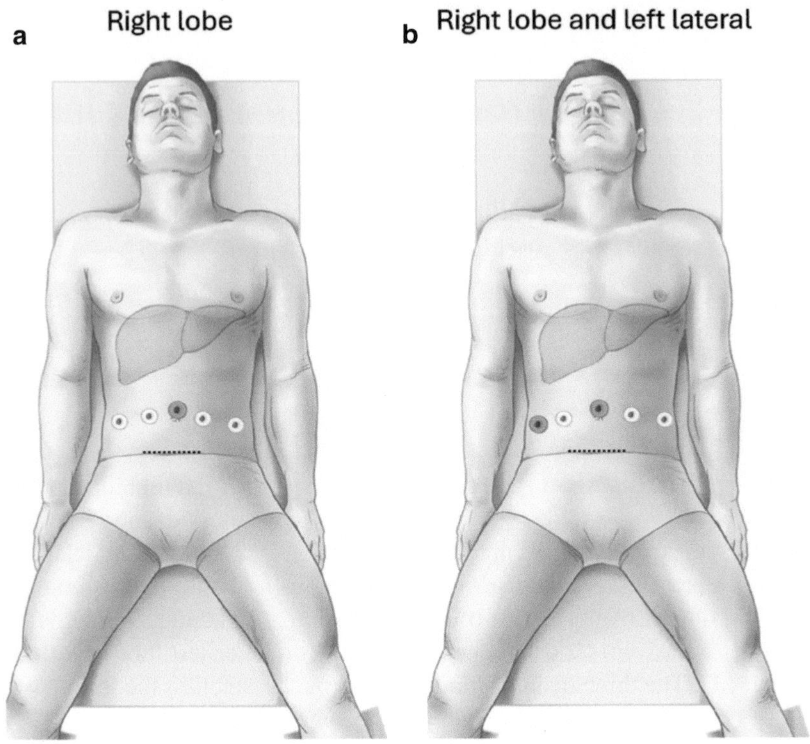

Fig. 27.1 Port placement for (**a**) right lobe hepatectomy, (**b**) left lobe and left lateral hepatectomy. White trocars, 8 mm; purple trocars, 12 mm; dotted line, Pfannenstiel incision

grasper grasps the gallbladder and retracts the attached right lobe to the left, thereby facilitating easier retraction of the right lobe with the ProGrasp™ in arm number 4. A step-by-step "piggyback" dissection of the right lobe is achieved, exposing the retro-hepatic inferior vena cava (IVC). The caudate branches as well as the caval ligament are dissected and clipped or ligated in between ligatures or transected with advanced bipolar energy (Vessel Sealer, Intuitive Surgical, Sunnyvale, USA), in order to expose the circumference of the lower IVC up to 3 o'clock inferiorly, and the upper IVC up to 12 o'clock position. In case of the presence of major right inferior hepatic veins, they should be preserved until the final clamping of the major vessels. A Foley catheter is inserted as a guide from the anterior surface of the IVC toward the upper surface of the liver, in the space between the RHV and the MHV, as a "hanging maneuver" of the liver.

27.2.2.2 Hilum Dissection

After this is completed, cholecystectomy is performed, and the cystic duct stump is secured with 2–0 Vicryl (Ethicon, Inc., Somerville, USA) in between two Hem-o-lok clips (Teleflex Incorporated, Wayne, USA) for assistant traction and better exposure. Then attention is turned to the hilum. Initially the portal vein is identified and dissected from the hilum, identifying the portal bifurcation (second fixed landmark of the operation) and isolating the right portal vein (RPV) with a vessel loop. The second structure to be dissected and isolated is the right hepatic artery (RHA). The dissection of the right hepatic artery should be done closely to the arterial wall and should stop at the right side of the main hepatic bile duct. Only in case of early branching of the right hepatic artery is the dissection continuous behind and occasionally to the left of the main hepatic duct. The order of dissection can vary according to the individual anatomy of the

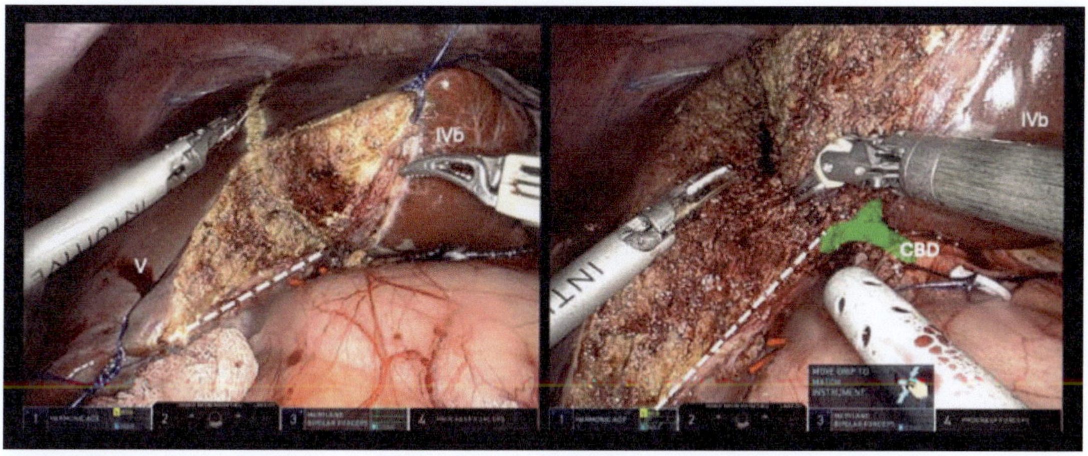

Fig. 27.2 Right lobe transection line with anatomical landmarks. Abbreviations: *IVb* segment IVb; *V* segment V; *CBD* common bile duct. Bile duct bifurcation identified in green. Dashed line with transection line planned based on anatomical landmarks

donor. Especially in Nakamura type C portal vein bifurcations (2), a separate isolation of the right anterior and right posterior portal vein is recommended. Finally, the hilar plate containing the right bile duct is left undissected until a later stage of the operation (Fig. 27.2).

Once this is completed, robotic applied bulldog vascular clamps are placed on the RPV and RHA, and indocyanine green (ICG) (25 mg, Verdye®, Diagnostic Green GmbH, Aschheim, Germany) is administered for better right lobe demarcation identification. The monopolar hook (Intuitive Surgical, Sunnyvale, USA) is then used to mark the parenchymal transection line from the dome of the liver (previously marked groove between RHV and MHV) to the gallbladder fossa. The vascular clamps are then removed, and liver transection is initiated.

27.2.2.3 Parenchymal Transection

To achieve better exposure, 2–0 Prolene (Ethicon, Inc., Somerville, USA) sutures are applied in the limit of the demarcation line in each side of the liver edge. They are exteriorized through the abdominal wall with a Cater-Thomason device (CooperSurgical, Trumbull, USA), lateral to the most lateral ports, and secured with a mosquito clamp outside of the abdomen.

Parenchymal transection is performed with the use of the built-in harmonic ultrasonic energy device initially (for the first few centimeters of transection) via robotic arm number 3 and consequently via arm number 1. Once deep in the parenchyma and especially closed to major tributaries of the MHV, two Maryland bipolar dissectors in arms number 1 and number 3 can be used sequentially by the clamp-crushing technique. When major vessels (segment V or segment VIII veins) are identified, the Maryland bipolar is used as a right-angle instrument to loop the vessels, and metallic clips (Challinger, Karl Storz Germany) are applied with the help of a 2–0 Vicryl (Ethicon, Inc., Somerville, USA) suture to expose the vessel. When vessels are too large for metallic clip ligation, hemostatic sutures using 5–0 PDS (Ethicon, Inc., Somerville, USA) are applied in a figure-of-eight fashion to the segment V veins. For the larger segment VIII veins, an 8 mm Endo GIA™ Universal Stapler (originally under Covidien, now Medtronic, Minneapolis, USA) is used for stapling.

The parenchymal transection progresses in layers from the dome to the hilar plate, increasing the exposure as it proceeds deeper on the right side of the middle hepatic vein (MHV), which is preserved in the donor. The angle of the transec-

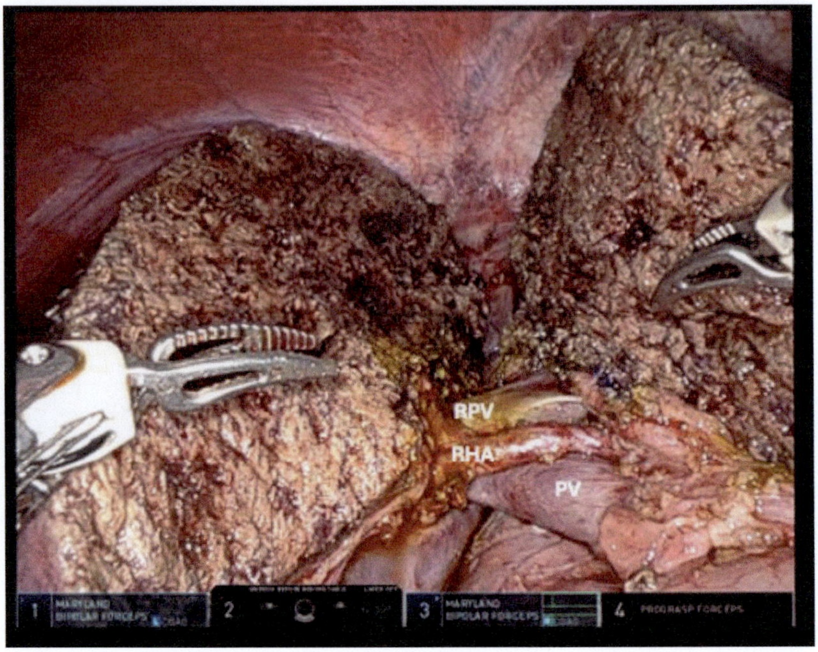

Fig. 27.3 Right lobe transection line prior to vascular clamping. Abbreviations: *RPV* right portal vein; *RHA* right hepatic artery; *PV* portal vein

tion line varies depending on the individual anatomy of the right lobe; however, it is typically slightly oblique (approximately 30%) to the left, following a line from the middle of the gallbladder fossa at the liver edge to the bifurcation of the main hepatic duct (Fig. 27.3). When the hilar plate is identified, transection is discontinued. The biliary anatomy is then delineated using ICG cholangiography and the Firefly™ Fluorescence Imaging Camera (Intuitive Surgical, Sunnyvale, USA).

27.2.2.4 Biliary Transection

After identifying the appropriate biliary anatomy and classifying it according to the Nakamura type, enhanced by real-time ICG cholangiography, the edge of the right bile duct (RBD) is transected using curved scissors in a blunt cut without the use of energy devices. This is followed by biliary probing to identify the bifurcation of the common bile duct (CBD) and the takeoff of the left hepatic duct (LHD). Once this is achieved, a "corridor" line is planned for the complete transection of the RBD with enough stump to allow later safe closure of the orifice and always preserving the donor side (in Nakamura II, III, or IV biliary anatomies (2), multiple ducts will be resulted in the liver graft). It is important to avoid using energy during this transection and to accept some oozing from the peri-biliary vascular plexus. Once the transection is completed, hemostasis is achieved using 6–0 PDS (Ethicon, Inc., Somerville, USA) sutures (Fig. 27.4).

27.2.2.5 Caudate Lobe Division (Segments I and IX)

After biliary transection, the caudate lobe is divided either with the harmonic or with Maryland bipolar forceps inferior to the hilum, under the vessels. The Foley catheter is then transferred from the anterior surface of the IVC to anteriorly to the hilum, completing the "hanging maneuver." The liver parenchyma is then transected in between segments I and IX with the harmonic or the use of the two Maryland bipolar instruments, with the crash-clump technique, from inferior to superior, lateral to the Foley catheter (this will ensure that no injury is caused to the IVC branches). This completes the parenchymal transection, and the closure of the donor bile ducts stumps as well as the donor hilar plate stump is performed, which is an important part of the procedure.

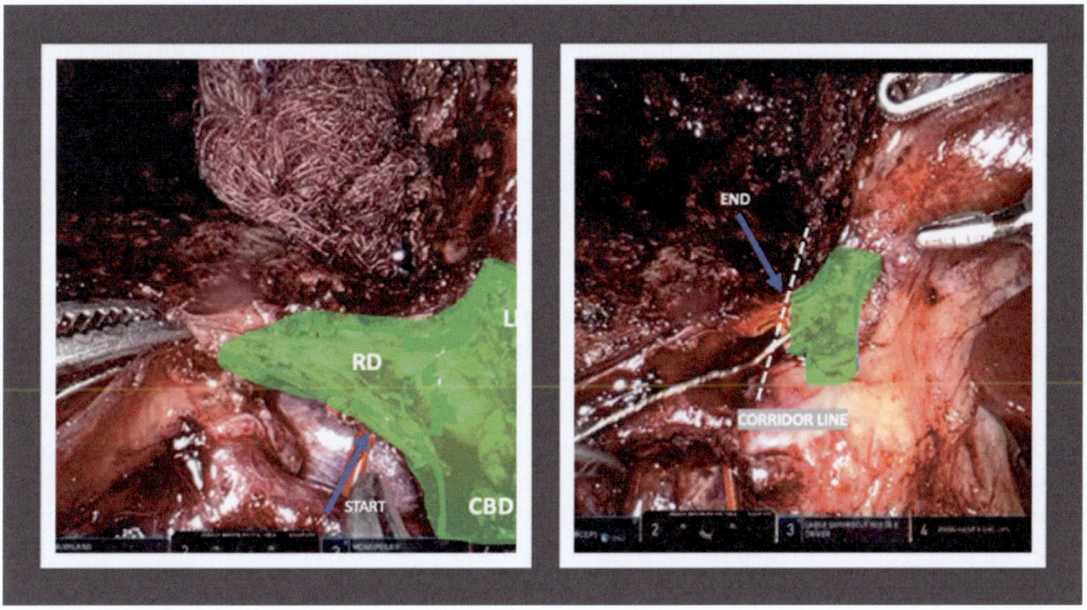

Fig. 27.4 Corridor line for biliary transection in Nakamura I anatomy. Abbreviations: *RD* right duct; *LD* left duct; *CBD* common bile duct

The closure of the donor's right hepatic duct stump starts with a single 6–0 PDS stitch at the upper end of the duct opening. This stitch is kept long and is used for pulling up and thus exposing the right bile duct stump with the robotic arm number 4. Then the right bile duct stump is closed with multiple interrupted 6–0 PDS sutures. The small caudate bile duct stumps are also selectively closed with 6–0 PDS interrupted sutures and additionally secured with small metal clips. The remaining hilar plate stump is secured with a running 6–0 PDS suture for complete hemostasis and biolostasis.

27.2.2.6 Vascular Division and Graft Removal

Once communication with the recipient team is established and the back table is prepared for the graft, the artery is secured with a 5–0 PDS suture followed by the placement of two Hem-o-lok clips (Teleflex Incorporated, Wayne, USA). It is then divided using round-tip scissors. This step is followed by the stapling of the right portal vein (RPV) with an 8-mm Endo GIA™ Universal Stapler (Medtronic, Minneapolis, USA) to preserve the length of the vessel, and the right hepatic vein (RHV) is stapled with a standard vascular laparoscopic stapler. If additional right inferior veins are present, they are also stapled. The graft is then placed in a large Endo Catch™ Gold specimen bag (Medtronic, Minneapolis, USA) and removed through the Pfannenstiel incision, which measures approximately 8–12 cm.

27.2.3 Left Lobe Donor Hepatectomy

27.2.3.1 Liver Mobilization

The first step is to mobilize the left lobe of the liver by detaching the coronary and triangular ligaments, identifying the groove between the RHV and MHV. Then this is marked with monopolar cautery as the first fixed landmark of the operation. The right lobe of the liver is not mobilized, and no piggyback or hanging maneuvers are performed. Then the left lateral segment is dissected free from the hepato-gastric ligament, with special attention to arterial anatomic variations to the left hepatic artery (LHA). The edge of the left hepatic vein (LHV) is dissected and exposed, as well as the caudate lobe and the Arantius ligament, dissecting down to the hilum.

27.2.3.2 Hilar Dissection

Cholecystectomy is performed, and the cystic duct stump is secured with two Hem-o-loks (Teleflex Incorporated, Wayne, USA). After this is completed, attention is turned to the hilum, dissecting the left portal vein (LPV) at the end of the Arantius ligament, toward the portal bifurcation (second fixed landmark of the operation). The LPV is then isolated and encircled with a vessel loop. The left hepatic artery (LHA) is then dissected down to the bifurcation of the common hepatic artery (CHA), and the right hepatic artery (RHA) needs to be identified (attention to arterial anatomic variations of the replaced RHA). Liver demarcation is then achieved in the same fashion as in right lobe robotic donor hepatectomy (described above).

27.2.3.3 Parenchymal Transection

The initial parenchymal transcction is performed in a similar fashion as for the right lobe donor hepatectomy. The same technique and strategies are applied, as well as the same landmarks until exposure of segment V vein. Once the segment V vein is transected and secured with either clips or stitching, the parenchymal transection lines continuous horizontal toward the line of Arantius'. The horizontal parenchymal plane follows superiorly the angulation of the middle hepatic vein toward the cava (Fig. 27.5).

27.2.3.4 Biliary Transection

For the left hilar transection, the parenchyma behind the left hilar plate should be freed using Maryland bipolar forceps with minimal energy. Additionally, the left lobe should be lifted anteriorly using robotic arm number 4. After appropriately exposing the left hilar plate and identifying the main biliary anatomy using the ICG-enhanced FireFly™ technique, the left bile duct (LBD) is transected with curved scissors. This cut is made away from the main bile duct bifurcation but before the confluence between segment IV and the left lateral duct, using a blunt technique without energy. After probing of the bile duct, the further angulation of the left hilar plate cutting is determined thus trying to preserve the segment I duct and achieving one bile duct stump on the graft. In cases of variant biliary anatomy, particu-

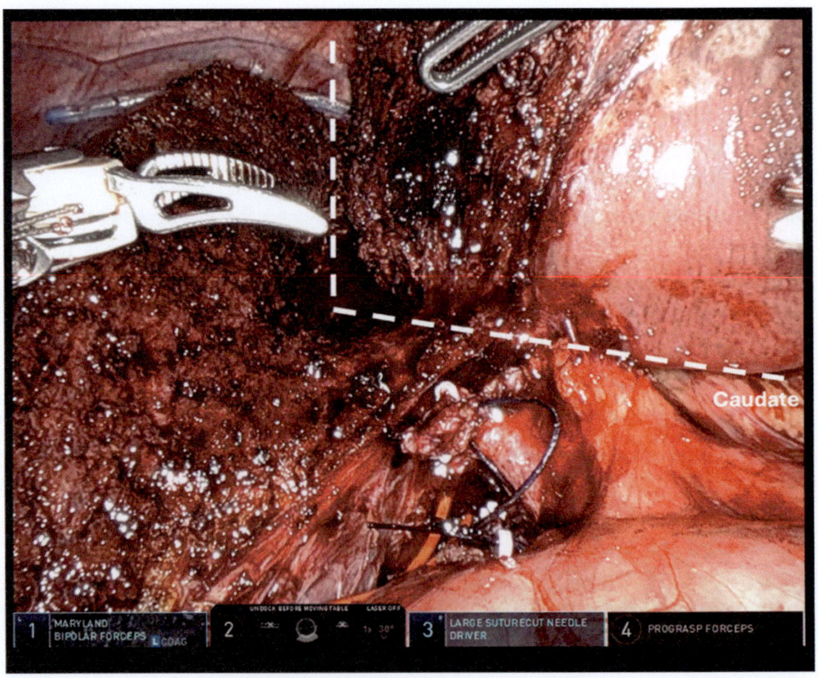

Fig. 27.5 Left lobe parenchymal transection line with landmarks

larly when the right anterior duct drains far into the left hepatic duct, intraoperative cholangiography may be necessary to safely identify the transection point of the left bile duct. It is very important to preserve the anatomy of the donor, and as a core principle never perform any biliary reconstruction to a donor.

27.2.3.5 Segments I and IV Division

After transecting the left hilar plate, the caudate lobe is identified. The Arantius ligament serves as a reference for the level of parenchymal transection between the left lobe and the paracaval portions of the liver on both the left and right sides, preserving caudate segments I and IX for the donor. This completes the dissection, extending from the hilum up to the common trunk of the left hepatic vein (LHV) and middle hepatic vein (MHV). Special attention is required for the large segment VIII veins and caudate branches that drain into the back of the common trunk to prevent bleeding. In most cases, large segment VIII veins need to be secured with stapling or large clips (metal or Hem-o-lok™). However, caution is advised when using clips near the upper part of the common trunk, as they can interfere with the stapling of the common trunk. Once the parenchymal transection is complete, the common trunk can be mobilized from the cava and diaphragm, thus maximizing the length of the venous outflow.

27.2.3.6 Vascular Division and Graft Removal

Once communicated with the recipient team and back table is ready for the graft, the artery is ligated with 5–0 PDS (Ethicon, Inc., Somerville, USA) suture followed by two Hem-o-lok™ clips (Teleflex Incorporated, Wayne, USA) and divided with round-tip scissors. This is followed by stapling of the LPV with the Signia™ Stapling System (Medtronic, Minneapolis, USA) (to preserve length of the vessel) and stapling of the LHV/MHV with robotic vascular stapler (inserted through the robotic arm number 1, hence the need for a 12-mm robotic port). The graft is then placed in a large Endo Catch™ Gold specimen bag (Medtronic, Minneapolis, USA) and removed through the Pfannenstiel incision (approximately 5–8 cm).

27.2.4 Left Lateral Donor Hepatectomy

27.2.4.1 Liver Mobilization

The first step is to mobilize the left lobe of the liver as described for left lobe donation except that the marking on the upper surface of the liver is between the left and the middle hepatic vein.

27.2.4.2 Hilar Dissection

Cholecystectomy is not performed routinely. The hilar dissection begins by identifying the origin of the left hepatic artery from the proper hepatic artery and freeing it from surrounding nerves and lymphatics. This dissection should be limited to a few millimeters. During this phase in most donors the right gastric artery needs to be ligated. The further opening of the hilar peritoneum extends along the segment IV artery toward segment IV. The subsequent dissection occurs in the umbilical triangular area (between the main bile duct, the left hepatic artery, and the left hilar plate), thereby identifying the left portal vein as well as segment IV artery. Small portal branches to segments I and IV need to be transected to facilitate proper access to the left main portal vein and the left hilar plate, but this should be minimized to preserve portal perfusion to liver segments IV and I. The left portal vein (LPV) is then isolated but not encircled. Attention is required for arterial anatomical variations of the hepatic artery (HA); branches to segment IV may need to be sacrificed if the diameter of the branches to segments II/III is too large. No liver demarcation is performed, as anatomical landmarks are sufficient for parenchymal transection. Distal portal branches to segment IV are carefully divided from the portal bifurcation up to the umbilical plate.

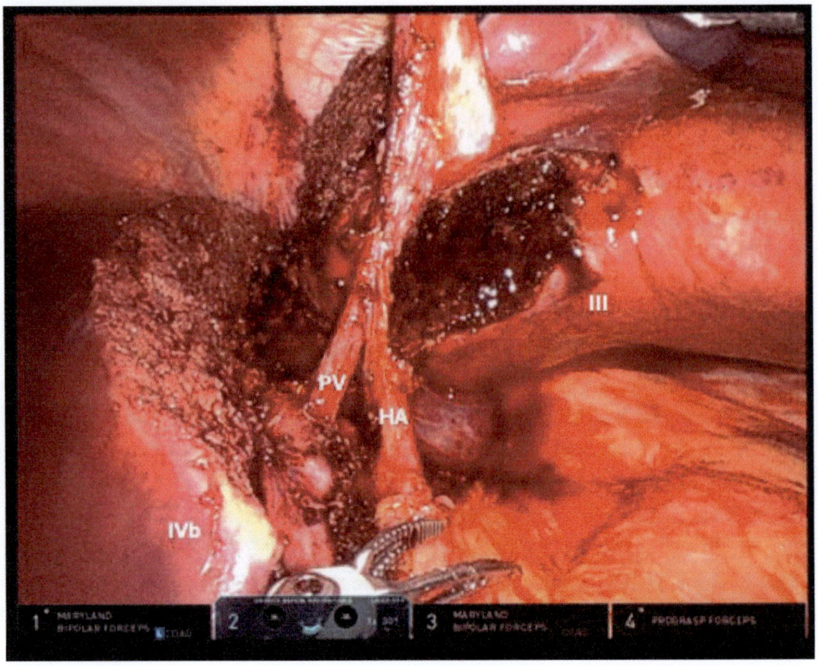

Fig. 27.6 Left lateral segment transection line prior to vascular clamping. Abbreviations: *PV* portal vein; *HA* hepatic artery; *III* segment III; *IVb* segment IVb

27.2.4.3 Parenchymal Transection

Parenchymal transection is performed in a similar fashion as the right and left lobe hepatectomy. The same technique and strategies are applied, as well as same landmarks until exposure of the umbilical plate. Since no liver demarcation is pursued, the transection line is carried out on the right aspect of the falciform ligament, toward to the groove of the LHV/MHV. In most cases the initial parenchymal transection (40%) down to the tip of the umbilical plate is performed before the hilar dissection to allow better exposure to the hilum (Fig. 27.6).

27.2.4.4 Biliary Transection

After appropriate biliary anatomy identification, the umbilical plate is transected with monopolar curved scissors with a blunt cut without the use of energy devices. The "corridor" line is also planned in the same manner for the right and left lobe hepatectomies. Segment IV duct should be preserved to minimize biliary complications in the donor, but in few cases it may have to be sacrificed at this stage in order to achieve adequate biliary anatomy for the graft (segment II and III) (Fig. 27.7).

27.2.4.5 Segment I and Left Lateral Division

After umbilical plate transection, the caudate lobe is identified, and the Arantius ligament is used as a reference for the level of parenchymal transection for the completion of the dissection, from the hilum up to the LHV. Attention is dedicated to crossing branches closer to the hepatic veins, dissected with the clamp-crush technique, and avoiding the use of energy devices in non-dissected structures. This completes the parenchymal transection, and Pfannenstiel incision can be performed.

27.2.4.6 Vascular Division and Graft Removal

Once communicated with the recipient team and back table is ready for the graft, the left artery is clamped with two Hem-o-lok clips (Teleflex Incorporated, Wayne, USA) (no tie is used) and divided with round-tip scissors. This is followed by stapling of the LPV with the Signia™ Stapling System (Medtronic, Minneapolis, USA) (to preserve length of the vessel) and stapling of the LHV with robotic vascular stapler (inserted via robotic arm number 1, hence the need for a

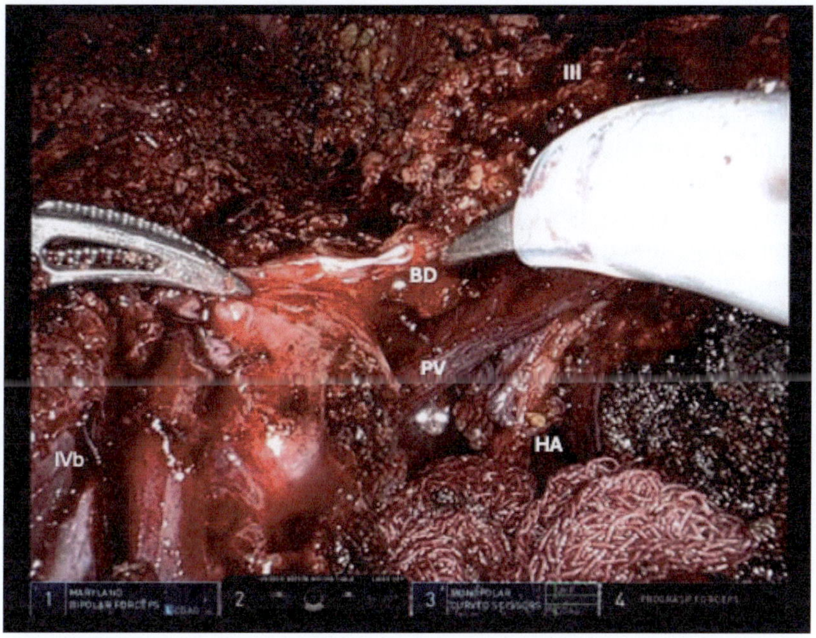

Fig. 27.7 Left lateral donor hepatectomy with illustrating the beginning of the transection of the left bile duct (BD). Abbreviations: *HA* hepatic artery; *PV* portal vein; *III* segment 3; *IVb* segment 4b of the liver

12-mm robotic port). The graft is placed in a medium Endo Catch™ Gold specimen bag (Medtronic, Minneapolis, USA) and removed through the Pfannenstiel incision (approximately 6 cm).

27.3 Tips and Tricks

- Careful identification of fixed anatomical landmarks as described is the key reference for these operations.
- Parenchymal dissection should be performed with harmonic ultrasonic energy device and two Maryland bipolar forceps with the crush-clump technique in the deeper planes of dissection.
- Biliary transection should be carefully caried out after appropriate identification of the biliary anatomy and "corridor" line planning.
- Biliary transection should be always performed with blunt scissors without energy and following the principle of preserving the anatomy of the donor as a priority over the graft.
- Left lobe and left lateral hepatectomies do not require right lobe mobilization or piggyback and hanging maneuver techniques.

27.4 Literature Overview

Living donor hepatectomy is performed in most of centers worldwide using an open surgical approach (3). The evolution of minimally invasive surgery has introduced laparoscopic techniques, which are associated with reduced postoperative pain and shorter hospital stays. However, laparoscopic approaches have not shown superiority over open surgery in reducing morbidity (4), and they present technical challenges such as limited instrument articulation and surgeon fatigue, potentially impacting surgical outcomes. Robotic donor hepatectomy has emerged as a promising development, offering enhanced precision, superior ergonomics, and better 3D visualization, which are critical for the delicate maneuvers required during donor hepatectomy (5) (6). A comprehensive study conducted at our center (Organ Transplant Center of Excellence, King Faisal Specialist Hospital and Research Centre, Riyadh, Saudi Arabia) provided important data comparing outcomes from robotic, laparoscopic, and open approaches (1). Analyzing over 1700 cases over a decade, the study demonstrated that robotic donor hepatectomy significantly reduced donor morbidity compared to both laparoscopic and open approaches. For instance, overall morbidity was 4% for robotic, compared to 8% for laparoscopic and 16% for open donor hepatectomies. Additionally, while recipient mortality did not differ significantly across approaches, major (Clavien-Dindo grade ≥ 3a) (7) recipient morbidity rates were significantly lower following robotic retrieved donor grafts. These findings suggest that the robotic approach may offer a safer alternative for both donors and recipients, highlighting its potential to enhance clinical outcomes in living donor liver transplantation.

27.5 Conclusions

Robotic donor hepatectomy can be safely performed in high-volume experienced centers with excellent outcomes for donors. Careful technical details must be followed to assure success of the procedure.

References

1. Raptis DA, Elsheikh Y, Alnemary Y, Marquez KAH, Bzeizi K, Alghamdi S, et al. Robotic living donor hepatectomy is associated with superior outcomes for both the donor and the recipient compared to laparoscopic or open – a single center prospective registry study of 3,448 cases. Am J Transplant Off J Am Soc Transplant Am Soc Transpl Surg. 2024:S1600-6135(24)00293-4. https://doi.org/10.1016/j.ajt.2024.04.020.
2. Nakamura T, Tanaka K, Kiuchi T, Kasahara M, Oike F, Ueda M, et al. Anatomical variations and surgical strategies in right lobe living donor liver transplantation: lessons from 120 cases. Transplantation. 2002;73(12):1896–903.
3. Lee SG. A complete treatment of adult living donor liver transplantation: a review of surgical technique and current challenges to expand indication of patients. Am J Transplant Off J Am Soc Transplant Am Soc Transpl Surg. 2015;15(1):17–38.
4. Hong SK, Kim JY, Lee J, Kim J, Choi HH, Lee S, et al. Pure laparoscopic donor hepatectomy: experience of 556 cases at Seoul National University Hospital. Am J Transplant Off J Am Soc Transplant Am Soc Transpl Surg. 2024;24(2):222–38.
5. Giulianotti PC, Tzvetanov I, Jeon H, Bianco F, Spaggiari M, Oberholzer J, et al. Robot-assisted right lobe donor hepatectomy. Transpl Int Off J Eur Soc Organ Transplant. 2012;25(1):e5–9.
6. Broering D, Sturdevant ML, Zidan A. Robotic donor hepatectomy: a major breakthrough in living donor liver transplantation. Am J Transplant Off J Am Soc Transplant Am Soc Transpl Surg. 2022;22(1):14–23.
7. Clavien PA, Barkun J, de Oliveira ML, Vauthey JN, Dindo D, Schulick RD, et al. The Clavien-Dindo classification of surgical complications: five-year experience. Ann Surg. 2009;250(2):187–96.

Safe Unplanned Open Conversion During Robotic Hepatectomy for Major Bleeding

28

Mariano C. Giglio, Gianluca Rompianesi, and Roberto I. Troisi

28.1 Introduction

The worldwide diffusion of minimally invasive liver surgery (MILS) over the last three decades [1] has led to the expansion of the number of centers performing laparoscopic and robotic resections as well as the increase of the possible indications and is no longer limited to small peripheral lesions located in the anterior segments [2–4]. Robotic-assisted surgery overcame several limitations of the laparoscopic technique [5, 6], allowing more surgeons to manage more complex liver resections with a minimally invasive approach. MILS has been demonstrated to be able to offer significant advantages both in minor and major hepatectomies [1], but despite the technical progress and the increased skills, in 4–26% of cases, the surgical team opts to convert from minimally invasive to open surgery [7]. This decision can be the result of a lack of progression and therefore be performed electively or determined by an intraoperative complication and therefore being performed as an emergency procedure. In both scenarios, the timing of the conversion is of paramount importance to minimize the entity of the negative impact that can affect patients' outcomes [8]. Intraoperative bleeding represents the most frequently reported cause of open conversion during laparoscopic hepatectomies, accounting for approximately one-third of cases, followed by oncological concerns and adhesions [8–11]. Intraoperative bleedings during liver resections most frequently originate from major vessels during dissection or mobilization maneuvers, from the liver parenchyma during the transection, or from other non-hepatic structure injuries. At higher risk of experiencing intraoperative bleeding are patients affected by cirrhosis, due to impaired coagulation, the diseased liver parenchyma, the distorted anatomy, and the presence of portocaval shunts in case of portal hypertension. Similarly to laparoscopic liver resections, robotic-assisted hepatectomies have bleeding and oncological reasons as the most frequent indications for open conversions [12]. Compared to laparoscopy, the robotic approach has shown to reduce by 60% the risk of open conversion during major hepatectomies, with rates as low as 5% [11, 13], but analyses performed on minor, low-difficulty resections failed to demonstrate these superior results [14, 15]. It needs to be pointed out that the available data on robotic liver resection are far more limited than the ones on laparoscopic resections, and that may be affected by a selection bias [7].

M. C. Giglio · G. Rompianesi · R. I. Troisi (✉)
Division of HBP, Minimally Invasive and Robotic Surgery, Transplantation Center, Federico II University Hospital, Naples, Italy

M. D'Hondt, I. Sucandy (eds.), *Textbook of Robotic Liver Surgery*,
https://doi.org/10.1007/978-3-031-76536-0_28

28.2 Risk Factors for Unplanned Conversion to Open Surgery

The relatively low proportion of converted patients and consequent small sample sizes and the different inclusion criteria adopted by reports in the literature led to diverse factors identified as associated with open conversion during MILS. Several demographic characteristics have been described as risk factors such as patient age, male gender, diabetes, hypertension, BMI, and ASA score, as well as oncological and surgical characteristics such as tumor size, presence of cirrhosis or portal hypertension, resections involving three or more segments or posterosuperior segments, high difficulty score, previous abdominal surgery, and concomitant surgical procedures [8–11, 14, 16]. Bleeding represents the most frequently observed cause of open conversion in most published series, and consequently, all factors and events that can increase intraoperative blood loss constitute risk factors for this unfavorable event. In case of major and difficult resections or the presence of cirrhosis and portal hypertension, exceptional caution should be taken, and the patients should be managed in experienced, high-volume centers to reduce the conversion rate and achieve optimal postoperative outcomes [17].

28.3 Technical Management of an Unplanned Conversion for Major Bleeding

Several methods for hemostasis can be used during a minimally invasive liver resection, including bipolar energy, clipping, sutures, and hemostatic devices. Examining all of them in details is beyond the scope of this chapter.

Once these strategies have failed and open conversion appears to be the best solution for bleeding control, it should be performed in the safest and most efficient way possible. An emergency conversion performed with ongoing and uncontrolled bleedings has higher chances to result in life-threatening situations, considering the time required for the undocking of the robotic system and the abdominal opening. Therefore, before proceeding with the open conversion, every attempt should be made to obtain the best temporary control of the bleeding sources.

At any time point during the liver resection, the surgical team should be prepared for an unexpected and sudden uncontrolled hemorrhage. The aspirator must work properly and be able to reach the bleeding site, the suction system canister must hold sufficient capacity, and one or more gauzes should be available in the abdominal cavity to compress the bleeding site.

The best temporary control of bleeding can be achieved as a result of mechanical maneuvers and adjuvant maneuvers.

It should be kept in mind that the temporary occlusion must operate efficiently for the entire time necessary for the robot to be released and the abdomen opened. Therefore, it must be kept in place by a laparoscopic instrument controlled by the table assistant, since robotic instruments will stop working at some point and will have to be removed since opening the abdomen with the robot in position is impractical and unsafe.

For the authors, the best method to obtain temporary control of major bleedings is the mechanical compression using gauze (Fig. 28.1).

This is effective especially in case of bleeding from the hepatic veins. In case of major bleeding, the table assistant will use the aspirator to keep a clean field and expose the bleeding site, while the console surgeon will compress the point of bleeding by handling a gauze with the robotic instruments. For this reason, it is better to have a gauze into the abdomen during the parenchymal transection, to be ready in case of bleeding. Once compression has effectively reduced the bleeding, the gauze should be kept in place by the laparoscopic instrument of the table assistant. Then, the robotic instruments can be removed and the robot undocked from the patient.

Another option for temporary bleeding control which can follow gauze's compression is vessel occlusion by grasping it at the bleeding point. However, this maneuver should be carefully evaluated since it can worsen the bleeding especially in case of very fragile veins; therefore, it is suggested to be performed only by experi-

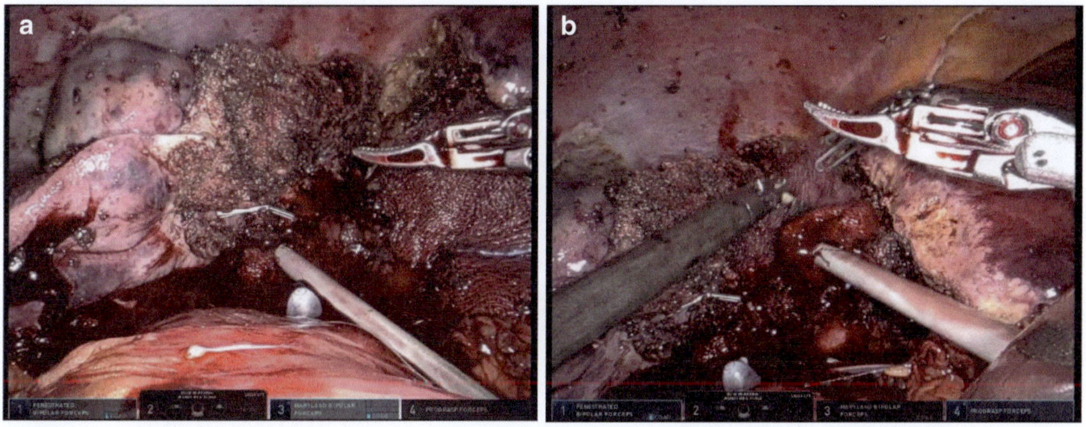

Fig. 28.1 (**a**) intraoperative bleeding during a robotic right hepatectomy for large hepatocellular carcinoma after SIRT; (**b**) temporary bleeding control with compression by gauze placed by the console surgeon (fenestrated bipolar on the first robotic arm) with the aid of the table assistant (suction device)

enced surgeons. During robotic liver resection, this maneuver at first will be performed by the console surgeon using a robotic instrument; however, to allow a safe conversion, this maneuver must be then continued by the table assistant using a laparoscopic instrument, to allow robot undocking. This switch of instruments seems unpractical and unsafe, and therefore this method of hemostasis appears as less convenient than vessel compression.

Some other actions can be helpful to control the bleeding. If the source of bleeding is a major portal branch or a Glissonean pedicle, a Pringle maneuver can be performed and maintained for the needed time [9]. In case of bleeding from hepatic veins, lowering the ventilator positive end-expiratory pressure (PEEP) can result in a significant reduction of the bleeding [18]. Also, a method to achieve a fast lowering of the central vein pressure is the adoption of a reverse Trendelenburg position [9]. However, it should be kept in mind that this position may increase the risk of air embolism and alter the patient's hemodynamics, so it should be discussed and agreed with the anesthesiological team.

The type of abdominal incision should be planned in order to achieve the best possible exposure of the bleeding site and to complete the liver resection. The authors prefer to perform the incision while the pneumoperitoneum is maintained, therefore leaving the robotic trocars in place since the intra-abdominal pressure can contribute in reducing the bleeding [9] and the robotic camera, controlled by the assistant surgeon, can monitor the effectiveness of the bleeding control during the abdominal incision.

The robotic platform offers an undoubted advantage of performing very precise sutures, especially in the case of bleeding. With this in mind, the assessment of the site of hemorrhage and the decision as to whether use the compression technique or vascular suturing first, as opposed to temporary closure using a clamp, are up to the operator. The management of sudden hemorrhages should be considered in a rapid and effective "conversion protocol" algorithm where, in addition to the techniques used (and specific tools at the table such as long vascular clamps or bulldogs), the roles of each surgical, medical, and nursing team member should be clearly defined and agreed beforehand in order to maximize the success rates and outcomes in such a rapidly evolving and stressful scenario. The console and table surgeons, the scrub nurse, and the OR nurses need to coordinate for the management of the temporary bleeding control, the undocking process, the instrument setup, and the open conversion, while the anesthetic team manages works toward maintaining the best possible hemodynamic balance during the critical phases and in preparation for the conversion. The approval of a "conversion protocol" is of funda-

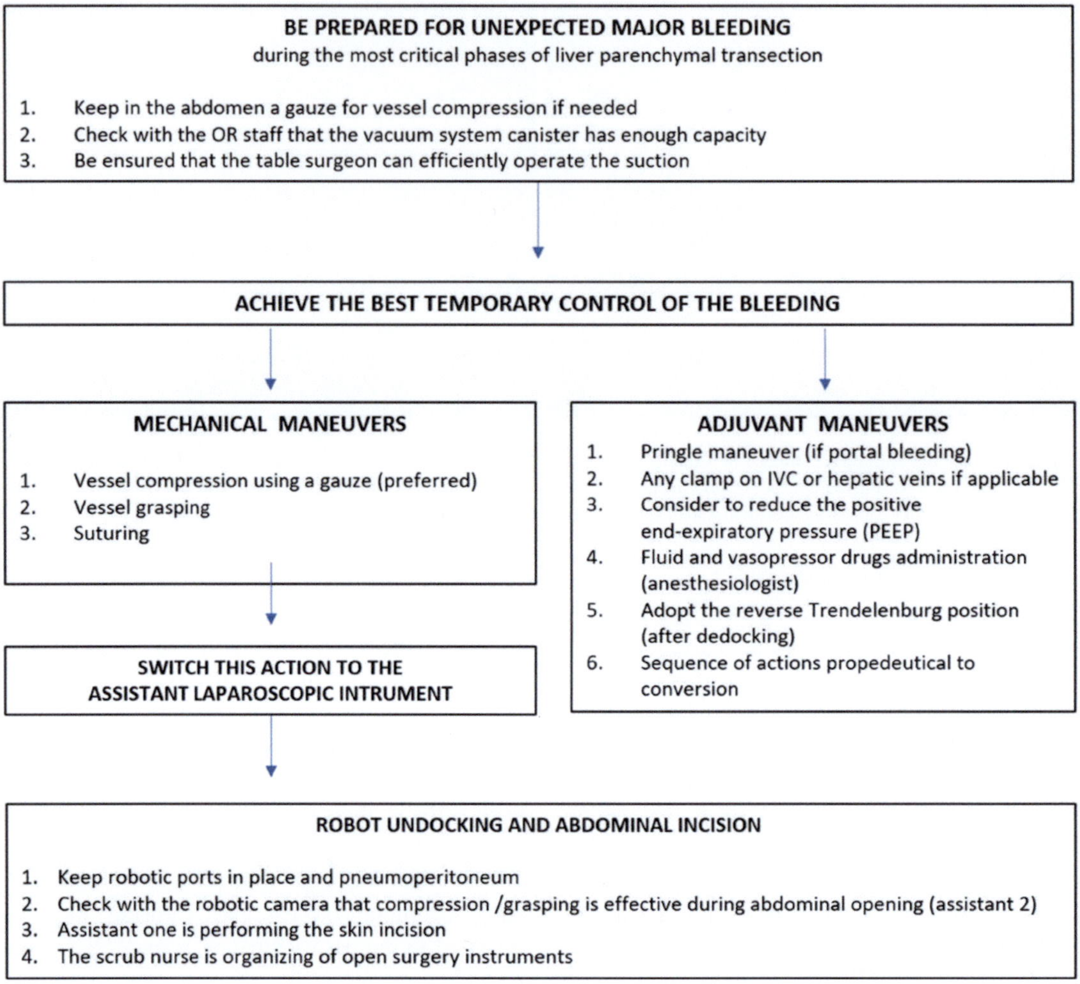

Fig. 28.2 Essential actions for temporary bleeding control during a robotic liver resection allowing safe conversion to open surgery

mental importance in robotic surgery programs, especially for complex cases.

The recommended maneuvers and sequence of events for an unplanned open conversion for hemorrhage during robotic liver resections are shown in Fig. 28.2.

28.4 Outcomes of Conversions During Robotic Liver Resection

A recent multicenter retrospective study involving 42 centers investigated the outcomes of patients undergoing robotic major liver resections requiring conversion to open surgery [11]. To date, this study represents the largest evidence on this topic. Conversion to open surgery was necessary in 4.9% of robotic major hepatectomies, as opposed to the 11.3% conversion rate observed in laparoscopic major hepatectomies. On multivariable analysis, taking into account resection difficulty, type of liver resection, presence of liver cirrhosis, and other patient characteristics, the robotic approach emerged as an independent protective factor against conversion to open surgery (OR = 0.39). As expected, patients requiring conversion had inferior outcomes compared to non-converted patients, with increased blood loss, transfusion rate, postoperative major morbidity, reoperation rate, and 30- and 90-day mortality. Indeed, this was expected since converted patients are negatively selected patients (higher ASA score, presence of liver cirrhosis and portal hypertension, larger tumors requiring more complex resections) who also lose the benefits of the minimally invasive approach. In addition, patients converted for bleeding showed increased postoperative morbidity and mortality compared to patients converted for other reasons.

Finally, when comparing outcomes of patients converted during robotic surgery with those of patients converted during laparoscopic surgery, after correcting for potential confounders, the analysis showed that robotic converted procedures were associated with higher blood losses, higher intraoperative transfusion rates, major complications (Clavien–Dindo >3) and reoperations, and higher 30-day and 90-day mortality. This suggests that emergency conversion may be more challenging during robotic than laparoscopic surgery, because of the time needed for robotic undocking and the difficulty for the first surgeon to achieve rapid full control of the operative field (inability of the surgeon to achieve rapid full control?). However, the inferior outcomes of converted robotic cases may be also due to the fact that only the most serious intraoperative incidents, leading to the worse outcomes, were converted during robotic surgery, while other events were more easily controlled with robotics avoiding a conversion as opposed to laparoscopy. Last but not least, it could be possible that robotic surgeons may have persisted with the minimally invasive approach much longer to avoid conversion and only converted to the open approach in the very late phase.

Further prospective studies analyzing in detail the causes of conversions will help in clarifying if robotic conversions are associated with poorer outcomes.

References

1. Ciria R, et al. Comparative short-term benefits of laparoscopic liver resection: 9000 cases and climbing. Ann Surg. 2016;263(4):761–77.
2. Buell JF, et al. The international position on laparoscopic liver surgery: The Louisville Statement, 2008. Ann Surg. 2009;250(5):825–30.
3. Wakabayashi G, et al. Recommendations for laparoscopic liver resection: a report from the second international consensus conference held in Morioka. Ann Surg. 2015;261(4):619–29.
4. Abu Hilal M, et al. The Southampton Consensus Guidelines for Laparoscopic Liver Surgery: From Indication to Implementation. Ann Surg. 2018;268(1):11–8.
5. Troisi RI, et al. The democratizing effects of robotic surgery: nine HPB manoeuvres exactly reproduced by the da vinci system. Surg Oncol. 2022;44:101822.
6. Troisi RI, et al. Robotic approach to the liver: open surgery in a closed abdomen or laparoscopic surgery with technical constraints? Surg Oncol. 2020;33:239–48.
7. Stiles ZE, et al. Predictors and implications of unplanned conversion during minimally invasive hepatectomy: an analysis of the ACS-NSQIP database. HPB (Oxford). 2017;19(11):957–65.
8. Halls MC, et al. Conversion for unfavorable intraoperative events results in significantly worse outcomes during laparoscopic liver resection: lessons learned from a multicenter review of 2861 cases. Ann Surg. 2018;268(6):1051–7.
9. Troisi RI, et al. Risk factors and management of conversions to an open approach in laparoscopic liver resection: analysis of 265 consecutive cases. HPB (Oxford). 2014;16(1):75–82.
10. Cauchy F, et al. Risk factors and consequences of conversion in laparoscopic major liver resection. Br J Surg. 2015;102(7):785–95.
11. Montalti R, et al. Risk factors and outcomes of open conversion during minimally invasive major hepatectomies: an international multicenter study on 3880 procedures comparing the laparoscopic and robotic approaches. Ann Surg Oncol. 2023;30(8):4783–96.
12. Gheza F, et al. Reasons for open conversion in robotic liver surgery: a systematic review with pooled analysis of more than 1000 patients. Int J Med Robot. 2019;15(2):e1976.
13. Tsung A, et al. Robotic versus laparoscopic hepatectomy: a matched comparison. Ann Surg. 2014;259(3):549–55.
14. Wang HP, et al. Factors associated with and impact of open conversion on the outcomes of minimally invasive left lateral sectionectomies: an international multicenter study. Surgery. 2022;172(2):617–24.
15. Zureikat AH, et al. Minimally invasive hepatopancreatobiliary surgery in North America: an ACS-NSQIP analysis of predictors of conversion for laparoscopic and robotic pancreatectomy and hepatectomy. HPB (Oxford). 2017;19(7):595–602.
16. Silva JP, et al. Minimally invasive hepatectomy conversions: an analysis of risk factors and outcomes. HPB (Oxford). 2018;20(2):132–9.
17. Goh BK, et al. Factors associated with and outcomes of open conversion after laparoscopic minor hepatectomy: initial experience at a single institution. Surg Endosc. 2015;29(9):2636–42.
18. Sand L, et al. Effect of patient position and PEEP on hepatic, portal and central venous pressures during liver resection. Acta Anaesthesiol Scand. 2011;55(9):1106–12.

Part VII

New Technology, Financial Impact, Registries and the Future of Robotic Liver Surgery

29 AI and XR in Robotics

Pieter De Backer, Rania Matthys, and Niki Rashidian

29.1 Introduction

29.1.1 What Is Artificial Intelligence (AI)?

Artificial intelligence (AI) is a subset of computer sciences focused on building smart algorithms capable of performing tasks that typically require human intelligence. Machine learning (ML) is a subtype of AI that can learn from data to identify, predict, and make decisions by recognizing novel nonlinear patterns within large datasets. Deep learning (DL) is defined as an ML algorithm that uses multiple layers to progressively extract higher-level features from the raw data. In the DL domain, a convolutional neural network (CNN) is a class of neural networks, most applied to analyzing visual imagery. While in ML, human rules are applied during the extraction of information, also referred to as feature engineering, DL and more specifically CNNs have the abstract capability of learning new patterns without any human intervention. As such, CNNs can analyze medical images based on complex patterns not detectable by the human eye. Computer vision is a field of ML that trains computers to interpret and understand the visual world [1, 2]. Machines can accurately identify and locate objects and then react to what they "see" using digital images from cameras, videos, and DL models. Trained algorithms can predict an outcome based upon its "experience" when presented with a novel yet similar data [3]. Robotic liver surgery, as a high-stake data-intensive process, is an ideal domain to benefit from AI technologies. Examples include using AI models to identify clinical diagnoses, support decision-making, improve surgical training, interpret medical images, navigate preoperative planning and intraoperative guidance, and predict postoperative course of patients.

29.1.2 Three-Dimensional (3D) Models and Extended Reality (XR)

Over the past decades, surgeons performed preoperative planning by interpreting two-dimensional (2D) CT or MRI images while cognitively reconstructing this information to a

P. De Backer (✉)
Orsi Academy, Melle, Belgium

IBiTech-bioMMeda, Ghent University, Ghent, Belgium
e-mail: pieter.de.backer@orsi.be

R. Matthys
Faculty of Medicine and Health Sciences, Ghent University, Ghent, Belgium

N. Rashidian
Department of General, HPB Surgery and Liver Transplantation, Ghent University Hospital, Ghent, Belgium
e-mail: nikdokht.rashidian@ugent.be

M. D'Hondt, I. Sucandy (eds.), *Textbook of Robotic Liver Surgery*,
https://doi.org/10.1007/978-3-031-76536-0_29

3D presentation of the patient's anatomy as would be encountered during surgery [4]. However, liver anatomy is complex with a high degree of variation in the hepatic vessels and bile ducts anatomy. As such, it is often difficult to reconstruct a 3D presentation based on 2D images alone [4–6]. To overcome this bottleneck, 3D models, reconstructed from planar 2D imaging, have found their way into preoperative planning in recent years. These 3D liver reconstructions have become an important clinical tool in preoperative planning of complex liver surgeries. This modality helps the surgeon to visualize and better understand the spatial relationships between vascular and biliary anatomy of liver and the surrounding parenchyma. In comparison to conventional CT or MRI, these models allow the surgeon to easily notice variations in liver anatomy [5, 7, 8]. The 3D reconstructions may as well add value to oncological liver surgery as they help the surgeon to determine the 3D relationship between the tumor and surrounding biliary and vascular structures and to estimate the tumor invasion into hilar vessels, hepatic veins, and bile ducts. This in turn leads to safer resection plane planning, an optimal surgical approach, and the prediction of possible complications [5, 8, 9]. In this way, 3D imaging ultimately improves preoperative planning for complex surgeries and can be considered superior to conventional 2D imaging for the visualization of anatomical structures [7]. In the clinical practice, 3D models have been found to be beneficial in decision-making for determining resectability or radicality, in calculation of future liver remnant and extent of resection, and in determining indications for hepatic vein resection/reconstruction or R1 vascular resections [10, 11].

3D reconstructions are made by a process called segmentation, which is defined by Ganguli et al. as the process of isolating an area of interest in the data files, which are in this case a series of DICOM images. When all these isolated areas are stacked together, they form the complete structure [12]. The final 3D image is converted to a stereolithography (STL) format [12, 13]. Slice thickness plays an important role in this process since thinner slices increase accuracy of the reconstructed 3D image [14]. The 3D images can be reconstructed either from CT or MRI images, where CT images are clearly more suitable than MRI images as MRI is of lower resolution and acquires thicker slices, resulting in a lower accuracy of the 3D visualization [9, 14]. The segmentation workflow can be distinguished based on the input of the user, in manual, semiautomatic, and automatic segmentation. Automatic segmentation requires no user input, while manual segmentation is entirely performed by the user [8, 15–17]. During manual segmentation, the user manually delineates the boundaries of liver parenchyma, hepatic and portal vasculature, hepatic artery, inferior vena cava, biliary ducts, and gallbladder in a slice-by-slice manner using software. Semiautomatic segmentation is an intermediate form where limited user input is required, e.g., marking seed points on top and bottom parts in every liver lobe or providing an initial liver contour in some slices [18]. Multiple segmentation software packages have been developed to do manual and semiautomated segmentation, both as open source (3D slicer as the most powerful solution at present) and as proprietary software. 3D models can also be offered as a service, in which the clinician only provides a CT or MRI and is returned a virtual file or a 3D-printed model [5, 16, 17, 19]. However, manual segmentation depends on the skills and experience of the user, making it subject to high intra- and interobserver variability. Further, manual segmentation is a time-consuming and tedious process for which automatic segmentation can offer a solution [15, 16, 18]. Despite the recent advances in computing methods, automatic liver segmentation has remained a challenging task, due to indefinite boundary, intensity inhomogeneity, and anatomical variations across subjects [15, 18].

Postoperative analysis of databases showed that 3D reconstructions can contribute to safer liver surgery. Surgeries performed with the help of a 3D reconstruction tend to have shorter operating times, lower estimated blood loss and complication rates, and positive surgical margins [19–21]. The reduction of blood loss is an important finding, as blood loss is closely related to postoperative recurrence in liver malignancies [21]. These 3D reconstructions also ensure a smaller difference between the predicted and the actual resected liver volume, making them a suitable tool for the estimation of the postoperative liver volume [21].

While 3D liver reconstructions clearly improve surgical planning, displaying a 3D image on a 2D monitor before or during surgery provides the surgeon with a limited sense of depth [7, 13, 22]. Creating a physical 3D model of the patient's liver by 3D printing may offer a solution to this drawback. 3D printing is the process of converting a 3D virtual model to a physical model by layer-by-layer fabrication [12, 22]. Physical 3D-printed models provide the surgeon preoperatively with a more comprehensive geometry and feeling and a better presentation of the spatial relationships between intrahepatic vessels, tumor, and parenchyma, making it more accessible for the surgeon to rehearse the planned surgical route, to discuss the case with the team, and to explain the procedure to the patient [12, 22–24]. The elasticity and softness of these models can be chosen to mimic the texture of real liver parenchyma [19, 23]. Preoperative dissection of this model can contribute to better understanding and confirming the operative plan and vision [23]. Intraoperatively, the 3D model can be positioned in the desired way to guide the surgeon trough the operative plan, and it can be easily reoriented to provide the surgeon with an optimal vision of the liver anatomy after changes in liver orientation, e.g., mobilization and rotation of the liver [7, 12, 23, 25]. Despite the physical 3D models improving pre- and intraoperative assessment of the liver anatomy compared to visual 3D models, these models are only reserved for complex surgeries and teaching anatomy, as the 3D printing process comes with a high cost and a long time of printing [13, 19, 25]. Printing one 3D model costs several hundreds of dollars, and the estimating printing time can take up to multiple days [13, 19, 25]. In comparison to printed 3D models, digital solutions offer a sustainable and available modality for individualized 3D presentation of liver models. Extended reality (XR) is defined as any technology that alters reality by adding digital elements to the physical or real-world environment including virtual reality (VR), augmented reality (AR), and mixed reality (MR). In complex liver surgery, these are evolving technologies enabling improved preoperative planning and intraoperative navigation [26, 27].

29.1.3 AI in Robotic Liver Surgery

29.1.3.1 Preoperative 3D Model Making

As mentioned above, 3D reconstructions of the liver are essential to multiple aspects of management of liver diseases, as these can be used to plan liver resections and transplantations and are also used in image-guided percutaneous ablation therapies [16, 28]. Since manual segmentation comes with a high workload and high intra- and interobserver variability, these liver reconstructions can also be generated automatically, which brings us to the field of AI [15, 16, 29]. Multiple ML algorithms have been developed to automatically generate 3D liver reconstructions from CT/MRI imaging [29]. Although many of these algorithms can successfully delineate most parts of the liver surface and are proven to be accurate and effective by quantitative validations, further research is still required before these algorithms can be applied to clinical practice in an automated way [16, 28]. A hybrid option where the AI model predicts parts of the 3D model and the surgeon alters part of the segmentation can be an intermediate solution.

29.1.3.2 Intraoperative Tracking and Monitoring

Artificial intelligence in robotic surgery is a field in full development where the focus currently lies on surgical instrument (SI) detection and tracking intraoperative events [30–32]. SI tracking and detection serves multiple goals, which include long-term goals such as robotic automation [33]. A more short-term goal is automatic and quantitative methods to determine where a surgeon stands on the steep learning curve of robotic surgery, since up to this day, there only exist subjective evaluation methods for this purpose [30, 31]. A DL algorithm was trained to identify surgical instruments and track them during the surgery. Especially the trajectory of the SI was defined as the movements of SI might be associated with surgical skills [31]. Based on these movements, nine motion metrics were calculated to estimate the skills of the surgeon: time to completion of surgery, instruments out of view, instrument collision, economy of motion, average speed, number of movements, economic factors, surgical instrument changes, and laparoscopy usage [31]. Compared to conventional methods, this new quantitative assessment shows 83% accuracy [31]. Nevertheless, other research shows that instrument tracking only is not suitable for surgical skill assessment, and other factors such as objectively defined errors and proficiency-based assessment are more transparent and objective. The aim of proficiency-based surgical skill assessment, compared to time-based assessment, is to achieve predefined expert-level performance and provide training tailored to individual practice needs that results in a more uniform skillset [34, 35]. As such, AI-powered SI tracking should be focused on assisting the efficient implementation of PBP methodologies (see chapter 11), rather than trying to derive skill from pure motion metrics.

The research of AI in laparoscopic liver surgery is still more extensive than in robot-assisted liver surgery. A large proportion of this research is conducted in laparoscopic cholecystectomy surgeries with the goal of preventing major bile duct injuries (BDI) by correctly identifying anatomical landmarks based on AI identification [9, 36–39]. A first deep learning (DL) algorithm is based on the mental model surgeons maintain during this surgery to identify safe and dangerous zones of dissection to minimize the risk of injuries. This algorithm was trained to classify every pixel of the surgical video in the *go* zone, where dissection could be performed with a low risk of a major BDI or the *no-go* zone, a deeper region in the hepatobiliary triangle with an unacceptable probability of causing major BDI [36, 37]. The output of this model can be displayed as an overlay over the video of the surgical field, either binary, presenting the *go* and *no-go* zone with exact boundaries, or displayed by topographical heatmaps, where each pixel represents a probability of belonging to the *go* or the *no-go* zone [36, 37]. This second option tends to be more consistent to fit the mental models of the surgeons, as they seldom make exact distinctions between safe and dangerous zones [36, 37]. These DL algorithms already show a high accuracy between the true manual annotations and annotations predicted by these models [36]. This form of deep learning where objects are identified and highlighted along their boundaries pixel by pixel as an overlay is called semantic segmentation. The aim of these semantic segmentation algorithms is to incorporate these in robotic platforms to take the next step in automated surgical tasks [33, 36].

Another DL algorithm in the field of laparoscopic cholecystectomy was trained to assess the critical view safety (CVS) criteria, consisting of the hepatocystic triangle being cleared from fat, the lower 1/3 of the gallbladder being separated from the liver, and only two tubular structures entering the gallbladder [38, 40]. The goal of fulfilling the CVS criteria is just as with the *go* and *no-go* zones, to prevent major BDI. This DL algorithm is already capable of recognizing the two tubular structures entering the gallbladder but has still some difficulties with fulfilling the first criterium of the CVS [38].

Lastly, current research also focuses on surgical phase and tool recognition [32, 41]. A DL algorithm is developed that can identify laparoscopic instruments during surgery with high accuracy and fast speed in surgical environments with low and medium levels of complexity. This algorithm works by classifying every pixel in either the foreground (being the instrument) or the background [42]. However, in more complex surgical environments, large regions are still segmented incorrectly. An example of these complex environments is when the specimen bag is inserted, as this bag has almost the same color as the surroundings and has a great overlap with neighboring structures [42]. Accurate tool detection can be used to extrapolate surgical skills metrics such as tool usage patterns, range, and economy of movements [43]. For surgical phase recognition, other DL algorithms are being worked on. The challenge of these applications is computers lacking a memory and context understanding. In this way, they miss knowledge of prior surgical steps, making it a great challenge to correctly identify the surgical phase [41]. To overcome this hurdle, long short-term memory (LSTM) networks and more recently transformers, which originated from natural language processing, were deployed in surgical phase recognition [44]. By giving the algorithms the equivalent of a memory cortex, they acquire knowledge of previous surgical events leading to the significant improvement of their surgical phase recognition performances [41, 43, 45]. Accurate phase recognition can give important feedback to the surgeon about how long each phase took, which instruments were used per surgical phase, and how much time was spent performing specific tasks [41]. Next to this, it can play a role in OR planning and logistics.

29.1.4 XR in Robotic Liver Surgery

29.1.4.1 Preoperative

While 3D models can improve preoperative planning of liver surgeries, a potential drawback of desktop-based visualization of 3D models is its representation on 2D monitors. Applying extended reality (XR) to these models is a possible solution to improve preoperative liver surgery planning. Furthermore, preoperative surgical planning can be merged into the operative room and even the intraoperative field. Prior research has shown the benefits of VR compared to desktop visualization of 3D models for preoperative surgical planning and decision-making in liver surgery [46]. Figure 29.1 shows a 2D CT scan of a patient with liver tumor converted to the 3D and VR models.

Mixed reality is defined at the result of merging the real and virtual world so that the virtual 3D models can be visualized in the real world, overlaying the real 3D objects or even anatomy [47, 48]. This mixed reality view can be achieved by wearable devices classified under head-mounted displays (HMD). They can be used everywhere and can thus be taken to the operating room to check the surgical plan before or during the surgery [47]. These visualizations trough mixed reality can provide the surgeon with a better anatomical overview, which could potentially lead to better decision-making in preoperative surgical planning. Mixed reality also allows the surgeon to preoperatively define a resection plan as deformable surfaces can be added to the 3D model that deform alongside the requested resection margins either by interactive manipulation or by changing the coordinates of control points that are distributed on a grid [47]. As such, 3D visualization trough mixed reality has already shown to significantly decrease the time for spatial understanding of the organ [47, 48]. For preoperative planning, VR technology is most often used to its

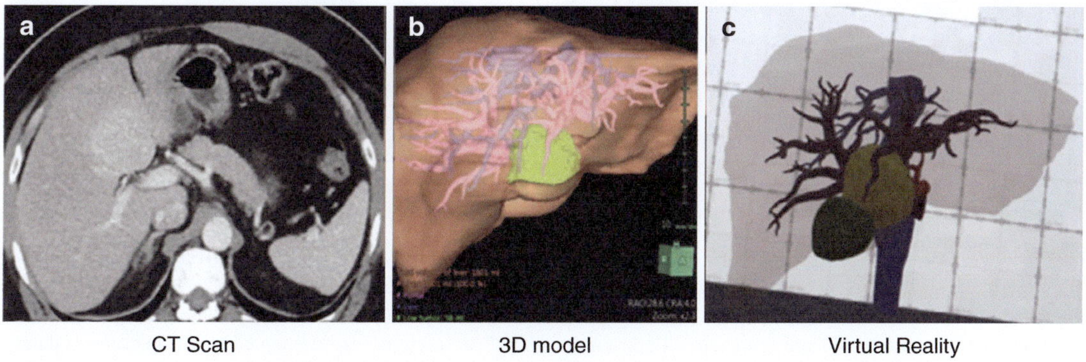

Fig. 29.1 Case of hepatocellular carcinoma. (**a**) CT showing the location of the tumor in the segment 4B/5. (**b**) 3D model based on CT. (**c**) 3D model displayed in virtual reality

lower cost, more user-friendliness, and more controlled environment for convenient education. Also most importantly, here we do not need to overlap the reality.

29.1.4.2 Intraoperative

Head-mounted displays allow for sterile interaction when hand movements are tracked. As such, mixed reality is particularly useful for intraoperative guidance for open surgery. Other items that could be displayed on the HMD during surgery are patient characteristics such as blood pressure, heart rate, ECG, or if needed a CT scan or 3D model to scroll through. Whereas mixed reality could potentially offer a benefit in the preoperative planning of liver resection and during open surgery, AR can make the greatest contribution during robot-assisted and laparoscopic liver surgery. Augmented reality provides additional information inside the operative field by alignment of virtual structures over the real anatomical structures. The surgical field is augmented with, e.g., a 3D model, providing extra information in the surgeon's view on target vessels or lesions of interest. In this case, AR can be defined as the fusion of live surgical images provided by a robotic camera, with preoperative patient-specific imaging, using different landmarks. This combination gives the surgeon the intraoperative robotic camera view overlaid with the preoperative 3D model [6, 14, 49, 50]. In this way, AR allows the surgeon to look inside closed cavities or dens organs, such as the liver [51]. This can then subsequently be taken one step further, where an instrument segmentation algorithm tracks the instruments throughout the surgery, allowing interaction with the overlayed 3D model [52]. Figure 29.2 shows an AR example, depicting real-time overlay of the portal vein 3D model during a robotic right hemihepatectomy for a cholangiocarcinoma on the Intuitive TilePro™.

Centrally, classical AR is shown, while the right image depicts the added benefit of AI-assisted instrument detection.

Another useful application of AR applications for robotic liver resection is the placement of the robotic ports. By positioning a beamer above the patient, a 3D model of the abdominal cavity can be projected on the skin of the patient, thus once more augmenting the surgical field with new information. The projection is correctly registered in relation to external landmarks [51, 53]. This helps the surgeon in strategic port and camera placement, adapted to the patient's anatomy, which can lead to less complications and a decrease in operating time [53]. Likewise, AR could potentially offer important benefits in the oncological liver surgery. As AR contributes to visualizing internal liver structures (veins, arteries, bile ducts, and tumor), it easily shows intrahepatic lesions, their real-time location, their relation to major vascular structures, and variants in hepatic anatomy [6, 14, 49, 51, 53, 54]. In this way, AR could lead the surgeon in deciding the transection planes in complex tumor resections close to main structures, leading to a correct dis-

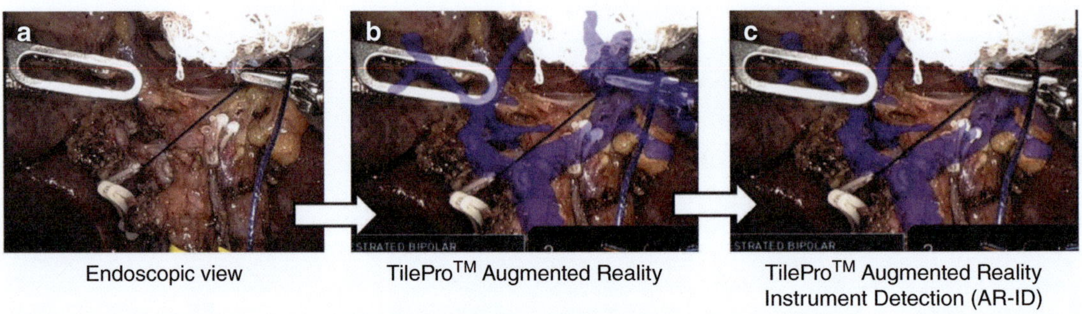

Fig. 29.2 Case of robotic right hemihepatectomy for a large intrahepatic cholangiocarcinoma. (**a**) Endoscopic view showing dissection of the right portal vein. (**b**) Augmented reality view. 3D patient-specific portal vein, derived from MRI, is projected onto the operative field inside the robotic console by means of the Intuitive TilePro™ feature. (**c**) Augmented reality instrument detection (AR-ID), where AI-assisted instrument detection causes the 3D model to no longer obstruct the robotic surgical instruments which enhances AR interaction

section of the tumor with precise recognition of all major vascular structures, with less blood loss, decreased OR time, and maximal liver parenchymal preservation [49, 51, 53, 55].

Nevertheless, augmented reality has several roadblocks ahead to be used in daily clinical practice. Up to now, AR can only offer the benefit of showing vanishing lesions in the liver parenchyma as the overlay accuracy is too inaccurate to allow the surgeon to perform a resection relaying solely on the AR visualization [49, 50]. To tackle this drawback, ongoing research focuses on further improving the registration process and the preoperative 3D visualizations [14]. The registration process in AR is defined as the process in which corresponding data points in virtual and real images are linked into one coordinate system for the purpose of a perfect overlay of the 3D visualization on the surgical images [14]. Due to the frequent anatomic variation in hepatobiliary anatomy, the overlay accuracy must be 100% to ensure a safe and effective surgery, making registration the key to a reliable navigation tool [14, 49, 55]. This process is acquired by defining and marking landmarks on the preoperative 3D images and corresponding points on the liver or in the abdominal cavity, such as the ribs, xiphoid process, liver boundaries, porta hepatis, gallbladder fossa, and inferior vena cava [14, 49, 51, 56]. This procedure can be performed manually, called interactive registration, or automatically, making it automatic registration [14, 50, 55, 56]. Additional complexity arises from the liver deformation during resection, where the anatomy and flexure of the liver combined with the pneumoperitoneum result in a different anatomy and then present during CT or MRI. Other challenges of current registration methods are the surgical manipulation before resection and respiratory movements ensuring that the surgery needs to be interrupted multiple times to repeat the registration process [49, 53, 57]. To adapt for these bottlenecks, the 3D models, which are mainly static at present, should become flexible and dynamic, which means they are able to deform alongside the organ as the surgery progresses [51, 58]. Consequently, it is critical that the 3D reconstructions are adapted to these intraoperative changes in order to enlarge the AR overlay accuracy and provide the surgeon with a reliable surgical navigation tool [51].

29.2 Conclusions

Surgical practice in general, and liver surgery in particular, is evolving through rapid advances in computer technologies and data sciences. AI in robotic surgery and image-guided navigation in liver surgery is affecting all areas of surgical practice, including pre-, intra-, and postoperative areas, to achieve individualized treatments and to improve patient outcomes. Augmented reality is an important modality for preoperative surgical planning and for intraoperative guidance through targeting the lesion, identifying the relation of the tumor with main anatomical structures, and defining resection margins. Robotic platforms can facilitate the use and integration of image navigation during the surgery. However, technological development is still ongoing, and full clinical implementation is to be awaited. Ideally, we need a flexible modeling process with dynamic registration capable of providing real-time feedback to the surgeon with minimal impact on surgical workflow.

References

1. Hashimoto DA, Rosman G, Rus D, Meireles OR. Artificial intelligence in surgery: promises and perils. Ann Surg. 2018;268(1):70–6.
2. Gholinejad M, A JL, Dankelman J. Surgical process modelling strategies: which method to choose for determining workflow? Minim Invasive Ther Allied Technol. 2019;28(2):91–104.
3. Jiang J, Trundle P, Ren J. Medical image analysis with artificial neural networks. Comput Med Imaging Graph. 2010;34(8):617–31.
4. Bijlstra OD, Broersen A, Oosterveer TTM, Faber RA, Achterberg FB, Hurks R, et al. Integration of three-dimensional liver models in a multimodal image-guided robotic liver surgery cockpit. Life (Basel). 2022;12(5):667.
5. Ivashchenko OV, Smit JN, Nijkamp J, Ter Beek LC, Rijkhorst EJ, Kok NFM, et al. Clinical implementation of in-house developed MR-based patient-specific 3D models of liver anatomy. Eur Surg Res. 2020;61(4–5):143–52.

6. Veerankutty FH, Jayan G, Yadav MK, Manoj KS, Yadav A, Nair SRS, et al. Artificial intelligence in hepatology, liver surgery and transplantation: emerging applications and frontiers of research. World J Hepatol. 2021;13(12):1977–90.
7. Zein NN, Hanouneh IA, Bishop PD, Samaan M, Eghtesad B, Quintini C, et al. Three-dimensional print of a liver for preoperative planning in living donor liver transplantation. Liver Transpl. 2013;19(12):1304–10.
8. Virzì A, Muller CO, Marret JB, Mille E, Berteloot L, Grévent D, et al. Comprehensive review of 3D segmentation software tools for MRI usable for pelvic surgery planning. J Digit Imaging. 2020;33(1):99–110.
9. Bari H, Wadhwani S, Dasari BVM. Role of artificial intelligence in hepatobiliary and pancreatic surgery. World J Gastrointest Surg. 2021;13(1):7–18.
10. Radtke A, Sotiropoulos GC, Molmenti EP, Schroeder T, Peitgen HO, Frilling A, et al. Computer-assisted surgery planning for complex liver resections: when is it helpful? A single-center experience over an 8-year period. Ann Surg. 2010;252(5):876–83.
11. Procopio F, Cimino M, Viganò L, Colombo AE, Franchi E, Costa G, et al. Prediction of remnant liver volume using 3D simulation software in patients undergoing R1vasc parenchyma-sparing hepatectomy for multiple bilobar colorectal liver metastases: reliability, clinical impact, and learning curve. HPB (Oxford). 2021;23(7):1084–94.
12. Ganguli A, Pagan-Diaz GJ, Grant L, Cvetkovic C, Bramlet M, Vozenilek J, et al. 3D printing for preoperative planning and surgical training: a review. Biomed Microdevices. 2018;20(3):65.
13. Kuroda S, Kobayashi T, Ohdan H. 3D printing model of the intrahepatic vessels for navigation during anatomical resection of hepatocellular carcinoma. Int J Surg Case Rep. 2017;41:219–22.
14. Tang R, Ma LF, Rong ZX, Li MD, Zeng JP, Wang XD, et al. Augmented reality technology for preoperative planning and intraoperative navigation during hepatobiliary surgery: A review of current methods. Hepatobiliary Pancreat Dis Int. 2018;17(2):101–12.
15. Beichel R, Bornik A, Bauer C, Sorantin E. Liver segmentation in contrast enhanced CT data using graph cuts and interactive 3D segmentation refinement methods. Med Phys. 2012;39(3):1361–73.
16. Hu P, Wu F, Peng J, Liang P, Kong D. Automatic 3D liver segmentation based on deep learning and globally optimized surface evolution. Phys Med Biol. 2016;61(24):8676–98.
17. Gotra A, Sivakumaran L, Chartrand G, Vu KN, Vandenbroucke-Menu F, Kauffmann C, et al. Liver segmentation: indications, techniques and future directions. Insights Imaging. 2017;8(4):377–92.
18. Le DC, Chinnasarn K, Chansangrat J, Keeratibharat N, Horkaew P. Semi-automatic liver segmentation based on probabilistic models and anatomical constraints. Sci Rep. 2021;11(1):6106.
19. Gavriilidis P, Edwin B, Pelanis E, Hidalgo E, de'Angelis N, Memeo R, et al. Navigated liver surgery: state of the art and future perspectives. Hepatobiliary Pancreat Dis Int. 2022;21(3):226–33.
20. Hallet J, Gayet B, Tsung A, Wakabayashi G, Pessaux P. Systematic review of the use of pre-operative simulation and navigation for hepatectomy: current status and future perspectives. J Hepatobiliary Pancreat Sci. 2015;22(5):353–62.
21. Zhang S, Huang Z, Cai L, Zhang W, Ding H, Zhang L, et al. Three-dimensional versus two-dimensional video-assisted hepatectomy for liver disease: a meta-analysis of clinical data. Wideochir Inne Tech Maloinwazyjne. 2021;16(1):1–9.
22. Soon DS, Chae MP, Pilgrim CH, Rozen WM, Spychal RT, Hunter-Smith DJ. 3D haptic modelling for preoperative planning of hepatic resection: A systematic review. Ann Med Surg (Lond). 2016;10:1–7.
23. Kuroda S, Kihara T, Akita Y, Kobayashi T, Nikawa H, Ohdan H. Simulation and navigation of living donor hepatectomy using a unique three-dimensional printed liver model with soft and transparent parenchyma. Surg Today. 2020;50(3):307–13.
24. Liu JP, Lerut J, Yang Z, Li ZK, Zheng SS. Three-dimensional modeling in complex liver surgery and liver transplantation. Hepatobiliary Pancreat Dis Int. 2022;21(4):318–24.
25. Berardi G, Colasanti M, Meniconi RL, Ferretti S, Guglielmo N, Mariano G, et al. The applications of 3D imaging and Indocyanine green dye fluorescence in laparoscopic liver surgery. Diagnostics (Basel). 2021;11(12):2169.
26. Quero G, Lapergola A, Soler L, Shahbaz M, Hostettler A, Collins T, et al. Virtual and augmented reality in oncologic liver surgery. Surg Oncol Clin N Am. 2019;28(1):31–44.
27. Balci D, Kirimker EO, Raptis DA, Gao Y, Kow AWC. Uses of a dedicated 3D reconstruction software with augmented and mixed reality in planning and performing advanced liver surgery and living donor liver transplantation (with videos). Hepatobiliary Pancreat Dis Int. 2022;21(5):455–61.
28. Alirr OI, Rahni AAA, Golkar E. An automated liver tumour segmentation from abdominal CT scans for hepatic surgical planning. Int J Comput Assist Radiol Surg. 2018;13(8):1169–76.
29. Gul S, Khan MS, Bibi A, Khandakar A, Ayari MA, Chowdhury MEH. Deep learning techniques for liver and liver tumor segmentation: A review. Comput Biol Med. 2022;147:105620.
30. Sarikaya D, Corso JJ, Guru KA. Detection and localization of robotic tools in robot-assisted surgery videos using deep neural networks for region proposal and detection. IEEE Trans Med Imaging. 2017;36(7):1542–9.
31. Lee D, Yu HW, Kwon H, Kong HJ, Lee KE, Kim HC. Evaluation of surgical skills during robotic surgery by deep learning-based multiple surgical instrument tracking in training and actual operations. J Clin Med. 2020;9(6):1964.
32. De Backer P, Eckhoff JA, Simoens J, Müller DT, Allaeys C, Creemers H, et al. Multicentric exploration

of tool annotation in robotic surgery: lessons learned when starting a surgical artificial intelligence project. Surg Endosc. 2022;36(11):8533–48.
33. Gruijthuijsen C, Garcia-Peraza-Herrera LC, Borghesan G, Reynaerts D, Deprest J, Ourselin S, et al. Robotic endoscope control via autonomous instrument tracking. Front Robot AI. 2022;9:832208.
34. Gallagher AG. Metric-based simulation training to proficiency in medical education:– what it is and how to do it. Ulster Med J. 2012;81(3):107–13.
35. N. R. Proficiency-based stepwise training for liver surgery. 2022.
36. Madani A, Namazi B, Altieri MS, Hashimoto DA, Rivera AM, Pucher PH, et al. Artificial intelligence for intraoperative guidance: using semantic segmentation to identify surgical anatomy during laparoscopic cholecystectomy. Ann Surg. 2022;276(2):363–9.
37. Laplante S, Namazi B, Kiani P, Hashimoto DA, Alseidi A, Pasten M, et al. Validation of an artificial intelligence platform for the guidance of safe laparoscopic cholecystectomy. Surg Endosc. 2023;37(3):2260–8.
38. Mascagni P, Vardazaryan A, Alapatt D, Urade T, Emre T, Fiorillo C, et al. Artificial intelligence for surgical safety: automatic assessment of the critical view of safety in laparoscopic cholecystectomy using deep learning. Ann Surg. 2022;275(5):955–61.
39. Kitaguchi D, Takeshita N, Hasegawa H, Ito M. Artificial intelligence-based computer vision in surgery: recent advances and future perspectives. Ann Gastroenterol Surg. 2022;6(1):29–36.
40. Pakhomov D, Premachandran V, Allan M, Azizian M, Navab N, editors. Deep residual learning for instrument segmentation in robotic surgery. International Workshop on Machine Learning in Medical Imaging, Springer; 2019.
41. Ward TM, Mascagni P, Ban Y, Rosman G, Padoy N, Meireles O, et al. Computer vision in surgery. Surgery. 2021;169(5):1253–6.
42. Zhang J, Gao X. Object extraction via deep learning-based marker-free tracking framework of surgical instruments for laparoscope-holder robots. Int J Comput Assist Radiol Surg. 2020;15(8):1335–45.
43. Kennedy-Metz LR, Mascagni P, Torralba A, Dias RD, Perona P, Shah JA, et al. Computer vision in the operating room: opportunities and caveats. IEEE Trans Med Robot Bionics. 2021;3(1):2–10.
44. Czempiel T, Paschali M, Ostler D, Kim ST, Busam B, Navab N, editors. Opera: attention-regularized transformers for surgical phase recognition. International Conference on Medical Image Computing and Computer-Assisted Intervention, Springer; 2021.
45. Cheng K, You J, Wu S, Chen Z, Zhou Z, Guan J, et al. Artificial intelligence-based automated laparoscopic cholecystectomy surgical phase recognition and analysis. Surg Endosc. 2022;36(5):3160–8.
46. Rashidian N, Giglio MC, Van Herzeele I, Smeets P, Morise Z, Alseidi A, et al. Effectiveness of an immersive virtual reality environment on curricular training for complex cognitive skills in liver surgery: a multicentric crossover randomized trial. HPB (Oxford). 2022;24(12):2086–95.
47. Kumar RP, Pelanis E, Bugge R, Brun H, Palomar R, Aghayan DL, et al. Use of mixed reality for surgery planning: assessment and development workflow. J Biomed Inform. 2020;112s:100077.
48. Pelanis E, Kumar RP, Aghayan DL, Palomar R, Fretland ÅA, Brun H, et al. Use of mixed reality for improved spatial understanding of liver anatomy. Minim Invasive Ther Allied Technol. 2020;29(3):154–60.
49. Prevost GA, Eigl B, Paolucci I, Rudolph T, Peterhans M, Weber S, et al. Efficiency, accuracy and clinical applicability of a new image-guided surgery system in 3D laparoscopic liver surgery. J Gastrointest Surg. 2020;24(10):2251–8.
50. Zygomalas A, Kehagias I. Up-to-date intraoperative computer assisted solutions for liver surgery. World J Gastrointest Surg. 2019;11(1):1–10.
51. Pessaux P, Diana M, Soler L, Piardi T, Mutter D, Marescaux J. Towards cybernetic surgery: robotic and augmented reality-assisted liver segmentectomy. Langenbeck's Arch Surg. 2015;400(3):381–5.
52. De Backer P, Van Praet C, Simoens J, Peraire Lores M, Creemers H, Mestdagh K, Allaeys C, Vermijs S, Piazza P, Mottaran A, Bravi CA, Paciotti M, Sarchi L, Farinha R, Puliatti S, Cisternino F, Ferraguti F, Debbaut C, De Naeyer G, Decaestecker K, Mottrie A. Improving augmented reality through deep learning: real-time instrument delineation in robotic renal surgery. Eur Urol. 2023;84(1):86–91. https://doi.org/10.1016/j.eururo.2023.02.024. Epub 2023 Mar 21. PMID: 36941148.
53. Giannone F, Felli E, Cherkaoui Z, Mascagni P, Pessaux P. Augmented reality and image-guided robotic liver surgery. Cancers (Basel). 2021;13(24):6268.
54. Haouchine N, Dequidt J, Berger MO, Cotin S. Deformation-based augmented reality for hepatic surgery. Stud Health Technol Inform. 2013;184:182–8.
55. Katić D, Wekerle AL, Görtler J, Spengler P, Bodenstedt S, Röhl S, et al. Context-aware Augmented Reality in laparoscopic surgery. Comput Med Imaging Graph. 2013;37(2):174–82.
56. Labrunie M, Ribeiro M, Mourthadhoi F, Tilmant C, Le Roy B, Buc E, et al. Automatic preoperative 3d model registration in laparoscopic liver resection. Int J Comput Assist Radiol Surg. 2022;17(8):1429–36.
57. Zhang W, Zhu W, Yang J, Xiang N, Zeng N, Hu H, et al. Augmented reality navigation for stereoscopic laparoscopic anatomical hepatectomy of primary liver cancer: preliminary experience. Front Oncol. 2021;11:663236.
58. Bertrand LR, Abdallah M, Espinel Y, Calvet L, Pereira B, Ozgur E, et al. A case series study of augmented reality in laparoscopic liver resection with a deformable preoperative model. Surg Endosc. 2020;34(12):5642–8.

30 Economic Cost and Value in Robotic Liver Surgery

Alessandro Parente and Robert P. Sutcliffe

Abbreviations

LDLT	Live donor liver transplantation
LLR	Laparoscopic liver resection
LLS	Left lateral sectionectomy
OLR	Open liver resection
RLR	Robotic liver resection

30.1 Introduction

There has been a global increase in the uptake of minimally invasive techniques in the field of liver surgery in recent decades, and many studies have demonstrated that both laparoscopic and robotic approaches are safe, feasible and associated with short-term advantages compared to open liver resection (OLR) [1–5]. The potential advantages of robotic over laparoscopic surgery are currently under investigation and the topic of several retrospective studies [6–10], although there is currently a lack of prospective data. There is emerging data to indicate that robotic liver resection (RLR) may increase the feasibility of a minimally invasive approach for technically difficult resections in comparison with laparoscopic liver resection (LLR), potentially by shortening the learning curve and reducing the rate of conversion to open surgery [10–12]. Implementation and widespread adoption of RLR programmes have been hindered by high capital and procedural costs of surgical robots, particularly in low-middle income countries, combined with a lack of robust evidence of a benefit over LLR. Estimation of the economic cost of a new surgical procedure, such as RLR, is highly complex and influenced by various factors including capital expenditure, cost of disposable instruments, operating room and hospital resources and postoperative patient outcomes. At the time of writing, several companies have launched novel surgical robots, and market forces are likely to significantly improve the cost-effectiveness of robotic procedures in the near future. In this chapter we provide a comprehensive overview of the current literature focusing on the economic costs and value of robotic liver resection compared to open and laparoscopic approaches.

30.2 Robotic Versus Open Liver Resection

The first research evaluating the costs of RLR was published by the University of Washington, USA, in 2016 [13]. Herein, the authors retrospectively compared 71 RLR with 88 OLR for benign and malignant liver diseases, using an adjusted generalized linear regression model, and they included both major and minor hepatectomies.

A. Parente
Institute of Liver Studies, Kings College Hospital, London, UK

R. P. Sutcliffe (✉)
Liver Unit, Queen Elizabeth Hospital, Birmingham, UK
e-mail: robert.sutcliffe@uhb.nhs.uk

M. D'Hondt, I. Sucandy (eds.), *Textbook of Robotic Liver Surgery*,
https://doi.org/10.1007/978-3-031-76536-0_30

Due to prolonged operating time and disposables costs, intraoperative costs were significantly higher for RLR (RLR: \$6026, OLR: \$5479; $p = 0.047$), but these were offset by reduced postoperative length of hospital stay after RLR, and the overall cost of RLR was significantly less than OLR (\$14,754 vs. \$18,998; $p = 0.001$).

In 2017, researchers at University of Illinois (Chicago, USA) illustrated the results of their study where they compared 68 RLR and 55 OLR for any indication where an independent company performed the financial analysis [14]. In a well-conducted study, these authors demonstrated that RLR had less overall morbidity and shorter hospital stay compared to OLR. The average total costs, including readmissions, were \$37,518 and \$41,948 for RLR and OLR, respectively. Of note, the authors found that this difference was not statistically significant. However, they have shown that RLR is financially comparable to OLR without sharply affecting hospital expenditures.

In an interesting report, Chinese authors evaluated the cost of robotic surgery in patients with hepatolithiasis [15]. Using a propensity scored-matched analysis in a 1:2 ratio, the authors compared the costs of 26 patients who underwent robotic surgery to 52 patients who underwent open surgery. There were no differences in postoperative complication rates (46.2% vs. 63.5%; $p = 0.145$), intraoperative stone clearance rates (96.2% vs. 90.4%; $p = 1.000$) or final stone clearance rates (100% vs. 98.1%; $p=0.652$) between the two groups. The robotic group had less blood loss and shorter postoperative hospital stays (13.54 ± 6.54 vs. 17.81 ± 7.49 days; $p = 0.016$) than the open group. They concluded that robotic surgery has advantages over open technique in terms of intraoperative blood loss and duration of hospital stay. However, the cost of robotic surgery was higher than open surgery. Notably, the differences in costs between the two groups were related to the start-up expenditure for the robotic system and the value of disposables which were not covered by medical insurance [15].

In 2019, a group from Taipei at the National Taiwan University conducted a study dividing perioperative and inpatient care costs. The robotic approach showed higher perioperative costs which were related to robotic system expenditures and higher anaesthesia fee due to longer operation time. However, patients who underwent RLR had lower cost of inpatient care due to shorter hospital stay as well as associated pharmacy and laboratory costs. Although the robotic approach was associated with decreased cost of inpatient care, the total cost remained high [16]. In the setting of living donor liver transplantation (LDLT), the first report was published in 2016 by the same group. They published their experience in robotic procurement of adult liver right lobe grafts and showed higher costs related to the robotic procedure when compared to open (\$13,436 vs \$5019; $p < 0.001$) [17].

In 2021, a preliminary experience from MedStar Georgetown University Hospital, USA, calculated the costs of 36 open resections versus 26 RLR by totalling the cost of disposables for open and robotic cases, cumulative operative time cost for their institution and cost of perioperative transfusion. The hospital cost analysis was calculated by totalling the cost of ward and ICU cost per day multiplied by the length of stay. They showed that operative cost was significantly higher in the robotic group; however, overall admission cost was comparable between the two. The total costs which included hospital readmission were \$17,288 for open resections compared to \$16,149 for RLR ($p = 0.357$) [18].

Considering minor hepatectomies only, a recent study compared 87 patients who underwent minor liver resections and found that RLR was associated with fewer complications and shorter hospital stay compared to OLR, with similar operating time and blood loss [19]. The clinical benefits of RLR compared to OLR in this study negated the additional costs of robotic instruments and translated into significantly lower overall costs after RLR (\$3597 vs. \$4131; $p = 0.03$).

A recent US study identified 267 RLR and compared the outcomes with 104 open hepatectomies [20]. Indications for resection included both benign and malignant disorders. They illustrated that the robotic group had significantly less blood loss ($p < 0.0001$), smaller lesion size ($p = 0.0052$), shorter length of stay ($p < 0.0001$) and decreased 90-day mortality ($p = 0.0028$). There were no significant differences in costs between the groups. The open group received higher reimbursement (\$29,297 versus \$19,102; $p < 0.001$) and profit (\$5005 versus \$6682; $p < 0.001$). The authors concluded that within their healthcare system, there was no greater

Table 30.1 Robotic versus open hepatectomy

Author	Country	N (RLR/OLR)	Intraoperative cost ($)			Postoperative cost ($)			Overall cost ($)			Hospital stay (days)			Morbidity		
			Robotic	Open	P value	Robotic	Open	P value	Robotic	Open	P value	Robotic	Open	P value	Robotic	Open	P value
Sham et al. [13]	US	71/88	6026	5479	0.047	8570	13,425	0.001	14,754	18,998	0.001	4.2	6.5	0.001	14.3%	23.9%	N/A
Daskalaki [14]	US	68/55	16,901	14,390	N/A	11,579	19,335	N/A	37,518	41,948	N/A	6.8	9.2	0.05	22%	40%	0.047
Wu [16]	Taipei	50/50	10,951	2735	0.001	1546	2185	0.002	12,498	4920	0.001	7.5	1[illegible].3	0.001	6%	14%	0.32
Chen [17]	Taipei	13/54	–	–	–	–	–	–	13,436	5019	0.001	7	7.3	0.18	7.7%	9.3%	0.68
Hawksworth [18]	US	26/36	12,021	7146	0.001	4425	8112	0.001	16,149	17,288	0.357	2.5	5	0.001	7.7%	22.6%	0.159
Stewart [19]	US	46/41	–	–	–	–	–		3597	4131	0.03	2.2	6.2	0.001	7%	24%	0.02
Shapera [20]	US	267/104	–	–	–	–	–	–	27,912	28,314	0.76	4	6	0.001	–	–	–

N/A not available; *RLR* robotic liver resection; *OLR* open liver resection cost associated with the robotic platform despite reduced reimbursement and profit [20].

A summary of studies which compared the costs of RLR and OLR is shown in Table 30.1. Variation in cost between studies is probably a reflection of differences in study methodology in respect of how intraoperative and postoperative costs were calculated, and some studies provided greater detail than others. Variation was also influenced by differences in healthcare systems in different countries. Interestingly, the overall cost of RLR was higher than OLR in both studies from Taipei, whilst the overall costs of RLR were either similar or less than OLR in the US studies.

30.3 Robotic Versus Laparoscopic Liver Resection

The first study to compare RLR and LLR was conducted at the University of Pittsburgh, USA, in 2012 [21]. Packiam et al. analysed 11 RLR and 18 LLR for left lateral sectionectomy (LLS) and found no differences in operative times, blood loss, major complications or 90-day mortality [21]. However, patients undergoing robotic LLS had more minor complications, more intensive care admission and longer hospital stay. Direct costs related to the procedures were \$5130 for RLR and \$4408 (p = 0.401) for LLR. However, robotic LLS expenditures were significantly higher when considering indirect costs which were calculated to be \$6553 versus \$4408 (p = 0.021) for RLR and LLR, respectively [21].

In 2016, a study from Asan Medical Center, South Korea, compared RLR and LLR in left-sided liver resections (LLS and left hepatectomy) [22]. The authors analysed 13 RLR (11 LLS and 3 left hepatectomies) and 17 LLR (6 LLS and 11 left hepatectomies). There were no significant differences in mean operative time (RLR versus LLR; 291.5 ± 85.1 vs. 240.9 ± 68.6 min; p = 0.082) and mean estimated blood loss (RLR versus LLR; 388.5 ± 65.0 vs. 342.6 ± 84.7 ml; p = 0.114) between groups. Total medical costs were significantly higher in the robotic group \$11,475 compared to \$6762 (p = 0.001) in the laparoscopic group [22].

A study from Yonsei University, South Korea, found similar results when comparing robotic and laparoscopic LLS [23]. By analysing robotic (n = 12) and laparoscopic (n = 31) LLS, they demonstrated no significant differences in perioperative outcomes (blood loss, major complications, length of stay), although operating time was longer in robotic than in laparoscopic LLS (391 versus 196 min, respectively). Overall, robotic LLS costs were higher than laparoscopic LLS ones (\$8183 versus \$5190; p = 0.009) [23].

A French study in 2017 compared outcomes of robotic and laparoscopic LLS [24] and found no differences in perioperative outcomes between groups, as well as similar overall costs (RLR €5522 vs. LLR €6035; $p = 0.7$). After matching, the surgical and economic outcomes remained similar, except longer operating time in the robotic group [24]. In contrast, the overall cost of robotic LLS was higher in robotic LLS compared to laparoscopic LLS in a Chinese study published in 2019, whilst perioperative outcomes were similar [25]. In this study it was noteworthy that robotic LLS was associated with reduced blood loss compared to laparoscopic LLS in a subgroup of complex cases (defined by tumour size >10 cm, BMI > 30, combined lymphadenectomy or choledochoscopy), whilst operating time and postoperative outcomes were similar between groups, and overall costs were higher for robotic LLS ($13,536.9 vs. $9186.7; $p = 0.006$) [25]. Another recent study from China also confirmed that robotic LLS had higher overall costs ($11,897 vs. $8004; $p < 0.001$) without any clinical benefits compared to laparoscopic LLS [26].

The cost and outcomes of minimally invasive hepatectomy were evaluated in a US study and found that a full laparoscopic approach was associated with the lowest direct hospital costs compared to either robotic or laparoscopic hand-assisted minor hepatectomy [27]. In this study, it was not possible to compare costs for major hepatectomy due to small sample size.

Interestingly, a matched comparison of almost 500 patients derived from the American NSQIP database indicated that the higher costs associated with RLR ($29,800 vs. $24,000; $p < 0.001$) compared to LLR were due to differences in blood transfusion rates, length of stay and operating time rather than cost of disposables [28].

Several meta-analyses have been published comparing RLR and LLR and have confirmed the findings that RLR is more expensive than LLR without any significant improvement in short-term clinical outcomes [29–35], although blood loss and conversion rates did favour the robotic approach in some studies [30, 31]. However, it is important to note that most studies are focused on minor hepatectomy, whilst data on the cost-

Table 30.2 Robotic versus laparoscopic hepatectomy

Author	Country	N (RLR/ LLR)	Intraoperative cost ($)			Postoperative cost ($)			Overall cost ($)			Hospital stay (days)			Morbidity		
			Robotic	Lap	P value	Robotic	Lap	P value	Robotic	Lap	P value	Robotic	Lap	P value	Robotic	Lap	P value
Packiam et al. [21]	USA	11/18	5130	4408	0.401	–	–	–	–	–	–	4	3	0.031	27%	0%	0.019
Yu et al. [22]	S Korea	13/17	–	–	–	–	–	–	11,475	6762	0.001	7.8	9.5	0.053	0%	11.7%	0.208
Kim et al. [23]	S Korea	12/31	–	–	–	–	–	–	8183	5190	0.009	7	7	0.491	25%	22.6%	0.668
Salloum et al. [24]	France	16/80	–	–	–	–	–	–	5522[a]	6035[a]	0.71	6	7	0.74	13	11	0.77
Hu et al. [25]	China	58/54	–	–	–	–	–	–	12,786	7974	0.001	4.3	4.4	0.448	1.7%	3.7%	0.517
Miller et al. [28]	USA	227/227	–	–	–	–	–	–	29.800	24.000	0.001	4	3	0.65	–	–	–
Zhu et al. [26]	China	176/82	4852	2490	0.001	–	–	–	12,621	8285	0.001	4	4	0.075	0.6%	2.4%	0.238

N/A not available; *RLR* robotic liver resection; *LLR* laparoscopic liver resection
[a]Values reported in Euros (€)

benefit profile of robotic major hepatectomy is currently limited [29, 31].

A summary of studies which compared economic costs between RLR and LLR is presented in Table 30.2.

30.4 Future Directions

In the future, global dissemination of robotic liver resection is likely to be influenced by clinical benefit, learning curve and affordability. The available data indicate that robotic liver resection is cost-effective compared to open liver resection, whilst the clinical benefits of robotic liver resection are less clear when compared with laparoscopic liver resection. In the absence of prospective trials, the best available evidence to date suggests that RLR may increase the feasibility of minimally invasive liver resection (less blood loss and lower conversion rates) compared to LLR, particularly for complex procedures such as extended hepatectomy [9], resection of posterosuperior segments [8] and central hepatectomy [6]. The learning curve for RLR may also be shorter than for LLR [11, 12], which may help to accelerate widespread adoption of minimally invasive liver resection. Lastly, the excess costs of robotic surgery compared to laparoscopic surgery may be counteracted by increased feasibility for complex procedures, but it is also expected that the cost differences are likely to reduce as market forces lead to reduced capital and maintenance costs of surgical robots in the future.

References

1. Buell JF, Cherqui D, Geller DA, O'Rourke N, Iannitti D, Dagher I, et al. Position on laparoscopic liver surgery. Ann Surg. 2009;250:825–30.
2. Wakabayashi G, Cherqui D, Geller DA, Buell JF, Kaneko H, Han HS, et al. Recommendations for laparoscopic liver resection: a report from the second international consensus conference held in Morioka. Ann Surg. 2015;261:619–29.
3. Mirnezami R, Mirnezami AH, Chandrakumaran K, Abu Hilal M, Pearce NW, Primrose JN, Sutcliffe RP. Short- and long-term outcomes after laparoscopic and open hepatic resection: systematic review and meta-analysis. HPB (Oxford). 2011;13(5):295–308.
4. Ratti F, Fiorentini G, Cipriani F, Catena M, Paganelli M, Aldrighetti L. Laparoscopic vs open surgery for colorectal liver metastases. JAMA Surg. 2018;153(11):1028–35.
5. Fretland ÅA, Dagenborg VJ, Bjørnelv GMW, Kazaryan AM, Kristiansen R, et al. Laparoscopic versus open resection for colorectal liver metastases: the OSLO-COMET randomized controlled trial. Ann Surg. 2018;267(2):199–207.
6. Yang HY, Choi GH, Chin KM, Choi SH, Syn NL, Cheung TT, The International Robotic and Laparoscopic Liver Resection Study Group Investigators, et al. Robotic and laparoscopic right anterior sectionectomy and central hepatectomy: multicentre propensity score-matched analysis. Br J Surg. 2022;109(4):311–4.
7. Willems E, D'Hondt M, Kingham TP, Fuks D, Choi GH, Syn NL, Sucandy I, Marino MV, Prieto M, Chong CC, International Robotic and Laparoscopic Liver Resection Study Group Investigators, et al. Comparison between minimally invasive right anterior and right posterior sectionectomy vs right hepatectomy: an international multicenter propensity score-matched and coarsened-exact-matched analysis of 1,100 patients. J Am Coll Surg. 2022;235(6):859–68.
8. D'Silva M, Han HS, Liu R, Kingham TP, Choi GH, Syn NLX, Prieto M, Choi SH, Sucandy I, Chiow AKH, International Robotic and Laparoscopic Liver Resection Study Group Investigators, et al. Limited liver resections in the posterosuperior segments: international multicentre propensity score-matched and coarsened exact-matched analysis comparing the laparoscopic and robotic approaches. Br J Surg. 2022;109(11):1140–9.
9. Chong CC, Fuks D, Lee KF, Zhao JJ, Choi GH, Sucandy I, Chiow AKH, Marino MV, Gastaca M, Wang X, International Robotic and Laparoscopic Liver Resection Study Group Investigators, et al. Propensity score-matched analysis comparing robotic and laparoscopic right and extended right hepatectomy. JAMA Surg. 2022;157(5):436–44.
10. Kadam P, Sutcliffe RP, Scatton O, Sucandy I, Kingham TP, Liu R, International Robotic and Laparoscopic Liver Resection Study Group Investigators, et al. An international multicenter propensity-score matched and coarsened-exact matched analysis comparing robotic versus laparoscopic partial liver resections of the anterolateral segments. J Hepatobiliary Pancreat Sci. 2022;29(8):843–54.
11. D'Hondt M, Devooght A, Willems E, Wicherts D, De Meyere C, Parmentier I, Provoost A, Pottel H, Verslype C. Transition from laparoscopic to robotic liver surgery: clinical outcomes, learning curve effect, and cost-effectiveness. J Robot Surg. 2022;17(1):79–88.
12. Chua D, Syn N, Koh YX, Goh BKP. Learning curves in minimally invasive hepatectomy: systematic review and meta-regression analysis. Br J Surg. 2021;108(4):351–8.

13. Sham JG, Richards MK, Seo YD, et al. Efficacy and cost of robotic hepatectomy: is the robot cost-prohibitive? J Robotic Surg. 2016;10:307–13.
14. Daskalaki D, Gonzalez-Heredia R, Brown M, Bianco FM, Tzvetanov I, Davis M, et al. Financial impact of the robotic approach in liver surgery: a comparative study of clinical outcomes and costs between the robotic and open technique in a single institution. J Laparoendosc Adv Surg Tech A. 2017;27(4):375–82.
15. Shu J, Wang X, Li J, et al. Robotic-assisted laparoscopic surgery for complex hepatolithiasis: a propensity score matching analysis. Surg Endosc. 2019;33:2539–47.
16. Wu CY, Chen PD, Chou WH, Liang JT, Huang CS, Wu YM. Is robotic hepatectomy cost-effective? In view of patient-reported outcomes. Asian J Surg. 2019;42(4):543–50.
17. Chen PD, Wu CY, Hu RH, Ho CM, Lee PH, Lai HS, Lin MT, Wu YM. Robotic liver donor right hepatectomy: a pure, minimally invasive approach. Liver Transpl. 2016;22(11):1509–18.
18. Hawksworth J, Llore N, Holzner ML, Radkani P, Meslar E, Winslow E, Satoskar R, He R, Jha R, Haddad N, Fishbein T. Robotic hepatectomy is a safe and cost-effective alternative to conventional open hepatectomy: a single-center preliminary experience. J Gastrointest Surg. 2021;25(3):825–8.
19. Stewart C, Wong P, Warner S, Raoof M, Singh G, Fong Y, Melstrom L. Robotic minor hepatectomy: optimizing outcomes and cost of care. HPB (Oxford). 2021;23(5):700–6.
20. Shapera E, Sucandy I, Syblis C, Crespo K, Ja'Karri T, Ross S, Rosemurgy A. Cost analysis of robotic versus open hepatectomy: is the robotic platform more expensive? J Robot Surg. 2022;16(6):1409–17.
21. Packiam V, Bartlett DL, Tohme S, Reddy S, Marsh JW, Geller DA, Tsung A. Minimally invasive liver resection: robotic versus laparoscopic left lateral sectionectomy. J Gastrointest Surg. 2012;16(12):2233–8.
22. Yu YD, Kim KH, Jung DH, Namkoong JM, Yoon SY, Jung SW, Lee SK, Lee SG. Robotic versus laparoscopic liver resection: a comparative study from a single center. Langenbeck's Arch Surg. 2014;399(8):1039–45.
23. Kim JK, Park JS, Han DH, Choi GH, Kim KS, Choi JS, Yoon DS. Robotic versus laparoscopic left lateral sectionectomy of liver. Surg Endosc. 2016;30(11):4756–64.
24. Salloum C, Lim C, Lahat E, Gavara CG, Levesque E, Compagnon P, Azoulay D. Robotic-assisted versus laparoscopic left lateral sectionectomy: analysis of surgical outcomes and costs by a propensity score matched cohort study. World J Surg. 2017;41(2):516–24.
25. Hu M, Liu Y, Li C, Wang G, Yin Z, Lau WY, Liu R. Robotic versus laparoscopic liver resection in complex cases of left lateral sectionectomy. Int J Surg. 2019;67:54–60.
26. Zhu L, Liu Y, Hu M, Zhao Z, Li C, Zhang X, Tan X, Wang F, Liu R. Comparison of robotic and laparoscopic liver resection in ordinary cases of left lateral sectionectomy. Surg Endosc. 2022;36(7):4923–31.
27. Mejia A, Cheng SS, Vivian E, Shah J, Oduor H, Archarya P. Minimally invasive liver resection in the era of robotics: analysis of 214 cases. Surg Endosc. 2020;34(1):339–48.
28. Miller HP, Hakim A, Kellish A, Wozniak M, Gaughan J, Sensenig R, Atabek UM, Spitz FR, Hong YK. Cost-benefit analysis of robotic vs. laparoscopic hepatectomy: a propensity-matched retrospective cohort study of American College of Surgeons National Surgical Quality Improvement Program Database. Am Surg. 2022;88(12):2886–92.
29. Guan R, Chen Y, Yang K, Ma D, Gong X, Shen B, Peng C. Clinical efficacy of robot-assisted versus laparoscopic liver resection: a meta-analysis. Asian J Surg. 2019;42(1):19–31.
30. Gavriilidis P, Roberts KJ, Aldrighetti L, Sutcliffe RP. A comparison between robotic, laparoscopic and open hepatectomy: a systematic review and network meta-analysis. Eur J Surg Oncol. 2020;46(7):1214–24.
31. Zhang L, Yuan Q, Xu Y, Wang W. Comparative clinical outcomes of robot-assisted liver resection versus laparoscopic liver resection: a meta-analysis. PLoS One. 2020;15(10):e0240593.
32. Wang JM, Li JF, Yuan GD, He SQ. Robot-assisted versus laparoscopic minor hepatectomy: a systematic review and meta-analysis. Medicine (Baltimore). 2021;100(17):e25648.
33. Ziogas IA, Evangeliou AP, Mylonas KS, Athanasiadis DI, Cherouveim P, Geller DA, Schulick RD, Alexopoulos SP, Tsoulfas G. Economic analysis of open versus laparoscopic versus robotic hepatectomy: a systematic review and meta-analysis. Eur J Health Econ. 2021;22(4):585–604.
34. Ciria R, Berardi G, Alconchel F, Briceño J, Choi GH, Wu YM, Sugioka A, Troisi RI, Salloum C, Soubrane O, Pratschke J, Martinie J, Tsung A, Araujo R, Sucandy I, Tang CN, Wakabayashi G. The impact of robotics in liver surgery: a worldwide systematic review and short-term outcomes meta-analysis on 2,728 cases. J Hepatobiliary Pancreat Sci. 2022;29(2):181–97.
35. Hajibandeh S, Hajibandeh S, Dosis A, Qayum MK, Hassan K, Kausar A, Satyadas T. Level 2a evidence comparing robotic versus laparoscopic left lateral hepatic sectionectomy: a meta-analysis. Langenbeck's Arch Surg. 2022;407(2):479–89.

31 Current Evidence on Robotic Liver Resections: Results from the International Robotic and Laparoscopic Liver Resection Study Group

Darren W. C. Chua, Hwee-Leong Tan, and Brian K. P. Goh

The introduction of robotic surgery began as early as the late 1990s, where Himpens et al. reported the first successful clinical application of a robotic surgical system (nicknamed "Mona") to perform the world's first robotic cholecystectomy [1]. Since then, technological improvements in the robotic platform have seen its adoption in a wide range of abdominal surgeries. However, its implementation in liver surgery was encumbered by the anatomic complexity and challenging operative exposure inherent to any liver resection [2]. Over the years, the continual development of advanced energy/bipolar devices, linear staplers and robotic technologies in parallel with the technical understanding of liver anatomy allowed robotic surgery to expand into all forms of liver resection, including major and technical major hepatectomies.

The robotic platform allows for certain advantages which are deficient in conventional laparoscopic surgery. The most striking advantage is the improved dexterity afforded by the elimination of intrinsic tremors and the seven degrees of rotational freedom permissible in each robotic arm. In addition, the availability of quality high definition, three-dimensional images, a surgeon-controlled camera and the use of adjunctive visualization tools such as near-infrared fluorescence using indocyanine green (ICG) allow for superior visualization of anatomic boundaries and biliovascular elements during parenchymal transection [3]. Its major limitation lies in its cost and absence of haptic feedback, a critical element in minimally invasive surgery which alerts the proceduralist on potential dangers via kinaesthetic and cutaneous feedback, a lack of which may contribute to an increase in the incidence of intraoperative injury [4]. Nevertheless, the overall benefit and comfort with the robot have seen its expansion to numerous forms of abdominal and extra-abdominal surgeries, with its adoption likely to be progressively increased with time as robot technology continually transforms modern surgery.

D. W. C. Chua · H.-L. Tan
Department of Hepatopancreatobiliary and Transplant Surgery, Singapore General Hospital and National Cancer Centre, Singapore, Singapore

B. K. P. Goh (✉)
Department of Hepatopancreatobiliary and Transplant Surgery, Singapore General Hospital and National Cancer Centre, Singapore, Singapore

Duke-National University of Singapore Medical School, Singapore, Singapore
e-mail: brian.goh@singhealth.com.sg

M. D'Hondt, I. Sucandy (eds.), *Textbook of Robotic Liver Surgery*,
https://doi.org/10.1007/978-3-031-76536-0_31

31.1 Learning Curves in Robotic Liver Surgery

In recent years, numerous studies have repeatedly demonstrated the safety and feasibility of minimally invasive liver resection (MILR), boasting not only improvements in perioperative morbidity but also oncologic outcomes that are equivalent to their open counterparts [5–8]. One of the inherent issues with MILR is the learning curve required to surmount the technical obstacles. Studies evaluating learning curves have yielding diverse results given the great deal of study heterogeneity, from the methods used to assess learning curves to the outcomes chosen as surrogates for operative proficiency [9]. Consequently, the cited learning curves in MILR ranges from 20 to over 80 cases [10–12], with laparoscopic liver resection (LLR) forming the bulk of contributing studies. Comparatively, studies evaluating learning curves for robotic liver resection (RLR) are significantly fewer but generally demonstrate a shorter learning curve [13]. A recent systematic review of learning curves in MILR reaffirms this, with a caseload required to surmount the learning curve for RLR being demonstrably less than that of laparoscopic liver resection (25 vs 50 cases), even after adjustment for year of adoption of MILR [12]. Beyond the apparent advantages of the robot platform, part of the reason for these findings lies in the period and relative experience of surgeons embarking in RLR at the point in time. The IDEAL framework proposed by Barkun et al. [14] describes the phases of surgical innovation. Briefly, the framework consists of five stages: idea (IDEAL stage 1), development (IDEAL stage 2a), exploration (IDEAL stage 2b), assessment (IDEAL stage 3) and long-term (IDEAL stage 4). Each stage thus provides a platform for critical appraisal in every step of the innovation process and allows for a set of high-quality recommendations on how evaluation should be performed in each phase [15]. The progression of LLR through the IDEAL stages began in the 1990s and has since developed over the last three decades. Its trajectory was hampered by numerous factors, including the initial scepticism to LLR and the deficiencies of available technology and knowledge of laparoscopic techniques [16]. This contrasts with the trajectory of RLR, which follows a much faster course and adoption. Part of the reasons beyond the advantages of the robot is due to the period in which RLR was adopted. It is likely that the surgeons who adopted RLR already had a clear understanding and familiarity with the anatomic and technical challenges and could therefore leverage on the advantage of the robot to help surmount the learning curve quicker. In fact, it is well understood that most surgeons who embark on RLR already have some prior experience with LLR [13, 17].

31.2 Difficulty Scoring Systems in Robotic Liver Surgery

In 2014, the Second International Consensus Conference on Laparoscopic Liver Resection (ICCLLR) which convened in Morioka highlighted the need for a stepwise approach in the adoption of MILR [18]. Thc introduction of difficulty scoring systems (DSS) helped bridge the gap between the case and surgeon by providing a score commensurate to a surgeon's current level of experience and skill. The first of such DSS, proposed by Ban et al. [19], consisted of a three-level difficulty score which was subsequently modified to a four-level system what is now known as the Iwate DSS [18]. Since then, numerous other scoring systems have been proposed, of which the Ban, Iwate, Hasegawa and IMM systems being the most validated, demonstrating robustness in predicting operative difficulty and post-operative outcomes [20, 21]. One of the issues with existing DSS validation studies is its predominant focus on laparoscopic liver resection. In a recent systematic review evaluating DSS among laparoscopic and robotic liver resections, only the Ban and Iwate DSS were found to be externally validated for robotic liver resections [22, 23]. However, the findings of these two studies yielded conflicting results, with Coletta et al. being unable to demonstrate the utility of a DSS in the context of robotic liver resection. Although it is arguable that the study may be possibly underpowered given the limited sample size, it may be possible that many of the variables included in DSS may not be applicable to RLR given the advantages a robotic platform confers. It is possible that operative factors within these DSS constructs such as tumour location, size and proximity to vessels may be easily overcome using the robot given the added advantage of operative dexterity and superior visualization. This highlights the need for further modifications to existing DSS for better discrimination among the difficulty strata.

31.3 Robotic Versus Laparoscopic Liver Surgery

The theoretical technical advantages afforded by RLR over LLR with resultant shorter learning curves were critically evaluated through a series of propensity score-matched comparative studies performed by the International Robotic and Laparoscopic Liver Resection Study Group, an international multicentre collaboration comprising over 20 institutions which have been performing MILR for more than a decade [24–29]. Two major interrelated variables influencing the technical difficulty of MILR cutting across the various DSS are the location and type of liver resection performed [20, 21]. As such, each of these comparative studies focused on a select group of MILR which share common anatomic and technical challenges so as to evaluate the comparative perioperative performance of RLR against LLR in each. The key findings are summarized in the following table:

Type of MILR	Patient cohort	Key findings (matched analysis)
Right and extended right hepatectomy [25]	Unmatched: 220 RLR vs 769 LLR Matched: 220 RLR vs 220 LLR	RLR had lower conversion rate (8.6% vs 15.9%, $p = 0.03$) RLR had shorter post-operative hospital stay (median [IQR] 7.0 [5.0–10.0] vs 7.0 [7.75–10.0] days, $p = 0.48$)
Right posterior sectionectomy [24]	Unmatched: 96 RLR vs 244 LLR Matched: 66 RLR vs 132 LLR	RLR had lower blood loss (median [IQR] 200 [100–400] vs 450 [200–800] mL, $p < 0.001$) RLR had lower need for intraoperative blood transfusion (9.1% vs 22.0%, $p = 0.026$) RLR had lower conversion rate (3.0% vs 9.8%, $p = 0.001$)
Left and extended left hepatectomy [28]	Unmatched: 190 RLR vs 390 LLR Matched: 164 RLR vs 164 LLR	RLR had lower blood loss (median [IQR] 100 [200] vs 200 [235] mL, $p = 0.029$) RLR had lower conversion rate (2.4% vs 7.9%, $p = 0.043$) RLR had shorter post-operative hospital stay (median [IQR] 6 [3] vs 7 [3.3] days, $p = 0.009$)
Posterosuperior segments (segments IVa, VII and VIII) [26]	Unmatched: 159 RLR vs 824 LLR Matched: 127 RLR vs 381 LLR	RLR had lower blood loss (median [IQR] 100 [40–200] vs 200 [100–500] mL, $p = 0.003$) RLR had lower need for intraoperative blood transfusion (3.1% vs 10.0%, $p = 0.025$) RLR had shorter duration of Pringle manoeuvre (median [IQR] 30 [20–46] vs 40 [25–58] min, $p = 0.012$) RLR had lower conversion rate [0/8% vs 7/9%, $p = 0.022$] RLR had shorter operating time (median [IQR] 175 [130–255] vs 224 [155–300] min, $p < 0.001$)
Anterolateral segments (segments II, III, IVB, V and VI) [27]	Unmatched: 358 RLR vs 1868 LLR Matched: 261 RLR vs 783 LLR	RLR had lower blood loss (median [IQR] 50 [20–200] vs 100 [50–300] mL, $p < 0.001$) RLR had longer operating time (median [IQR] 170 [125–225] vs 160 [112–224] min, $p = 0.036$) RLR had lower conversion rate (1.5% vs 6.8%, $p = 0.003$)
Right anterior sectionectomy and central hepatectomy [29]	Unmatched: 48 RLR vs 185 LLR Matched: 34 RLR vs 68 LLR	RLR had lower blood loss (median [IQR] 200 [100–500] vs 300 [192–700] mL, $p = 0.020$)

Taken together, these studies by our group clearly demonstrate a consistent trend in improved perioperative outcomes associated with RLR when compared with LLR, especially in reducing intraoperative blood loss and lowering rates of open conversion. Of note, RLR was associated with a longer median operating duration for the anterolateral segments [27], which may be attributable to the relatively lower technical demands for these resections diminishing the comparative advantage RLR may have over LLR in spite of additional time needed for robot docking and undocking.

While these series represent the largest comparative experience in MILR in literature to date, several limitations need to be considered. The retrospective nature of these multicentre analyses lends itself to issues of missing data, selection bias and reduced internal validity due to differing institutional and surgical practices. Additionally, due to the higher barrier of entry to robotic surgery in most centres, RLR is more likely to be performed by a smaller pool of more experienced minimally invasive liver surgeons, while LLR may be performed by a larger pool of surgeons with greater disparity in technical expertise and caseload. Furthermore, it is likely that many surgeons performing RLR had prior experience with LLR before embarking on RLR.

31.4 Conclusion

While MILR has transferred the pain of liver surgery from the patient to the surgeon, RLR holds promise for mitigating several negative aspects of LLR, which has translated into tangible benefits in terms of perioperative outcomes for the patient and a shorter learning curve for the surgeon.

References

1. Himpens J, Leman G, Cadiere GB. Telesurgical laparoscopic cholecystectomy. Surg Endosc. 1998;12(8):1091.
2. Ho CM, Wakabayashi G, Nitta H, Ito N, Hasegawa Y, Takahara T. Systematic review of robotic liver resection. Surg Endosc. 2013;27(3):732–9.
3. Lim HJ, Chiow AKH, Lee LS, Tan SS, Goh BK, Koh YX, et al. Novel method of intraoperative liver tumour localisation with indocyanine green and near-infrared imaging. Singapore Med J. 2021;62(4):182–9.
4. Okamura AM. Haptic feedback in robot-assisted minimally invasive surgery. Curr Opin Urol. 2009;19(1):102–7.
5. Kabir T, Tan ZZ, Syn NL, Wu E, Lin JD, Zhao JJ, et al. Laparoscopic versus open resection of hepatocellular carcinoma in patients with cirrhosis: meta-analysis. Br J Surg. 2021;109(1):21–9.
6. Syn NL, Kabir T, Koh YX, Tan HL, Wang LZ, Chin BZ, et al. Survival Advantage of Laparoscopic Versus Open Resection For Colorectal Liver Metastases: A Meta-analysis of Individual Patient Data From Randomized Trials and Propensity-score Matched Studies. Ann Surg. 2020;272(2):253–65.
7. Wang ZY, Chen QL, Sun LL, He SP, Luo XF, Huang LS, et al. Laparoscopic versus open major liver resection for hepatocellular carcinoma: systematic review and meta-analysis of comparative cohort studies. BMC Cancer. 2019;19(1):1047.
8. Zhang KJ, Liang L, Diao YK, Xie YM, Wang DD, Xu FQ, et al. Short- and long-term outcomes of laparoscopic versus open liver resection for large hepatocellular carcinoma: a propensity score study. Surg Today. 2023;53(3):322–31.
9. Harrysson IJ, Cook J, Sirimanna P, Feldman LS, Darzi A, Aggarwal R. Systematic review of learning curves for minimally invasive abdominal surgery: a review of the methodology of data collection, depiction of outcomes, and statistical analysis. Ann Surg. 2014;260(1):37–45.
10. Vigano L, Laurent A, Tayar C, Tomatis M, Ponti A, Cherqui D. The learning curve in laparoscopic liver resection: improved feasibility and reproducibility. Ann Surg. 2009;250(5):772–82.
11. Cannon RM, Brock GN, Marvin MR, Buell JF. Laparoscopic liver resection: an examination of our first 300 patients. J Am Coll Surg. 2011;213(4):501–7.
12. Chua D, Syn N, Koh YX, Goh BKP. Learning curves in minimally invasive hepatectomy: systematic review and meta-regression analysis. Br J Surg. 2021;108(4):351–8.
13. Efanov M, Alikhanov R, Tsvirkun V, Kazakov I, Melekhina O, Kim P, et al. Comparative analysis of learning curve in complex robot-assisted and laparoscopic liver resection. HPB (Oxford). 2017;19(9):818–24.
14. Barkun JS, Aronson JK, Feldman LS, Maddern GJ, Strasberg SM, Balliol C, et al. Evaluation and stages of surgical innovations. Lancet. 2009;374(9695):1089–96.
15. McCulloch P, Cook JA, Altman DG, Heneghan C, Diener MK, Group I. IDEAL framework for surgical innovation 1: the idea and development stages. BMJ. 2013;346:f3012.
16. Halls MC, Alseidi A, Berardi G, Cipriani F, Van der Poel M, Davila D, et al. A comparison of the learning curves of laparoscopic liver surgeons in differing

stages of the IDEAL paradigm of surgical innovation: standing on the shoulders of pioneers. Ann Surg. 2019;269(2):221–8.
17. Chen PD, Wu CY, Hu RH, Chen CN, Yuan RH, Liang JT, et al. Robotic major hepatectomy: is there a learning curve? Surgery. 2017;161(3):642–9.
18. Wakabayashi G. What has changed after the Morioka consensus conference 2014 on laparoscopic liver resection? Hepatobiliary Surg Nutr. 2016;5(4):281–9.
19. Hasegawa Y, Wakabayashi G, Nitta H, Takahara T, Katagiri H, Umemura A, et al. A novel model for prediction of pure laparoscopic liver resection surgical difficulty. Surg Endosc. 2017;31(12):5356–63.
20. Goh BKP, Prieto M, Syn N, Koh YX, Teo JY, Lee SY, et al. Validation and comparison of the Iwate, IMM, Southampton and Hasegawa difficulty scoring systems for primary laparoscopic hepatectomies. HPB (Oxford). 2021;23(5):770–6.
21. Lee SY, Goh BKP, Sepideh G, Allen JC, Merkow RP, Teo JY, et al. Laparoscopic liver resection difficulty score-a validation study. J Gastrointest Surg. 2019;23(3):545–55.
22. Chong CCN, Lok HT, Fung AKY, Fong AKW, Cheung YS, Wong J, et al. Robotic versus laparoscopic hepatectomy: application of the difficulty scoring system. Surg Endosc. 2020;34(5):2000–6.
23. Coletta D, Ossola P, Parrino C, Oddi A, D'Annibale M, Perri P, et al. Robotic liver resections: application of difficulty score systems to an initial experience. Is a specific robotic difficulty score necessary? J Laparoendosc Adv Surg Tech A. 2020;30(11):1177–82.
24. Chiow AKH, Fuks D, Choi GH, Syn N, Sucandy I, Marino MV, et al. International multicentre propensity score-matched analysis comparing robotic versus laparoscopic right posterior sectionectomy. Br J Surg. 2021;108(12):1513–20.
25. Chong CC, Fuks D, Lee KF, Zhao JJ, Choi GH, Sucandy I, et al. Propensity score-matched analysis comparing robotic and laparoscopic right and extended right hepatectomy. JAMA Surg. 2022;157(5):436–44.
26. D'Silva M, Han HS, Liu R, Kingham TP, Choi GH, Syn NLX, et al. Limited liver resections in the posterosuperior segments: international multicentre propensity score-matched and coarsened exact-matched analysis comparing the laparoscopic and robotic approaches. Br J Surg. 2022;109(11):1140–9.
27. Kadam P, Sutcliffe RP, Scatton O, Sucandy I, Kingham TP, Liu R, et al. An international multicenter propensity-score matched and coarsened-exact matched analysis comparing robotic versus laparoscopic partial liver resections of the anterolateral segments. J Hepatobiliary Pancreat Sci. 2022;29(8):843–54.
28. Sucandy I, Rayman S, Lai EC, Tang CN, Chong Y, Efanov M, et al. Robotic versus laparoscopic left and extended left hepatectomy: an international multicenter study propensity score-matched analysis. Ann Surg Oncol. 2022;29(13):8398–406.
29. Yang HY, Choi GH, Chin KM, Choi SH, Syn NL, Cheung TT, et al. Robotic and laparoscopic right anterior sectionectomy and central hepatectomy: multicentre propensity score-matched analysis. Br J Surg. 2022;109(4):311–4.

The Future and Hurdles of Robotic Liver Surgery

32

Paolo Magistri, Ning Qi Pang, Stefano Di Sandro, and Fabrizio Di Benedetto

32.1 Introduction

Since the very first series report by Giulianotti and colleagues on robotic approach for liver resections published in 2011 [1], the role of robotics in liver surgery evolved significantly moving from a pure resective to a reconstructive tool. As a matter of fact, in just 10 years the community of surgeons expanded the opportunities of application of the robotic platform with its pioneering work, achieving sensational results like the first liver transplantation with robotic approach [2–4]. In between, technical refinements and growing experience led to the demonstration of comparable short-term and long-term outcomes versus the standard open approach. Nonetheless there are still margins for the optimization of robotic surgical platforms in the treatment of liver diseases.

P. Magistri · S. Di Sandro · F. Di Benedetto (✉)
Hepato-Pancreato-Biliary Surgery and Liver Transplantation Unit, University of Modena and Reggio Emilia, Modena, Italy
e-mail: fabrizio.dibenedetto@unimore.it

N. Q. Pang
Division of Hepatobiliary & Pancreatic Surgery, Department of Surgery, National University Hospital, Singapore, Singapore

32.2 Progresses of Robotic Liver Surgery

Data published so far showed that robotic approach in the field of liver surgery is safe and feasible not only for intrahepatic benign and malignant tumors [5–7] but also for perihilar cholangiocarcinoma [8, 9]. Advantages of the robotic approach for liver surgery pertain to both the field of oncologic efficacy and technical characteristics of the device, as well as some of its drawbacks. We report below some crucial aspects for each domain.

32.2.1 Challenging Scenarios

Robotic approach showed significant advantages over conventional laparoscopy in the setting of complex liver resections. In particular, it reduced the rate of conversion to open in patients undergoing high difficulty liver resections according to the IWATE difficulty score in an Italian multicenter series from experienced robotic HPB centers [10]. These results were also confirmed by a recent international multicenter study comparing laparoscopic versus robotic right and extended right hepatectomies, showing after propensity score matching that robotic approach reduces the rate of conversion to open and postoperative hospital stay [11]. Therefore, it is not surprising that some authors have speculated that the use of

M. D'Hondt, I. Sucandy (eds.), *Textbook of Robotic Liver Surgery*,
https://doi.org/10.1007/978-3-031-76536-0_32

robotic may actually mitigate the difficulty of the resection evaluated according to tumor position and segment location [12]. Under this perspective major liver resection may seem more feasible compared to posterior segments due to the straight transection plane. Nonetheless, the manipulation of the right lobe, hepatic vein cuff, and hilar dissection requires an adequate learning curve to guarantee adequate outcomes for patients requiring a major hepatectomy. A study from 2017 reported that the learning curve for robotic major hepatectomies can be divided into 3 phases: initial (phase 1, 15 patients), intermediate (phase 2, 25 patients), and mature (phase 3, 52 patients). After phase 1 the effects of learning are expressed by the reduction of operative time and hospital stay, after phase 2 reduction of blood loss, while phase 3 represented the mature phase of general improvement [13]. Subsequent studies suggested a total of 45 and 35 cases to gain proficiency with robotic right and left hepatectomy, respectively [14] and that robotic major hepatectomy is associated to the less use of Pringle maneuver, blood loss, and conversion to open compared to standard laparoscopy [15]. It is clear that the inner characteristics of the robotic platform, namely, stability and magnification of the view, immersive 3D vision, tremor filtration, and motion scaling, improve the learning curve and allow to translate the standard open approach gesture in a closed abdomen [16] and make robotic approach potentially feasible in the most complex scenarios. Moreover, the use of indocyanine green (ICG)-based fluoroscopy allows a clear visualization of the anatomical segments of the liver and assessment of the vitality and perfusion of the residual parenchyma (Fig. 32.1). Therefore, there is no actual limit to the potential indications to robotic surgery, and the inclusion and exclusion criteria rely only on the experience of the operating surgeon and HPB center. Even complex cases like redo-surgery, liver resections after previous open surgery, and even hepaticojejunostomy for biliary strictures after liver transplantation have been proven to be feasible in expert hands [17–19].

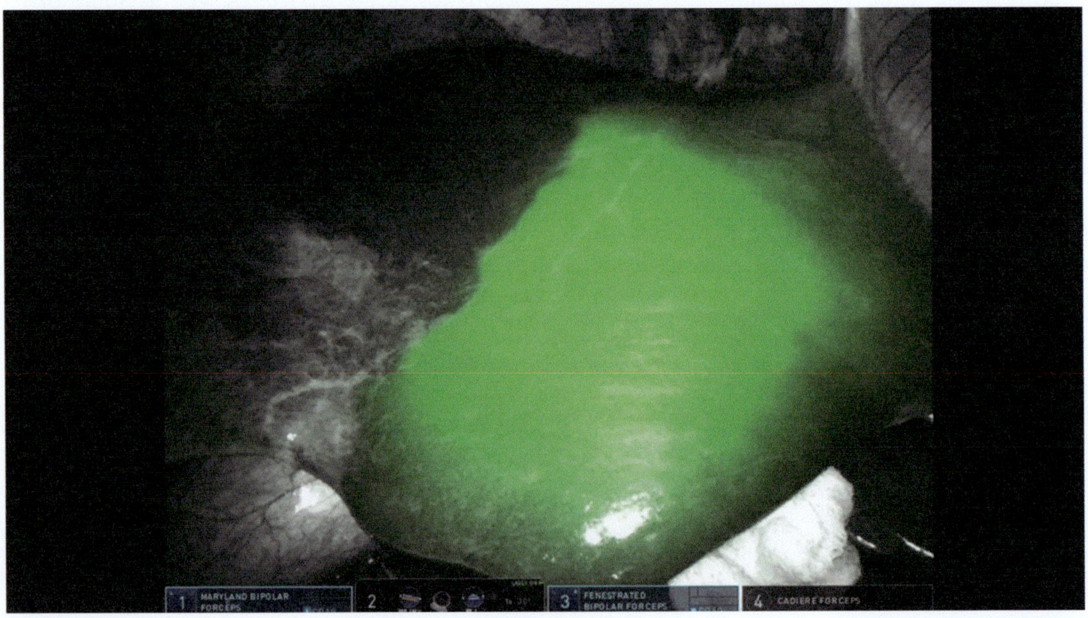

Fig. 32.1 Negative staining for a right hepatectomy

32.2.2 HCC and Portal Hypertension

Two major series were published comparing the results of robotic approach versus the standard open surgery for the treatment of hepatocellular carcinoma (HCC). In 2017 Chen and colleagues [20] reported the very first series that compared oncologic results between open and robotic cases. The results showed that besides longer operative time, the robotic approach provided shorter length of hospital stay, less use of analgesics, and comparable surgical margins, without differences in the disease-free survival (DFS) and overall survival (OS) rates. Consistently, our recent western multicenter series comparing robotic cases from high volume HPB centers versus open cases from a center not performing robotic surgery not only confirmed these results but also showed that robotic approached reduced the incidence of post-hepatectomy liver failure (PHLF) and the length of stay in the intensive care unit [6]. Besides the well-known advantages offered by minimally invasive surgery in patient with liver cirrhosis and even Child B liver decompensation [21], few data are available on the impact of robotic liver resection in patients with clinically significant portal hypertension [22, 23]. The presence of liver cirrhosis in a series analyzing patients undergoing left lateral sectionectomy seems not to increase the risk of conversion, although increases operative time, blood loss, need for blood transfusion, and hospital stay. Nonetheless, it did not influence the incidence and severity of postoperative morbidity despite the perioperative outcomes [22]. Conversely, when it comes to complex procedures like posterosuperior segments, postoperative morbidity got worse in patients with cirrhosis compared to non-cirrhotic cases [23].

32.2.3 Perihilar Cholangiocarcinoma

The most recent development of robotic liver surgery is probably represented by the integration of the reconstructive phase, which is the most characteristic aspect of the surgical treatment for biliary tumors [24]. Few data are available in literature so far since this kind of approach is still considered pioneering. The largest series published in literature is from Li and colleagues, including 48 cases, 23 of whom were associated biliary resections and major hepatectomies and no Bismuth-Corlette type 4 cases, showing 72.9% R0 resection rate, with 4.2% incidence of bile leak, and low readmission rate [25]. We recently published our experience from Modena University with robotic approach for perihilar cholangiocarcinoma including all type of tumors from Bismuth-Corlette 1–4 [9]. The series included consecutive cases of biliary tumors treated with robotic approach. In fact, starting from the very first case performed, we subsequently approached Klatskin tumors robotically except those cases that presented with locally advanced disease and suspect arterial infiltration. Interestingly only one case was converted to open due to lack of compliance to the pneumoperitoneum. Our results, although preliminary and ranging from isolated biliary resections to ALPPS, showed no inferiority of the robotic approach to the standard open according to the most recent benchmark outcomes [26]. Similarly, our multicenter study showed reduced estimated blood loss, length of stay, and complications in the robotic group compared to the open cases from the same time interval, with similar mortality rates [8]. Even though the caudal view allows a direct view of the duct during the reconstructive phase, approaching biliary surgery with minimally invasive strategies requires tools to compensate the lack of haptic feedback to assess biliary anatomy and tumor extension. This can be obtained in the robotic platform thanks to the abovementioned characteristics of the machine and the TilePro function that permits the integration of ultrasound and endoscopic imaging, as, for example, the use of choledocoscopy to assess biliary anatomy, potential residual disease, and treating intrahepatic lithiasis (Fig. 32.2).

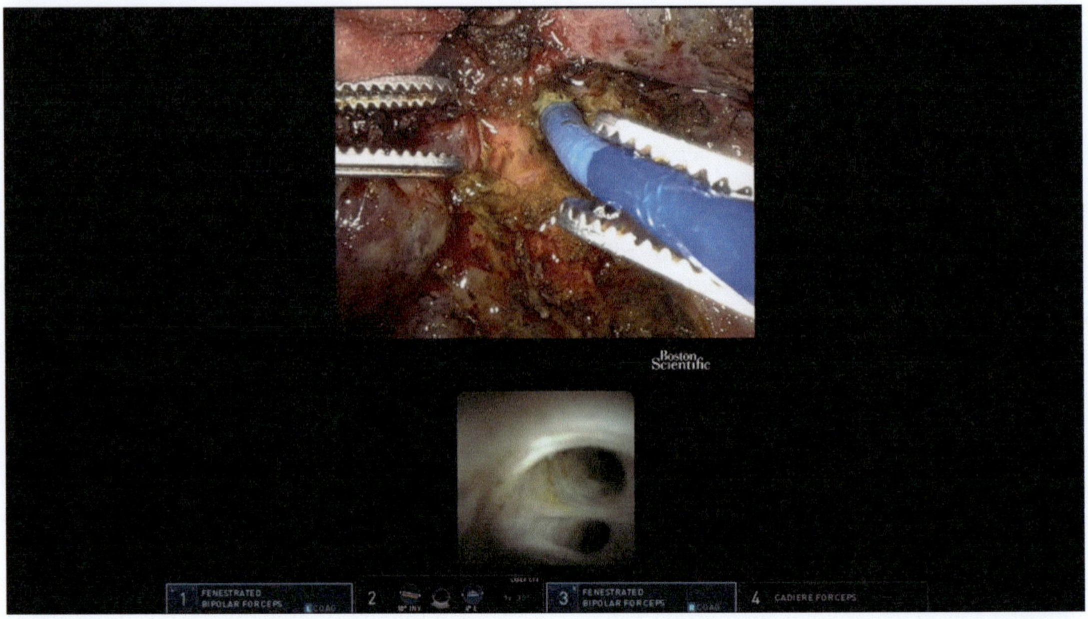

Fig. 32.2 Intraoperative robotic choledocoscopy

32.2.4 Donor Hepatectomy and Liver Transplantation

Donor hepatectomy is probably one of the most delicate procedures to be carried out since it involves a healthy individual who is volunteering the greatest possible sign of affection and solidarity toward a patient waiting for a life-saving transplantation. Minimally invasive approach for donor hepatectomy is already a reality in the most advanced living donor liver transplantation (LDLT) programs, and consensus is now obtained about the safety and feasibility of the procedure [27]. In fact, due to the expansion of LDLT and its growing indication in the field of transplant oncology, more donor candidates will probably ask for minimally invasive strategies to minimize the impact of the procedure on their health, and robotic approach is rising as an improvement of standard laparoscopy. The group of King Faisal Hospital in Riyadh is currently leading the use of robotic approach for donor hepatectomy, having showed a quick expansion of the program with excellent results for both donors and recipients [28]. Notably, even complex cases with anatomical vascular and biliary variants can be safely approached with the robot in expert centers and after an adequate learning curve (Fig. 32.3). Interestingly, robotic approach has been successfully applied also for more advanced LDLT strategies like single-segment procurement and reduction graft [29]. It is not surprising that in this context of technical evolution also liver implantation has been carried out successfully with a hybrid or full robotic approach [2, 30]. Some controversies are open on this latter application of the robotic technology, mainly due to the vascular management and especially clamping of the vena cava. Nonetheless, this technique may transfer some of the well-known advantages of minimally invasive surgery to liver transplantation recipients that may get an extreme benefit from a reduced surgical impact, improving recovery and social reintegration. It is not likely to be a solution for all patients undergoing liver transplantation in the next future, but it will definitely represent a front of implementation of robotic approach.

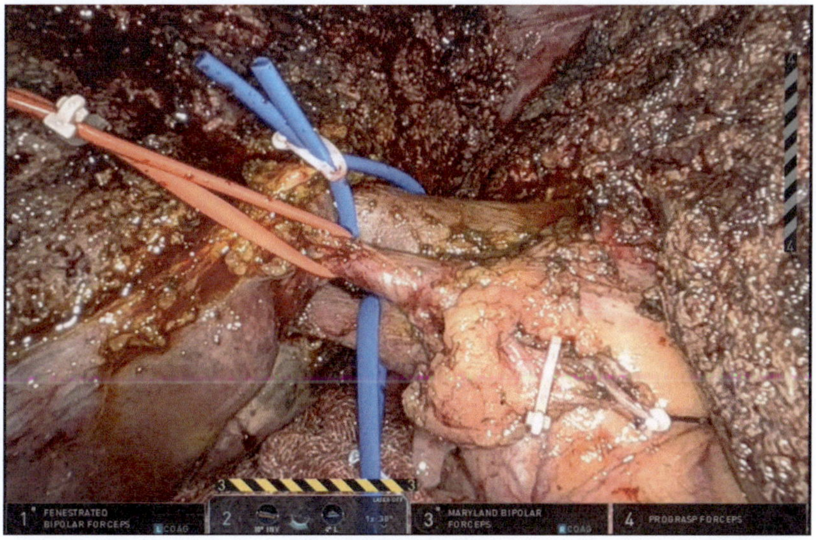

Fig. 32.3 Type 3 portal vein anatomy preparation during donor hepatectomy

32.3 Hurdles of Robotic Liver Surgery

Instrumentation is one of the most hotly debated topics, with a particular interest on the lack of CUSA among robotic tools. Some authors claim that an integration of robotic approach with laparoscopic CUSA from the bedside may increase safety and reduce morbidity and most importantly reproduces the technique classically adopted in the open and laparoscopic approach. From our perspective the use of CUSA from the table side becomes detrimental from the cost-effectiveness point of view. Moreover, during robotic surgery, the table side surgeon may become physically uncomfortable due to the narrow robotic trocar lines and encumbrance of robotic arms. The presence of a robotic CUSA would be helpful to further increase the performance of robotic liver resection; however, ultrasonic instruments cannot fulfill the robotic concept because they lack a flexible design, like the Harmonic ACE (Intuitive Surgical, Sunnyvale, CA, USA). Therefore, bipolar-mediated parenchymal dissection still represents the best option because of the agility of its movements and the degrees of freedom [6]. Another technical aspect that is still under-investigated is the feasibility of vascular resections with robotic approach in the field of liver surgery. In fact, while some reports of robotic pancreatoduodenectomies with portal vein resection have proved the safety and efficacy of the technique [31], very few cases are reported in literature on vascular management during robotic liver surgery [9, 32]. The underlying reasons are not immediately evident since the flexibility of robotic instrument and stability of the platform favor an easy reconstruction with the same technique as for open approach (Fig. 32.4). Safety concerns regarding vascular clamping in a closed abdomen and learning curve in vascular management even in open setting are potential

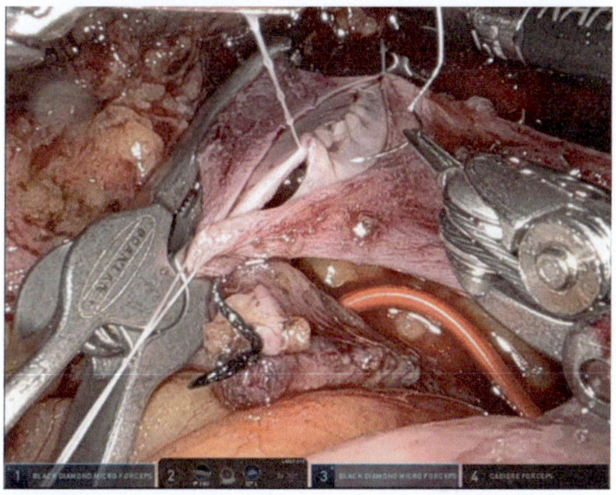

Fig. 32.4 End-to-end portal vein anastomosis during left hepatectomy for Klatskin 3b

factors that influence the slow diffusion of this technical solution. Finally, several analyses on costs of robotic surgery have been reported so far, with controversial results about the balance between costs and benefits. Taken together, data in literature are too heterogeneous so far to clearly solve the issue of costs comparison between robotic surgery and other techniques [33]. Lastly, formal learning curve analysis on the most complex scenarios like biliary and vascular reconstruction and donor hepatectomy is yet to be defined and requires novel high-quality data. Most importantly, we will face the need to find innovative strategies to train the younger generations of surgeons that are native minimally invasive, to be confident in the open when an emergency conversion is needed, given the increasing indication to minimally invasive surgery even for extremely complex cases.

32.4 Conclusions and Future Perspectives

Robotic approach can achieve safe outcomes in patients with cirrhosis and clinically significant portal hypertension and is currently expanding toward more technically complex cases that include reconstructive phases, like perihilar cholangiocarcinoma and liver transplantation. The benefits in terms of health-related quality of life for patients undergoing robotic surgery have been demonstrated for every indication, showing that the only limit to the adoption of the robotic approach is the experience of the center and surgeons' proficiency. Moreover, the technical characteristics of the platform offer resources for the implementation with advanced imaging, fluorescence-based guidance, integrated choledocoscopy, 3D reconstruction, and potentially in the next future AI-based navigation.

References

1. Giulianotti PC, Coratti A, Sbrana F, Addeo P, Bianco FM, Buchs NC, et al. Robotic liver surgery: results for 70 resections. Surgery. 2011;149(1):29–39.
2. Suh KS, Hong SK, Lee S, Hong SY, Suh S, Han ES, et al. Purely laparoscopic explant hepatectomy and hybrid laparoscopic/robotic graft implantation in living donor liver transplantation. Br J Surg. 2022;109(2):162–4.
3. Pinto-Marques H, Sobral M, Magistri P, Gomes da Silva S, Guerrini GP, Mega R, Guidetti C, Coelho JS, Di Sandro S, Di Benedetto F. Full Robotic Whole Graft Liver Transplantation: A Step Into The Future. Ann Surg. 2024. https://doi.org/10.1097/SLA.0000000000006420. Epub ahead of print. PMID: 38953528.
4. Khan AS, Scherer M, Panni R, Cullinan D, Martens G, Kangarga I, King CR, Benzinger R, Wellen JR, Chapman WC, Doyle MB. Total robotic liver transplant: the final frontier of minimally invasive surgery. Am J Transplant. 2024;24(8):1467–72. https://doi.org/10.1016/j.ajt.2024.03.030. Epub 2024 Mar 29. PMID: 38556089.
5. Sucandy I, Schlosser S, Bourdeau T, Spence J, Attili A, Ross S, et al. Robotic hepatectomy for benign and malignant liver tumors. J Robot Surg. 2019;14(1):75–80.
6. Di Benedetto F, Magistri P, Di Sandro S, Sposito C, Oberkofler C, Brandon E, et al. Safety and efficacy of robotic vs open liver resection for hepatocellular carcinoma. JAMA Surg. 2023;158(1):46–54.
7. Magistri P, Pecchi A, Franceschini E, Pesi B, Guadagni S, Catellani B, et al. Not just minor resections: robotic approach for cystic echinococcosis of the liver. Infection. 2019;47(6):973–9.
8. Sucandy I, Marques HP, Lippert T, Magistri P, Coelho JS, Ross SB, et al. Clinical outcomes of robotic resection for Perihilar cholangiocarcinoma: a first, multicenter, trans-Atlantic, expert-center. Collaborative Study Ann Surg Oncol. 2023;31(1):81–9.
9. Magistri P, Pang NQ, Guidetti C, Caracciolo D, Odorizzi R, Catellani B, et al. Robotic approach for perihilar cholangiocarcinoma: from Bismuth 1 to vascular resection. Eur J Surg Oncol J Eur Soc Surg Oncol Br Assoc Surg Oncol. 2023;49(11):107002.
10. Cipriani F, Fiorentini G, Magistri P, Fontani A, Menonna F, Annecchiarico M, et al. Pure laparoscopic versus robotic liver resections: multicentric propensity score-based analysis with stratification according to difficulty scores. J Hepato-Biliary-Pancreat Sci. 2021;29(10):1108–23.
11. Chong CC, Fuks D, Lee KF, Zhao JJ, Choi GH, Sucandy I, et al. Propensity score-matched analysis comparing robotic and laparoscopic right and extended right hepatectomy. JAMA Surg. 2022;157(5):436–44.
12. Luberice K, Sucandy I, Modasi A, Castro M, Krill E, Ross S, et al. Applying IWATE criteria to robotic hepatectomy: is there a "robotic effect"? HPB. 2021;23(6):899–906.
13. Chen PD, Wu CY, Hu RH, Chen CN, Yuan RH, Liang JT, et al. Robotic major hepatectomy: is there a learning curve? Surgery. 2017;161(3):642–9.

14. Liu Q, Zhang T, Hu M, Zhao Z, Zhao G, Li C, et al. Comparison of the learning curves for robotic left and right hemihepatectomy: a prospective cohort study. Int J Surg Lond Engl. 2020;81:19–25.
15. Liu Q, Zhang W, Zhao JJ, Syn NL, Cipriani F, Alzoubi M, et al. Propensity-score matched and coarsened-exact matched analysis comparing robotic and laparoscopic major hepatectomies: an international multicenter study of 4822 cases. Ann Surg. 2023;278(6):969–75.
16. Troisi RI, Pegoraro F, Giglio MC, Rompianesi G, Berardi G, Tomassini F, et al. Robotic approach to the liver: open surgery in a closed abdomen or laparoscopic surgery with technical constraints? Surg Oncol. 2020;33:239–48.
17. Machado MA, Surjan RC, Basseres T, Makdissi F. Robotic repeat right hepatectomy for recurrent colorectal liver metastasis. Ann Surg Oncol. 2019;26(1):292–5.
18. Feldbrügge L, Ortiz Galindo SA, Frisch O, Benzing C, Krenzien F, Riddermann A, et al. Safety and feasibility of robotic liver resection after previous abdominal surgeries. Surg Endosc. 2022;36(5):2842–9.
19. Hawksworth J, Radkani P, Nguyen B, Aguirre O, Winslow E, Kroemer A, et al. Robotic hepaticojejunostomy for late anastomotic biliary stricture after liver transplantation: technical description and case series. Ann Surg. 2022;275(6):e801–3.
20. Chen PD, Wu CY, Hu RH, Chou WH, Lai HS, Liang JT, et al. Robotic versus open hepatectomy for hepatocellular carcinoma: a matched comparison. Ann Surg Oncol. 2016;24(4):1021–8.
21. Troisi RI, Berardi G, Morise Z, Cipriani F, Ariizumi S, Sposito C, et al. Laparoscopic and open liver resection for hepatocellular carcinoma with Child-Pugh B cirrhosis: multicentre propensity score-matched study. Br J Surg. 2021;108(2):196–204.
22. Coelho FF, Herman P, Kruger JAP, Wu AGR, Chin KM, Hasegawa K, et al. Impact of liver cirrhosis, the severity of cirrhosis, and portal hypertension on the outcomes of minimally invasive left lateral sectionectomies for primary liver malignancies. Surgery. 2023;174(3):581–92.
23. Lim C, Scatton O, Wu AGR, Zhang W, Hasegawa K, Cipriani F, et al. Impact of liver cirrhosis and portal hypertension on minimally invasive limited liver resection for primary liver malignancies in the posterosuperior segments: an international multicenter study. Eur J Surg Oncol J Eur Soc Surg Oncol Br Assoc Surg Oncol. 2023;49(10):106997.
24. D'Hondt M, Wicherts DA. Pure robotic major hepatectomy with biliary reconstruction for hepatobiliary malignancies: first European results. Surg Endosc. 2023;37(6):4396–402.
25. Li J, Tan X, Zhang X, Zhao G, Hu M, Zhao Z, et al. Robotic radical surgery for hilar cholangiocarcinoma: a single-centre case series. Int J Med Robot Comput Assist Surg MRCAS. 2020;16(2):e2076.
26. Mueller M, Breuer E, Mizuno T, Bartsch F, Ratti F, Benzing C, et al. Perihilar cholangiocarcinoma – novel benchmark values for surgical and oncological outcomes from 24 expert centers. Ann Surg. 2021;274(5):780–8.
27. Cherqui D, Ciria R, Kwon CHD, Kim KH, Broering D, Wakabayashi G, et al. Expert consensus guidelines on minimally invasive donor hepatectomy for living donor liver transplantation from innovation to implementation: A joint initiative from the International Laparoscopic Liver Society (ILLS) and the Asian-Pacific Hepato-Pancreato-Biliary Association (A-PHPBA). Ann Surg. 2021;273(1):96–108.
28. Broering D, Sturdevant ML, Zidan A. Robotic donor hepatectomy: a major breakthrough in living donor liver transplantation. Am J Transplant Off J Am Soc Transplant Am Soc Transpl Surg. 2022;22(1):14–23.
29. Rela M, Rajalingam R, Shetty G, Cherukuru R, Rammohan A. Robotic monosegment donor hepatectomy for pediatric liver transplantation: first report. Pediatr Transplant. 2022;26(1):e14110.
30. Lee KW, Choi Y, Hong SK, Lee S, Hong SY, Suh S, et al. Laparoscopic donor and recipient hepatectomy followed by robot-assisted liver graft implantation in living donor liver transplantation. Am J Transplant Off J Am Soc Transplant Am Soc Transpl Surg. 2022;22(4):1230–5.
31. Beane JD, Zenati M, Hamad A, Hogg ME, Zeh HJ, Zureikat AH. Robotic pancreatoduodenectomy with vascular resection: outcomes and learning curve. Surgery. 2019;166(1):8–14.
32. Machado MA, Mattos BH, Lobo Filho MM, Makdissi FF. Robotic right hepatectomy with portal vein Thrombectomy for colorectal liver metastasis (with video). J Gastrointest Surg Off J Soc Surg Aliment Tract. 2021;25(7):1932–5.
33. Di Benedetto F, Petrowsky H, Magistri P, Halazun KJ. Robotic liver resection: hurdles and beyond. Int J Surg [Internet]. 2020 Jun 3 [cited 2020 Jun 5]; Available from: http://www.sciencedirect.com/science/article/pii/S1743919120304532.

MIX
Papier aus verantwortungsvollen Quellen
Paper from responsible sources
FSC® C105338

FSC
www.fsc.org

If you have any concerns about our products,
you can contact us on
ProductSafety@springernature.com

In case Publisher is established outside the EU,
the EU authorized representative is:
Springer Nature Customer Service Center GmbH
Europaplatz 3, 69115 Heidelberg, Germany

Printed by Libri Plureos GmbH
in Hamburg, Germany